Preparation and Application of Polymer Nanocomposites

Preparation and Application of Polymer Nanocomposites

Topic Editors

Hongbo Gu
Zilong Deng
Donglu Fang
Xianhu Liu
Kai Sun
Hu Liu

MDPI

Basel • Beijing • Wuhan • Barcelona • Belgrade • Novi Sad • Cluj • Manchester

Topic Editors

Hongbo Gu
School of Chemical Science and Engineering
Tongji University
Shanghai
China

Zilong Deng
School of Environmental Science and Engineering
Tongji University
Shanghai
China

Donglu Fang
State Key Laboratory for the Development and Utilization of Forest Food Resources
Nanjing Forestry University
Nanjing
China

Xianhu Liu
National Engineering Research Center for Advanced Polymer Processing Technology
Zhengzhou University
Zhengzhou
China

Kai Sun
School of Marine Science and Engineering
Shanghai Maritime University
Shanghai
China

Hu Liu
National Engineering Research Center for Advanced Polymer Processing Technology
Zhengzhou University
Zhengzhou
China

Editorial Office
MDPI AG
Grosspeteranlage 5
4052 Basel, Switzerland

This is a reprint of the Topic, published open access by the journals *Applied Nano* (ISSN 2673-3501), *Materials* (ISSN 1996-1944), *Molecules* (ISSN 1420-3049), *Nanomaterials* (ISSN 2079-4991), *Polymers* (ISSN 2073-4360) and *Solids* (ISSN 2673-6497), freely accessible at: https://www.mdpi.com/topics/34GN7IBW15.

For citation purposes, cite each article independently as indicated on the article page online and as indicated below:

Lastname, Firstname, Firstname Lastname, and Firstname Lastname. Article Title. *Journal Name* **Year**, *Volume Number*, Page Range.

ISBN 978-3-7258-7064-6 (Hbk)
ISBN 978-3-7258-7065-3 (PDF)
https://doi.org/10.3390/books978-3-7258-7065-3

Contents

About the Editors

Hongbo Gu

Hongbo Gu is currently an associate professor and doctoral supervisor at the School of Chemical Science and Engineering, Tongji University. She received his bachelor's degree from Harbin Institute of Technology in 2008 and his doctoral degree from the same university in 2014. During this period, she received funding from the China Scholarship Council to conduct joint training at Lamar University in the United States from 2011 to 2013. She joined Tongji University in September of the same year. Her research focuses on the integrated design, preparation, and application of polymer nanocomposites, encompassing electrofunctional polymers, polymer crystallization, thermoelectric and magnetothermoelectric properties, electromagnetic properties, and their applications in flexible electronics, environmental remediation, and sensors. She has published over 100 SCI papers in internationally journals such as *Chemical Society Reviews*, *Nature Communications*, and *Advanced Materials*, including 65 as first or corresponding author. In total, 22 of these papers are ESI highly cited, and 8 are ESI hot papers, with a total citation count exceeding 8400 and an h-index of 53. She holds 10 authorized national invention patents and has been selected for various talent programs, including the China Association for Science and Technology's Young Talent Support Project, the Shanghai Youth Science and Technology Rising Star Program, the Shanghai Youth Science and Technology Elite Sailing Program, and the Shanghai Association for Science and Technology's "Flying Plan". She also serves as an executive member of the Youth Working Committee of the Chinese Society for Composite Materials, a member of the Micro/Nano Composite Materials and Intelligent Composite Materials Professional Committee, co-editor of *Emerging Materials Research*, and executive editor and editorial board member of journals such as *Engineered Science*. She serves as a reviewer for over 100 international journals.

Zilong Deng

Zilong Deng is currently an Associate Professor, Doctoral Supervisor and Assistant Dean at the School of Environmental Science and Engineering, Tongji University. From July 2024 to August 2025, he served as Deputy General Manager and Party Committee Member at COSCO Shipping Group/Shanghai Ship Research Institute/China Shipping Environmental Technology (Shanghai) Co., Ltd. From April 2019 to November 2023, he was an Assistant Professor and Master's Supervisor at the School of Environmental Science and Engineering, Tongji University. From May 2011 to July 2013, he worked at Cott Beverages in the UK, holding positions such as Technical Assistant, Senior Experimentalist, and Production Supervisor. His main research interests are the environmental and agricultural applications of bio-based nanocellulose composite materials. He was selected as a Shanghai Oriental Talent (Youth) and has published over 50 SCI papers, including over 30 JCR Q1 papers as first/corresponding author. He has been granted and published over 10 patents as the first inventor. He serves as an expert in the expert database of the Chinese Society for Environmental Sciences and the Jiangxi Provincial Association for Science and Technology, and is a member of the Shanghai Rural Revitalization Youth Talent Association.

Donglu Fang

Donglu Fang is an Associate Research Fellow and Master's Supervisor in the Department of Horticulture, College of Forestry and Grassland and College of Soil and Water Conservation, Nanjing Forestry University. He is a member of the Jiangsu Provincial Agricultural Society and one of the first experts of the Jiangsu Provincial Non-Major Crop Variety Identification Committee. His main research areas are forest food storage, preservation and high-value utilization, and intelligent packaging materials for fruits. He conducts research based on the National Key Laboratory of Forest Food Resource Exploration and Utilization. He has presided over projects such as the Jiangsu Provincial Excellent Youth Fund, and was selected for the Jiangsu Higher Education "Blue Project" and the Jiangsu Provincial Association for Science and Technology's Young Scientific and Technological Talent Support Project. He has published relevant academic achievements in journals such as *Food Hydrocolloids*. In teaching, he won the second prize in the college's young teacher teaching competition, presided over university-level higher education research projects, guided students to win national gold medals in subject competitions, and served as a class teacher in the Department of Food Science and Engineering.

Xianhu Liu

Xianhu Liu is a directly appointed professor and Doctoral Supervisor at the National Engineering Research Center for Rubber and Plastic Molds, Zhengzhou University. He holds a PhD from the University of Erlangen-Nuremberg, Germany, and is the deputy director of the Henan Provincial International Joint Laboratory for Polymer Material Molding and Molds. His research focuses on polymer material molding and functionalization, multifunctional composite materials, porous polymer materials, multifunctional coatings, and spectrally selective polymer composite films. He is a Fellow of the Royal Society of Chemistry, a Highly Cited Researcher, and a recipient of the Henan Provincial Outstanding Youth Science Fund. He has led more than 10 projects, including National Natural Science Foundation of China projects and military projects. He has published over 100 papers as first/corresponding author in journals such as *Science* and *Nature Communications*, with over 15,000 citations and an H-index of 39. He serves on the editorial boards of several international journals and is an expert reviewer for domestic and international projects.

Kai Sun

Kai Sun, born in December 1988, is a member of the Communist Party of China. He is a professor and Doctoral Supervisor at the School of Marine Science and Engineering, Shanghai Maritime University. He received his Ph.D. in Materials Science and Engineering from Shandong University. From 2016 to 2017, he received funding from the China Scholarship Council to conduct joint training at the University of Tennessee in the United States. He joined Shanghai Maritime University in March 2018 and was promoted to professor and Doctoral Supervisor in July 2024. His research focuses on ship and marine engineering equipment detection technology and protective materials, electromagnetic metamaterials, and electro-magnetic–thermal multifunctional composite materials. He has presided over more than ten projects, including the National Natural Science Foundation of China. He has published over 100 SCI papers in journals such as *Advanced Functional Materials*, with a total citation count of over 9,700. He has received numerous awards, including the Shanghai Science and Technology Award and the Ministry of Education Natural Science Award. He has been selected for several talent programs of the China Association for Science and Technology and Shanghai Municipality, and is among the top 2% of scientists globally recognized by Stanford University. He has mentored students who have won numerous national and provincial-level awards.

Hu Liu

Hu Liu is a professor and Doctoral Supervisor at the National Engineering Research Center for Rubber and Plastic Molds, Zhengzhou University, and a permanent staff member of the National Key Laboratory of Industrial Equipment Structural Analysis and Optimization and CAE Software. He received his PhD in Materials Processing Engineering from Zhengzhou University in 2017 and was a joint supervisor at the University of Tennessee, USA, from 2015 to 2016. His research focuses on polymer-based functional nanocomposites such as flexible force-sensitive sensors, electromagnetic wave absorption, and thermal insulation aerogels. He has published over 90 SCI papers as first/corresponding author in journals such as *Advanced Functional Materials, Science Bulletin,* and *Nano Research*, with over 20 papers selected as ESI highly cited papers. He has led 8 projects, including National Natural Science Foundation of China (NSFC) projects, Henan Provincial Excellent Youth Fund projects, and China Postdoctoral Science Foundation projects. He was selected for the Central Plains Elite Talent Program—Central Plains Science and Technology Innovation Young Top Talent, Henan Provincial Excellent Youth Fund recipient, Henan Provincial Excellent Doctoral Dissertation recipient, and was listed in the 2022–2025 Global Top 2% Scientists' "Lifetime Scientific Impact" and "Annual Scientific Impact" lists. He also serves as a member of the Advanced Composites and an Editorial Board member of international journals such as *Hybrid Materials* and *Journal of Composites Science*.

nanomaterials

MDPI

Article

Effect of Rapid Thermal Annealing on the Characteristics of Micro Zn-Doped Ga_2O_3 Films by Using Mixed Atomic Layer Deposition

Jiajia Tao [1,2,3], Xishun Jiang [4,*], Aijie Fan [2], Xianyu Hu [3], Ping Wang [4], Zuoru Dong [1] and Yingjie Wu [1]

1 Shanghai Microwave Technology Research Institute (No. 50 Research Institute of China Electronics Technology Group Corporation), Shanghai 200331, China
2 Zhangjiang Laboratory, Shanghai 201210, China
3 State Key Laboratory of ASIC and System, School of Microelectronics, Fudan University, Shanghai 200433, China
4 School of Mechanical and Electronical Engineering, Chuzhou University, Chuzhou 239000, China
* Correspondence: jxs@chzu.edu.cn

Abstract: In this work, micro Zn-doped Ga_2O_3 films (GZO) were deposited by one-step mixed atomic layer deposition (ALD) followed by post-thermal engineering. The effects of Zn doping and post-annealing temperature on both structure characteristics and electric properties were investigated in detail. The combination of plasma-enhanced ALD of Ga_2O_3 and thermal ALD of ZnO can realize the fast growth rate (0.62 nm/supercyc.), high density (4.9 g/cm^3), and smooth interface (average R_q = 0.51 nm) of Zn-doped Ga_2O_3 film. In addition, the thermal engineering of the GZO was achieved by setting the annealing temperature to 400, 600, 800, and 1000 °C, respectively. The GZO film annealed at 800 °C exhibits a typical crystalline structure (Ga_2O_3: β phase, ZnO: hexagonal wurtzite), a lower roughness (average R_q = 2.7 nm), and a higher average breakdown field (16.47 MV/cm). Notably, compared with the pure GZO film, the breakdown field annealed at 800 °C increases by 180%. The O_V content in the GZO after annealing at 800 °C is as low as 34.8%, resulting in a remarkable enhancement of electrical properties. These research findings offer a new perspective on the high-quality ALD-doped materials and application of GZO in high-power electronics and high-sensitive devices.

Keywords: micro Zn-doped Ga_2O_3 film; atomic layer deposition; thermal annealing effect; electrical property

1. Introduction

In recent years, the next-generation wide bandgap semiconductor material gallium oxide (Ga_2O_3) has shown great potential applications in power electronic devices and solar-blind ultraviolet photo-detectors due to its superior characteristics, including a larger bandgap (4.6–4.9 eV), a higher breakdown field strength (~8 MV/cm), and superior chemical and thermal stability [1–3]. At present, researchers have improved the optical, electrical, photoelectric, or optoelectrical properties of Ga_2O_3 materials through doping, noble metal modification, or energy band engineering. Among the doping, modification, and energy band engineering, doping engineering is the simplest and most effective method to enhance the performance of Ga_2O_3-based devices [4]. However, the related precise control of electrical properties for Ga_2O_3 thin films by doping engineering is still a huge challenge. It was found that Si-doped β-Ga_2O_3 thin films developed through MOCVD can only non-uniform line dope controlled by employing pulse width modulation [5]. In addition, the correlation between structure, chemistry, and dielectric properties of iron-doped Ga_2O_3 were investigated roughly by the standard high-temperature solid-state chemical reaction method [6]. Moreover, the Zr-doping strategy for Cu_2O/Ga_2O_3 (Ga_2O_3/ZrO_2 = 98.7:1.3 wt%) was investigated to achieve a high-performance solar-blind ultraviolet photodetector through magnetron sputtering deposition [7]. In summary, conventional synthesis methods face significant challenges in achieving large-area fabrication, highly uniform deposition, and precise trace doping of Ga_2O_3 thin films due to inherent limitations in traditional equipment capabilities and intricate reaction kinetics, which are essential for effectively tailoring their electrical and optoelectronic properties.

Atomic layer deposition (ALD) has emerged as a suitable technique for achieving precise control over both thickness and compositional uniformity in surface engineering, particularly in conformal coating applications and in situ doping processes of nanostructured semiconductors. While numerous studies have demonstrated the successful deposition of Ga_2O_3 thin films using ALD methodology, research focusing on controlled elemental doping strategies in Ga_2O_3 via ALD remains comparatively limited. Moreover, the window temperature of the ALD growth of Ga_2O_3 is below 400 °C, in which Ga_2O_3 films are in an amorphous or polycrystalline state [8]. These amorphous or polycrystal states contain a large number of uncertain grain boundaries, defects, and oxygen vacancies, which greatly hinder the electric or photoelectric performance of the device [9]. Accordingly, the thermal engineering after the ALD processing of Ga_2O_3 film is an important issue for improving photoelectric performance [10]. In addition, Zn atoms have similar atomic radii to Ga atoms, so they can be used for the most efficient doping control [11]. Tao et al. investigated the morphological, electrical, and optical properties of the Zn-doped Ga_2O_3 amorphous film, including the change of film density and resistivity, roughness, transmittance, and bandgap [12]. Singh et al. achieved green light emission and p-type characteristics of ZnGaO films [13]. Hu et al. prepared Zn: Ga_2O_3 films by co-injecting sALD with predictable doping content and realized a high-performance photodetector [14]. However, the post-thermal engineering of Zn element micro doped-Ga_2O_3 during the ALD process is still lacking systematic research. Moreover, the basic mechanisms of the doping and post-thermal effect are still absent, which hinders the basic understanding of the ALD doping effect and the exploration of new performances of Ga_2O_3-based devices.

Herein, the Zn-doped Ga_2O_3 thin films were grown by mixed ALD, in which the Zn: Ga were precisely adjusted by a fixed cycle of thermal processing of ZnO and a plasma-enhanced process of Ga_2O_3. The subsequent post-thermal engineering of thin films grown by ALD was achieved by rapid annealing at 400 °C, 600 °C, 800 °C, and 1000 °C for 5 min in an N_2 atmosphere. The ALD micro-doping of the Zn element and thermal engineering effects on optical, structural, and electrical properties of Ga_2O_3 film were disused in detail. The annealed GZO film shows superior crystallinity, high transparency, low surface roughness, and a high breakdown field. This micro dopant and post-thermal treatment process successfully improved the optical and electrical properties of GZO film, which broadens its potential applications in photodetection and high-power devices.

2. Materials and Methods

Zn-doped Ga_2O_3 films were deposited on Si substrates by using the ALD system (BENEQ TFS200, Espoo, Finland). Trimethylgallium (TMG) and diethylzinc (DEZn) were used as Ga and Zn precursors. Oxygen plasma and H_2O were used as the precursors for Ga_2O_3 and ZnO. All the precursors in the experiment were weighed using a METTLER TOLEDO ME 204 balance. The deposited temperatures during all ALD processes were kept at 200 °C. The specific parameters of ALD-deposited films are listed in the Supplementary Information. After ALD processing, the samples were annealed at 400, 600, 800, and 1000 °C, respectively, in an N_2 atmosphere for 5 min through rapid thermal processing (Annealsys, Montpellier, France, AS-One). The thickness of the deposited GZO films was measured by spectroscopic ellipsometry (SE, SOPRA GES-5E). X-ray diffraction (XRD, D8 Advance, Bruker, Billerica, MA, USA) was used to examine the crystallinity of the films. Meanwhile, the film density was fitted using X-ray reflectivity (XRR, D8-Advance, Bruker). The surface roughness of GZO films was checked with atomic force microscopy (AFM, Bruker). X-ray photoelectron spectroscopy (XPS, Thermo Scientific, Waltham, MA, USA) was conducted to analyze the chemical composition and state. The MOS device fabrication details are shown in the Supplementary Information. The current-voltage (I–V) characteristics of the GZO-based MOS devices were measured by a Keithley 4200 semiconductor parameters test system.

3. Results and Discussion

The cycle ratio of Ga_2O_3 and ZnO with the same film thickness was precisely regulated and compared by using the unique atomic layer film deposition characteristics of ALD, and a super cycle of Zn-doped Ga_2O_3 films containing seven cycles of Ga_2O_3 and one cycle of ZnO was designed. For better differentiation, the sample prepared after several super cycles is named GZO. A supercycle process and the structure diagram of GZO deposited by ALD are shown in Figure 1a,b. Accordingly, the structure of the GZO based on the MOS device after photolithography, metal deposition, and stripping is shown in Figure 1c.

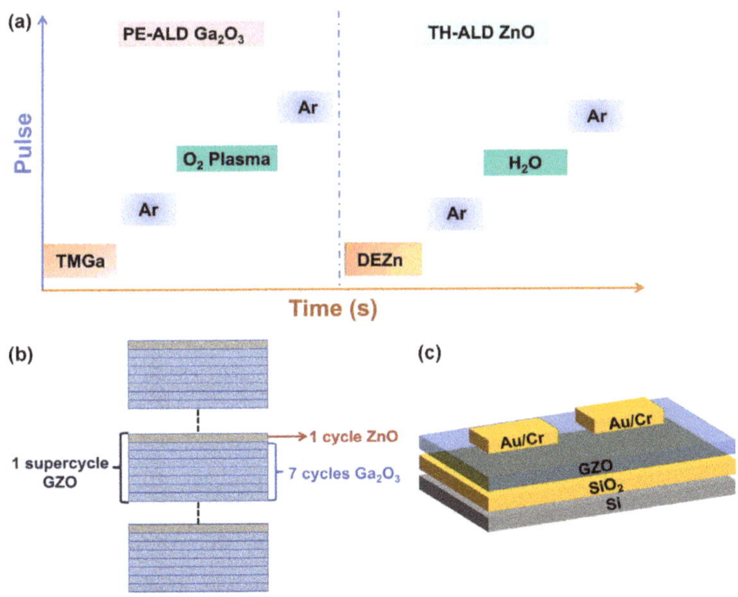

Figure 1. (**a**) GZO thin film deposition process, including Ga_2O_3 and ZnO monolayer deposition process; (**b**) schematic diagram of GZO laminated film structure consisting of seven layers of Ga_2O_3 and one layer of ZnO, and (**c**) MOS device structure based on GZO structure.

In order to obtain the thickness, refractive index (n), extinction coefficient (k), and density of the films, SE was used to systematically measure the micro Zn-doped Ga_2O_3 films deposited by ALD. The corresponding results are shown in Figure 2a–c. Figure 2a shows the SE test and fitting curves of the polarization angle and phase angle of the GZO thin film. Experimental results are highly consistent with calculated results in the test wavelength range, indicating the high accuracy of fitting results. The thickness of the GZO film is

20 nm through the calculation and fitting of the Cauchy model. As shown in Figure 2b, in the wavelength range of 230–800 nm, with the increase in wavelength, the refractive index of GZO film became smaller and maintained normal dispersion characteristics. At the wavelength of 632.8 nm, the n value of the film is 2.26, which is higher than that of GZO films prepared by traditional methods [15,16], indicating that GZO films prepared by ALD have the advantages of dense and high electron density. At the same time, the extinction coefficient of all films decreases with the increase in wavelength and is closer to zero at 500–800 nm, indicating that GZO has a high transmittance in the visible light region [17]. This suggests that ALD shows great advantages in depositing transparent electrode films due to their high density and good transmittance. Figure 2c shows the growth rate for Ga_2O_3, ZnO, and GZO films. By comparison, it is found that the deposition rate of GZO film (0.76 nm/cyc.) is much higher than the superposition deposition rate of the ZnO and Ga_2O_3 thin films (0.62 nm/supercyc.). The design of GZO is based on a supercycle consisting of seven PE-ALD deposition Ga_2O_3 cycles and one TH-ALD deposition ZnO cycle. The estimated and measured thickness of ZGO film are listed in Table S1. It shows that the scheme realizes the rapid growth of Zn-doped Ga_2O_3 thin films. In addition, Figure 2c also shows the different film density values obtained after SE fitting. The densities of Ga_2O_3, GZO, and ZnO films are 5.3, 4.9, and 5.6 g/cm^3, respectively. The GZO thin film shows an amorphous state. It can be concluded that the decrease of thin film density is mainly attributed to the corresponding defects in the internal Ga_2O_3 thin films caused by Zn doping [18].

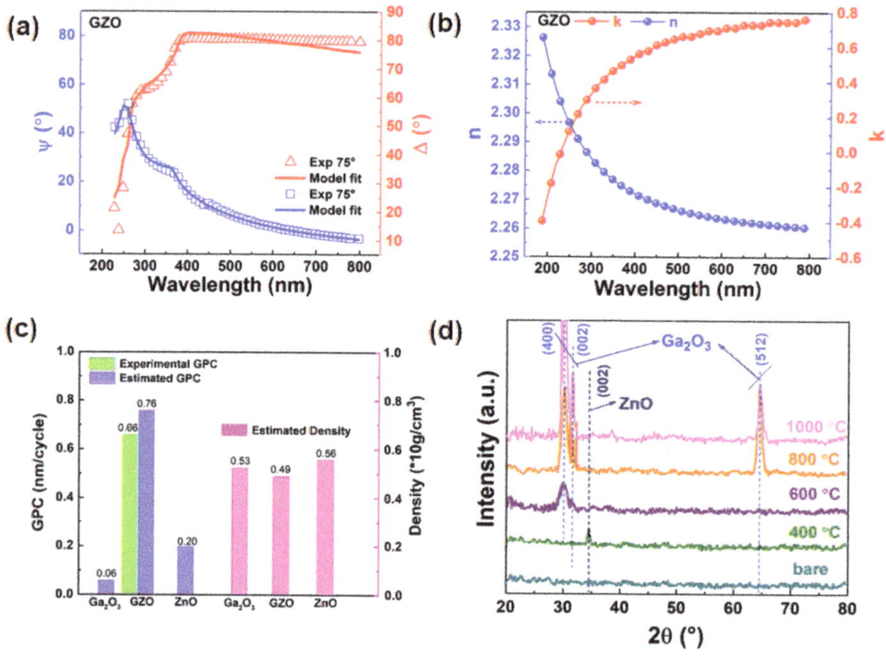

Figure 2. (**a**) SE test and fitting curves of polarization angles and phase angles, (**b**) the change of refractive index n and extinction coefficient k with wavelength of original GZO films, (**c**) a supercyclic theoretical superposition and measured deposition rate of GZO, Ga_2O_3, and ZnO films, (**d**) XRD patterns of GZO films with different rapid annealing temperatures.

Figure 2d shows the XRD pattern of GZO films on Si substrate before and after the annealing process. In the bare GZO films, no other diffraction peaks are found, which indicates that the Zn-doped Ga_2O_3 film is amorphous. The peaks of the Si substrate are shown in Figure S1 for better comparison. However, according to a previous study [12], the ZnO films deposited by TH-ALD have a hexagonal wurtzite crystal structure, unlike the amorphous state of deposited Ga_2O_3 film by PE-ALD. Obviously, the amorphous state of Ga_2O_3 film leads to the poor crystallinity of the GZO film. In particular, the change of crystallinity of Zn-doped Ga_2O_3 films deposited in this mixed ALD mode is different from that of other ALD-doped composite films. For example, during the ALD process of Al-doped ZnO, the crystallinity of AlZnO films depends on the Al content [19]. The crystalline properties of doped films depend on the preparation method, intrinsic property, and post-thermal engineering. Accordingly, the (400) and (002) peaks of β-Ga_2O_3 in GZO film exhibit the highest diffraction intensity after a 600 °C annealing process [20]. As the annealing temperature increases, the peak intensity of Ga_2O_3 rises, and the FWHM becomes narrower. This indicates that high temperature promotes the crystallization and growth of the film. The reason is that the high-temperature annealing treatment at the appropriate temperature enables the gallium atoms and oxygen atoms inside the film to obtain sufficient energy to migrate to the appropriate positions and eliminate the oxygen vacancies inside the film. In addition, a typical crystalline hexagonal wurtzite structure of ZnO with (002) peak is shown in the GZO before and after annealing. At low annealing temperatures ($\leq$ 400 °C), Zn may act as a nucleation site, lowering the energy barrier for crystallization. This can lead to an earlier onset of crystallization compared to undoped Ga_2O_3. In addition, Zn might stabilize metastable Ga_2O_3 polymorphs at lower temperatures due to lattice strain or defect interactions, delaying the transition to the stable β-phase. At high annealing temperatures ($\geq$ 600 °C), Zn doping could accelerate β-Ga_2O_3 formation by facilitating atomic rearrangement despite initial strain from the larger Zn^{2+} ions. Also, exceeding Zn solubility limits may result in ZnO precipitates, which could either pin grain boundaries or act as templates for epitaxial growth, depending on interfacial compatibility [21].

In order to study the effect of rapid annealing temperature on the surface morphology of Ga_2O_3 films, AFM was used to test the films' roughness. The results are shown in Figure 3. The average roughness (R_q) of GZO films annealed at 0, 400, 600, 800, and 1000 °C is 0.51, 1.28, 3.32, 2.70, and 3.89 nm, respectively. With the increase of annealing temperature, the roughness first increases, then decreases, and finally increases, which is related to the rapid annealing process and island regrowth mechanism of the film [22]. With the temperature rising from 0 °C to 800 °C, the active site of film island regrowth continues to approach saturation, the regrowth rate tends to be stable, and the roughness of the film decreases. On the contrary, at the high temperature of 1000 °C, a small number of active sites are activated, grow, and aggregate again, resulting in a faster growth rate and maximum roughness of the film.

Figure 4 shows the XPS spectra of GZO film at different rapid annealing temperatures. The survey spectra in Figure 4a exhibits peaks of Ga (2s, 2p, 3p, and 3d), Zn (2p), and O (1s), as well as Auger peaks from gallium (Ga LM1, LM2) and oxygen (O KL1), which is consistent with previous reports on unannealed Ga_2O_3 and ZnO [23,24]. In the survey spectrum, no peaks of other impurities were detected, indicating that the GZO film retained its high purity throughout the atomic layer deposition and rapid thermal annealing processes. In Figure 4b, the characteristic doublet of Ga 2p is clearly visible, with binding energies of 1144.91 eV for Ga 2p1/2 and 1118.07 eV for Ga 2p3/2, yielding an energy separation of 26.84 eV, which is consistent with previously reported values [25]. Furthermore, as depicted in Figure 4c, the peak positions of Ga 3d remained unchanged after annealing at different temperatures, suggesting that Ga atoms exhibited excellent stability after annealing. For the Zn dopant, the two peaks of Zn 2p1/2 and Zn 2p3/2 are located at 1044.97 and 1021.99 eV, respectively, with a consistent energy difference of 22.98 eV [26]. However, the peak signal of Zn2p in the sample annealed at 1000 °C is weak, implying potential transformations of ZnO at elevated temperatures, such as the formation of ZnO_2 or carbonization [27]. Collectively, the above XPS analysis indicates that after multiple rapid high-temperature annealing cycles, the Zn element in the doped film remained relatively stable.

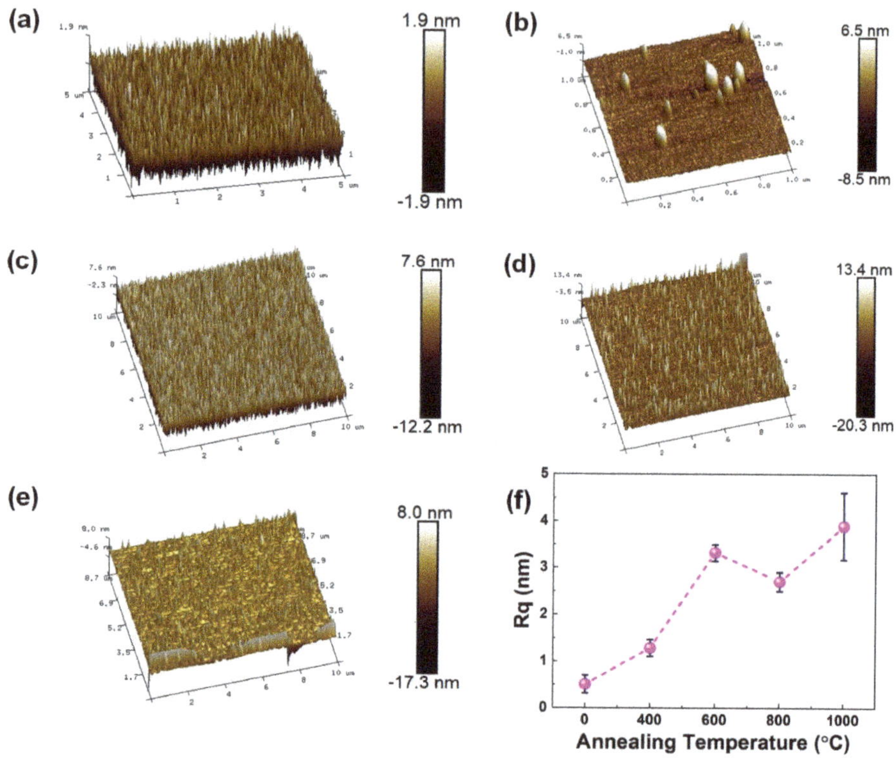

Figure 3. AFM surface morphologies of the GZO films with rapid annealing (**a**) as deposited, at (**b**) 400 °C, (**c**) 600 °C, (**d**) 800 °C, and (**e**) 1000 °C, and (**f**) the R_q changes of GZO film with rapid annealing temperature.

The O1s of the GZO film before and after the rapid annealing process can be divided into three peaks [24,28,29], namely, the peak of the lattice oxygen (O_L), the peak of the oxygen vacancy (O_V), and the peak related to the hydroxyl groups (O_H), as shown in Figure 5a–e. The corresponding peak positions of the three types of O1s are 530.5 eV of the O_L, 531.2 eV of the O_V, and 531.9 eV of the O_H. At the same time, the contents of different types of O 1s components are also calculated and shown in the corresponding insets. As the annealing temperature increases from 0 °C to 800 °C, the O_H content decreases from 19.0% to 4.35%, indicating that the annealing-induced growth reaction significantly purifies the film. Additionally, the surface roughness first increases and then decreases at this temperature range, further confirming the substantial improvement in film quality post-annealing. From 800 °C to 1000 °C, the O_H content exhibits a slight increase, which can be attributed to the crystallization process occurring at higher temperatures. Notably, even after annealing at 1000 °C, the maximum O_H content remains at 11.3%, substantially lower than the bare 19%. This suggests that annealing effectively reduces residual unreacted and adsorbed organic groups within the film during deposition, thereby enhancing its density, reducing surface roughness, and improving electrical properties [30].

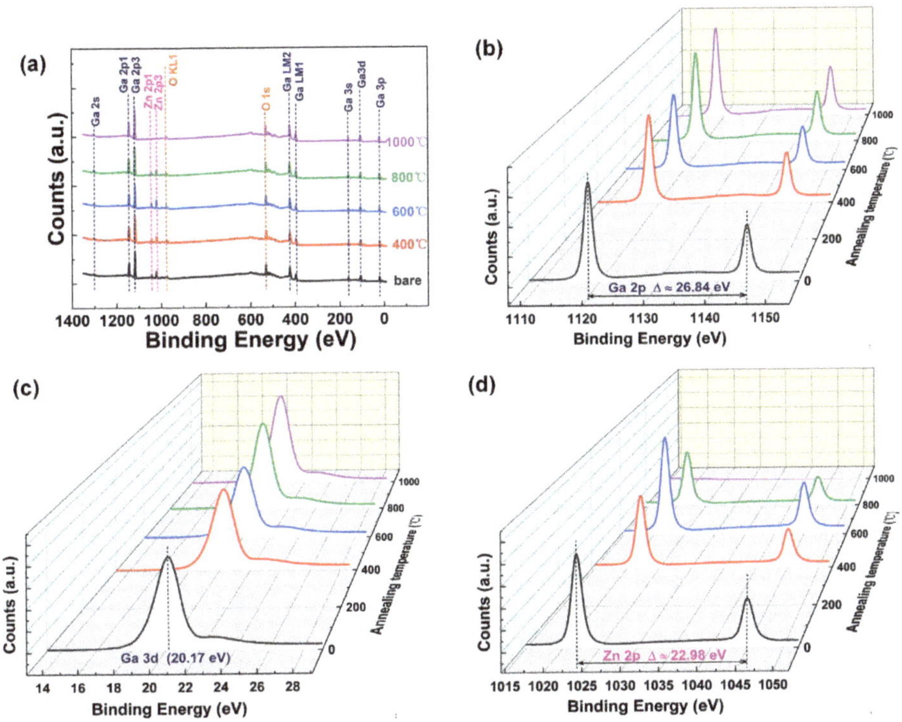

Figure 4. XPS spectra of GZO films after different rapid annealing temperatures. (**a**) Survey spectra, (**b**) Ga 2p spectra, (**c**) Ga 3d spectra, and (**d**) Zn 2p spectra.

Most notably, the oxygen vacancy (O_V) content in GZO films exhibits a dynamic evolution during thermal annealing across different temperature regimes, as demonstrated in Figure 5f. The O_V concentration initially increases from 38.0% at 0 °C to a maximum of 46.8% at 400 °C, followed by a progressive decline, with further temperature elevation to 800 °C. Notably, the minimum O_V content of 25% is achieved at 800 °C, while subsequent annealing at 1000 °C restores the O_V concentration to 39%. This behavior corresponds to the characteristic growth-crystallization-regrowth transition occurring during the 0–1000 °C annealing sequence [31]. The observed O_V concentration fluctuations critically govern the electrical performance of the films. Particularly significant is the 34.8% reduction in O_V content following 800 °C annealing compared to the bare state, which correlates with enhanced electrical properties in the processed films.

The electrical properties of annealed Zn-doped Ga_2O_3 films were achieved using the MOS devices described above. The typical I–V characteristic curves of the annealed Zn-doped Ga_2O_3 film-based MOS structure are shown in Figure 6a. At 0.5 V, the leakage current of all devices was less than 5.5×10^{-9} A, which suggests micro Zn-doped Ga_2O_3 maintains good insulation properties. When the applied voltage gradually increases to the breakdown voltage, the leakage current of the device suddenly increases, which accords with the hard breakdown characteristic of the device. The Zn-doped Ga_2O_3 film prepared by ALD has the characteristics of high density, and the regrowth of crystal during annealing will affect the electrical properties of the device. This is different from the soft breakdown characteristics of bare Ga_2O_3-based films [32]. The average breakdown voltage fields are 5.88, 9.12, 10.31, 16.47, and 12.90 MV/cm for Zn-doped Ga_2O_3 film with annealed temperatures of 0, 400, 600, 800, and 1000 °C, respectively. The average breakdown field of MOS devices based on Zn-doped Ga_2O_3 films with different annealing temperatures are listed in Table S2. By increasing the annealed temperature, the E_g value decreases, which is the main intrinsic breakdown voltage [33]. On the other hand, the main breakdown characteristics are due to the internal structural characteristics of the film. Accordingly, by taking into account the film thickness, the breakdown electric field of the film was calculated and is shown in Figure 6b. Similar to the variation of breakdown voltage, the breakdown field strength first increases and then decreases with the increase of annealing temperature. This could be due to the changes of oxygen vacancies and defects in internal Ga_2O_3 and ZnO. Under a certain electric field strength, internal defects and vacancies can move and rearrange, and then act as conductive channels, making the film more easily broken down, as exhibited in Figure 7a. The concentration of oxygen defects, as shown in Figure 5, first increases, then decreases, and finally increases with the annealing temperature increasing. The GZO film annealed at 800 °C has the lowest concentration of oxygen defects and thus requires a larger breakdown electric field. Additionally, the breakdown field strength of the GZO film annealed at 800 °C is higher than that of Ga_2O_3 and ZnO [34,35]. This indicates that trace Zn doping and annealing engineering can significantly enhance the electrical properties of Ga_2O_3, promoting its potential application in wide bandgap and high-power devices.

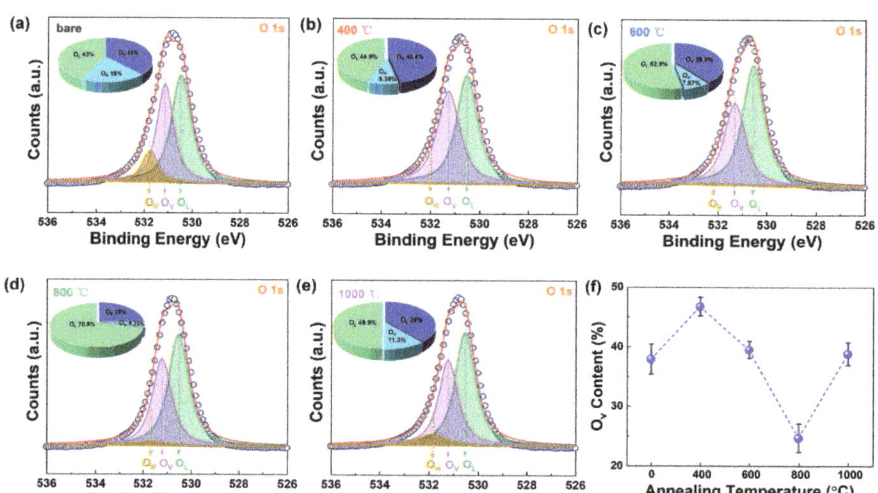

Figure 5. (**a**–**e**) The O 1s spectra of GZO film after different rapid annealing temperatures. The ratio of different oxygen components is obtained after calculation and shown in the inset. (**f**) The ratio of oxygen vacancy as a function of annealing temperature.

Figure 7a shows the breakdown mechanism of GZO thin film. As described above, under the action of an electric field, oxygen vacancies and other defect states aggregate with each other to form conductive channels, thus causing the thin film to be broken down under a certain voltage [36]. After annealing, obvious changes occurred inside the GZO film that increased the breakdown voltage, as shown in Figure 7b,c. First, the unstable defects in Ga_2O_3 and ZnO grow again after annealing, thus reducing the density of the defect states. Second and third, after high-temperature annealing, the unstable oxygen vacancy in the interior and at the interface will be transformed into a stable oxygen vacancy so that the carrier concentration involved in the breakdown will be reduced. Fourth, the defect states with different charge characteristics will also compound with each other, forming faults and failing to form a complete conductive channel. In addition, oxygen vacancies create mid-gap states that facilitate trap-assisted tunneling or impact ionization, lowering the effective breakdown field by providing pathways for premature current flow. Further, the interfaces between different phases can concentrate electric fields, accelerating carrier multiplication. All of the above significantly reduced internal defect states of GZO films caused by annealing and significantly enhanced the breakdown characteristics of GZO films.

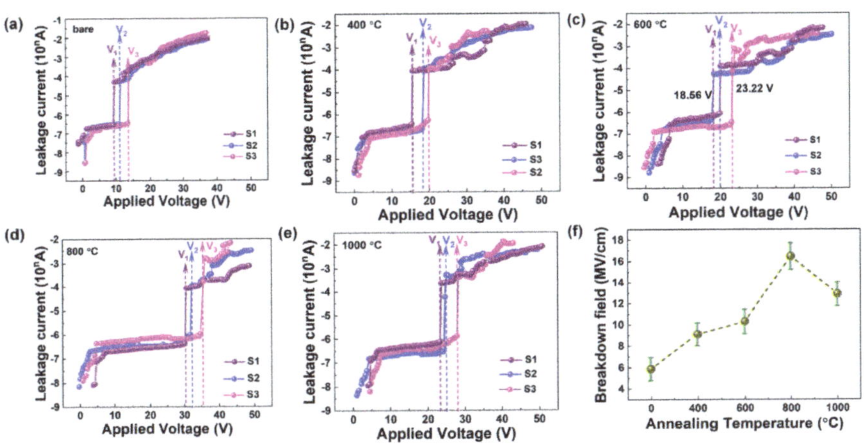

Figure 6. (**a**–**e**) I–V curves of the Zn-doped Ga_2O_3 thin films with different annealed temperatures; (**f**) the breakdown field as a function of annealed temperature.

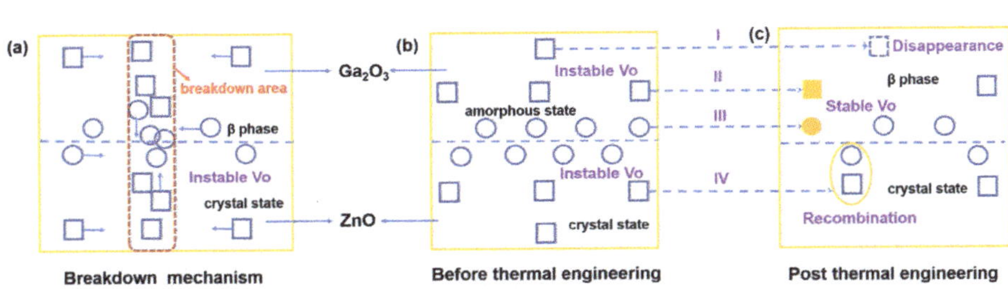

Figure 7. (**a**) The breakdown mechanism of GZO film, (**b**) the movement and redistribution process of oxygen vacancies before and (**c**) after thermal engineering under an external electric field.

4. Conclusions

In this study, Zn-doped Ga_2O_3 thin films with optimized post-deposition thermal engineering were successfully fabricated through ALD followed by rapid thermal annealing. The amorphous GZO films with desirable morphological and structural characteristics were achieved through precise control of the layer-by-layer ALD deposition process. Subsequent thermal treatment within the 400–1000 °C range induced controlled crystallization of the GZO films. Notably, the 800 °C annealed GZO demonstrated an enhancement in the breakdown field compared to both as-deposited films and those annealed at other temperatures. The electrical property of films is closely related to the behavior of the internal oxygen vacancy. The lower oxygen vacancy concentration and excellent structure characteristics of GZO annealing at 800 °C significantly improved the breakdown characteristics of the MOS device. Moreover, the breakdown mechanism of GZO thin film was also discussed in detail. The developed GZO material system demonstrates significant potential for advanced optoelectronic applications, particularly in UV-C/UV-B flexible photodetectors, as well as in high-power electronic devices through implementation in n-n junctions and heterojunction transistors.

Supplementary Materials: The following supporting information can be downloaded at: https://www.mdpi.com/article/10.3390/nano15070499/s1, Figure S1: XRD patterns of the grown Ga_2O_3, ZnO, and GZO films grown on Si (100) substrate; Table S1: Summary of the details of the parameters for one growth supercycle during the reaction process, estimated and measured thickness of GZO films; Table S2: The breakdown voltages and average breakdown field of MOS devices based on Zn-doped Ga_2O_3 films with different annealing temperature. The experimental methods of ALD deposition of GZO and MOS device fabrication and measurement are also described in the Supplementary Materials.

Author Contributions: J.T. and P.W. designed the research; X.H., Z.D. and Y.W. contributed to device fabrication; X.J. and A.F. assisted with data analysis; J.T. wrote the paper. All authors have read and agreed to the published version of the manuscript.

Funding: The presented work is supported by the Shanghai Sailing Program (Nos. 22YF1446200, 23YF1444300), Shanghai Super Postdoctoral Plans Encouragement Programs (No. 2023078). This work is also supported by the Anhui University Provincial Natural Science Research Project, China (KJ2021A1089), Research Initiation Fund Project of Chuzhou University (2023qd94).

Data Availability Statement: Data are contained within the article and Supplementary Materials.

Conflicts of Interest: The authors declare no conflict of interest.

References

1. Mazumder, S.K.; Voss, L.F.; Dowling, K.M.; Conway, A.; Hall, D.; Kaplar, R.J.; Pickrell, G.W.; Flicker, J.; Binder, A.T.; Chowdhury, S.; et al. Overview of Wide/Ultrawide Bandgap Power Semiconductor Devices for Distributed Energy Resources. *IEEE J. Emerg. Sel. Top. Power Electron.* **2023**, *11*, 3957–3982. [CrossRef]
2. Huang, Z.; Zhou, S.; Chen, L.; Zheng, Q.; Li, H.; Xiong, Y.; Ye, L.; Kong, C.; Fan, S.; Zhang, H.; et al. Fully Transparent Amorphous Ga_2O_3-Based Solar-Blind Ultraviolet Photodetector with Graphitic Carbon Electrodes. *Crystals* **2022**, *12*, 1427. [CrossRef]
3. Lee, D.Y.; Jeong, S.; Moon, S.; Yang, M.; Kim, S.H.; Kim, D.; Lee, S.Y.; Lee, I.S.; Jeon, D.W.; Park, J.H.; et al. Giant Colloidal Quantum Dot/α-Ga_2O_3 Heterojunction for High Performance UV-Vis-IR Broadband Photodetector. *ACS Nano* **2024**, *18*, 34741–34749. [CrossRef]
4. Lee, S.; Bielinski, A.R.; Fahrenkrug, E.; Dasgupta, N.P.; Maldonado, S. Macroporous *p*-GaP Photocathodes Prepared by Anodic Etching and Atomic Layer Deposition Doping. *ACS Appl. Mater. Interfaces* **2016**, *8*, 16178–16185. [CrossRef]
5. Hernandeza, A.; Islama, M.M.; Saddatkiaa, P.; Codding, C.; Dulal, P.; Agarwal, S.; Janover, A.; Novak, S.; Huang, M.B.; Dang, T.; et al. MOCVD growth and characterization of conductive homoepitaxial Si-doped Ga_2O_3. *Result Phys.* **2021**, *25*, 104167. [CrossRef]
6. Roy, S.; Mallesham, B.; Zade, V.B.; Martinez, A.; Shutthanandan, V.; Thevuthasan, S.; Ramana, C.V. Correlation between Structure, Chemistry, and Dielectric Properties of Iron-Doped Gallium Oxide ($Ga_{2-x}Fe_xO_3$). *J. Phys. Chem. C* **2018**, *122*, 27597–27607. [CrossRef]
7. Jiang, J.; Wu, S.M.; Li, X.Y.; Xin, Q.; Tian, Y. Zr-Doping Strategy of High-Quality Cu_2O/β-Ga_2O_3 Heterojunctionfor Ultrahigh-Performance Solar-Blind Ultraviolet Photodetection. *ACS Appl. Mater. Interfaces* **2024**, *16*, 64136–64145. [CrossRef]
8. Comstock, D.J.; Elam, J.W. Atomic Layer Deposition of Ga_2O_3 Films Using Trimethylgallium and Ozone. *Chem. Mater.* **2012**, *24*, 4011–4018. [CrossRef]
9. Gebert, M.; Bhattacharyya, S.; Bounds, C.C.; Syed, N.; Daeneke, T.; Fuhrer, M.S. Passivating Graphene and Suppressing Interfacial Phonon Scattering with Mechanically Transferred Large-Area Ga_2O_3. *Nano Lett.* **2023**, *23*, 363–370. [CrossRef]
10. Cheng, Z.; Shi, J.J.; Yuan, C.; Kim, S.; Graham, S. Chapter Four-Thermal science and engineering of β-Ga_2O_3 materials and devices. *Semicond. Semimet.* **2021**, *107*, 77–99.

11. Fan, H.C.; Wang, C.; Ruan, Y.J.; Shen, K.C.; Wu, W.Y.; Wuu, D.S.; Lai, F.M.; Lien, S.Y.; Zhu, W.Z. Enhanced responsivity of solar blind ultraviolet photodetector by PEALD deposited Zn-doped Ga_2O_3 thin films. *IEEE Trans. Electron. Dev.* **2024**, *71*, 664–669. [CrossRef]
12. Tao, J.J.; Lu, H.L.; Gu, Y.; Ma, H.P.; Li, X.; Chen, J.X.; Liu, W.J.; Zhang, H.; Feng, J.J. Investigation of growth characteristics, compositions, and properties of atomic layer deposited amorphous Zn-doped Ga_2O_3 films. *Appl. Surf. Sci.* **2019**, *476*, 733–740.
13. Singh, A.K.; Yen, C.C.; Chang, K.P.; Wuu, D.S. Structural and photoluminescence properties of co-sputtered p-type Zn-doped β-Ga_2O_3 thin films on sapphire substrates. *J. Lumin.* **2023**, *260*, 119836.
14. Hsu, C.H.; Wu, W.B.; Yue, X.X.; Wu, W.Y.; Gao, P.; Wuu, D.S.; Cho, Y.S.; Huang, C.J.; Lien, S.Y. Improved performance of solar blind ultraviolet photodetectors by spatial ALD Zn-doped Ga_2O_3 film and post-annealing. *Surf. Coat. Technol.* **2025**, *497*, 131798.
15. Lee, S.H.; Lee, J.H.; Choi, S.J.; Park, J.S. Studies of thermoelectric transport properties of atomic layer deposited gallium-doped ZnO. *Ceram. Int.* **2017**, *43*, 7784–7788. [CrossRef]
16. Lee, Y.L.; Chen, S.F.; Ho, C.L.; Wu, M.C. Effects of oxygen plasma post-treatment on Ga-doped ZnO films grown by thermal-mode ALD. *ECS J. Solid State Sci. Technol.* **2013**, *2*, 316–320.
17. Dasgupta, N.P.; Neubert, S.; Lee, W.; Trejo, O.; Lee, J.R.; Prinz, F.B. Atomic layer deposition of Al-doped ZnO films: Effect of grain orientation on conductivity. *Chem. Mater.* **2010**, *22*, 4769–4775.
18. Yuan, N.Y.; Wang, S.Y.; Tan, C.B.; Wang, X.Q.; Chen, G.G.; Ding, J.N. The influence of deposition temperature on growth mode, optical and mechanical properties of ZnO films prepared by the ALD method. *J. Cryst. Growth* **2013**, *366*, 43–46. [CrossRef]
19. Lung, C.; Toma, M.; Pop, M.; Pop, A. Characterization of the structural and optical properties of ZnO thin films doped with Ga, Al and (Al plus Ga). *J. Alloys Compd.* **2017**, *725*, 1238–1243.
20. Dong, L.P.; Jia, R.X.; Xin, B.; Zhang, Y.M. Effects of post-annealing temperature and oxygen concentration during sputtering on the structural and optical properties of β-Ga_2O_3 films. *J. Vac. Sci. Technol. A* **2016**, *34*, 060602. [CrossRef]
21. Liu, X.J.; Sun, H.F.; Wu, Y.Z.; Sun, S.Y.; Ren, Y.; Tan, X.; Jia, H.L. Theoretical Study of Sulphur Atoms' Adsorption and Migration Behaviors on Diamond (001) Surface. *Coatings* **2019**, *9*, 184. [CrossRef]
22. Zheng, H.; Zhang, R.J.; Xu, J.P.; Wang, S.X.; Zhang, T.N.; Sun, Y.; Zheng, Y.X.; Wang, S.Y.; Chen, X.; Chen, L.Y.; et al. Thickness-Dependent Optical Constants and Annealed Phase Transitions of Ultrathin ZnO Films. *J. Phys. Chem. C* **2016**, *120*, 22532–22538.
23. Ramachandran, R.K.; Dendooven, J.; Botterman, J.; Pulinthanathu Sree, S.; Poelman, D.; Martens, J.A.; Poelman, H.; Detavernier, C. Plasma enhanced atomic layer deposition of Ga_2O_3 thin films. *J. Mater. Chem. A* **2014**, *2*, 19232–19238.
24. Yu, C.F.; Chen, S.H.; Sun, S.J.; Chou, H. Influence of the grain boundary barrier height on the electrical properties of Gallium doped ZnO thin films. *Appl. Surf. Sci.* **2011**, *257*, 6498–6502.
25. Chandiran, A.K.; Tetreault, N.; Humphry-Baker, R.; Kessler, F.; Baranoff, E.; Yi, C.; Nazeeruddin, M.K.; Grätzel, M. Subnanometer Ga_2O_3 tunnelling layer by atomic layer deposition to achieve 1.1 V open-circuit potential in dye-sensitized solar cells. *Nano Lett.* **2012**, *12*, 3941–3947. [CrossRef]
26. Guziewicz, E.; Krajewski, T.A.; Przezdziecka, E.; Korona, K.P.; Czechowski, N.; Klopotowski, L.; Terziyska, P. Zinc Oxide Grown by Atomic Layer Deposition: From Heavily n-Type to p-Type Material. *Phys. Status Solidi B* **2019**, *257*, 1900472.
27. Suhariadi, I.; Itagaki, N.; Shiratani, M. Improved Nanoscale Al-Doped ZnO with a ZnO Buffer Layer Fabricated by Nitrogen-Mediated Crystallization for Flexible Optoelectronic Devices. *ACS Appl. Nano Mater.* **2020**, *3*, 2480–2490.
28. Gao, Z.; Banerjee, P. Review Article: Atomic layer deposition of doped ZnO films. *J. Vacuum Sci. Technol. A* **2019**, *37*, 050802.
29. Jiao, S.; Lu, H.; Wang, X.; Nie, Y.; Wang, D.; Gao, S.; Wang, J. The structural and photoelectrical properties of gallium oxide thin film grown by radio frequency magnetron sputtering. *ECS J. Solid State Sci. Technol.* **2019**, *8*, Q3086–Q3090.
30. Jeong, S.W.; Lee, H.J.; Kim, K.S.; You, M.T.; Roh, Y.; Noguchi, T.; Xianyu, W.; Jung, J. Effects of annealing temperature on the characteristics of ALD-deposited HfO_2 in MIM capacitors. *Thin Solid Films* **2006**, *515*, 526–530.
31. Ylivaara, O.; Langner, A.; Ek, S.; Malm, J.; Julin, J.; Laitinen, M.; Ali, S.; Sintonen, S.; Lipsanen, H.; Sajavaara, T.; et al. Thermomechanical properties of aluminum oxide thin films made by atomic layer deposition. *J. Vac. Sci. Technol. A* **2022**, *40*, 062414.
32. Shen, Y.; Ma, H.P.; Gu, L.; Zhang, J.; Huang, W.; Zhu, J.T.; Zhang, Q.C. Atomic-Level Sn Doping Effect in Ga_2O_3 Films Using Plasma-Enhanced Atomic Layer Deposition. *Nanomaterials* **2022**, *12*, 4256. [CrossRef] [PubMed]
33. Shen, Y.; Ma, H.P.; Wang, Z.Y.; Gu, L.; Zhang, J.; Li, A.; Yang, M.; Zhang, Q.C. Influence of Rapid Thermal Annealing on the Characteristics of Sn-Doped Ga_2O_3 Films Fabricated Using Plasma-Enhanced Atomic Layer Deposition. *Crystals* **2023**, *13*, 301. [CrossRef]
34. Mohamed, H.F.; Xia, C.; Sai, Q.; Cui, H.; Pan, M.; Qi, H. Growth and fundamentals of bulk β-Ga_2O_3 single crystals. *J. Semicond.* **2019**, *40*, 011801.

35. Liu, S.Y.; Liu, S.; Zhou, Y.Y.; Piao, Y.J.; Li, G.J.; Wang, Q. Transparent ZnO: Al_2O_3 films with high breakdown voltage and resistivity. *Appl. Phys. Lett.* **2018**, *113*, 032102. [CrossRef]
36. Dong, L.P.; Jia, R.X.; Xin, B.; Peng, B.; Zhang, Y.M. Effects of oxygen vacancies on the structural and optical properties of β-Ga_2O_3. *Sci. Rep.* **2017**, *7*, 40160.

polymers

MDPI

Article

Cellulose/Aminated Multi-Walled Carbon Nanotube Nanocomposite Aerogels for Oil Adsorption

Runlin Han [1,†], Zihan Liu [2,†], Faxiang Feng [1], Shi Su [1], Guilin Dong [1], Xiaobing Liu [1] and Hongbo Gu [2,*]

1 Key Laboratory of Jiangxi Province for Special Optoelectronic Artificial Crystal Materials, School of Chemistry and Chemical Engineering, Jinggangshan University, Ji'an 343009, China; hanrunlin@163.com (R.H.); fengfaxiangfj@163.com (F.F.); 178885271280@163.com (S.S.); dong2990455445@163.com (G.D.); liuxiaobing805@163.com (X.L.)

2 Key Laboratory of Chemical Assessment and Sustainability, School of Chemical Science and Engineering, Tongji University, Shanghai 200092, China; 15638112382@139.com

* Correspondence: hongbogu2014@tongji.edu.cn

† These authors contributed equally to this work.

Abstract: At present, the oil extraction and chemical industry and other industries produce a large amount of oily wastewater and organic sewage, and the world is suffering from oil spills and organic wastewater pollution. As a porous material, aerogels are promising in the field of oil adsorption. In this work, the nanocellulose/aminated multi-walled carbon nanotube (NC-MWCNT-NH_2) nanocomposite aerogel with a high porosity of up to 97.80% is prepared by varying the weight percentage of MWCNTs-NH_2, among which the nanocomposite aerogel with 0.1% weight percentage of MWCNTs-NH_2 exhibits the best adsorption performance with the adsorption capacity to cyclohexane, ethyl acetate, anhydrous ethanol, methylene dichloride, acetone, kerosene, pump oil, and used pump oil of 39.77 ± 0.82, 44.54 ± 1.67, 43.03 ± 1.06, 62.13 ± 0.36, 39.92 ± 1.09, 39.37 ± 0.27, 43.48 ± 0.06, and 38.45 ± 0.84 $g \cdot g^{-1}$, respectively. Compared with pure nanocellulose aerogel, the adsorption capacity of the NC-MWCNT-NH_2 aerogel to pump oil is improved by up to 93.9%, which exhibits excellent adsorption properties and could be utilized in the field of oil adsorption.

Keywords: aerogel; cellulose; aminated multi-walled carbon nanotubes (MWCNTs-NH_2); composites; oil adsorption

1. Introduction

The rapid development of industries such as petrochemicals and production has promoted the rapid growth of the country's economy, but it has also brought a lot of environmental problems [1–3]. Oily wastewater refers to wastewater discharged from industrial production containing oil, like petroleum-related products, tar and its fractions, fats, and edible animal and vegetable oils [4]. There is an urgent need to develop innovative technologies and materials that are sustainable and scalable for the efficient collection and separation of spilled oils and organic solvents from contaminated water sources. Chemical methods such as flocculation and precipitation are often expensive and complicated. Also, these methods are not environmentally friendly [5]. The adsorption method uses solid porous adsorbents such as activated carbon, polypropylene fibers, and polyurethane foams that are commonly used to clean up oil spills. It is one of the most promising methods for its high efficiency, simple operation, relatively lower cost, less secondary pollution, and higher recyclability [1,6].

Aerogel is a nanoscale porous solid material formed by replacing the liquid phase of a gel with gas, and is the lowest-density solid material known to exist on earth [7,8]. In addition to this, it has a compressive capacity equivalent to thousands of times its own weight, extremely high thermal and insulating properties, and a low refractive index. Aerogels have extremely high porosity, a high specific surface area, and an ultra-high pore volume ratio [9].

Cellulose is one of the world's most abundant and renewable resources. It is a linear homopolysaccharide composed of β-D-glucopyranose units linked by β-1-4 linkages, with highly ordered crystalline phases as well as amorphous domains such as lignin, pectin, hemicellulose, and so on [10]. Cellulose aerogels combine the advantages of aerogels with low density, high porosity, large specific surface area, and excellent biodegradability [11]. It also has good adsorption properties, which could adsorb oil substances and greatly benefit energy saving and carbon emission reduction [12]. Cellulose nanofibrils (CNFs) are composed of a long, flexible, and interconnected network of cellulose nanofibers, with a diameter of 2–60 nm and a length of several micrometers, containing both crystalline and amorphous cellulose domains. CNF aerogels have a broad range of functional applications, such as efficient adsorption, thermal insulation, and energy storage [13–15]. Zhang et al. prepared a highly compressible and hydrophobic aerogel by freeze-drying of the aqueous suspensions of CNFs, which showed a hexane absorption capacity of 45 $g{\cdot}g^{-1}$ [16]. Zhang et al. [17] used nanocellulose as a carrier and chloropropyltriethoxysilane as a crosslinking agent to make a hydrophobic nanocellulose hydrogel with a low density ($\leq$0.086 $g{\cdot}cm^{-3}$) and high porosity ($\geq$95.2%).

Carbon nanotubes (CNTs) are widely regarded as one-dimensional nanomaterials because they have an inner diameter of about 1 nm, an outer diameter of a few to several tens of nanometers, a tube length in the micrometer range, and an aspect ratio of more than 100. Among them, multi-walled carbon nanotubes (MWCNTs) and MWCNT-based adsorbents have been explored for the removal of heavy metals from industrial effluents with high adsorption performance [18]. Jiang et al. [19] designed a double-layer aerogel structure with cellulose nanofibrils as the matrix and CNTs as the light-absorbing material, which could achieve a 76.3% solar energy conversion efficiency under the standard solar radiation. Zou et al. [20] prepared an ultralight free-standing MWCNT aerogel with an anisotropic macroporous honeycomb structure and a surface area of 580 $m^2{\cdot}g^{-1}$.

In this work, cellulose/MWCNT-NH_2 nanocomposite (NC-MWCNT-NH_2) aerogels are prepared from cellulose nanofibers and MWCNTs-NH_2 using a freeze-drying method for adsorption of oil and organic solvents. The different MWCNT contents are selected to evaluate the effect of the NC-MWCNT-NH_2 structure on the adsorption property. The structure of as-prepared aerogels is analyzed using scanning electron microscope (SEM), Fourier transform infrared spectroscopy (FTIR), thermogravimetric analyzer (TGA), and specific surface and porosity analyzer (BET). The adsorption property of these aerogels for various oils and organics is measured in detail. It is found that the adsorption performance of NC-MWCNT-NH_2 nanocomposite aerogels is increased first and then decreased with the increasing MMWCNT-NH_2 content. The aerogel with 0.1 wt.% MWCNTs-NH_2 shows the highest adsorption capacities for all the oils and organic solvents. The research demonstrates that the introduction of MWCNTs-NH_2 can significantly improve the adsorption properties of aerogels and offers prospects for the application of cellulose aerogels in oil spill treatment.

2. Materials and Methods

2.1. Materials

Nanocellulose (CNF, diameter: 4–10 nm, length: 1–3 μm, 0.8 wt.% aqueous solution) was purchased from Guilin Qihong Technology Co., Ltd, Guilin, China. Aminated MWCNTs (outer diameter: 8–15 nm, length: ~50 μm) were provided by Jiangsu XFNANO Materials Tech Co., Ltd., Nanjing, China. Cyclohexane ethyl acetate, ethanol, and dichloromethane are all of analytical purity and were supplied by Shanghai Macklin Biochemical Technology Co., Ltd., Shanghai, China. Other chemicals were used as received without purification.

2.2. Preparation of Nanocellulose Aerogel

The preparation of pure NC aerogel was divided into three steps as a whole. First, a certain amount of aqueous nanocellulose solution was put in the de-ionized (DI) water to dilute the nanocellulose solution to 0.4 wt% of NC. Then, the above solution was poured into a high-speed mixer (4000 $r\cdot min^{-1}$) for 10 min. Finally, the solution was taken out into the centrifuge tube (7 mL). Then, the tube was put into the freezing dryer with the temperature set to −56 °C and freeze-dried for 24 h. The pure NC aerogel was marked as A0.

2.3. Preparation of NC-MWCNT-NH_2 Nanocomposite Aerogels

The same procedure was applied to prepare NC-MWCNT-NH_2 nanocomposite aerogels, in which the MWCNTs were firstly added into the NC solution and then put in the high-speed mixer for mixing. The aminated MWCNT content in the aqueous solution was 0.05, 0.1, and 0.15% in mass fraction, and the corresponding samples were named A1, A2, and A3, respectively. Contents of raw materials in nanocomposite aerogels are shown in Table 1.

Table 1. Contents of raw materials in nanocomposite aerogels.

NO	MWCNT-NH_2 Content (wt.%)	Cellulose Content (wt.%)	Mass Ratio of MWCNT-NH_2 and Cellulose
A1	0.05	0.4	1:8
A2	0.1	0.4	2:8
A3	0.15	0.4	3:8
A0	0	0.4	0

2.4. Characterization of Aerogels

The surface morphology of aerogels was observed on SEM (Hitachi S-4800, Tokyo, Japan) with an acceleration voltage of 5 kV. The chemical structure of aerogels was conducted on the FTIR (IS20, Thermo Scientific, Thermo Nicolet NEXUS, Waltham, MA, USA) with an attenuated total reflection (ATR) accessory over the range from 4000 to 400 cm^{-1}. Thermal degradation of aerogels was investigated by thermogravimetric analysis (TGA, Q50, TA Instruments, New Castle, DE, USA) with a heating rate of 10 °C min^{-1} in the air. The specific surface area of aerogels was determined using the Specific Surface Area Analyzer (Tristar3020, McMurray Tick Instruments Co., Ltd., Shanghai, China). The porosity (s) of aerogels was calculated using Equation (1) [21]:

$$s = \frac{m_2 - m_1}{m_2 - m_1 + m} \quad (1)$$

where m_1 is the mass of the sealed weighing vial filled with anhydrous ethanol after adsorption, m_2 is the mass of the sealed weighing vial filled with anhydrous ethanol before adsorption, and m is the mass of the aerogel used.

2.5. Adsorption Test

The adsorption tests were conducted as follows. Firstly, a piece of aerogel was weighed as m_3. Secondly, the small weighing bottle with a lid containing 4 mL of ethyl acetate, anhydrous ethanol, dichloromethane, acetone, kerosene, pump oil, and used pump oil was weighed as m_4. After that, the aerogel was put into oil for adsorption for 5 min without agitation. Then, the aerogel was quickly taken out and the mass of the bottle after adsorption was weighed as m_5. The adsorption capacity (A) of oils could be computed as the weight difference between m_5 and m_4 divided by aerogel mass m_3 shown in Equation (2). The reported value is the average of three measurements with the error bars.

$$A = \frac{m_4 - m_5}{m_3} \quad (2)$$

3. Results

3.1. Structure Characterization

The morphology of MWCNTs-NH_2 used in this work is characterized by SEM, and the results are shown in Figure S1. It presents a typical carbon nanotube morphological structure, showing a highly entangled tube structure with a diameter of about 10–20 nm. The as-prepared NC-MWCNT-NH_2 aerogel with a density of about 6 mg/cm^3, as shown in Figure S2, could be placed on the stamen of a flower without the stamen being deformed.

The morphology of pure NC aerogel and NC-MWCNT-NH_2 aerogel is carried out by SEM. In Figure 1a,b, the SEM images of the A0 sample show many fibrous thread-like structures with radial diameters of about 2–3 μm, which are coiled and entangled with each other, forming a three-dimensional network structure. There is a certain lamellar structure under the tread-like structure. The SEM images of the A1 sample reveal an obvious typical lamellar structure (Figure 1c,d). The structure of the A2 sample is relatively porous, and it can be observed that some filamentary fibers are still retained in its structure, and these filamentary fibers and MWCNTs-NH_2 together build up the laminated structure of this composite aerogel (Figure 1e,f). The A3 sample has an overall layered structure (Figure 1g,h), which is also reported in the polyvinyl alcohol and CNF aerogel with a laminated structure [16,19].

Figure 2 shows the FTIR spectra of A0, A1, A2, and A3. As illustrated in Figure 2, the absorption peaks at 1060, 1325, 1417, 1593, and 2923 cm^{-1} are assigned to the stretching vibration of C-O, -OH, CH_2-OH, hydrogen bonding, and the saturated C-H bond in the cellulose structure, respectively [22]. These characteristic absorption peaks can also be observed in the NC-MWCNT-NH_2 nanocomposite aerogels [23]. The FTIR spectrum of MWCNTs-NH_2 is shown in Figure S3. The peak around 3394 cm^{-1} belongs to the -NH stretching vibration of MWCNTs-NH_2 [24].

BET analysis of pure NC aerogel and NC-MWCNT-NH_2 nanocomposite aerogels with mass fractions of 0.05%, 0.1%, and 0.15%, respectively, was carried out. The nitrogen adsorption–desorption curves and pore size distribution curves of these aerogel samples are shown in Figure 3. In the inset of Figure 3, in the pore size distribution curves, the pore sizes of these samples are in the range of about 2–55 nm, but most of the pore sizes are distributed in the range of 2–15 nm. The nitrogen adsorption–desorption curves of these aerogels are in accordance with the type-IV adsorption isotherm, exhibiting mesoporous properties. As shown in Table S1, the resulting BET specific surface areas of A0, A1, A2, and A3 samples are 11.09, 13.42, 14.95, and 38.86 m^2·g^{-1}, respectively. With the gradual increase in the mass fraction of MWCNTs-NH_2 in the nanocomposite aerogel, its BET specific surface area is also gradually increased. The porosity of A0, A1, A2, and A3 samples is 96.27, 97.55,

97.80, and 96.81%, respectively, in which the NC-MWCNT-NH_2 nanocomposite aerogel with an MWCNT-NH_2 mass fraction of 0.1%, i.e., the A2 sample, has the largest porosity.

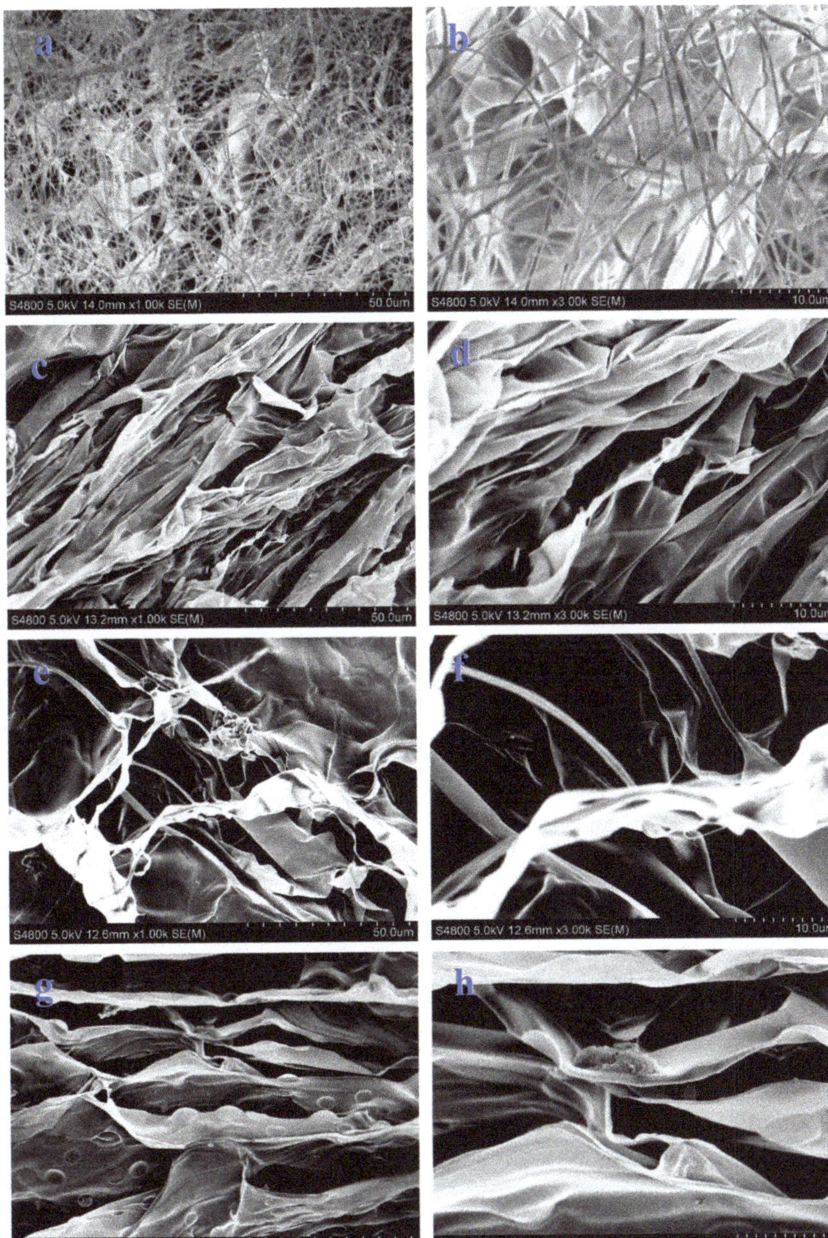

Figure 1. SEM images of (**a**,**b**) A0, (**c**,**d**) A1, (**e**,**f**) A2, and (**g**,**h**) A3.

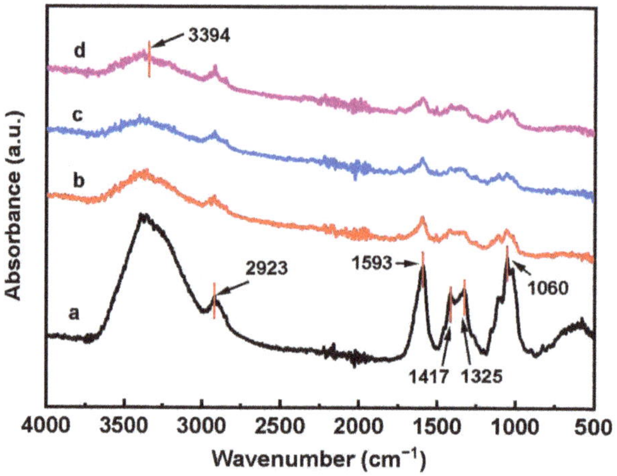

Figure 2. FTIR spectra of the aerogel samples: (a) A0, (b) A1, (c) A2, and (d) A3.

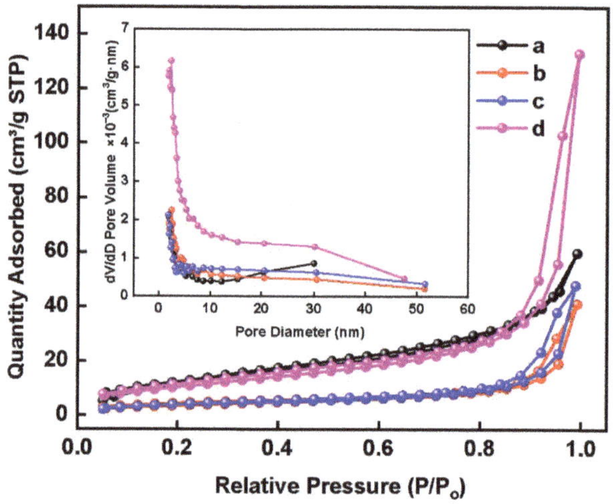

Figure 3. Nitrogen adsorption–desorption curves and pore size distribution curves (inset) of aerogel samples: (a) A0, (b) A1, (c) A2, and (d) A3.

To analyze the thermal stability of aerogels, TGA analysis was performed in the air condition in Figure 4. All four curves show about 10% weight loss before rising to 150 °C, arising from the evaporation of adsorbed water in the samples. The first stage of weight loss for the samples from 260 to 320 °C is due to the degradation of the 3D network structure of the nanofibrillar cellulose, and the variation in weight loss of pure NC aerogel at this stage is higher than that of the other three samples. The second stage of weight loss for the samples is in the range of 380 to 520 °C, which is caused by the thermal decomposition of the carbon skeleton of the aerogel. The variation in weight loss at this stage of the three nanocomposite aerogels is obviously higher than that of the pure NC aerogel for the addition of MWCNTs-NH_2 as listed in Table S3. We also investigated the thermal stability of MWCNTs in air, and as seen in Figure S4, the MWCNTs-NH_2 are thermally stable and begin to show rapid decomposition at temperature higher than 500 °C. Both A1 and A3 show a typical lamellar structure, so the TGA curves of samples are very similar, while the small amount of fiber filaments in sample A2 not only increases the porosity of sample, but also alters its weight loss.

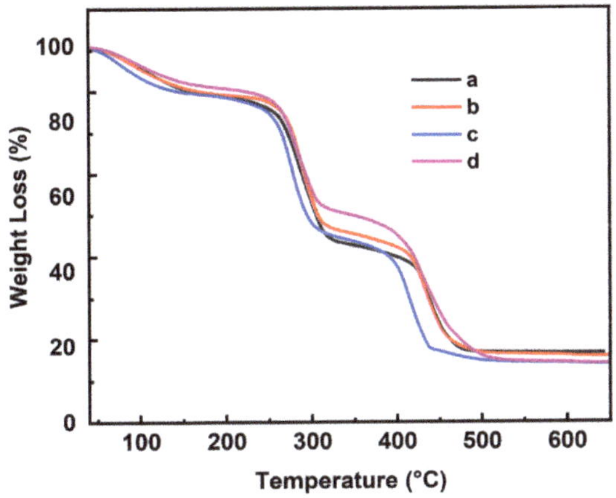

Figure 4. TGA curves of aerogel samples: (a) A0, (b) A1, (c) A2, and (d) A3.

3.2. Adsorption Performance

To investigate the adsorption properties of pure NC aerogel and NC-MWCNT-NH_2 nanocomposite aerogels on oils and organic solvents, the cyclohexane, ethyl acetate, ethanol, dichloromethane, acetone, kerosene, pump oil, and wasted pump oil were selected for testing. As shown in Figure 5a, the adsorption capacity of the pure NC aerogel on cyclohexane, ethyl acetate, ethanol, dichloromethane, acetone, kerosene, pump oil, and wasted pump oil is 34.78 ± 0.87, 28.99 ± 0.08, 24.83 ± 0.92, 39.48 ± 0.9, 30.36 ± 1.59, 36.26 ± 0.36, 22.43 ± 0.48, and 30.65 ± 0.93 $g{\cdot}g^{-1}$, respectively. From the data in Figure 5b, NC-MWCNT-NH_2 nanocomposite aerogels with 0.1% mass fraction of NC-MWCNT-NH_2 have the best adsorption capacity, and its adsorption amounts of cyclohexane, ethyl acetate, ethanol, dichloromethane, acetone, kerosene, pump oil, and wasted pump oil are 39.77 ± 0.82, 44.54 ± 1.67, 43.03 ± 1.06, 62.13 ± 0.36, 39.92 ± 1.09, 39.37 ± 0.28, 43.48 ± 0.06, and 38.45 ± 0.84 $g{\cdot}g^{-1}$, respectively. The results are also listed in Table S4 for clarity. The adsorption capacity of pump oil is increased significantly, almost doubling. Overall, the adsorption performance of NC-MWCNT-NH_2 nanocomposite aerogels was basically higher than that of the pure NC aerogel. It is noticed that the adsorption performance of NC-MWCNT-NH_2 nanocomposite aerogels is increased first and then decreased with the increasing MWCNT-NH_2 content. This may be the reason that, according to the adsorption data, sample A2 shows the highest adsorption capacities for all the oils and organic solvents. The results are consistent with the highest porosity of sample A2, and the laminated structure with filamentary fibers are remained in the nano composite aerogel. At the same time, all aerogels show a higher adsorption capacity for dichloromethane because the adsorption capacities of aerogels for organic solvents are affected by their density and surface tension. The greater the density and surface tension, the larger the adsorption capacities will be obtained [25].

As shown in Table 2, compared to other reported oil adsorbents, our NC-MWCNT-NH_2 nanocomposite aerogels exhibit relatively higher adsorption capacities than those of sponges, foams, and inorganic aerogels [26–29], but lower than nanocellulose aerogels prepared by crosslinking individual cellulose nanofibers and the hydrophobic modification of the surface wettability [30].

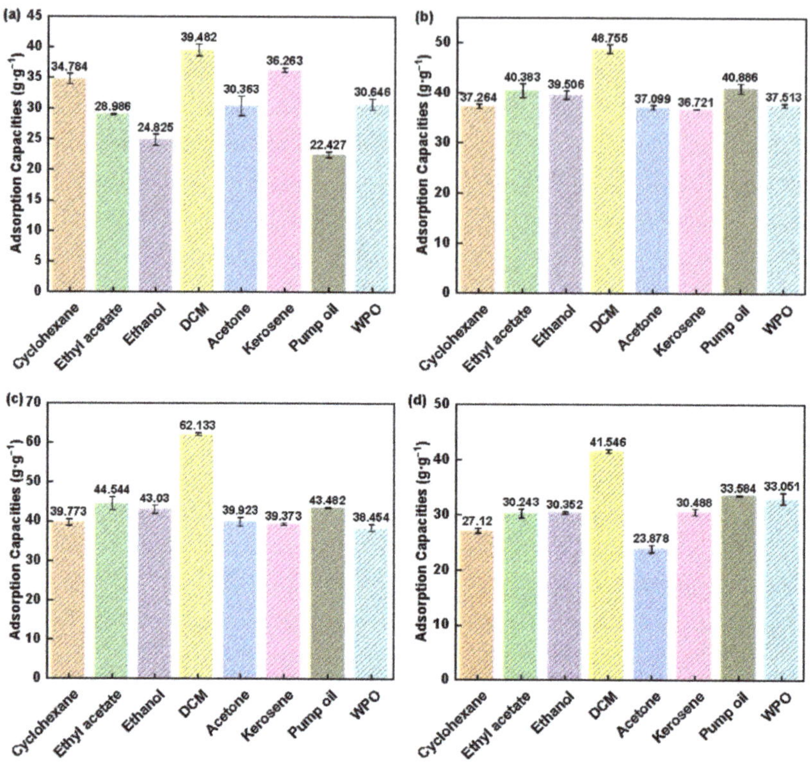

Figure 5. Adsorption of oils and organic solvents by different aerogels: (**a**) A0, (**b**) A1, (**c**) A2, and (**d**) A3, where DCM means dichloromethane, and WPO means wasted pump oil.

Table 2. Comparison of properties for various oil-absorbing materials in the literature.

Materials	Oil Type	Adsorption Capacity (g·g^{-1})	Literature
BN/SiO_2@PU	Chloroform	3	[26]
F-rGO@WS	Pump oil	3.63	[27]
Graphene aerogels	Diesel oil	25	[28]
Silica aerogels	Motor oil	15.1	[29]
Nanocellulose aerogels	Hexane	80	[30]
NC-MWCNT-NH_2 aerogel	Pump oil	43.48	This work

4. Conclusions

In this work, the NC-MWCNT-NH_2 nanocomposite aerogels with different contents of MWCNTs-NH_2 are prepared, and the adsorption performance is studied in detail. The NC-MWCNT-NH_2 nanocomposite aerogel with a MWCNT-NH_2 mass fraction of 0.1% has the best adsorption performance with the adsorption capacity to cyclohexane, ethyl acetate, anhydrous ethanol, dichloromethane, acetone, kerosene, pump oil, and wasted pump oil of 39.77 ± 0.82, 44.54 ± 1.67, 43.03 ± 1.06, 62.13 ± 0.36, 39.92 ± 1.09, 39.37 ± 0.28, 43.48 ± 0.06, and 38.45 ± 0.84 g·g^{-1}, respectively. These results are increased by 14.3, 53.7, 73.3, 57.4, 31.5, 8.6, 93.9, and 25.5%, respectively, compared with that of the pure NC aerogel. It can be found that the adsorption performance of the NC-MWCNT-NH_2 nanocomposite aerogel increases and then decreases with the increase in MWCNT-NH_2 content. The A2 sample exhibits the highest adsorption capacity for all oils and organic solvents. This result is consistent with the highest porosity and the retention of the lamellar structure of filamentous fibers in the nanocomposite aerogel.

In a word, the introduction of MWCNTs-NH_2 into nanocellulose aerogels could significantly improve the structure of aerogels, which is more conducive to the adsorption of oils and organic matter.

Supplementary Materials: The following supporting information can be downloaded at: https://www.mdpi.com/article/10.3390/polym17070869/s1, Figure S1: The morphology of MWCNTs-NH_2; Figure S2: The resulting nanocomposite aerogel placed on the stamen of a flower; Figure S3: The FTIR spectrum of MWCNTs-NH_2; Figure S4: The TGA curve of MWCNTs-NH_2; Figure S5. SEM image of pure nanocellulose. Table S1: BET specific surface area and porosity of sample; Table S2: Adsorption of oils and organic solvents by different aerogels; Table S3. Variation of weight percentage of aerogels in TGA. (First stage, from 260 to 320 °C; Second stage, from 380 to 520 °C); Table S4 Adsorption of oils and organic solvents by different aerogels.

Author Contributions: Conceptualization, methodology, H.G.; investigation, formal analysis, data curation, R.H. and Z.L.; data curation, writing—original draft preparation, R.H. and H.G.; funding acquisition, H.G. and R.H.; formal analysis, F.F. and S.S.; data curation, G.D. and X.L. All authors have read and agreed to the published version of the manuscript.

Funding: This research was funded by the National Natural Science Foundation of China, grant number 22366021, 22268023, Jiangxi Province Double Thousand Talents Plan, grant number jxsq2020101049, and China Postdoctoral Science Foundation, grant number 2021M691964.

Institutional Review Board Statement: Not applicable.

Data Availability Statement: No data were used for the research described in the article.

Conflicts of Interest: The authors declare no conflicts of interest.

References

1. Laitinen, O.; Suopajärvi, T.; Österberg, M.; Liimatainen, H. Hydrophobic, Superabsorbing Aerogels from Choline Chloride-Based Deep Eutectic Solvent Pretreated and Silylated Cellulose Nanofibrils for Selective Oil Removal. *ACS Appl. Mater. Interfaces* **2017**, *9*, 25029–25037. [CrossRef] [PubMed]
2. Xu, X.; Pan, Y.; Ge, L.; Chen, Y.; Mao, X.; Guan, D.; Li, M.; Zhong, Y.; Hu, Z.; Peterson, V.K.; et al. High-Performance Perovskite Composite Electrocatalysts Enabled by Controllable Interface Engineering. *Small* **2021**, *17*, 2101573. [CrossRef]
3. Abdelghafar, F.; Xu, X.; Guan, D.; Lin, Z.; Hu, Z.; Ni, M.; Huang, H.; Bhatelia, T.; Jiang, S.P.; Shao, Z. New Nanocomposites Derived from CationNonstoichiometric Bax(Co, Fe, Zr, Y)$O_{3-\delta}$ as Efficient Electrocatalysts for Water Oxidation in Alkaline Solution. *ACS Mater. Lett.* **2024**, *6*, 2985–2994. [CrossRef]
4. Akhlamadi, G.; Goharshadi, E.K. Sustainable and Superhydrophobic Cellulose Nanocrystal-Based Aerogel Derived from Waste Tissue Paper as a Sorbent for Efficient Oil/Water Separation. *Process Saf. Environ. Prot.* **2021**, *154*, 155–167. [CrossRef]
5. Yang, X.; Cranston, E.D. Chemically Cross-Linked Cellulose Nanocrystal Aerogels with Shape Recovery and Superabsorbent Properties. *Chem. Mater.* **2014**, *26*, 6016–6025. [CrossRef]
6. Rao, R.; Pint, C.L.; Islam, A.E.; Weatherup, R.S.; Hofmann, S.; Meshot, E.R.; Wu, F.; Zhou, C.; Dee, N.; Amama, P.B.; et al. Carbon Nanotubes and Related Nanomaterials: Critical Advances and Challenges for Synthesis toward Mainstream Commercial Applications. *ACS Nano* **2018**, *12*, 11756–11784. [CrossRef]
7. Zhao, S.; Siqueira, G.; Drdova, S.; Norris, D.; Ubert, C.; Bonnin, A.; Galmarini, S.; Ganobjak, M.; Pan, Z.; Brunner, S.; et al. Additive Manufacturing of Silica Aerogels. *Nature* **2020**, *584*, 387–392. [CrossRef]
8. Gu, H.; Gao, C.; Zhou, X.; Du, A.; Naik, N.; Guo, Z. Nanocellulose Nanocomposite Aerogel towards Efficient Oil and Organic Solvent Adsorption. *Adv. Compos. Hybrid Mater.* **2021**, *4*, 459–468. [CrossRef]
9. Li, M.; Wu, Z.; Chen, X.; Gan, F.; Teng, C.; Li, X.; Dong, J.; Zhao, X.; Zhang, Q. Continuous and Strong Polyimide Aerogel Fibers Enhanced by *Para*-Aramid Nanofibers Prepared via a "Reaction Spinning" for Thermal Insulation. *Chem. Eng. J.* **2024**, *486*, 150255. [CrossRef]
10. Abouzeid, R.E.; Khiari, R.; El-Wakil, N.; Dufresne, A. Current State and New Trends in the Use of Cellulose Nanomaterials for Wastewater Treatment. *Biomacromolecules* **2019**, *20*, 573–597. [CrossRef]

11. Zhang, Q.; Li, L.; Wu, H.; Cheng, Y.; Liu, C.; Fang, C. Recent Advances in Cellulose Based Aerogels with Various Dimensions: Design, Functionalization, and Applications. *Cellulose* **2024**, *32*, 1–27. [CrossRef]
12. Zhao, Y.; Zeng, Q.; Lai, X.; Li, H.; Zhao, Y.; Li, K.; Jiang, C.; Zeng, X. Multifunctional Cellulose-Based Aerogel for Intelligent Fire Fighting. *Carbohyd. Polym.* **2023**, *316*, 121060. [CrossRef]
13. Chen, Y.; Zhang, L.; Yang, Y.; Pang, B.; Xu, W.; Duan, G.; Jiang, S.; Zhang, K. Recent Progress on Nanocellulose Aerogels: Preparation, Modification, Composite Fabrication, Applications. *Adv. Mater.* **2021**, *33*, 2005569. [CrossRef] [PubMed]
14. Idumah, C.I.; Ezika, A.C.; Okpechi, V.U. Emerging Trends in Polymer Aerogel Nanoarchitectures, Surfaces, Interfaces and Applications. *Surf. Interfaces* **2021**, *25*, 101258. [CrossRef]
15. Gu, H.; Zhou, X.; Lyu, S.; Pan, D.; Dong, M.; Wu, S.; Ding, T.; Wei, X.; Seok, I.; Wei, S.; et al. Magnetic nanocellulose-magnetite aerogel for easy oil adsorption. *J. Colloid Interface Sci.* **2020**, *560*, 849–856.
16. Zhang, X.; Wang, H.; Cai, Z.; Yan, N.; Liu, M.; Yu, Y. Highly Compressible and Hydrophobic Anisotropic Aerogels for Selective Oil/Organic Solvent Absorption. *ACS Sustain. Chem. Eng.* **2019**, *7*, 332–340. [CrossRef]
17. Zhang, Y.; Yin, M.; Lin, X.; Ren, X.; Huang, T.-S.; Kim, I.S. Functional Nanocomposite Aerogels Based on Nanocrystalline Cellulose for Selective Oil/Water Separation and Antibacterial Applications. *Chem. Eng. J.* **2019**, *371*, 306–313. [CrossRef]
18. Sengupta, A.; Gupta, N.K. MWCNTs Based Sorbents for Nuclear Waste Management: A Review. *J. Environ. Chem. Eng.* **2017**, *5*, 5099–5114. [CrossRef]
19. Jiang, F.; Liu, H.; Li, Y.; Kuang, Y.; Xu, X.; Chen, C.; Huang, H.; Jia, C.; Zhao, X.; Hitz, E.; et al. Lightweight, Mesoporous, and Highly Absorptive All-Nanofiber Aerogel for Efficient Solar Steam Generation. *ACS Appl. Mater. Interfaces* **2018**, *10*, 1104–1112. [CrossRef]
20. Zou, J.; Liu, J.; Karakoti, A.S.; Kumar, A.; Joung, D.; Li, Q.; Khondaker, S.I.; Seal, S.; Zhai, L. Ultralight Multiwalled Carbon Nanotube Aerogel. *ACS Nano* **2010**, *4*, 7293–7302. [CrossRef]
21. Zhou, X.; Fu, Q.; Liu, H.; Gu, H.; Guo, Z. Solvent-Free Nanoalumina Loaded Nanocellulose Aerogel for Efficient Oil and Organic Solvent Adsorption. *J. Colloid Interface Sci.* **2021**, *581*, 299–306. [CrossRef] [PubMed]
22. Jiang, F.; Hsieh, Y.-L. Cellulose Nanofibril Aerogels: Synergistic Improvement of Hydrophobicity, Strength, and Thermal Stability via Cross-Linking with Diisocyanate. *ACS Appl. Mater. Interfaces* **2017**, *9*, 2825–2834. [CrossRef] [PubMed]
23. Yu, L.; Li, C.M.; Zhou, Q.; Gan, Y.; Bao, Q.L. Functionalized Multi-Walled Carbon Nanotubes as Affinity Ligands. *Nanotechnology* **2007**, *18*, 115614. [CrossRef]
24. Bora, D.K.; Bavdane, P.P.; Bhatt, B.; Nikumbe, D.Y.; Sethia, G.; Nagarale, R.K. Ruthenium-Anchored Aminated MWCNTs/Polyaniline Membrane Electrode Assembly for Alkaline Water Splitting. *J. Electroanal. Chem.* **2024**, *954*, 118027. [CrossRef]
25. Hu, Y.; Jiang, Y.; Ni, L.; Huang, Z.; Liu, L.; Ke, Q.; Xu, H. An elastic MOF/graphene aerogel with high photothermal efficiency for rapid removal of crude oil. *J. Hazard. Mater.* **2023**, *443*, 130339. [CrossRef]
26. Zhai, Y.; Yuan, X. One-Pot Fabrication of Hydrophobic, Superelastic, Harakeke-Derived Nanocellulose Aerogels with Excellent Shape Recovery for Oil Adsorption and Water-in-Oil Emulsion Separation. *Inter. J. Biol. Macromol.* **2024**, *280*, 135489. [CrossRef]
27. Sam, E.K.; Sam, D.K.; Chen, J.; Lv, X.; Liu, J. Hydrophobic Porous BN/SiO_2@PU as Ternary Adsorbents for Efficient Oil/Water Separation. *J. Porous Mater.* **2020**, *27*, 1149–1158. [CrossRef]
28. Huang, W.; Zhang, L.; Lai, X.; Li, H.; Zeng, X. Highly Hydrophobic F-rGO@wood Sponge for Efficient Clean-up of Viscous Crude Oil. *Chem. Eng. J.* **2020**, *386*, 123994. [CrossRef]
29. Huang, J.; Yan, Z. Adsorption Mechanism of Oil by Resilient Graphene Aerogels from Oil–Water Emulsion. *Langmuir* **2018**, *34*, 1890–1898. [CrossRef]
30. Wang, D.; McLaughlin, E.; Pfeffer, R.; Lin, Y.S. Adsorption of Oils from Pure Liquid and Oil–Water Emulsion on Hydrophobic Silica Aerogels. *Separ. Purif. Technol.* **2012**, *99*, 28–35. [CrossRef]

nanomaterials

MDPI

Review

Graphene-Based Impregnation into Polymeric Coating for Corrosion Resistance

Arti Yadav [1], Santosh Panjikar [2,3] and R. K. Singh Raman [1,4,*]

[1] Department of Chemical and Biological Engineering, Monash University, Melbourne, VIC 3800, Australia; arti.yadav@monash.edu
[2] Australian Nuclear Science and Technology Organisation, Australian Synchrotron, 800 Blackburn Road, Clayton, Melbourne, VIC 3168, Australia; santoshp@ansto.gov.au
[3] Department of Biochemistry and Molecular Biology, Monash University, Melbourne, VIC 3800, Australia
[4] Department of Mechanical and Aerospace Engineering, Monash University, Melbourne, VIC 3800, Australia
* Correspondence: raman.singh@monash.edu

Abstract: This review explores the development and application of the impregnation of graphene-based materials into polymeric coatings to enhance corrosion resistance. Derivatives of graphene, such as graphene oxide (GO) and reduced graphene oxide (rGO), have been increasingly integrated into polymer matrices to enhance polymers' mechanical, thermal, and barrier properties. Various synthesis approaches, viz., electrochemical deposition, chemical reduction, and the incorporation of functionalised graphene derivatives, have been explored for improving the dispersion and stability of graphene within polymers. These graphene-impregnated coatings have shown promising results in improving corrosion resistance by enhancing impermeability to corrosive agents and reinforcing mechanical strength under corrosive conditions. While the addition of graphene notably enhances coating performance, challenges remain in achieving uniform graphene dispersion and addressing the trade-offs between thickness and flexibility. This review highlights current advancements, limitations, and future directions, with a particular emphasis on optimising the synthesis techniques to maximise corrosion resistance while maintaining coating durability and economic feasibility.

Keywords: graphene-based nano fillers; polymer-matrix composites; corrosion resistance

1. Introduction

Corrosion is destructive chemical/electrochemical process that causes great harm to the economy, as well as safety challenges worldwide. The annual losses due to corrosion are estimated at approximately USD 2.5 trillion, i.e., 3.4% of the global Gross Domestic Product (GDP) [1,2]. Although corrosion is an inevitable process (for fundamental thermodynamic reasons), it can be significantly mitigated using advanced corrosion-resistant materials, advanced material design, effective cathodic protection systems, cutting-edge corrosion inhibitors, and high-performance coatings.

In recent years, the use of various types of corrosion inhibitors, including inorganic, organic, and natural inhibitors, has gained significant attention due to their efficiency in mitigating corrosion. Natural inhibitors derived from plant extracts, in particular, have been highlighted for their eco-friendliness and efficiency. For instance, the study *"Investigating the Adsorption and Corrosion Protection Efficacy and Mechanism of Marjoram Extract on Mild Steel"* demonstrates the effectiveness of marjoram extract as a natural inhibitor through adsorption mechanisms [3]. Similarly, the study *"Electrochemical, Structural and*

https://doi.org/10.3390/nano15070486

Thermodynamic Investigations of Methanolic Parsley Extract as a Green Corrosion Inhibitor for C37 Steel" underscores the potential of green inhibitors to provide sustainable corrosion protection [4]. Furthermore, the study *"Experimental and DFT Atomistic Insights into the Mechanism of Corrosion Protection of Low-Carbon Steel in an Acidic Medium by Polymethoxyflavones from Citrus"* offers a molecular-level understanding of the inhibition mechanisms provided by natural compounds, highlighting the role of flavonoids in enhancing adsorption and forming protective layers on metal surfaces [5]. These studies collectively emphasise the growing interest in eco-friendly corrosion inhibition strategies.

Though most means of ameliorating/mitigating corrosion may have been exhausted because of the age-old nature of corrosion and its huge economic implications, there have been recent advances that include the application of novel materials, such as graphene, for corrosion resistance [6,7]. In fact, the global market for graphene-based coatings is projected to grow significantly, reaching approximately USD 1 billion by 2028 [8], driven by a steady compound annual growth rate (CAGR) and demand across automotive, aerospace, and electronics applications.

In recent years, there has been a growing interest in eco-friendly corrosion inhibition approaches, particularly those involving graphene and biopolymer-based nanocomposites. As highlighted in the book *"Corrosion Protection of Metals and Alloys Using Graphene and Biopolymer-Based Nanocomposites"*, these materials offer significant advantages due to their non-toxic nature and exceptional barrier properties. The integration of biopolymers, such as chitosan, cellulose, and polyaniline, with graphene derivatives not only enhances the mechanical and anticorrosive properties of the coatings but also aligns with the increasing demand for sustainable and environmentally friendly corrosion protection solutions. This approach leverages the excellent impermeability of graphene to inhibit the ingress of corrosive species, while biopolymers contribute to enhanced adhesion and flexibility of the coating.

In recent years, the incorporation of graphene and its derivatives, such as graphene oxide (GO) and reduced graphene oxide (rGO), into polymer matrices has led to effective anticorrosion coatings that perform exceptionally even in aggressive corrosive conditions [9–22]. Incorporation of suitable derivatives of GO has been found to be particularly effective because the two-dimensional sheets of GO constitute a tortuous path for the inward transport of the corrosive species. Underpinned by this mechanism, Huang et al. [9] demonstrated that a nanocomposite coating of polyaniline (PANI) and GO nanosheets functionalised with polydopamine (PDA) considerably retarded the corrosion resistance of steel. Pu et al. [23] developed a composite coating of functionalised GO and a waterborne polyurethane (WPU) matrix, which forms a robust interpenetrating network, enhancing both hydrophobicity and adhesion to metal surfaces, and this resulted in an impressive 98% decrease in corrosion rate. Kumari et al. [24] showed that the incorporation of reduced GO (rGO) into a poly (4-vinylpyridine-co-butyl methacrylate) (PVPBM) matrix enhanced the corrosion protection efficiency to 95.4% while attaining high adhesion strengths. A study on GO–chitosan–silver composites demonstrated two orders of magnitude improvement in corrosion resistance [25]. Furthermore, the composites also enhanced thermal stability [21]. However, incorporation of GO into a polymeric matrix often raises the challenge of agglomeration of GO. Effective dispersion and compatibility of GO within the resin matrix are typically achieved through suitable covalent and non-covalent functionalisation of GO, such as through the incorporation of noble metals (e.g., Pt, Pd) or a semiconductor oxide (e.g., TiO_2, ZnO), as well as chemical bonding with the two primary sites on GO/rGO sheets, i.e., oxygenated groups (–OH, –COC, and –COOH) for condensation and esterification, and carbon double-bond (C=C) sites on the basal plane for free-radical bonding [26]. Such functionalisation facilitates bonding with the polymer matrix, further enhancing GO's ionic barrier properties [27]. Collectively, these studies underscore the versatility of the incorporation of graphene and GO-based coatings for varied industrial applications where both high corrosion resistance due to coatings and mechanical durability are of paramount importance [11,28].

This review discusses the incorporation of graphene and its derivatives into polymer composites as a cutting-edge approach to enhancing corrosion resistance. Figure 1 presents a schematic overview of the synthesis of graphene and its derivatives as fillers in polymer composites. By leveraging the excellent properties of graphene oxide (GO) and reduced graphene oxide (rGO), such as a high surface area, mechanical robustness, and impermeability, these composites considerably lengthen the transport pathways for corrosives within the coating, effectively and considerably retarding corrosion rates, making them highly attractive for industrial applications.

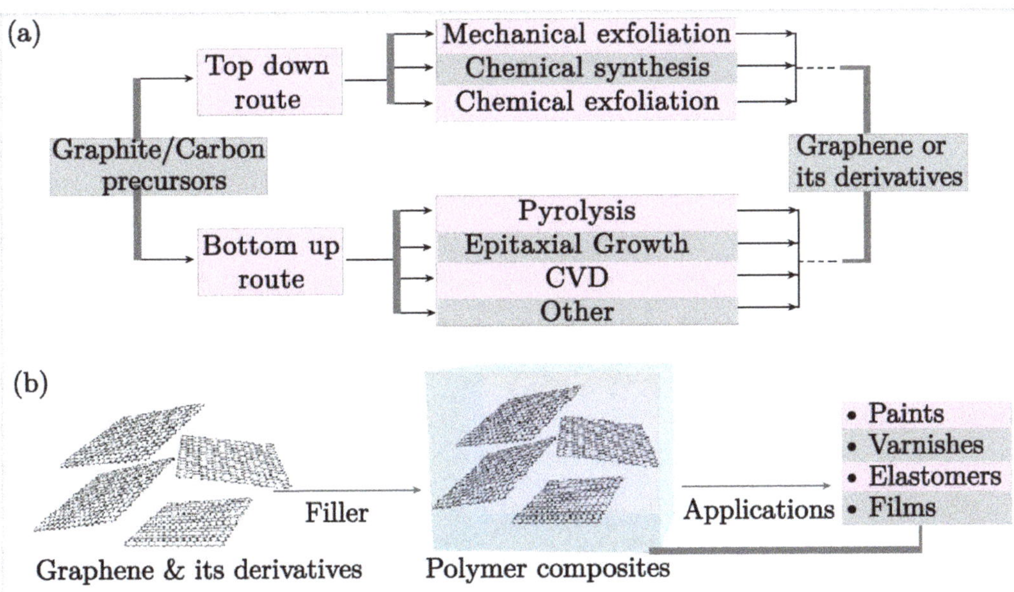

Figure 1. Schematic representation of (**a**) synthesis of graphene, (**b**) graphene and its derivatives as fillers in polymer composites.

2. Preparation of Graphene and Its Derivatives

Graphene demonstrates a unique combination of properties essential for an ideal coating material, i.e., outstanding impermeability, pronounced chemical inertness, and exceptional mechanical toughness; no other single material possesses these attributes simultaneously [29–32]. Graphene is a two-dimensional (2D) sheet of sp^2-hybridised carbon atoms arranged in a hexagonal lattice, as shown in Figure 2. This structure confers graphene with remarkable properties, making it the most investigated and attractive materials in recent years. Its exceptional characteristics include a remarkable electron mobility (2.5×10^5 $cm^2V^{-1}s^{-1}$), outstanding thermal conductivity (3000 WmK^{-1}), excellent mechanical properties (Young's modulus of 1 TPa), remarkable chemical stability, very high specific surface area (<2600 m^2g^{-1}), and exceptional optical transparency (97.4% transmittance at 550 nm) [33–36]. These properties can be tuned by adjusting factors such as the defect density, porosity, and number of graphene layers [33,37].

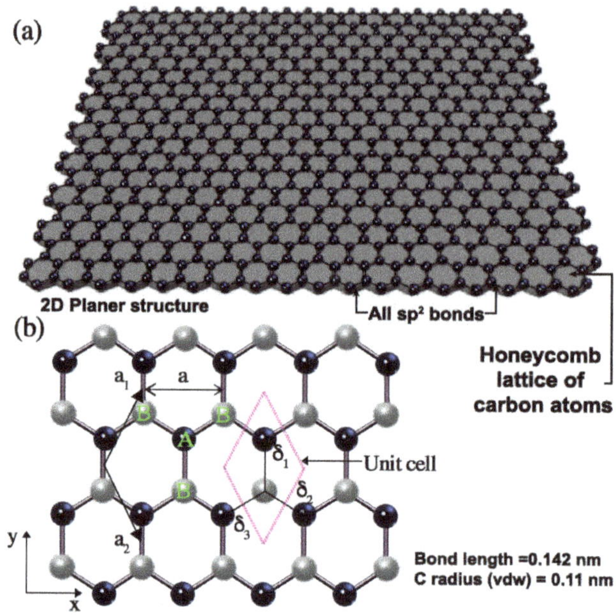

Figure 2. (**a**) Graphene lattice structure. (**b**) The honeycomb lattice configuration of monolayer graphene, where the grey and black circles represent carbon atoms located at specific lattice sites. Each atom in sublattice A (shown in green) has three nearest neighbours in sublattice B (also shown in green) and vice versa.

The π–π stacking interactions and strong van der Waals forces between graphene layers contribute to a tightly packed structure, which can enhance the barrier properties of graphene-based coatings. When graphene sheets aggregate, they form a dense layer that can effectively block the passage of corrosive agents like water, oxygen, and chloride ions, thereby protecting the underlying material from corrosion.

The synthesis of graphene includes mechanical exfoliation, epitaxial growth, liquid-phase exfoliation, electrochemical exfoliation, chemical oxidation and reduction processes, and chemical vapour deposition (CVD) [38–47]. The epitaxial growth of graphene on suitable substrates, such as silicon carbide (SiC), requires precise thermal control [46,47]. Liquid-phase exfoliation utilises specific solvents capable of dispersing and exfoliating individual layers, from graphite, thereby resulting in suspensions of graphene flakes within the solvent [48]. In chemical synthesis, graphite undergoes oxidation into graphene oxide (GO) flakes [49], which are subsequently chemically reduced to produce reduced graphene oxide (rGO) [45]. However, the graphene obtained through this method is not entirely pristine and contains numerous functional groups, which is why it is frequently designated as rGO, rather than graphene.

Electrodeposition of rGO has emerged as a promising technique due to its ability to provide precise control over the thickness of the deposited films and ensure better adherence and stability on metal substrates. This technique not only facilitates a uniform deposition but also enables the incorporation of functional materials such as nanoparticles to enhance the properties of rGO. For instance, the study *"Gold Nanoparticles Decorated Graphene as a High Performance Sensor for Determination of Trace Hydrazine Levels"* demonstrates the effectiveness of electrodeposition in achieving stable and uniform coatings with improved functional properties [50]. The enhanced adherence and stability provided by electrodeposition make it particularly suitable for applications requiring durable and high-performance anticorrosion coatings.

The chemical vapour deposition (CVD) technique for graphene synthesis entails the regulated generation of carbon atoms via catalytic dissociation of a hydrocarbon precursor on a substrate; the carbon atoms thus generated during this process self-organise on the substrate to form graphene [38,51,52]. Additionally, alternative methodologies such as spray coating, electrophoretic deposition, chemical conversion coating, flame spraying, electrochemical deposition, and laser-assisted transformation have also been explored for the synthesis of graphene.

CVD enables the production of high-quality, pristine graphene, rendering it the most prevalently utilised technique when highest-quality graphene is required [53–56]. CVD-produced graphene films are characterised by minimal defects and high purity. However, the substrates used for CVD graphene must be capable of catalytically decomposing the hydrocarbon precursor to generate nascent carbon atoms, as depicted in [51].

2.1. Preparation Methods of Graphene and Its Derivatives for Polymer Composites

The preparation methodologies significantly influence the composite structure, dispersion quality, and resultant properties. Common methods for preparing graphene-based polymer composites (Figure 3b–f) include solution mixing, in situ polymerisation, electrochemical techniques (electrospinning and electrodeposition), layer-by-layer (LbL) assembly, and melt blending. These techniques offer various advantages in terms of dispersion control, structural precision, scalability, and environmental considerations.

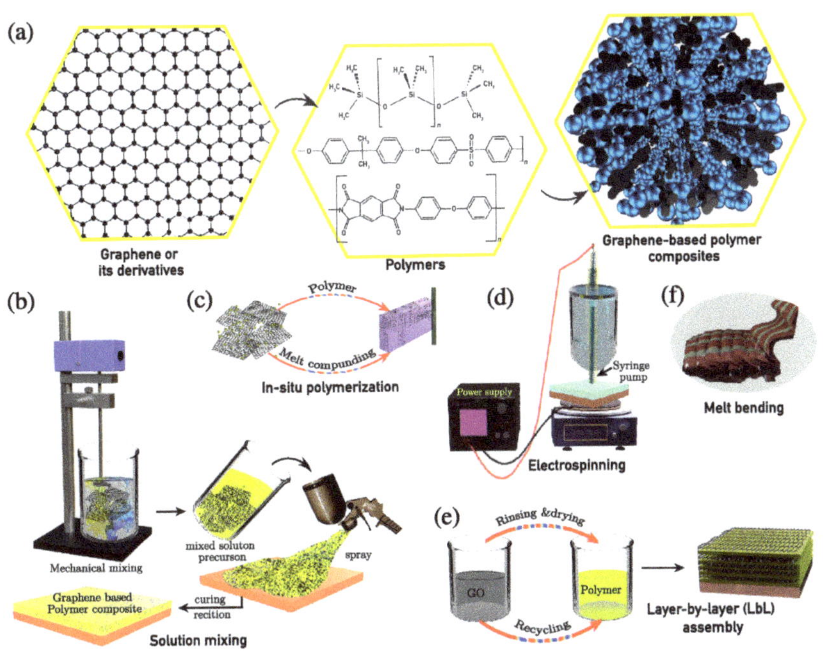

Figure 3. (**a**) Schematic illustration of a polymer composite with impregnated graphene or its derivatives. This illustrates various widely employed preparation methods of graphene and its derivatives for polymer composites: (**b**) solution mixing, (**c**) in situ polymerisation, (**d**) electrochemical reaction, (**e**) layer-by-layer (LbL) assembly, and (**f**) melt blending.

2.1.1. Solution Mixing

Solution mixing is a widely utilised technique for synthesising graphene and its derivative-based polymer composites due to its simplicity and efficiency. The process involves dispersing graphene or its derivatives (e.g., graphene oxide, reduced graphene oxide, functionalised graphene) in an appropriate solvent such as N,N-dimethylformamide (DMF), tetrahydrofuran (THF), chloroform, or ethanol through simple mixing, sheer/mechanical stirring, or ultrasonication to ensure uniform exfoliation (Figure 3b). The polymer is then introduced, followed by solvent removal via evaporation or vacuum distillation to obtain the final composite [57]. The quality of dispersion and interfacial interactions strongly influence the resulting properties. To enhance these interactions, surface functionalisation strategies such as polymer grafting or surfactant-assisted dispersion can be employed [13].

2.1.2. In Situ Polymerisation

In situ polymerisation ensures a uniform dispersion of graphene and its derivatives by initiating polymerisation in the presence of graphene sheets, allowing polymer chains to grow around the sheets, thereby preventing agglomeration (Figure 3c) [57–59]. The process starts with the dispersion of graphene in a monomer solution, followed by polymerisation using thermal, chemical, or photoinitiators. Controlled polymerisation methods such as atom transfer radical polymerisation (ATRP) and reversible addition–fragmentation chain transfer (RAFT) polymerisation allow molecular weight control and improved compatibility between graphene derivatives and the polymer matrix [13,60].

2.1.3. Electrochemical Reaction

Electrospinning is a scalable method for fabricating graphene and its derivatives with a high aspect ratio for polymer-nanofibre composites. The process employs a high-voltage (5–20 kV) electric field to create a charged polymer jet that elongates and solidifies into nanoscale fibres, as illustrated in Figure 3d [61,62]. Common solvents used in electrospinning include DMF, acetone, and chloroform, which help control polymer chain entanglement and jet stability. Advanced electrospinning techniques, such as coaxial electrospinning and emulsion electrospinning, enable the fabrication of core–shell and multi-layered nanofibres with tailored functionalities [63].

Electrodeposition is an efficient and precise technique for fabricating nanocomposites via electrochemical processes. This method facilitates the electro-polymerisation of a graphene/polymer nanocomposite from a monomer, dopant (if needed), and graphene oxide (GO), under controlled electrochemical conditions. The deposition occurs at a defined potential and ceases upon reaching the requisite charge transfer threshold [62]. As a result, the synthesised nanocomposite forms a uniform and adherent coating on the electrode surface.

2.1.4. Layer-by-Layer (LbL) Assembly

The LbL assembly technique is a highly controlled synthesis method for constructing multilayered graphene and its derivatives for polymer-fibre nanocomposites. The process involves sequential deposition of graphene-based materials and polymeric layers, utilising electrostatic interactions, hydrogen bonding, or van der Waals forces (Figure 3e) [64]. This method enables precise tuning of the thickness, surface chemistry, and composite properties. The stability and charge compatibility of graphene suspensions are crucial for achieving uniform film growth. Common solvents used in LbL assembly include water, ethanol, and methanol, ensuring the stable dispersion of graphene nanosheets [60].

2.1.5. Melt Blending

Melt blending is an industry-preferred, solvent-free synthesis method for fabricating graphene and its derivatives for polymer-fibre nanocomposites. The technique involves dispersing graphene into a molten polymer matrix under high shear forces, typically using twin-screw extrusion or injection moulding (Figure 3f) [62]. The effectiveness of melt blending depends on graphene's dispersion quality, which is often enhanced by surface modifications or compatibilisers. While this method offers scalability and environmental advantages, achieving a uniform dispersion remains challenging due to the high viscosity of molten polymers. Melt blending is widely used for fabricating conductive polymer composites, structural reinforcements, and lightweight functional materials [65].

3. Incorporation of Graphene and Its Derivatives into Polymer Composites

Incorporation of graphene and its derivatives into polymer composites has emerged as a promising strategy for enhancing corrosion resistance (schematic illustration shown in Figure 3a). Suitably functionalised graphene nanoplatelets have seamlessly been integrated into a wide range of polymeric matrices, which include epoxy [66], polyaniline [67], polyurethane [68], polyamide [69], polystyrene [70], polymethylmethacrylate [71], polypropylene [72], polydopamine [73], polyetherimide [74], polyvinyl butyral [75], polyvinyl alcohol [76], polypyrrole [77], biopolymer [78], silane [79], and polydimethylsiloxane [80], among others. These composites enhance properties like durability, flexibility, and corrosion resistance.

For instance, polyaniline/graphene composites (PAGCs) have demonstrated remarkable barrier properties against oxygen and water, outperforming traditional materials like neat polyaniline or even polyaniline/clay composites [67,81]. This enhanced performance is primarily due to the well-dispersed graphene nanoplatelets that inhibit the transport of corrosive molecules, thereby enhancing the corrosion protection capabilities of the coatings [81]. Functional groups of graphene oxide (GO) can be further functionalised to facilitate bonding with the polymer matrix. Incorporation of functionalised GO into polystyrene remarkably improved corrosion protection efficiency (from 37.90% to 99.53%, with just 2 wt% of the additive) [70]. Likewise, graphene/pernigraniline composites (GPCs) enhanced corrosion protection of copper substrates. These composites not only mitigate the corrosion but can also enhance other properties of polyaniline [74,82]. The use of 4-aminobensoyl group-functionalised graphene-like sheets as a conductive filler in these composites enables improved dispersion within the polymer matrix, which is crucial for forming an effective barrier against corrosive agents [11]. However, this requires the specific functionalisation of GO for effective bonding with the polymer matrix, and the incorporation of properly functionalised GO can improve both mechanical and barrier properties of the coating [13]. For instance, a study on PSS-PANI/rGO composite found the ultimate tensile strength and toughness of the epoxy composite with just 0.5 wt% PSS-PANI/rGO to be improved by 39% and 127%, respectively, when compared to the respective values of the plain epoxy [83]. Suitably functionalised GO facilitates the formation of a network that enables improved properties across the coating. With impressively improved properties of functionalised GO-polymer composites, as discusses earlier, these advanced materials hold significant potential for practical applications in various industries, particularly in corrosive environments [84]. Table 1 outlines the advancements in graphene-based polymer composites, highlighting the role of various polymer matrices and fabrication techniques in the enhancement of corrosion, mechanical, and barrier properties.

Table 1. Comparative efficiency of various polymer coatings with graphene nanosheets (GNSs).

Base Polymer	GNS Type	GNS Concentration	Fabrication Method	Observed Effect	Reference
Epoxy Resin	Graphene Oxide (GO)	0.5 wt%	Solution Mixing	Enhanced corrosion resistance and mechanical properties	[66]
Polyurethane	Functionalised GO	1.0 wt%	In situ Polymerisation	Improved barrier properties and thermal stability	[68]
Polyaniline	Reduced GO (rGO)	2.0 wt%	Electrophoretic Deposition	Increased electrical conductivity and anticorrosion performance	[67]
Polyvinyl Alcohol	GO-Silver Nanocomposite	0.3 wt%	Layer-by-Layer Assembly	Superior antibacterial activity and corrosion protection	[85]
Poly (methyl methacrylate)	GO	0.75 wt%	Melt Blending	Enhanced UV resistance and mechanical strength	[71]

The incorporation of few-layer graphene and graphene oxide (GO) as fillers in polymer composites has emerged as a promising strategy to enhance corrosion resistance [12]. Both graphene and GO serve to improve the barrier properties of these composites by effectively retarding the permeability of corrosive agents through the coating. Coatings of composites incorporated with platelets of functionalised GO produce a considerably superior corrosion resistance when compared to the composite coatings with other fillers [12]. This enhancement is attributed to the structure of graphene, which is an atomically thick 2D platelet with a remarkably high aspect ratio, which makes the path very tortuous for permeating corrosive agents, thereby providing a considerably more effective barrier against corrosion [19]. Besides providing an enhanced permeability barrier, the polymer-GO composite also possesses enhanced mechanical properties and overall durability. In this context, electroactive polyimide/graphene nanocomposite coatings are particularly attractive for anticorrosive properties. Such coatings are prepared through thermal imidisation to incorporate carboxyl-graphene nanosheets that significantly retard gas permeation, thereby enhancing corrosion resistance due to the coatings [28]. Further, the presence of amino-capped aniline trimer (ACAT) units within such composites is noted for their redox catalytic capabilities, which can induce the formation of passive metal oxide layers on the substrate, providing an additional layer of protection against corrosion [28].

3.1. Graphene/Polyaniline Composites

Chang et al. [11] demonstrated very impressive anticorrosion properties due to coatings of polyaniline/graphene composites (PAGCs). The PAGC coatings considerably reduce the permeability of gases such as oxygen (O_2) and water vapour (H_2O) compared to traditional polyaniline and polyaniline/clay composites (PACCs). Specifically, PAGCs with 0.1 wt.% ABF-G loading showed a reduction by approximately 55% and 60% in O_2 and H_2O permeabilities, respectively. This improvement is attributed to the high aspect ratio of the 4-aminobenzoyl group-functionalised graphene-like sheets (ABF-Gs) and a high aspect ratio of approximately 500 (compared to 220 for clay), which presents a more tortuous path for gas molecules, effectively enhancing the barrier properties of the composites. The study further reveals that increasing the ABF-G loading results in even greater enhancements in barrier properties, making PAGCs more effective gas barrier materials for anticorrosion coatings than PACCs. Figure 4a illustrates the synthesis process schematic for PANI/graphene composites (PAGCs), demonstrating the structural and functional characteristics of the composites, and the TEM images (Figure 4c,d) confirm that ABF-G is well-dispersed within the polyaniline matrix, indicating good compatibility and effective functionalisation. These structural modifications and the presence of nitrogen from the functional groups contribute to the enhanced barrier properties and electrical conductivity of PAGCs.

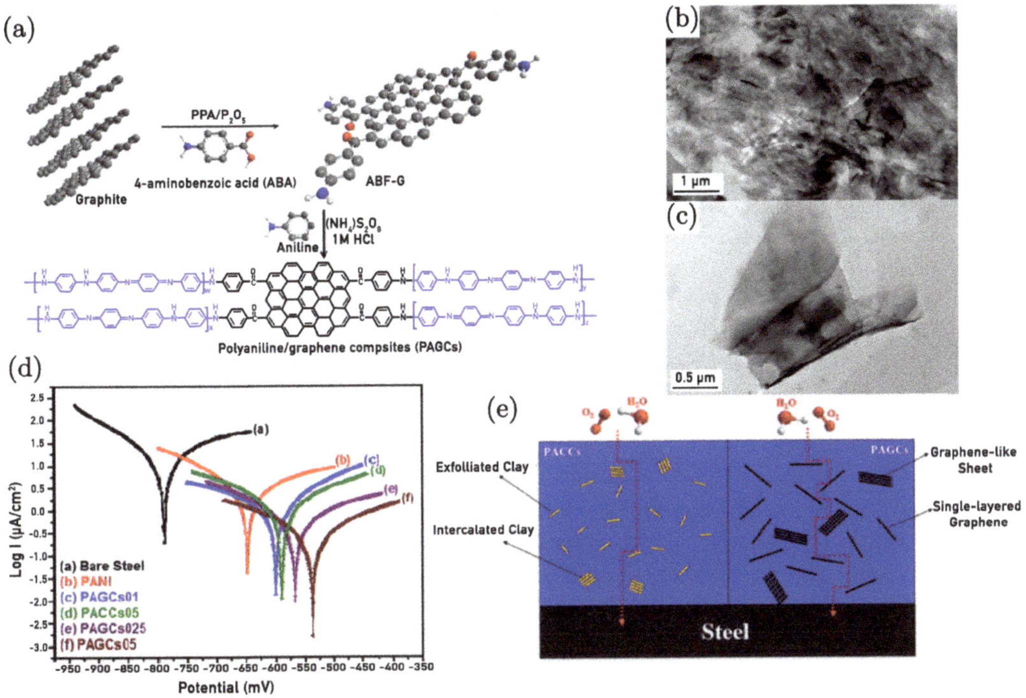

Figure 4. (**a**) Synthesis process for PANI/graphene composites (PAGCs). (**b**) TEM images of PAGCs05 at a low magnification and (**c**) high magnification. (**d**) Tafel plots comparing bare steel and PANI-coated, PAGCs01-coated, PACCs05-coated, PAGCs025-coated, and PAGCs05-coated electrodes, assessed in a 3.5 wt.% NaCl solution. (**e**) Conceptual illustration of O_2 and H_2O molecules traversing a tortuous path through PACCs and PAGCs [11].

Another study on GO/polyaniline composites by Gao et al. [86] investigated the enhancement of corrosion resistance due to waterborne epoxy coatings developed upon the incorporation of polyaniline (PANI) nanorods and nitrogen and fluorine dual-doped graphene oxide (NFGO) composites. XRD confirmed the crystalline structure of PANI and the integration of NFGO. TEM images reveal that PANI forms short rod-like structures with diameters of approximately 200–400 nm. When combined with NFGO, PANI rods grow on the graphene surface, increasing their aspect ratio due to additional nucleation sites provided by the graphene. They identified the morphology of PANI to be significantly influenced by the polymerisation time, which is crucial for the coating's performance. A shorter than optimal polymerisation time results in a coral-like particulate structure, while an excessively long time leads to agglomeration and an irregular morphology. The optimal polymerisation time results in a rod-like structure, which is essential for effective dispersion and interaction with nitrogen and fluorine dual-doped graphene oxide (NFGO). Electrochemical impedance spectroscopy (EIS) of PANI/NFGO/WEP coatings conferred the highest corrosion resistance, with R_{ct} values considerably higher than that for neat PANI coating. The salt spray test results show that PANI/NFGO/WEP coatings effectively inhibit the penetration of corrosive media, maintaining good adhesion even under wet conditions. This is attributed again to the two-dimensional GO blocking water molecule penetration, as a result of the effective bonding of GO with the polymeric matrix. The coated substrate demonstrated minimal rusting, even after extended exposure, highlighting its durability.

The enhanced corrosion resistance observed in the coatings can be attributed to the synergistic effect of polyaniline (PANI) and nitrogen and fluorine dual-doped graphene oxide (NFGO). The dual-doping of NFGO with nitrogen and fluorine not only improves electrical conductivity but also introduces additional active sites for the polymerisation of PANI. This facilitates a more uniform dispersion and stronger interaction between PANI and NFGO within the coating matrix. The high specific surface area and excellent conductivity of NFGO, combined with the passivation effect of PANI, create a robust barrier against the ingress of corrosive media.

In addition to the barrier protection mechanism, the incorporation of PANI provides anodic protection by forming a passivation layer on the substrate surface. This layer can shift the corrosion potential to more noble values, thereby reducing the anodic dissolution of the metal substrate. PANI, being a conductive polymer, also enables electron transfer processes that support the formation of passive oxide layers, further enhancing the protective capabilities of the coating. This dual protection mechanism—barrier and anodic protection—significantly improves the corrosion resistance compared to traditional polymer/clay composites.

The importance of this dual protection mechanism is further demonstrated in Figure 5d, which shows a substantial reduction in corrosion current density for the PANI/NFGO composite coatings compared to the control samples. The lower corrosion current density indicates a more effective suppression of electrochemical reactions at the metal surface, corroborating the protective role of both PANI and NFGO.

Both studies [11,86] underscore the potential of polyaniline/graphene composites as advanced anticorrosion coatings. The combination of a more tortuous path for diffusing gas molecules and the formation of a passive protective layer effectively prevent corrosion, making these composites a promising alternative to traditional coatings.

3.2. Graphene Oxide–Polymer Bilayer Coating

George et al. [85] studied a robust bilayer coating of graphene oxide (GO) and acrylic polymer on Cu-Ni alloys, for corrosion resistance in chloride environments. GO was prepared using a modified Hummers' method and deposited onto the alloy via electrophoretic deposition (EPD) at a constant voltage of 10 V for 60 s, followed by overnight drying at room temperature. The acrylic polymer layer was applied using a dip coating technique, where the GO-coated samples were immersed in a 10% PMMA solution in THF for 30 s, followed by lifting at a controlled speed. The process was repeated twice, and the samples were cured at 110 °C for 12 h. In their study, the surface characteristics of the GO-coated Cu-Ni alloy, the polymer-coated alloy, and the GO–polymer bilayer-coated alloy were examined by AFM. The GO-polymer bilayer coating was found to be 5–6 μm thick, which provided a robust barrier against corrosion, with a protection efficiency of 99.8%, attributed to the synergistic properties of GO and the polymer. As evidenced by the EIS results (Figure 6b), the bilayer coating shows a significantly higher coating resistance compared to the uncoated and single-layer coated samples, confirming the superior protective nature of the former in aggressive chloride environments. The GO layer enhances impermeability, while the polymer provides good film-forming, insulating, and adhesion properties.

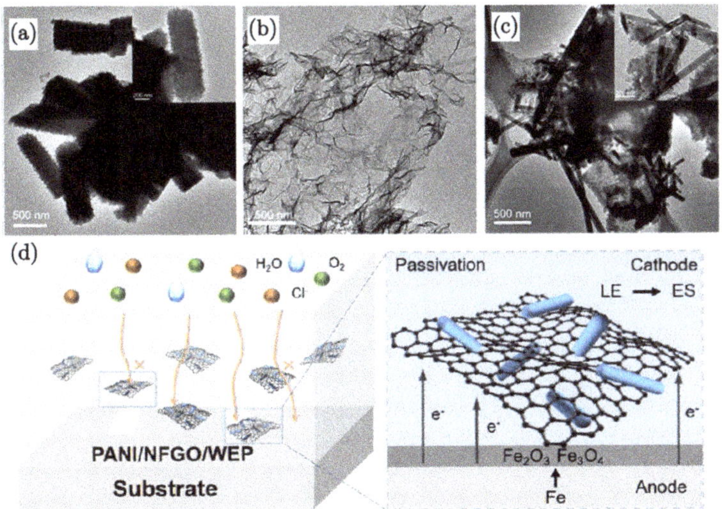

Figure 5. (**a**) TEM image of PANI, (**b**) TEM image of NFGO, and (**c**) TEM image of the PANI/NFGO composite. (**d**) Schematic illustration of the corrosion protection mechanism [86].

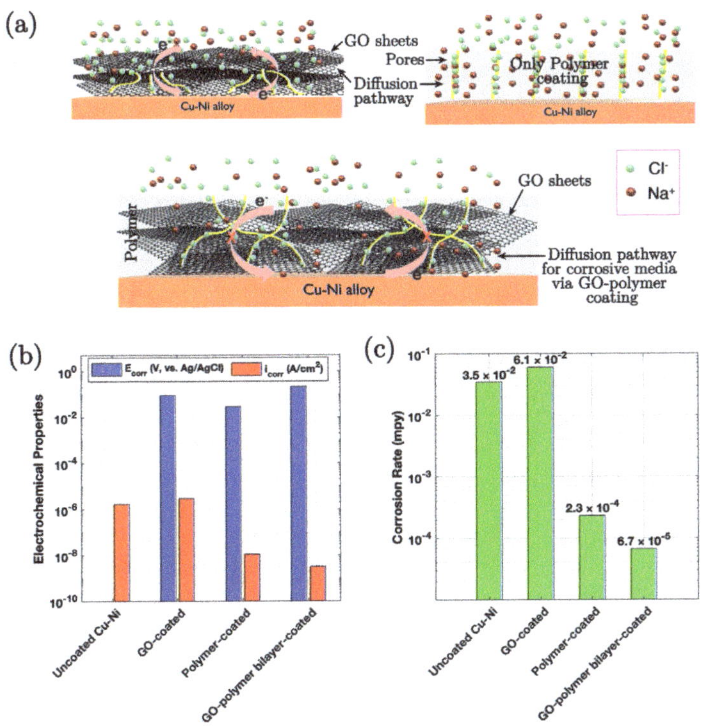

Figure 6. (**a**) Schematic representation of the corrosion protection mechanism for GO-coated, polymer-coated, and GO–polymer bilayer-coated Cu-Ni samples. (**b**) Bar chart depicting E_{corr} and i_{corr} values of the uncoated, GO-coated, polymer-coated, and GO–polymer bilayer-coated Cu-Ni samples in a 3.5% (w/v) NaCl solution after 1 h of stabilisation. (**c**) Corrosion rates of uncoated, GO-coated, polymer-coated, and GO–polymer bilayer-coated Cu-Ni samples in a 3.5% (w/v) NaCl solution, in mils per year [85].

3.3. Graphene–Epoxy Powder Coating

Zhang et al. [87] investigated the impact of graphene nanosheet addition to epoxy powder coatings on the wear and corrosion resistance. A schematic illustration of the wear improvement mechanism and corrosion resistance mechanism in the G/EP composite powder coating are given in Figure 7. The addition of 0.4 wt% graphene nanosheets significantly enhanced the coating properties. The graphene nanosheets enhanced lubrication of the wearing surface, thereby reducing wear. In addition, the dense structure of the nanosheet-containing surface improved adhesion, which suppressed deformation and microcracks, resulting in a shallow and narrow wear track, beneficial for wear resistance.

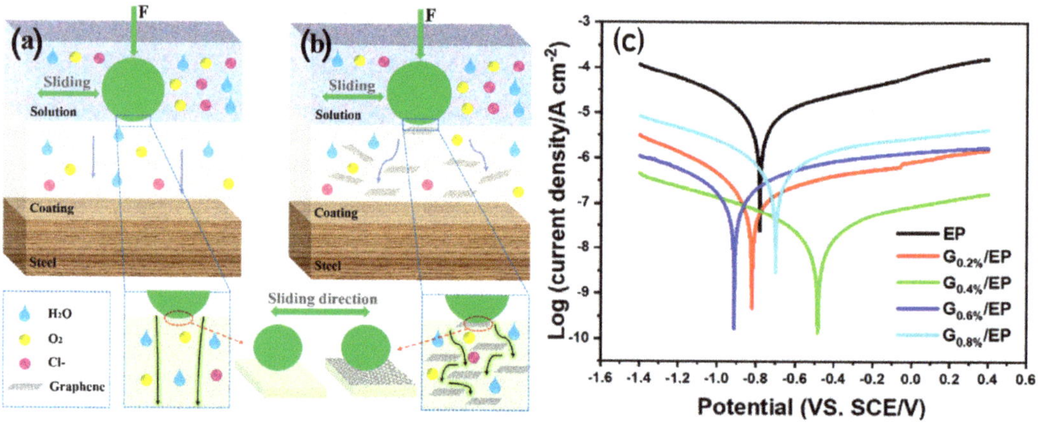

Figure 7. (**a**) Schematic illustration of the wear improvement mechanism and (**b**) corrosion resistance mechanism in the G/EP composite powder coating. (**c**) Potentiodynamic polarisation curves of the neat EP and G/EP composite coatings [87].

In terms of corrosion resistance, the incorporation of graphene nanosheets increases the compactness of the coating, which retards transport of the corrosive medium. This results in an excellent improvement in corrosion protection (Figure 7c), as evident from the high |Z| 0.01 Hz value of 9.65×10^5 Ω cm^2 after 50 days of immersion. The R_c and R_{ct} values, which reflect the physical barrier effect and severity of the corrosion reaction, respectively, also suggest that the G 0.4%/EP sample maintains the highest values, demonstrating enhanced corrosion resistance.

A previous study [78] also highlights the considerably inferior performance of epoxy powder coating without graphene nanosheets. The loose structure, inferior adhesion, and lack of a lubricating phase leave the coating prone to delamination or peeling during wear, leading to coating failure. Surface defects such as pinholes, voids, and microcracks can extend below the sliding surface, forming channels for corrosive medium penetration, which accelerates damage and coating peeling. The graphene-reinforced coating, however, mitigates these problems, as described earlier.

Overall, the findings suggest that the impregnation of graphene nanoplates into coatings offers a promising strategy for providing robust resistance to degradation due to combined corrosion and mechanically assisted degradation, such as corrosion–wear.

3.4. GO Modified with Polyamide–Epoxy/Zn Coating

Zn-rich polymer composite paint is a widely applied type of coating due to the effect of the Zn particles, which work as anodes to provide cathodic protection. In order to enhance the anticorrosion effect of epoxy/zinc (EP/Zn) coatings, graphene oxide (GO) can be added as filler. A recent study [88] showed GO modified with polyamide to further enhance the corrosion resistance on a carbon steel (Q235) substrate. A comprehensive study demonstrated an enhanced corrosion resistance of epoxy–zinc (EP/Zn) coatings upon the addition of graphene oxide (GO) and its modified form, polyamide-modified GO PGO [88]. The incorporation of 0.3 wt.% GO into the EP/Zn coating significantly improved corrosion resistance, as indicated by the increase in the R_p to 13,115 Ω cm^2 for PGO/EP/Zn, compared to 5386 Ω cm^2 for GO/EP/Zn. This enhancement is attributed to the grafting of polyamide curing agent (PA-651) onto PGO, which improves its dispersibility and interaction with the epoxy matrix, effectively plugging micropores and preventing the permeation of corrosive agents. The study also reported that the R_t, a critical parameter for evaluating corrosion resistance, was 64,520 Ω cm^2 for PGO/EP/Zn, i.e., more than double that of GO/EP/Zn at 31,830 Ω cm^2. This significant increase in R_t suggests a lower rate of electrochemical corrosion, indicating the superior protective properties of the PGO/EP/Zn coating. The enhanced performance is further supported by the visual evidence of the absence of red ferric compound on the surface of the GO/EP/Zn and PGO/EP/Zn coatings after immersion tests, i.e., unlike the neat EP/Zn coating (Figure 8), which showed extensive red corrosion products.

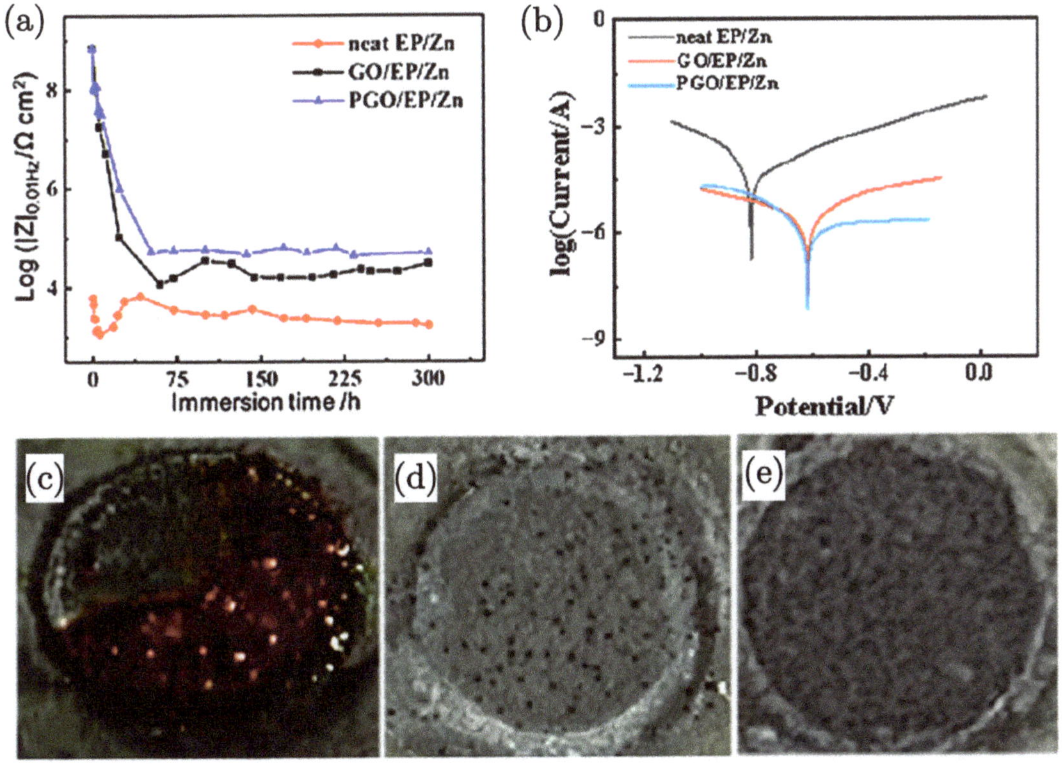

Figure 8. (**a**) Impedance dependence of neat EP/Zn, GO/EP/Zn, and PGO/EP/Zn coatings over immersion time in a 3.5 wt.% NaCl solution. (**b**) Tafel curves of neat EP/Zn, GO/EP/Zn, and PGO/EP/Zn samples after 300 h of immersion in 3.5 wt.% NaCl solution. Visual observations of coating samples exposed to 3.5 wt.% NaCl solution after 300 h of immersion: (**c**) neat EP/Zn, (**d**) GO/EP/Zn, and (**e**) PGO/EP/Zn, with a circle diameter of 1.1 cm in each image [88].

The mechanism behind the improved performance described above is linked to the modification of GO with PA-651. This modification results in a reduced O/C atomic ratio, indicating a decrease in oxygen-containing functional groups and an increase in re-aromatisation, which enhances the dispersion stability of PGO in organic solvents. This was confirmed by Raman spectroscopy, which revealed shifts in the D and G bands, suggesting charge transfer between the polyamide and GO, contributing to the improved properties of the composite.

In summary, the study effectively demonstrates that the in situ modification of GO with a curing agent significantly enhances the corrosion resistance of epoxy–zinc coatings. The findings highlight the potential of using modified GO in protective coatings to extend the lifespan of metal substrates in corrosive environments.

3.5. Graphene/Polyeugenol Composite

Adiwibawa Prasetya et al. [89] investigated the anticorrosion properties of a polyeugenol/graphene (PE/G) composite coating on copper metal. The results demonstrate that the addition of graphene significantly enhances the corrosion resistance of the coating. The electrochemical analysis shows a decrease in the corrosion rate with an increasing graphene content, as indicated by a reduction in the corrosion current density (I_{corr}), which is attributed to graphene acting as a physical barrier, preventing/hindering corrosive agents like oxygen and chloride ions from reaching the metal surface. The potentiodynamic polarisation revealed that the PE/G composite offers superior anticorrosion properties compared to pure polyeugenol. The corrosion protection efficiency increased from 37% to 78% with the incorporation of 1.25 wt% of graphene into the polyeugenol matrix (Figure 9h). This improvement highlights the effectiveness of graphene as a reinforcing agent in polymer matrices for corrosion protection. Morphological analysis using SEM before and after immersion (Figure 9a–f) further supported these findings. The mechanism underlying the enhanced corrosion resistance involves the formation of a physical barrier by graphene (Figure 9g). This is supported by the absence of corrosion products in the XRD analysis of the coated samples, indicating that the PE/G coating effectively prevents the formation of corrosion products such as Cu_2O and $Cu_2(OH)_3Cl$, which were observed in the case of the uncoated samples. The presence of graphene in the composite makes it challenging for corrosive species to penetrate the coating, thereby enhancing its protective capabilities.

SEM of the morphology also suggested that the PE/G-coated metal exhibited higher hydrophobicity compared to uncoated or PE-coated metal. This increased hydrophobicity contributes to the coating's ability to repel water and other corrosive agents, further enhancing its anticorrosion performance. Overall, the study provides strong evidence for the potential of PE/G composites as effective anticorrosion coatings, with significant potential for industrial applications.

3.6. Poly(m-Phenylenediamine)-Encapsulated Graphene

Zhang et al. [90] used a two-step process to prepare poly(m-phenylenediamine)-encapsulated graphene (G@PmPD); the preparation method is as illustrated in Figure 10a. High-energy ball milling provided impermeable graphene-encapsulated poly-m-phenylenediamine (via non-covalent π-π interactions), which reduced the conductivity to 1.2×10^{-7} S cm^{-1} and mitigated/ameliorated the electrochemical corrosion-promoting effect due to the cathodic nature of graphene. The amino-rich surface of G@PmPD improved dispersibility and compatibility with organic solvents and polymer matrices, facilitating the application of the composite coatings, thereby improving corrosion resistance. The incorporation of 0.5 wt% of G@PmPD into an epoxy matrix coating markedly improved the corrosion protection of steel substrates in a 60-day test. This enhancement is attributed to the impermeability of G@PmPD as a result of the enhanced compactness due to cross-linking between G@PmPD and the epoxy resin. Notably, even when the composite coating is damaged, the presence of G@PmPD effectively mitigates the damage (Figure 10b–d).

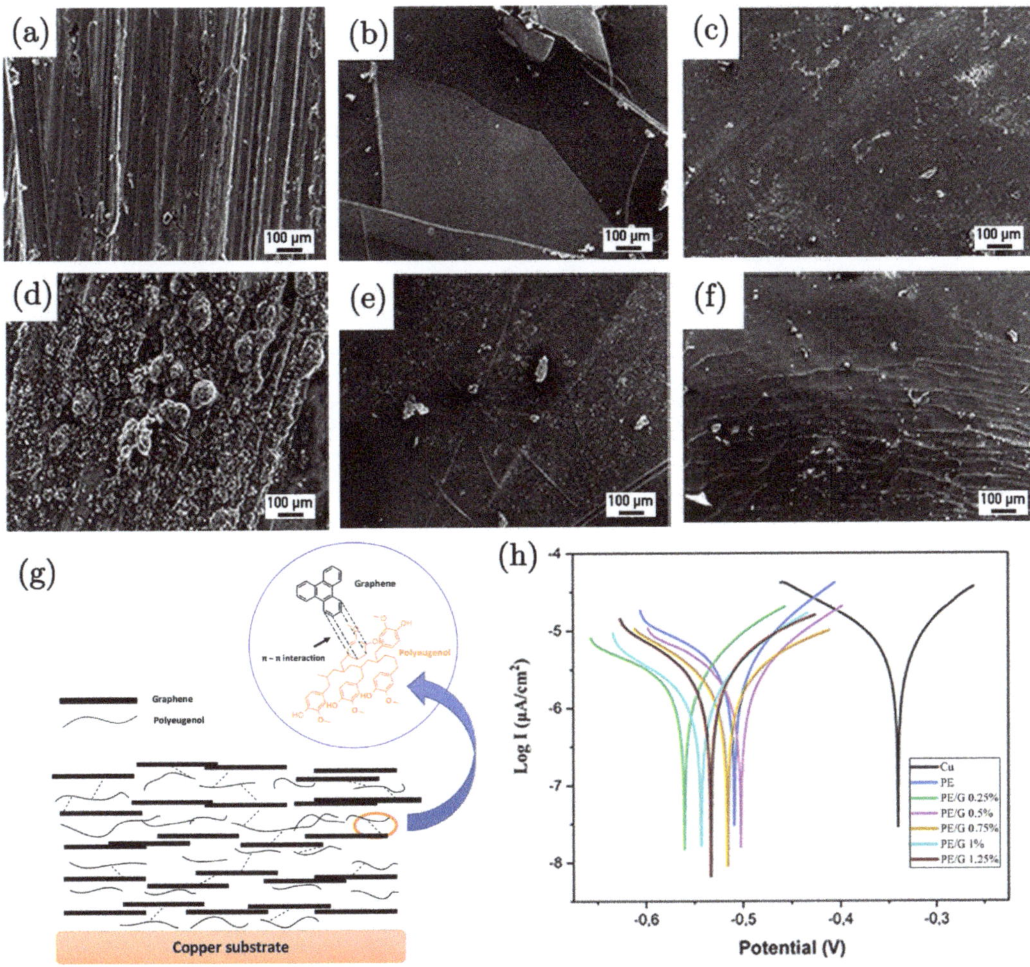

Figure 9. SEM images of bare Cu (**a**) before and (**d**) after immersion, PE-coated Cu (**b**) before and (**e**) after immersion, and PE/G 1.25%-coated Cu (**c**) before and (**f**) after immersion. (**g**) Schematic illustration of the PE/G composite coating on copper. (**h**) Tafel plots of Cu, PE, PE/G 0.25%, PE/G 0.5%, PE/G 0.75%, and PE/G 1% in 3.5 wt.% NaCl aqueous solution [89].

The morphological images revealed that the encapsulation of graphene by PmPD results in a thicker and less transparent structure compared to the original graphene. Elemental mapping by EDS confirms the uniform distribution of nitrogen from PmPD on the G@PmPD surface, verifying successful encapsulation, which contributes to the improved properties of the composite coating. The mechanism of enhanced corrosion protection involves several key factors. The insulating encapsulation of graphene by PmPD reduces its conductivity, thereby enhancing corrosion protection. The π-π interaction between PmPD and graphene ensures a stable and uniform dispersion within the epoxy matrix, which facilitates effective tortuosity of the diffusion path for corrosives, delaying their penetration. Additionally, the cross-linking between G@PmPD and the epoxy resin increases the compactness and cross-link density of the coating, further enhancing its barrier properties. This comprehensive approach also effectively addresses the challenges of graphene agglomeration, which is often a problem in the incorporation of graphene into a polymeric matrix.

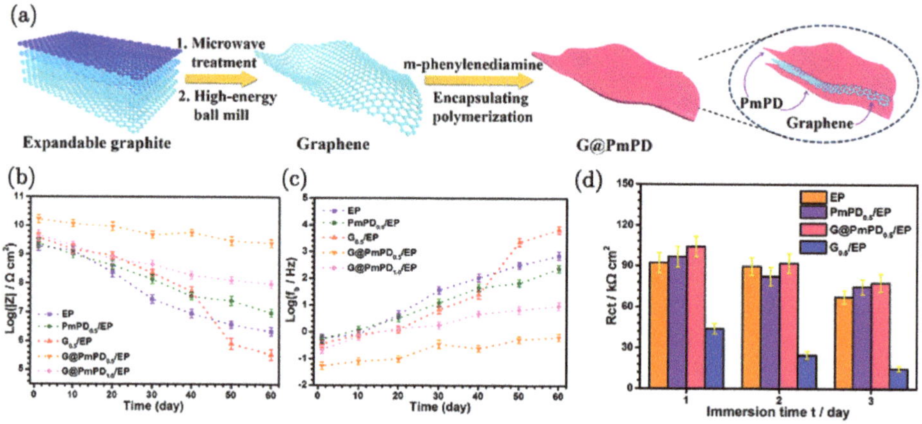

Figure 10. (**a**) Preparation process of G@PmPD. (**b**) $|Z|_{10\text{m Hz}}$ and (**c**) f_b values for EP, $PmPD_{0.5}/EP$, $G_{0.5}/EP$, $G@PmPD_{0.5}/EP$, and $G@PmPD_{1.0}/EP$ coatings as a function of immersion time. (**d**) Fitting results of R_{ct} for different scratched samples [90].

3.7. GO-Polyethylene Imine-Polyacrylic Composite

Zheng et al. [91] prepared a graphene/supramolecular polymer (PEI-PAA-GO) based on a graphene oxide (GO) composite with branched poly (ethylene imine) (PEI) and poly (acrylic acid) (PAA). Mild steel substrate coupons with a proper surface finish and cleaning were coated with PEI-PAA-GO coating. Coatings with increasing GO contents were named GO-0 wt%, GO-0.025 wt%, GO-0.05 wt%, and GO-0.1 wt%. EIS results (Figure 11c) indicated that the GO-containing coatings (GO-0.05–GO-0.15 wt%) exhibited superior corrosion resistance than the one without GO (GO-0 wt%). The self-repairing performance of the coatings was assessed by measuring the healing efficiency (η healing). GO-0.025 wt% showed a higher healing efficiency of 143.75 $\Omega\ h^{-1}$ compared to 90 $\Omega\ h^{-1}$ for GO-0 wt%. The enhanced self-repairing ability of GO-containing coatings is attributed to stronger intramolecular hydrogen bonding and electrostatic interactions.

The mechanism of protection involves the combination of the physical barrier effect of the 2D lamellar graphene oxide and the water-retaining property of the supramolecular polymer. This synergistic effect resulted in high corrosion resistance, with an impedance modulus reaching $8.0 \times 10^3\ \Omega\ cm^2$, which is four times higher than that of a pure PEI/PAA coating. The Bode-phase diagrams suggested the degradation characteristics of certain coatings over time, while others maintained high/durable corrosion resistance. The water uptake too decreases significantly, preventing swelling and enhancing the protective capability of the coating. Figure 11a illustrates a schematic of the degradation and self-repairing mechanisms. The self-repairing is attributed to the re-establishment of the polymer network at the fracture surface, facilitated by hydrogen bonding.

An overview illustrating the corrosion resistance performance of different graphene-based polymer composites, highlighting their material compositions, corrosive environments, and critical electrochemical parameters such as inhibition efficiency, Tafel slopes ($\beta a/\beta c$), corrosion current density (I_{corr}), and corrosion potential (E_{corr}), is provided in Table 2.

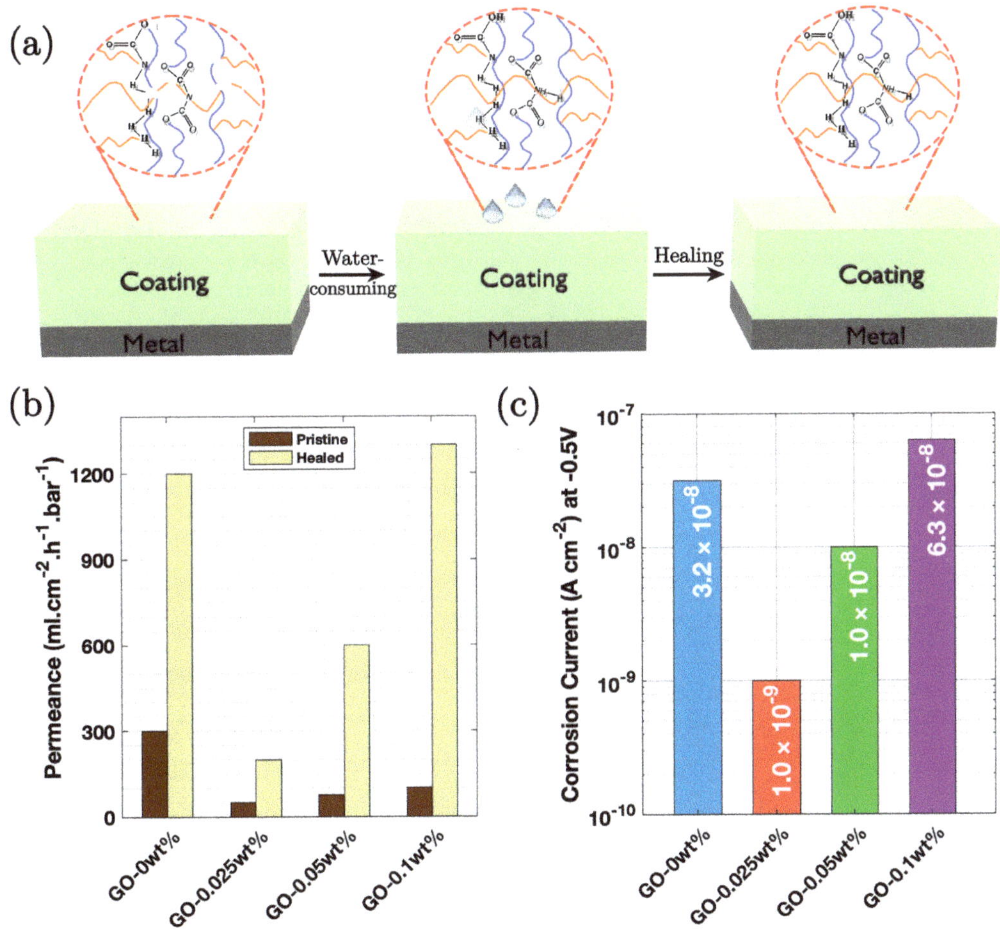

Figure 11. (**a**) Illustration of the self-repairing mechanism, where hydrogen bonds form upon water absorption, enabling healing. (**b**) Changes in water permeability for coatings in different states, highlighting the effect of the self-healing process. (**c**) EIS results of the PEI-PAA with GO-containing (GO-0 wt%, GO-0.025 wt%, GO-0.05 wt%, and GO-0.1 wt%) coatings exposed to air following 360 h of immersion in a 3.5 wt.% NaCl solution, demonstrating corrosion resistance [91].

Table 2. Corrosion resistance performance of graphene-integrated polymer composites.

Material Composition	Corrosive Medium	Inhibition Efficiency (%)	Tafel Slope ($\beta a/\beta c$ in mV/Decade)	Corrosion Current Density (I_{corr}, μA/cm^2)	Corrosion Potential (E_{corr}, V)	Ref.
PANI/NFGO	3.5% NaCl	93.5	95/110	0.12	−0.25	[86]
GO/Epoxy	3.5% NaCl	90.2	120/105	0.18	−0.3	[16]
rGO/Polypyrrole	1 M HCl	85.4	130/115	0.25	−0.35	[77]
Graphene/Polyurethane	3.5% NaCl	88.7	100/98	0.2	−0.28	[68]
GO/Chitosan	0.5 M NaCl	82.3	115/125	0.3	−0.32	[25]

4. Overview of Functionalised Graphene in Polymer Composites

Graphene, renowned for its exceptional mechanical and electrical properties, serves as an effective nanofiller in various polymer matrices, greatly enhancing their barrier properties against corrosion. Functionalised graphene and its derivatives when incorporated as nanofillers into polymeric coatings improve corrosion resistance. For instance, Yang et al. [81] demonstrated enhanced corrosion protection of waterborne epoxy (WEP) coatings when impregnated with functionalised graphene oxide derivatives, specifically GOEH. Figure 12a illustrates the syntheses of GOH, GOE, and GOEH. Incorporation of GOEH into WEP coatings provides a mechanism of both physical and chemical protective effects. The physical barrier is formed by well-dispersed graphene layers, which inhibit the diffusion of corrosive molecules/ions. The chemically mediated protection is primarily attributed to the presence of P-OH groups on GOEH, which can chelate with metal ions at the anode, forming insoluble metal–organic complexes. This reaction not only blocks the migration of Fe^{2+} ions but also plugs defects in the coating. Additionally, the P-OH groups enhance adhesion (as demonstrated through pull-off test results in Figure 12b), which is caused by reaction with the hydroxylated metal surface. This further retards the penetration of corrosive molecules along the GOEH/WEP composite interface.

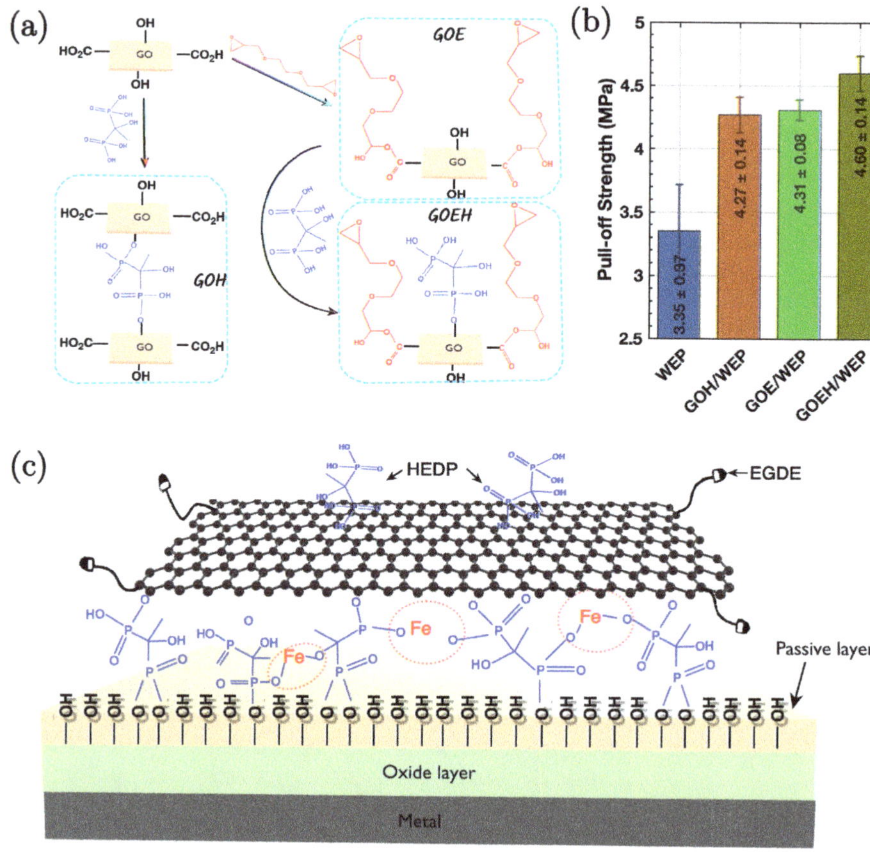

Figure 12. (**a**) Schematic illustration of the preparation process for GOH, GOE, and GOEH. (**b**) Adhesion strength of the specimens measured by pull-off testing. (**c**) Illustration of the corrosion protection mechanism for GOEH/WEP composite coatings [81].

The structural analysis of GOH, GOE, and GOEH reveals significant insights into their functional properties. FTIR and XPS spectroscopy indicate the successful grafting of HEDP onto GO, with characteristic peaks confirming the presence of P-O and P-OH groups. The reaction between the epoxide groups of EGDE and the carboxyl groups on GO results in the formation of GOE, which enlarges the interlayer spacing and improves compatibility with the WEP matrix, since structural modification is crucial for maintaining the dispersion of graphene layers, which is essential for the formation of an effective physical barrier.

Figure 12c illustrates the corrosion protection mechanism for GOEH/WEP composite coatings. The presence of EGDE molecules on GOEH allows for covalent bonding with the WEP matrix. This bonding is vital for preventing the penetration of corrosive molecules and ensuring the integrity of the coating. The structural characteristics of GOEH, such as the retention of P-OH groups, play a significant role in its ability to form strong bonds with metal substrates, contributing to the overall corrosion protection mechanism.

The performance of the GOEH/WEP coatings was evaluated through their impedance modulus, which stayed $> 1.5 \times 10^9\ \Omega\ cm^2$ for five months with the addition of 0.5 wt% graphene derivatives. The sustenance of high impedance indicates excellent barrier properties and long-term stability of the coatings. The formation of a dense passivation layer by GOEH layers and their bonding with Fe^{2+} and hydroxyl groups on metal substrates further enhances the corrosion protection performance. This passivation layer acts as an additional barrier, preventing the ingress of corrosive species and maintaining the coating's protective capabilities over time.

This review assessed the development of advanced corrosion protection coatings comprising graphene and its derivatives. The protection offered by such coatings can be leveraged to design protection systems with enhanced durability and performance in various industrial applications. Future research could explore the optimisation of graphene derivative concentrations and the potential for scaling up the production of these coatings for commercial use. Additionally, investigating the long-term environmental impact and sustainability of these coatings could provide valuable insights for their widespread adoption.

5. Conclusions

This review highlights the significant progress in utilising graphene-based composites, specifically graphene oxide (GO) and reduced graphene oxide (rGO), to enhance corrosion resistance in polymer matrices. Recent studies demonstrate that these composites provide superior barrier properties by extending ion diffusion paths and reducing corrosive agent permeability. Notably, some graphene-based polymer composites achieve an impressive corrosion protection efficiency of approximately 99% and a two- to four-order reduction in corrosion current density, underscoring the effectiveness of these materials in high-performance coatings. Additionally, advancements in functionalisation techniques have optimised GO dispersion within polymer matrices, enhancing compatibility, adhesion, and mechanical resilience.

However, despite these promising advancements, several key challenges remain that hinder the widespread practical application of graphene-based coatings. Issues such as the effective dispersion of graphene, achieving scalable production methods, and ensuring long-term environmental stability need to be addressed. For a comprehensive discussion of these challenges, please refer to Section 6.

Addressing these challenges will require targeted research efforts focusing on the development of scalable and cost-effective production techniques, innovating advanced functionalisation methods to maintain graphene's inherent properties, and adopting sustainable and eco-friendly synthesis approaches to minimise environmental impacts.

These developments address the growing demand for sustainable, cost-efficient materials in industry by combining long-term durability with a reduced environmental impact. In summary, while graphene-impregnated coatings hold immense promise for enhancing corrosion resistance in a wide range of industrial applications, addressing the existing challenges through innovative research and development is imperative for their successful implementation. As research progresses, graphene-based polymer composites are emerging as promising candidates for next-generation anticorrosion technologies, offering a balance between high protective performance and ecological and economic sustainability.

6. Challenges and Future Perspectives

Despite the significant advancements in graphene-based anticorrosion coatings, several challenges remain that need to be addressed for their widespread adoption.

Agglomeration of Graphene Sheets: One of the primary challenges is the tendency of graphene and its derivatives, such as graphene oxide (GO) and reduced graphene oxide (rGO), to agglomerate within polymer matrices. This agglomeration disrupts uniform dispersion, leading to compromised barrier properties and reduced corrosion protection. Effective dispersion techniques, including covalent and non-covalent functionalisation, as well as the use of surfactants, are essential to overcome this issue.

Functionalisation Challenges: Achieving effective functionalisation of graphene without compromising its intrinsic electrical and barrier properties remains a significant challenge. Advanced functionalisation methods, such as plasma treatment, silanisation, and the incorporation of nanoparticles, could enhance compatibility with polymer matrices and improve corrosion resistance.

Scalability and Cost: The scalable production of high-quality graphene with minimal defects is both technically challenging and cost-prohibitive. Techniques such as chemical vapour deposition (CVD) and electrochemical exfoliation, although capable of producing high-quality graphene, are still expensive for large-scale applications. Research into cost-effective and scalable production methods, such as liquid-phase exfoliation and microwave-assisted reduction, is essential.

Long-term Stability: The long-term stability of graphene-impregnated polymeric coatings under harsh environmental conditions, including exposure to UV radiation, high salinity, and fluctuating pH levels, needs further investigation. The development of hybrid composites with synergistic properties offers a promising solution to this challenge.

Author Contributions: A.Y.: planning and writing original and final drafts, illustrations; S.P.: writing and revision of final draft; R.K.S.R. and A.Y.: conceptualisation, planning and writing original and final drafts, supervision. All authors have read and agreed to the published version of the manuscript.

Funding: This research received no external funding.

Data Availability Statement: All data are available within the manuscript.

Conflicts of Interest: The authors declare no conflicts of interest.

References

1. Koch, G. 1—Cost of corrosion. In *Trends in Oil and Gas Corrosion Research and Technologies*; Woodhead Publishing Series in Energy; El-Sherik, A.M., Ed.; Woodhead Publishing: Boston, MA, USA, 2017; pp. 3–30.
2. Kumar Gupta, R.; Malviya, M.; Verma, C.K.; Gupta, N.A.; Quraishi, M. Pyridine-based functionalized graphene oxides as a new class of corrosion inhibitors for mild steel: An experimental and DFT approach. *RSC Adv.* **2017**, *7*, 39063–39074.
3. Sabiha, M.; Kerroum, Y.; El Hawary, M.; Boudalia, M.; Bellaouchou, A.; Hammani, O.; Amin, H.M.A. Investigating the Adsorption and Corrosion Protection Efficacy and Mechanism of Marjoram Extract on Mild Steel in HCl Medium. *Molecules* **2025**, *30*, 272. [CrossRef] [PubMed]
4. Boutoumit, A.; Elhawary, M.; Bellaouchou, A.; Boudalia, M.; Hammani, O.; José Garcia, A.; Amin, H.M.A. Electrochemical, Structural and Thermodynamic Investigations of Methanolic Parsley Extract as a Green Corrosion Inhibitor for C37 Steel in HCl. *Coatings* **2024**, *14*, 783. [CrossRef]

5. Najem, A.; Campos, O.S.; Girst, G.; Raji, M.; Hunyadi, A.; García-Antón, J.; Bellaouchou, A.; Amin, H.M.A.; Boudalia, M. Experimental and DFT Atomistic Insights into the Mechanism of Corrosion Protection of Low-Carbon Steel in an Acidic Medium by Polymethoxyflavones from Citrus Peel Waste. *J. Electrochem. Soc.* **2023**, *170*, 093512.
6. Singh Raman, R.K.; Chakraborty Banerjee, P.; Lobo, D.E.; Gullapalli, H.; Sumandasa, M.; Kumar, A.; Choudhary, L.; Tkacz, R.; Ajayan, P.M.; Majumder, M. Protecting copper from electrochemical degradation by graphene coating. *Carbon* **2012**, *50*, 4040–4045.
7. Raman, R.K.S.; Sanjid, A.; Banerjee, P.C.; Arya, A.K.; Parmar, R.; Amati, M.; Gregoratti, L. Remarkably Corrosion Resistant Graphene Coating on Steel Enabled Through Metallurgical Tailoring. *Small* **2024**, *20*, 2302498. [CrossRef]
8. Data Bridge Market Research. *Graphene-Based Coatings Market—Global Industry Trends and Forecast to 2028*; Data Bridge Market Research: Pune, India, 2023.
9. Huang, Y.; Zhang, B.; Wu, J.; Hong, R.; Xu, J. Preparation and characterization of graphene oxide/polyaniline/polydopamine nanocomposites towards long-term anticorrosive performance of epoxy coatings. *Polymers* **2022**, *14*, 3355. [CrossRef]
10. Krishnamoorthy, K.; Jeyasubramanian, K.; Premanathan, M.; Subbiah, G.; Shin, H.S.; Kim, S.J. Graphene oxide nanopaint. *Carbon* **2014**, *72*, 328–337.
11. Chang, C.-H.; Huang, T.-C.; Peng, C.-W.; Yeh, T.-C.; Lu, H.-I.; Hung, W.-I.; Weng, C.-J.; Yang, T.-I.; Yeh, J.-M. Novel anticorrosion coatings prepared from polyaniline/graphene composites. *Carbon* **2012**, *50*, 5044–5051.
12. Yi, M.; Shen, Z.; Zhao, X.; Liu, L.; Liang, S.; Zhang, X. Exploring few-layer graphene and graphene oxide as fillers to enhance the oxygen-atom corrosion resistance of composites. *Phys. Chem. Chem. Phys.* **2014**, *16*, 11162–11167.
13. Othman, N.H.; Che Ismail, M.; Mustapha, M.; Sallih, N.; Kee, K.E.; Ahmad Jaal, R. Graphene-based polymer nanocomposites as barrier coatings for corrosion protection. *Prog. Org. Coat.* **2019**, *135*, 82–99.
14. Qian, Y.; Zheng, W.; Chen, W.; Feng, T.; Liu, T.X.; Fu, Y.Q. Enhanced functional properties of CeO_2 modified graphene/epoxy nanocomposite coating through interface engineering. *Surf. Coat. Technol.* **2021**, *409*, 126819. [CrossRef]
15. Kausar, A.; Ahmad, I.; Eisa, M.H.; Maaza, M. Avant-Garde Polymer/Graphene Nanocomposites for Corrosion Protection: Design, Features, and Performance. *Corros. Mater. Degrad.* **2023**, *4*, 33–53. [CrossRef]
16. Zhou, Y.; Wang, H.; Zhang, C.; Zhou, Q.; Rodrigues, D.F. Graphene Oxide-Hybridized Waterborne Epoxy Coating for Simultaneous Anticorrosive and Antibiofilm Functions. *Front. Mater.* **2022**, *9*, 910152.
17. Guo, R.; Li, W.; Wang, X.; Lv, Y.; Chen, M.; Chen, Z.; Liu, Z.; Han, G.-C. Antimicrobial corrosion study of the epoxy coating with the graphene oxide supported Schiff base quaternary ammonium salt additives. *Mater. Today Commun.* **2023**, *35*, 105517. [CrossRef]
18. Lv, K.; Pan, R.; Zhang, L.; Tian, Y.; Sui, Y.; Wan, D. Synergistically assembled graphene/ZnO composite to enhance anticorrosion performance of waterborne epoxy coatings. *RSC Adv.* **2022**, *12*, 9069–9076. [CrossRef]
19. Li, J.; Zhang, H.; Sun, F.; Zhou, H.; Zhao, L.; Song, Y. The multiscale effects of graphene oxide on the corrosion resistance properties of waterborne alkyd resin coatings. *J. Mater. Res.* **2019**, *34*, 950–958. [CrossRef]
20. Kabeb, S.M.; Hassan, A.; Ahmad, F.; Mohamad, Z.; Sharer, Z.; Mokhtar, M. Synergistic effects of hybrid nanofillers on graphene oxide reinforced epoxy coating on corrosion resistance and fire retardancy. *J. Appl. Polym. Sci.* **2022**, *139*, 51640.
21. Khan, A.R.; Zhao, T.; Zheng, M. Composite fabrication of epoxy and graphene oxide coating to enrich the anticorrosion and thermal properties of carbon-steel. *Surf. Rev. Lett.* **2022**, *29*, 2250072. [CrossRef]
22. Tian, P.; Yang, S.; He, Y.; Liu, W.; Zhang, J. High-performance naphthalene-based photocurable epoxy resin for advanced printed circuit boards coatings by optimizing fluorodiamine-mediator extension chemistry. *Polymer* **2024**, *308*, 127421. [CrossRef]
23. Pu, Z.; Pang, J.; He, X.; Zhu, R.; Zhong, J.; Liu, X.; Jiang, D.; Ning, Z. Interpenetrating network structure design and corrosion resistance of high-performance waterborne polyurethane coating modified by epoxy resin and functionalized graphene oxide. *J. Appl. Polym. Sci.* **2023**, *140*, e53717.
24. Kumari, S.; Panigrahi, A.; Singh, S.K.; Pradhan, S.K. Enhanced corrosion resistance and mechanical properties of nanostructured graphene-polymer composite coating on copper by electrophoretic deposition. *J. Coat. Technol. Res.* **2018**, *15*, 583–592.
25. Jena, G.; Anandkumar, B.; Vanithakumari, S.C.; George, R.P.; Philip, J.; Amarendra, G. Graphene oxide-chitosan-silver composite coating on Cu-Ni alloy with enhanced anticorrosive and antibacterial properties suitable for marine applications. *Prog. Org. Coat.* **2020**, *139*, 105444.
26. Gong, X.; Liu, G.; Li, Y.; Yu, D.Y.W.; Teoh, W.Y. Functionalized-graphene composites: Fabrication and applications in sustainable energy and environment. *Chem. Mater.* **2016**, *28*, 8082–8118.
27. He, W.; Zhu, L.; Chen, H.; Nan, H.; Li, W.; Liu, H.; Wang, Y. Electrophoretic deposition of graphene oxide as a corrosion inhibitor for sintered NdFeB. *Appl. Surf. Sci.* **2013**, *279*, 416–423.
28. Chang, K.C.; Hsu, C.H.; Lu, H.I.; Ji, W.F.; Chang, C.H.; Li, W.Y.; Chuang, T.L.; Yeh, J.M.; Liu, W.R.; Tsai, M.H. Advanced anticorrosive coatings prepared from electroactive polyimide/graphene nanocomposites with synergistic effects of redox catalytic capability and gas barrier properties. *Express Polym. Lett.* **2014**, *8*, 243–255.
29. Berry, V. Impermeability of graphene and its applications. *Carbon* **2013**, *62*, 1–10.

30. Ren, S.; Cui, M.; Liu, C.; Wang, L. A comprehensive review on ultrathin, multi-functionalized, and smart graphene and graphene-based composite protective coatings. *Corros. Sci.* **2023**, *212*, 110939.
31. Stolyarova, E.; Stolyarov, D.; Bolotin, K.; Ryu, S.; Liu, L.; Rim, K.T.; Klima, M.; Hybertsen, M.; Pogorelsky, I.; Pavlishin, I. Observation of graphene bubbles and effective mass transport under graphene films. *Nano Lett.* **2009**, *9*, 332–337.
32. Prasai, D.; Tuberquia, J.C.; Harl, R.R.; Jennings, G.K.; Bolotin, K.I. Graphene: Corrosion-inhibiting coating. *ACS Nano* **2012**, *6*, 1102–1108.
33. Abbas, Q.; Shinde, P.A.; Abdelkareem, M.A.; Alami, A.H.; Mirzaeian, M.; Yadav, A.; Olabi, A.G. Graphene Synthesis Techniques and Environmental Applications. *Materials* **2022**, *15*, 7804. [CrossRef] [PubMed]
34. Gadipelli, S.; Guo, Z.X. Graphene-based materials: Synthesis and gas sorption, storage and separation. *Prog. Mater. Sci.* **2015**, *69*, 1–60.
35. Peigney, A.; Laurent Ch Flahaut, E.; Bacsa, R.R.; Rousset, A. Specific surface area of carbon nanotubes and bundles of carbon nanotubes. *Carbon* **2001**, *39*, 507–514.
36. Novoselov, K.S.; Colombo, L.; Gellert, P.R.; Schwab, M.G.; Kim, K. A roadmap for graphene. *Nature* **2012**, *490*, 192–200.
37. Kumar, N.; Salehiyan, R.; Chauke, V.; Joseph Botlhoko, O.; Setshedi, K.; Scriba, M.; Masukume, M.; Sinha Ray, S. Top-down synthesis of graphene: A comprehensive review. *FlatChem* **2021**, *27*, 100224.
38. Simonov, K.A.; Vinogradov, N.A.; Vinogradov, A.S.; Generalov, A.V.; Zagrebina, E.M.; Svirskiy, G.I.; Cafolla, A.A.; Carpy, T.; Cunniffe, J.P.; Taketsugu, T.; et al. From Graphene Nanoribbons on Cu(111) to Nanographene on Cu(110): Critical Role of Substrate Structure in the Bottom-Up Fabrication Strategy. *ACS Nano* **2015**, *9*, 8997–9011.
39. Seo, J.; Lee, J.; Jang, A.-R.; Choi, Y.; Kim, U.; Shin, H.S.; Park, H. Study of Cooling Rate on the Growth of Graphene via Chemical Vapor Deposition. *Chem. Mater.* **2017**, *29*, 4202–4208.
40. Batzill, M. The surface science of graphene: Metal interfaces, CVD synthesis, nanoribbons, chemical modifications, and defects. *Surf. Sci. Rep.* **2012**, *67*, 83–115.
41. Park, B.-J.; Choi, J.-S.; Eom, J.-H.; Ha, H.; Kim, H.Y.; Lee, S.; Shin, H.; Yoon, S.-G. Defect-Free Graphene Synthesized Directly at 150 °C via Chemical Vapor Deposition with No Transfer. *ACS Nano* **2018**, *12*, 2008–2016.
42. Huang, Y.; Pan, Y.-H.; Yang, R.; Bao, L.-H.; Meng, L.; Luo, H.-L.; Cai, Y.-Q.; Liu, G.-D.; Zhao, W.-J.; Zhou, Z.; et al. Universal mechanical exfoliation of large-area 2D crystals. *Nat. Commun.* **2020**, *11*, 2453.
43. Gao, L.; Ren, W.; Zhao, J.; Ma, L.-P.; Chen, Z.; Cheng, H.-M. Efficient growth of high-quality graphene films on Cu foils by ambient pressure chemical vapor deposition. *Appl. Phys. Lett.* **2010**, *97*, 183109. [CrossRef]
44. Park, S.; Ruoff, R.S. Chemical methods for the production of graphenes. *Nat. Nanotechnol.* **2009**, *4*, 217–224. [CrossRef] [PubMed]
45. Abdolhosseinzadeh, S.; Asgharzadeh, H.; Seop Kim, H. Fast and fully-scalable synthesis of reduced graphene oxide. *Sci. Rep.* **2015**, *5*, 10160.
46. Lu, X.; Yu, M.; Huang, H.; Ruoff, R.S. Tailoring graphite with the goal of achieving single sheets. *Nanotechnology* **1999**, *10*, 269. [CrossRef]
47. Novoselov, K.S.; Jiang, D.; Schedin, F.; Booth, T.J.; Khotkevich, V.V.; Morozov, S.V.; Geim, A.K. Two-dimensional atomic crystals. *Proc. Natl. Acad. Sci. USA* **2005**, *102*, 10451–10453.
48. Bourlinos, A.B.; Georgakilas, V.; Zboril, R.; Steriotis, T.A.; Stubos, A.K. Liquid-phase exfoliation of graphite towards solubilized graphenes. *Small* **2009**, *5*, 1841–1845. [CrossRef]
49. Marcano, D.C.; Kosynkin, D.V.; Berlin, J.M.; Sinitskii, A.; Sun, Z.; Slesarev, A.; Alemany, L.B.; Lu, W.; Tour, J.M. Improved synthesis of graphene oxide. *ACS Nano* **2010**, *4*, 4806–4814. [CrossRef]
50. Amin, H.M.A.; El-Kady, M.F.; Atta, N.F.; Galal, A. Gold Nanoparticles Decorated Graphene as a High Performance Sensor for Determination of Trace Hydrazine Levels in Water. *Electroanalysis* **2018**, *30*, 1757–1766. [CrossRef]
51. Reddy, K.M.; Gledhill, A.D.; Chen, C.-H.; Drexler, J.M.; Padture, N.P. High quality, transferrable graphene grown on single crystal Cu (111) thin films on basal-plane sapphire. *Appl. Phys. Lett.* **2011**, *98*, 113117.
52. Ishihara, M.; Koga, Y.; Kim, J.; Tsugawa, K.; Hasegawa, M. Direct evidence of advantage of Cu(111) for graphene synthesis by using Raman mapping and electron backscatter diffraction. *Mater. Lett.* **2011**, *65*, 2864–2867.
53. Banszerus, L.; Schmitz, M.; Engels, S.; Dauber, J.; Oellers, M.; Haupt, F.; Watanabe, K.; Taniguchi, T.; Beschoten, B.; Stampfer, C. Ultrahigh-mobility graphene devices from chemical vapor deposition on reusable copper. *Sci. Adv.* **2015**, *1*, e1500222.
54. Kong, W.; Kum, H.; Bae, S.-H.; Shim, J.; Kim, H.; Kong, L.; Meng, Y.; Wang, K.; Kim, C.; Kim, J. Path towards graphene commercialization from lab to market. *Nat. Nanotechnol.* **2019**, *14*, 927–938.
55. Munoz, R.; Gómez-Aleixandre, C. Review of CVD synthesis of graphene. *Chem. Vap. Depos.* **2013**, *19*, 297–322.
56. Wei, D.; Liu, Y.; Wang, Y.; Zhang, H.; Huang, L.; Yu, G. Synthesis of N-doped graphene by chemical vapor deposition and its electrical properties. *Nano Lett.* **2009**, *9*, 1752–1758. [PubMed]
57. Lawal, A.T. Recent progress in graphene based polymer nanocomposites. *Cogent Chem.* **2020**, *6*, 1833476.
58. Wang, X.; Hu, Y.; Song, L.; Yang, H.; Xing, W.; Lu, H. In situ polymerization of graphene nanosheets and polyurethane with enhanced mechanical and thermal properties. *J. Mater. Chem.* **2011**, *21*, 4222–4227.

59. Milani, M.A.; González, D.; Quijada, R.; Basso, N.R.S.; Cerrada, M.L.; Azambuja, D.S.; Galland, G.B. Polypropylene/graphene nanosheet nanocomposites by in situ polymerization: Synthesis, characterization and fundamental properties. *Compos. Sci. Technol.* **2013**, *84*, 1–7.
60. Zafar, M.; Muhammad Imran, S.; Iqbal, I.; Azeem, M.; Chaudhary, S.; Ahmad, S.; Kim, W.Y. Graphene-based polymer nanocomposites for energy applications: Recent advancements and future prospects. *Results Phys.* **2024**, *60*, 107655.
61. Lee, S.J.; Yoon, S.J.; Jeon, I.Y. Graphene/Polymer Nanocomposites: Preparation, Mechanical Properties, and Application. *Polymers* **2022**, *14*, 4733. [CrossRef]
62. Chee, W.K.; Lim, H.N.; Huang, N.M.; Harrison, I. Nanocomposites of graphene/polymers: A review. *RSC Adv.* **2015**, *5*, 68014–68051. [CrossRef]
63. Correa, E.; Moncada, M.E.; Gutiérrez, O.D.; Vargas, C.A.; Zapata, V.H. Characterization of polycaprolactone/rGO nanocomposite scaffolds obtained by electrospinning. *Mater. Sci. Eng. C* **2019**, *103*, 109773.
64. Kotov, N.A.; Dékány, I.; Fendler, J.H. Ultrathin graphite oxide–polyelectrolyte composites prepared by self-assembly: Transition between conductive and non-conductive states. *Adv. Mater.* **1996**, *8*, 637–641.
65. Guo, Y.; Xu, G.; Yang, X.; Ruan, K.; Ma, T.; Zhang, Q.; Gu, J.; Wu, Y.; Liu, H.; Guo, Z. Significantly enhanced and precisely modeled thermal conductivity in polyimide nanocomposites with chemically modified graphene via in situ polymerization and electrospinning-hot press technology. *J. Mater. Chem. C* **2018**, *6*, 3004–3015.
66. Pourhashem, S.; Vaezi, M.R.; Rashidi, A.; Bagherzadeh, M.R. Exploring corrosion protection properties of solvent based epoxy-graphene oxide nanocomposite coatings on mild steel. *Corros. Sci.* **2017**, *115*, 78–92.
67. Qiu, C.; Liu, D.; Jin, K.; Fang, L.; Xie, G.; Robertson, J. Electrochemical functionalization of 316 stainless steel with polyaniline-graphene oxide: Corrosion resistance study. *Mater. Chem. Phys.* **2017**, *198*, 90–98.
68. Haghdadeh, P.; Ghaffari, M.; Ramezanzadeh, B.; Bahlakeh, G.; Saeb, M.R. The role of functionalized graphene oxide on the mechanical and anti-corrosion properties of polyurethane coating. *J. Taiwan Inst. Chem. Eng.* **2018**, *86*, 199–212.
69. Kausar, A. Composite coatings of polyamide/graphene: Microstructure, mechanical, thermal, and barrier properties. *Compos. Interfaces* **2018**, *25*, 109–125.
70. Yu, Y.-H.; Lin, Y.-Y.; Lin, C.-H.; Chan, C.-C.; Huang, Y.-C. High-performance polystyrene/graphene-based nanocomposites with excellent anti-corrosion properties. *Polym. Chem.* **2013**, *5*, 535–550.
71. Chang, K.-C.; Ji, W.-F.; Lai, M.-C.; Hsiao, Y.-R.; Hsu, C.-H.; Chuang, T.-L.; Wei, Y.; Yeh, J.-M.; Liu, W.-R. Synergistic effects of hydrophobicity and gas barrier properties on the anticorrosion property of PMMA nanocomposite coatings embedded with graphene nanosheets. *Polym. Chem.* **2013**, *5*, 1049–1056.
72. Pan, Q.; Shim, E.; Pourdeyhimi, B.; Gao, W. Highly Conductive Polypropylene–Graphene Nonwoven Composite via Interface Engineering. *Langmuir* **2017**, *33*, 7452–7458.
73. Ou, J.; Liu, L.; Wang, J.; Wang, F.; Xue, M.; Li, W. Fabrication and Tribological Investigation of a Novel Hydrophobic Polydopamine/Graphene Oxide Multilayer Film. *Tribol. Lett.* **2012**, *48*, 407–415.
74. Alhumade, H.; Abdala, A.; Yu, A.; Elkamel, A.; Simon, L. Corrosion inhibition of copper in sodium chloride solution using polyetherimide/graphene composites. *Can. J. Chem. Eng.* **2016**, *94*, 896–904.
75. Zhu, G.; Cui, X.; Zhang, Y.; Chen, S.; Dong, M.; Liu, H.; Shao, Q.; Ding, T.; Wu, S.; Guo, Z. Poly (vinyl butyral)/Graphene oxide/poly (methylhydrosiloxane) nanocomposite coating for improved aluminum alloy anticorrosion. *Polymer* **2019**, *172*, 415–422.
76. Hikku, G.S.; Jeyasubramanian, K.; Venugopal, A.; Ghosh, R. Corrosion resistance behaviour of graphene/polyvinyl alcohol nanocomposite coating for aluminium-2219 alloy. *J. Alloys Compd.* **2017**, *716*, 259–269.
77. Liu, S.; Pan, T.J.; Wang, R.F.; Yue, Y.; Shen, J. Anti-corrosion and conductivity of the electrodeposited graphene/polypyrrole composite coating for metallic bipolar plates. *Prog. Org. Coat.* **2019**, *136*, 105237.
78. Fayyad, E.M.; Sadasivuni, K.K.; Ponnamma, D.; Al-Maadeed, M.A.A. Oleic acid-grafted chitosan/graphene oxide composite coating for corrosion protection of carbon steel. *Carbohydr. Polym.* **2016**, *151*, 871–878.
79. Zhu, H.; Yue, L.; Zhuang, C.; Zhang, Y.; Liu, X.; Yin, Y.; Chen, S. Fabrication and characterization of self-assembled graphene oxide/silane coatings for corrosion resistance. *Surf. Coat. Technol.* **2016**, *304*, 76–84.
80. Tang, J.; Yao, W.; Li, W.; Xu, J.; Jin, L.; Zhang, J.; Xu, Z. Study on a novel composite coating based on PDMS doped with modified graphene oxide. *J. Coat. Technol. Res.* **2018**, *15*, 375–383.
81. Cui, J.; Xiong, Z.; Qiu, H.; Li, J.; Yang, J. Functionalized graphene oxide: Carrier for corrosion inhibitor and barrier in waterborne epoxy coatings. *Compos. Part Appl. Sci. Manuf.* **2021**, *144*, 106354.
82. Sun, W.; Wang, L.; Wu, T.; Pan, Y.; Liu, G. Synthesis of low-electrical-conductivity graphene/pernigraniline composites and their application in corrosion protection. *Carbon* **2014**, *79*, 605–614.
83. Lin, Y.-T.; Don, T.-M.; Wong, C.-J.; Meng, F.-C.; Lin, Y.-J.; Lee, S.-Y.; Lee, C.-F.; Chiu, W.-Y. Improvement of mechanical properties and anticorrosion performance of epoxy coatings by the introduction of polyaniline/graphene composite. *Surf. Coat. Technol.* **2019**, *374*, 1128–1138.

84. Yang, S.; Zhu, S.; Hong, R. Graphene oxide/polyaniline nanocomposites used in anticorrosive coatings for environmental protection. *Coatings* **2020**, *10*, 1215. [CrossRef]
85. Jena, G.; Vanithakumari, S.C.; Polaki, S.R.; George, R.P.; Philip, J.; Amarendra, G. Electrophoretically deposited graphene oxide–polymer bilayer coating on Cu-Ni alloy with enhanced corrosion resistance in simulated chloride environment. *J. Coat. Technol. Res.* **2019**, *16*, 1317–1335.
86. An, H.; Liu, K.; Wang, S.; Jiang, C.; Liang, S.; Xiao, J.; Zhao, X.; Sun, Z.; Cao, C.; Gao, Y. Enhanced Corrosion Resistance of Waterborne Epoxy Coatings by Polyaniline Nanorods and Nitrogen and Fluorine Dual-Doped Graphene Oxide Composites. *ACS Appl. Nano Mater.* **2023**, *6*, 13250–13259.
87. Zhang, J.; Kong, G.; Li, S.; Le, Y.; Che, C.; Zhang, S.; Lai, D.; Liao, X. Graphene-reinforced epoxy powder coating to achieve high performance wear and corrosion resistance. *J. Mater. Res. Technol.* **2022**, *20*, 4148–4160.
88. He, S.; Wei, G.; Zhang, Z.; Yang, L.; Lin, Y.; Du, L.; Du, X. Incorporation of Graphene Oxide Modified with Polyamide Curing Agent into the Epoxy–Zinc Composite Coating for Promoting Its Corrosion Resistance. *Polymers* **2023**, *15*, 1873. [CrossRef]
89. Adiwibawa Prasetya, N.B.; Ibnu Ajizan, A.; Setiyo Widodo, D.; Ngadiwiyana, N.; Gunawan, G. A polyeugenol/graphene composite with excellent anti-corrosion coating properties. *Mater. Adv.* **2023**, *4*, 248–255.
90. Zhang, J.; Zheng, Y. Poly(m-phenylenediamine) encapsulated graphene for enhancing corrosion protection performance of epoxy coatings. *Nanotechnology* **2021**, *33*, 075705.
91. Zheng, M.; Ji, Q.; Ullah, Z.; Zhang, Y.; Chen, M.; Li, W.; Li, Q.; Liu, L. High protection performance based on corrosion media-consumption and barrier properties of the supramolecular polymer reinforced graphene oxide composite coatings. *J. Polym. Res.* **2021**, *28*, 426.

materials

MDPI

Article

Characterizing Six Percolation Cases in Flexible Electronic Composites: A Monte Carlo-Based 3D Compressive Percolation Model for Wearable Pressure Sensors

Sang-Un Kim [1] and Joo-Yong Kim [2,*]

[1] Department of Smart Wearable Engineering, Soongsil University, Seoul 06978, Republic of Korea; tkddnsl0723@naver.com
[2] Department of Materials Science and Engineering, Soongsil University, Seoul 06978, Republic of Korea
* Correspondence: jykim@ssu.ac.kr; Tel.: +82-10-8720-0631

Abstract: This study employs a Monte Carlo-based 3D compressive percolation model to systematically analyze the electrical behavior of flexible electronic composites under compressive deformation. By simulating the spatial distribution and connectivity of conductive particles, this study identifies six distinct percolation cases, each describing a unique connectivity evolution under strain. The model reveals that excessive initial connectivity leads to saturation effects, reducing sensitivity, while a high Poisson's ratio ($\geq$0.3) causes connectivity loss due to shear plane expansion. Notably, asymmetric particle shapes, such as cylinders and rectangles, exhibit superior percolation behavior, forming infinite clusters at lower strain thresholds (~0.4) compared to spherical particles (~0.5). Monte Carlo simulations with 3000 particles validate these findings, showing consistent trends in percolation behavior across different deformation states. By classifying and quantifying these six connectivity scenarios, this research provides a structured framework for optimizing flexible sensor designs, ensuring an optimal balance between conductivity and sensitivity. These findings contribute to advancing flexible electronics, particularly in wearable health monitoring, robotics, and smart textiles.

Keywords: 3D compressive percolation; flexible electronic composites; wearable pressure sensor; Monte Carlo simulation

1. Introduction

As nanoparticles with excellent electrical properties such as carbon black [1,2], graphene [3,4], carbon nanotube [5,6], and MXene [7–10] have been actively researched, these studies have extended to the development of flexible electronic materials through composite fabrication. Flexible electronic composites represent a cutting-edge material trend by integrating conductive particles into flexible substrates, thereby combining the advantages of both components. Unlike conventional rigid substrates used in many electronic devices, these flexible composites exhibit unique electrical properties alongside mechanical flexibility [11–14].

These materials have found applications in various advanced technological fields, such as wearable sensors, electronic skins, and flexible displays. Wearable sensors have garnered significant attention, leveraging the principle that the material's deformation and corresponding changes in electrical properties are influenced by human movement or physiological signals [15–18]. Flexible electronic composites have also been utilized as electrodes for collecting electrical signals, such as electrocardiograms (ECGs) and electromyograms (EMGs), by adhering closely to the skin [19,20]. Additionally, they serve as band-type sensors that monitor respiration by detecting changes in abdominal volume during inhalation and exhalation [21]. Other applications include sensors attached to joints or muscles [22], which measure stretching or bending degrees to recognize physical activities [23]. These examples highlight the extensive potential of flexible electronic composites in diverse applications.

One fundamental mechanism underlying the sensing of such physiological signals is piezo resistivity. When pressure deforms the material, the conductive particles embedded within its structure move closer together or further apart, resulting in measurable changes in the overall resistance of the composite [24,25]. This property is central to the functionality of flexible sensors designed for detecting biomechanical and physiological signals. The behavior of nanoscale conductive particles and the corresponding changes in a material's electrical properties are effectively described by percolation theory, a foundational framework in this study [26,27]. This theory evaluates the arrangement and interconnectivity of conductive particles embedded within a flexible matrix material, where the matrix primarily determines mechanical properties, and the particles define electrical characteristics [28].

As shown in Figure 1, the spatial distribution of conductive particles is quantified using the volume fraction probability p, which represents the fraction of the total material volume occupied by the conductive particles. When these particles are positioned sufficiently close to one another, they form conductive paths, collectively referred to as clusters. As the volume fraction p increases, the clusters grow in size and begin to coalesce, eventually forming a continuous network that spans from one side of the structure to the opposite side. This large-scale conductive network is known as the infinite cluster. The critical point at which this infinite cluster forms is defined by the critical volume fraction p_c. At p_c, the material transitions from an insulating to a conductive state due to the establishment of long-range connectivity. The electrical conductivity of the composite is significantly enhanced by the formation of the infinite cluster, and its strength is determined by the number and arrangement of conductive particles within it. This critical phenomenon is central to understanding and optimizing the electrical properties of conductive composites. The following equation can explain the percolation theory representing electrical resistance.

$$R(p) = \frac{1}{\sigma(p)} = \frac{1}{\sigma_0 (p - p_c)^t} \tag{1}$$

where the $\sigma(p)$ is the electrical conductivity at the particle volume fraction p, σ_0 is the proportionality constant, and t is the experimentally determined critical exponent. When $p < p_c$, insufficient connectivity results in conductivity approaching zero and resistance tending to infinity; when $p > p_c$, conductive pathways form, leading to a sharp increase in conductivity and a decrease in resistance. Near $p \approx p_c$, conductivity exhibits nonlinear growth governed by the critical exponent ttt, highlighting strong transition characteristics.

A related study by M. Åkerfeldt et al. used percolation theory to analyze the electrical properties of composites by adding the conductive polymer PEDOT-PSS to a polyurethane (PU) binder. The study showed a significant decrease in surface resistance at around 2 wt% PEDOT-PSS, indicating the formation of a conductive network. Beyond this concentration, the resistance plateaued, highlighting a critical threshold for achieving electrical connectivity [29]. Similarly, the study by Wang et al. demonstrated the occurrence of a percolation zone in mechanical sensing applications using flexible materials. This zone was characterized by a significant decrease in electrical resistance as the content of conductive fillers increased, highlighting the critical role of filler concentration in achieving electrical conductivity [30]. When fabricating flexible composite materials, numerous studies have shown that adjusting the volume fraction of conductive particles within the matrix to exceed the critical volume fraction p_c enhances electrical properties. However, for wearable sensors, sensitivity—arguably the most critical performance metric—requires significant changes in electrical properties to detect subtle variations effectively. Therefore, applying percolation theory to analyze and optimize the deformation behavior of flexible electronic composites is crucial for achieving high sensitivity.

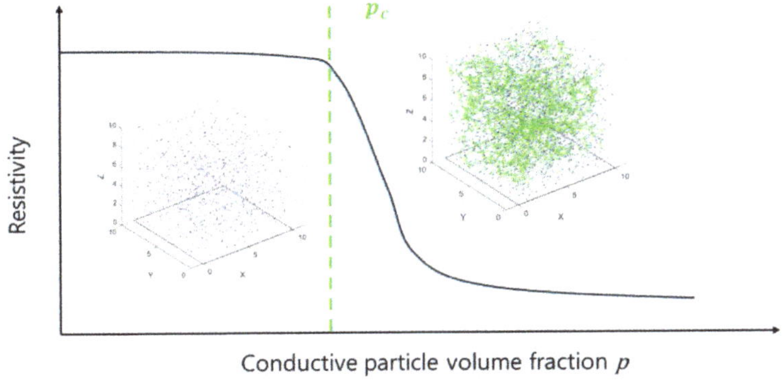

Figure 1. Percolation theory of conductive composites.

To address this, the present study references a Monte Carlo method [28,31,32], leveraging random number generation to simulate complex systems and extract data for probability-based numerical calculations. This approach considers the inherent challenges in controlling the distribution of nanoscale conductive particles within a flexible material matrix. Additionally, a 3D compressive percolation system was developed as an optimization simulation for wearable sensor sensitivity. This system enabled the analysis of electrical property changes under compression, providing insights into trends and behaviors for real-world applications.

2. Materials and Methods

2.1. Percolation Model

The percolation model used in this study was developed using MATLAB R2023a. The model simulates conductive particles randomly distributed within a defined cubic space, with particle sizes also varying randomly within a specified range. To account for the typical conductive fillers—such as carbon black, CNTs, graphene, and MXenes—the particle shapes were categorized as sphere (0D), cylinder (1D), and rectangular (2D), serving as key variables in the simulation. Connectivity within the percolation model was defined such that particles are considered connected when the distance between them falls below a specific threshold, termed the connectivity length, ξ. The relationship between the distance of two particles and the connectivity length determines the network formation, which can be expressed as follows:

$$r_{ij} = |r_i - r_j| \tag{2}$$

$$A_{ij} = \begin{cases} 1, & r_{ij} \leq \xi \\ 0, & r_{ij} > \xi \end{cases} \tag{3}$$

where the r_{ij} represents the distance between particles i and j; r_i and r_j are their position vectors; and A_{ij} is the adjacency matrix element indicating connection. In addition to the arrangement of conductive particles in Equation (2), an equation that considers the size of the particles can be expressed as follows:

$$A_{ij} = \begin{cases} 1, & r_{ij} - (R_i + R_j) \leq \xi \\ 0, & r_{ij} - (R_i + R_j) > \xi \end{cases} \quad (4)$$

where the R represents the radius of particles i and j. It indicates that a connection can be made if the distance between the centers of particles is within the critical distance minus the size of the particles. The final equations related to the alignment and connectivity of conductive particles, considering three distinct shapes (sphere, cylinder, and rectangular), can be expressed as follows:

$$P_{connect,sphere}(r_{ij}) = \exp\left(-\frac{r_{ij} - (R_i + R_j)}{\xi}\right) \quad (5)$$

$$P_{connect,cylinder}(r_{ij}, \theta_i, \theta_j) = \exp\left(-\frac{r_{ij} - (R_i + R_j + L_i cos\theta_i + L_j cos\theta_j)}{\xi}\right) \quad (6)$$

$$P_{connect,rectangular}(r_{ij}, \theta_i, \theta_j, a_i, a_j, b_i, b_j) = \exp\left(-\frac{r_{ij} - \sqrt{(a_i cos\theta_i + a_j cos\theta_j)^2 + (b_i sin\theta_i + b_j sin\theta_j)^2}}{\xi}\right) \quad (7)$$

where L_i and L_j are the effective length of cylinder particles i and j; θ_i and θ_j are the angular alignments i and j; a_i, a_j, b_i, and b_j are the length of major axis and minor axis; and $P_{connect,sphere}$, $P_{connect,cylinder}$, and $P_{connect,rectangular}$ are the connection probabilities. These equations describe the connectivity between particles by incorporating their size, alignment, and orientation for each geometry type.

The percolation model, described by Equations (5)–(7), defines the initial connectivity of a conductive network by incorporating the influence of different particle geometries—spherical, cylindrical, and rectangular—on electrical conductivity. In the case of spherical particles, as expressed in Equation (5), connectivity depends solely on the spatial proximity of two particles, meaning that a conductive path is formed if the center-to-center distance between two spheres is within a defined threshold ξ, with connectivity probability $P_{connect,sphere}$ being isotropic due to the uniform geometry of the spheres. However, in cylindrical particles, governed by Equation (6), connectivity is influenced by both spatial proximity and directional alignment, where two cylinders can only form a conductive link if they are not only within the threshold distance ξ but also aligned at favorable angles θ_i and θ_j, with their effective lengths L_i and L_j determining the extent of possible conductive interaction, thereby introducing anisotropic connectivity characteristics. In contrast, rectangular particles, as described in Equation (7), exhibit even more pronounced anisotropic behavior, where connectivity is affected by both the major axis a, which influences long-range connectivity, and the minor axis b, which affects local contact probability, while the orientation angles θ_i and θ_j dictate the likelihood of forming a conductive path, meaning that connectivity is highly dependent on both geometric proportions and spatial positioning.

The electrically connected group of conductive particles is referred to as a cluster. Within a cluster, the free electrons of the conductive particles facilitate charge transport, thereby forming an electrically conductive network. For a composite material to exhibit effective conductivity, it is not sufficient to have finite clusters that are isolated within the material. Instead, as the volume fraction of conductive particles increases, these finite clusters gradually merge, forming a continuous network that spans the entire structure from one side to the other. This ultimately leads to the formation of an infinite cluster, which can also establish connectivity with external electrodes, enabling the material to function as a conductive composite.

The fundamental percolation model in this study was designed to distinguish and analyze various parameters, including the number of conductive particles, particle size, shape, and connectivity length, as well as the definitions of finite and infinite clusters. The model calculates the number of clusters and the number of particles within the infinite cluster, providing insights into the electrical properties of the material. This approach enables a systematic evaluation of how these factors influence the transition from isolated clusters to a fully connected conductive network. This model allows for systematic analysis of how particle shape, size, and distribution influence the percolation threshold and overall network connectivity.

2.2. Three-Dimensional Compressive Percolation

From the basic percolation model, this study extends the analysis to model a 3D compressive percolation system, simulating the behavior of flexible electronic composites under compressive deformation when used as pressure sensors. The Poisson's ratio, which may vary depending on the flexible substrate, was also considered in this model. Instead, the focus was on analyzing the changes in clusters as the initial distribution of conductive particles proportionally shifted along the compression direction.

To incorporate the effects of compression, a new variable, compressive strain, was introduced into the percolation model. This variable accounts for the positional shifts in particles under compression, leading to closer connections between them. Using this modified model, the total number of clusters, the number of infinite clusters, and the number of particles constituting the infinite cluster can be expressed as follows:

$$L_{connection}(\varepsilon, r_{ij}) = \frac{N(N-1)}{2}\exp\left(-\frac{(1-\varepsilon)r_{ij,0}}{\xi}\right) \tag{8}$$

$$C(\varepsilon, r_{ij}) = \alpha L_{connection} \tag{9}$$

$$n_{i.c}(\varepsilon) = n_{max}(1 - \exp\left(-k(P_{connection}(\varepsilon) - P_c)\right)) \tag{10}$$

where N is the number of particles, ε represents the compressive strain, C is the total number of clusters, α is a proportionality constant, $L_{connection}$ determines the connection probability under compression, $n_{i,c}(\varepsilon)$ is the number of particles in the infinite cluster under compressive strain, and n_{max} is the maximum sensitivity. These equations effectively describe the behavior of particle connectivity and cluster formation under compressive strain, providing a framework to analyze and optimize the performance of flexible electronic composites.

While Equations (5)–(7) define the initial network structure of percolation model under static conditions, Equations (8)–(10) extend this framework by incorporating compressive strain ε to model dynamic changes in connectivity. Equation (8) describes how the total number of clusters N evolves as a function of strain, explaining that as compressive deformation increases, more particles transition from isolated states to connected conductive paths; Equation (9) focuses on the formation of the infinite cluster, which represents a fully connected conductive network spanning the material, showing that the connectivity probability P_c and the applied strain ε dictate whether a percolating network emerges or collapses; and, finally, Equation (10) defines maximum sensitivity n_{max} by quantifying the strain-dependent growth of electrical conductivity, demonstrating that the relationship between compressive strain, cluster formation, and connectivity probability determines the overall electrical performance of the material. It is possible to analyze how different particle shapes, orientations, and initial distributions influence the formation of conductive pathways and predict how external mechanical deformation affects electrical connectivity, thereby enabling the optimization of flexible electronic composites for high-performance applications.

2.3. Monte Carlo Method

In the developed 3D compressive percolation model, the Monte Carlo method was employed to collect data under random conditions, enabling statistical parameter estimation for the model. Among the input variables, factors such as the initial positions, sizes, and orientations of the conductive particles—which are uncontrollable in real flexible electronic composites—were assigned randomly. Meanwhile, controllable factors were carefully regulated. The composite's dimensions were fixed at a $10 \times 10 \times 10$ cubic structure, and the Poisson's ratio was varied between 0 and 0.3. The number of conductive particles ranged from 0 to 4500, while their shapes were categorized into spheres, cylinders, and rectangles. The connectivity length was defined as the threshold distance at which particles were considered to be in contact, and the compressive strain was set at 0.8.

The output data included the number of finite clusters, the connections in the infinite cluster, and the maximum number of conductive particles in the infinite cluster. These outputs provided a comprehensive understanding of the structural and electrical characteristics of the flexible electronic composite under compression.

3. Results

Case of 3D Compressive Percolation

Figure 2a shows a characteristic image of the 3D compressive infiltration model generated using the Monte Carlo method. The blue dots represent conductive particles randomly placed throughout space, and the image is scaled down to 1:10 of the size of the placed particles for visualization. The green lines represent connections between these particles. These connections indicate the formation of infinite clusters, and more green lines indicate increased connectivity, reflecting the enhanced electrical conductivity of the material. Figure 2b illustrates the three shapes of conductive particles used in the model: sphere, cylinder, and rectangular. These shapes were selected to study the impact of particle geometry on the percolation behavior and overall conductivity of the composite material.

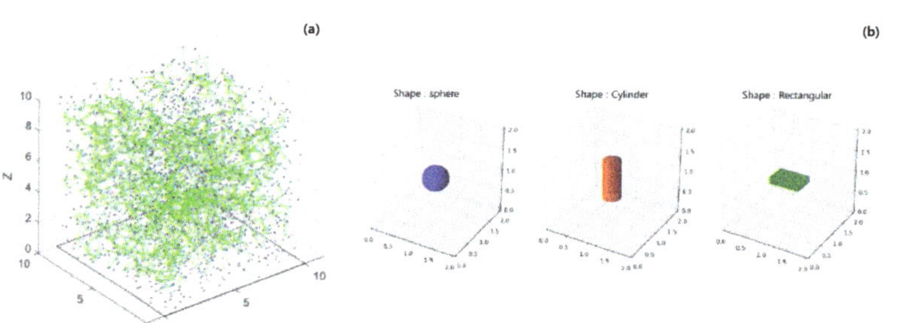

Figure 2. The results of the 3D modeling images: (**a**) 3D compressive percolation model, (**b**) the 3 shape of conductive particles.

Table 1 presents the results of the number of connected particles within the infinite cluster under the same compression conditions and within a given particle size range, where particle sizes and spatial distributions were randomly assigned. Despite conducting the Monte Carlo simulation under identical conditions—with 3000 particles and a Poisson's ratio of 0—the first result yielded an initial connectivity of 0.28, which was higher than the second result. However, during compression, the second result exhibited a greater number of connections. Nevertheless, the overall trend of increasing connectivity remained similar. This confirms that the Monte Carlo method successfully replicates aspects that should be considered in actual fabrication.

Table 1. The infinite cluster results from Monte Carlo-based 3D compressive percolation.

Strain	1st Result *	2nd Result *
0	0.28	0.2
0.11	0.62	0.68
0.23	0.84	0.85
0.34	0.93	0.92
0.46	0.96	0.95
0.57	0.98	0.98
0.69	0.99	1.00
0.80	1.00	1.00

* The condition of Monto Carlo is 3000 particles and Poisson's ratio is 0.

The analysis of changes in conductive particle connections, which determine the electrical performance of flexible electronic composites under compressive deformation, revealed six cases using the 3D compressive percolation model. Figure 3 illustrates these six cases of connection changes during compression, with each case representing a characteristic scenario from the overall conditions. Figure 3a represents cases where the number of conductive particles is insufficient both before and after compression, resulting in no infinite cluster formation and, consequently, no electrical conductivity. Figure 3b occurs when the number of conductive particles increases, and while no connections exist before compression, particles connect during compression as inter-particle distances decrease, leading to the formation of an infinite cluster. Figure 3c represents a case where the number of conductive particles is high enough that an infinite cluster exists before compression, and, at a final compressive strain of 0.8, all conductive particles are ideally connected, achieving optimal electrical conductivity. Figure 3d begins with a more extensive infinite cluster, but as all particles connect during compression, the system reaches equilibrium, and no further connections are formed despite continued deformation. Figure 3e is caused by a high Poisson's ratio in the flexible matrix, where connections initially increase during compression, but expansion in the shear plane increases the distance between particles, leading to the loss of connections and the disappearance of the infinite cluster. Finally, Figure 3f also results from a high Poisson's ratio, where initially high connectivity decreases during compression, resulting in the disappearance of the infinite cluster. These six cases highlight the diverse effects of compression and particle distribution on the conductive properties of flexible electronic composites.

Figure 4 illustrates the comparative analysis of finite and infinite cluster formation, and the differences in the number of nodes constituting the clusters, under three representative particle shapes: sphere, cylinder, and rectangular. To isolate the effects of particle shape, all other conditions were kept identical, with particle counts of 500, 1000, and 1500, and the expansion in the shear plane was restricted. In Figure 4a, which shows the results for sphere-shaped particles, infinite clusters did not form under compressive strains up to 0.4. For particle counts of 1000 and 1500, infinite clusters began to form at compressive strains exceeding 0.5 and 0.4, respectively. Figure 4b,c present the results for cylinder and rectangular particles under the same particle count and random volume distribution conditions. For the cylinders with a diameter-to-length ratio of 5 and rectangular particles with a major axis-to-minor axis ratio of 1.5 and a thickness of 0.2, infinite clusters formed more rapidly, with more particles connecting under compressive deformation.

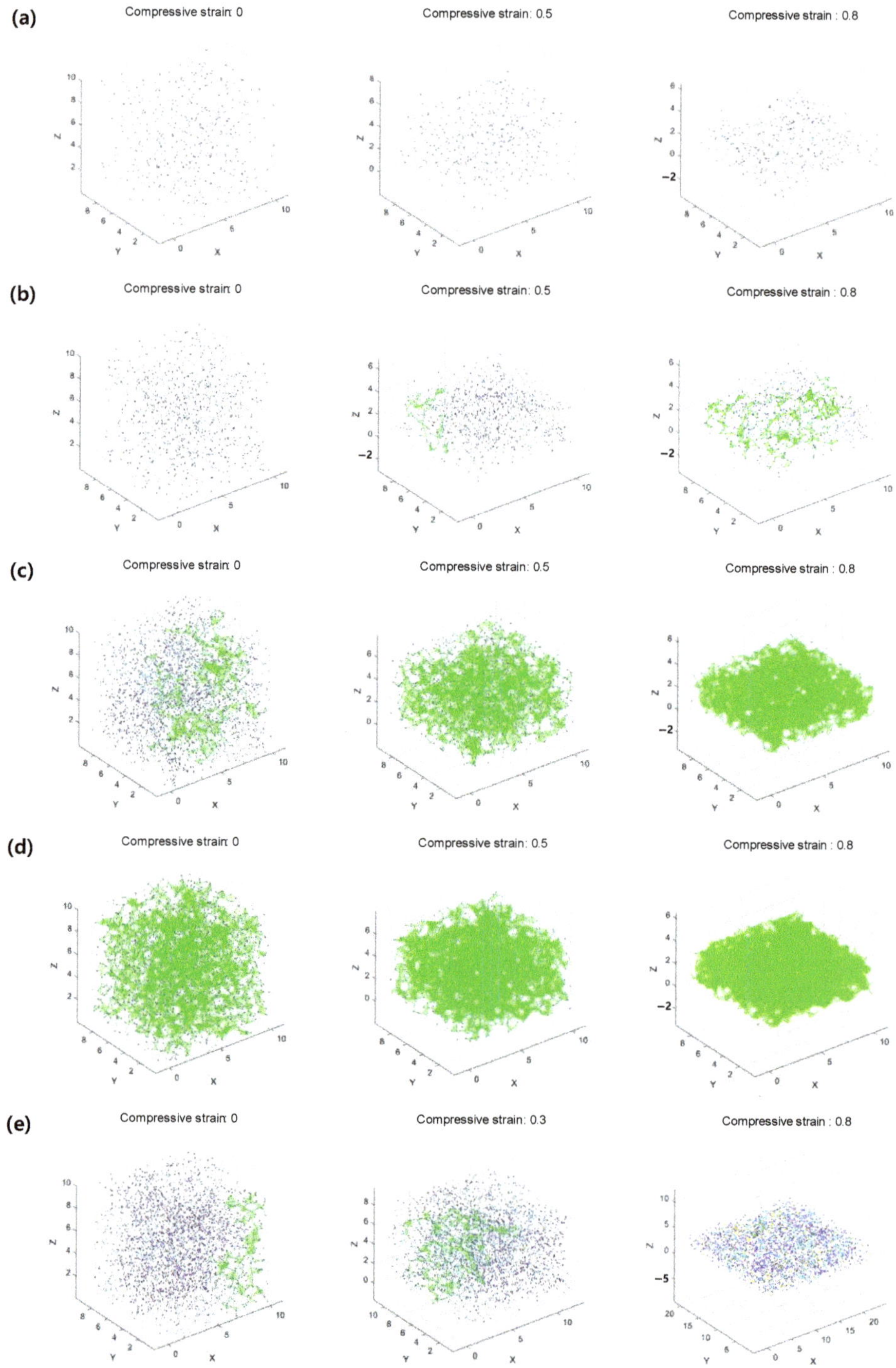

Figure 3. *Cont.*

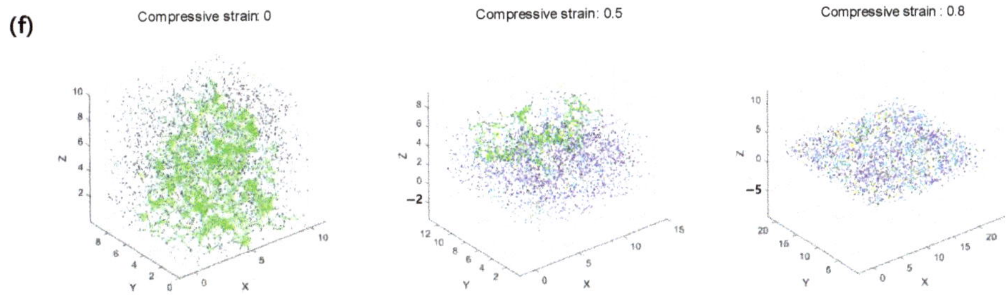

Figure 3. The six cases of the 3D compressive percolation model (**a**–**f**).

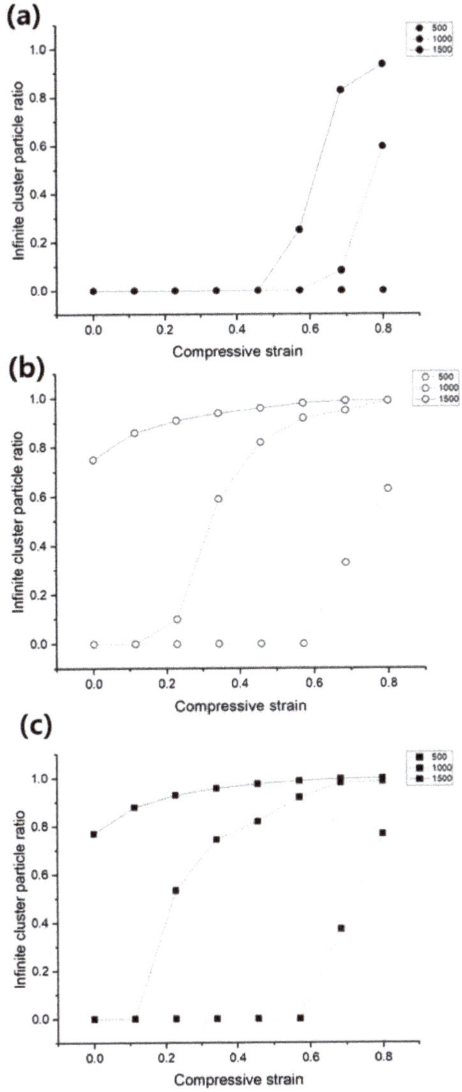

Figure 4. The infinite cluster results in 3 different shapes of conductive particles in 3D compressive percolation model. (**a**) Sphere, (**b**) cylinder, (**c**) rectangular.

4. Discussion

This study leveraged a Monte Carlo simulation approach to explore the behavior of conductive particles in flexible electronic composites under compressive deformation. Variables that are challenging to control experimentally—such as the number and shape of conductive particles, their spatial distribution, and the mechanical properties of the matrix—were systematically analyzed. By generating random distributions for particle positions, sizes, and orientations, the simulation replicated real-world irregularities, enabling robust statistical analysis. These results provided critical insights into the electrical property changes in flexible electronic composites and allowed the classification of deformation behaviors into six cases.

Among the six identified cases, the configuration with the largest variation in the number of connected particles from the initial state to the final compression was found to be the most suitable for achieving high sensitivity. This behavior allows the sensor to detect subtle changes in external forces, making it ideal for applications requiring precise measurement of compressive strain. However, many pressure sensors experience a decline in sensitivity within their operational range due to connection saturation. When the initial number of conductive particles is excessively high, overconnectivity occurs at the start of deformation, limiting the potential for additional connections as compression progresses. This saturation reduces the overall sensitivity of the composite, particularly in scenarios requiring nuanced detection across a broad range of compressive forces.

This study also revealed the influence of matrix properties, particularly Poisson's ratio, on the behavior of conductive composites. In materials with a high Poisson's ratio, compressive deformation induces expansion in the shear plane, increasing the distance between conductive particles. This phenomenon leads to the loss of connectivity, resulting in a decline in electrical properties. Such behavior can be analogously understood through the lens of tensile sensors, where stretching increases particle separation and electrical resistance. This finding underscores the need to carefully consider Poisson's ratio when designing flexible electronic composites for both compression- and tension-based applications.

Another significant aspect of this research involved analyzing the effect of particle geometry on cluster formation and connectivity. Conductive particles were categorized into three representative shapes—sphere, cylinder, and rectangular. By isolating particle geometry as a variable while keeping other parameters constant, this study demonstrated that asymmetry and directional randomness significantly enhance connectivity. Spherical particles, due to their symmetric shape, require higher compressive strain to achieve sufficient connectivity. In contrast, asymmetric shapes such as cylinders and rectangles form connections more readily under similar conditions. For cylindrical particles with a diameter-to-length ratio of 5 and rectangular particles with a major axis-to-minor axis ratio of 1.5, infinite clusters formed more rapidly, with more particles participating in connections during compression. This observation suggests that asymmetric and directionally random shapes are inherently more advantageous for forming robust conductive networks, making them preferable for applications requiring high electrical performance under deformation.

Overall, the findings in this 3D compressive model have direct implications for the design of wearable sensors. Effective sensitivity in these sensors depends on the ability of the material's electrical properties to respond proportionally to deformation. The results highlight the importance of optimizing the number and shape of conductive particles, as well as the mechanical properties of the matrix, to balance initial connectivity and sensitivity under operational conditions. Materials exhibiting excessive initial connectivity may suffer from saturation effects, reducing their performance in dynamic environments. Additionally, the phenomenon of shear plane expansion due to high Poisson's ratios provides a critical consideration for material design. For applications involving tensile deformation, strategies to mitigate the loss of connectivity must be employed, such as modifying the matrix material or adjusting particle geometry to enhance resilience against separation.

Future research will focus on incorporating a broader range of variables into the model to maximize sensor sensitivity within the desired sensing range. Examples include factors such as the agglomeration of conductive particles due to van der Waals forces, entanglement effects, particle rearrangement under repeated compression, and the inclusion of partial space matrices, such as fiber fabrics. These additional variables will enable more comprehensive simulations of realistic scenarios. Furthermore, to enhance the analysis, data extracted using the Monte Carlo method in this study will be integrated with artificial intelligence to develop advanced algorithmic models. This approach aims to predict and optimize the behavior of flexible electronic composites under various conditions, providing a more robust framework for sensor design and performance evaluation.

5. Conclusions

This study successfully employed a Monte Carlo-based 3D compressive percolation model to analyze the electrical performance of flexible electronic composites under compressive deformation. By simulating variables that are challenging to control experimentally, such as the number, shape, and spatial distribution of conductive particles, the research provided new insights into the behavior of these materials. The findings classified electrical property changes into six distinct cases, emphasizing the importance of factors like initial connectivity, saturation effects, and particle geometry. Additionally, the influence of matrix properties, such as Poisson's ratio, on conductive particle connectivity was demonstrated, providing a robust framework for optimizing wearable sensor sensitivity and performance.

The expected impact of this study is significant for the development of advanced wearable sensors and flexible electronic applications. For instance, in wearable health monitoring devices, a flexible sensor embedded in a wristband could monitor physiological signals such as heart rate, respiration, or joint movement. By applying the insights gained from this research, the sensor design could be optimized to enhance sensitivity and ensure accurate detection of subtle signals under varying conditions, improving device reliability. In industrial applications, pressure sensors for robotics or soft actuators could be developed using optimized flexible composites. By tailoring particle geometry and connectivity, sensors could achieve consistent performance across a wide range of forces, ensuring accurate and repeatable measurements in dynamic environments. In the field of electronic textiles, integrating these materials into smart fabrics could enable lightweight, flexible clothing that monitors wearer activity and environmental changes. The findings of this study provide a basis for designing textiles with enhanced electrical properties, allowing for efficient signal transmission even under strain.

Overall, the research offers a strong foundation for future advancements in flexible electronic materials, enabling new applications in healthcare, industry, and beyond. By incorporating additional variables and leveraging artificial intelligence for further optimization, the potential of these materials can be fully realized.

Author Contributions: As a corresponding author, J.-Y.K. was responsible for the whole structure construction, while S.-U.K. was responsible for reviewing, supervising, experiment design, and modeling. S.-U.K. was responsible for the data collection, processing, and material selection. J.-Y.K. and S.-U.K. drafted the manuscript. All authors have read and agreed to the published version of the manuscript.

Funding: This research was partly funded by the Technology Innovation Program (or Industrial Strategic Technology Development Program-Materials/Parts Package Type) (20016038, development of the textile-IT converged digital sensor modules for smart wear to monitor bio & activity signal in exercise, and KS standard), the Ministry of Trade, Industry & Energy (MOTIE, Republic of Korea), and the Korea Institute for Advancement of Technology (KIAT) grant funded by the Korean Government (MOTIE) (P0012770).

Institutional Review Board Statement: Not applicable.

Informed Consent Statement: Not applicable.

Data Availability Statement: Data are contained within the article.

Conflicts of Interest: The authors declare no conflicts of interest.

References

1. Wang, C.; Xia, K.; Wang, H.; Liang, X.; Yin, Z.; Zhang, Y. Advanced carbon for flexible and wearable electronics. *Adv. Mater.* **2019**, *31*, 1801072. [CrossRef]
2. Liu, L.; Shen, Z.; Zhang, X.; Ma, H. Highly conductive graphene/carbon black screen printing inks for flexible electronics. *J. Colloid Interface Sci.* **2021**, *582*, 12–21. [CrossRef] [PubMed]
3. Lee, S.-M.; Kim, J.-H.; Ahn, J.-H. Graphene as a flexible electronic material: Mechanical limitations by defect formation and efforts to overcome. *Mater. Today* **2015**, *18*, 336–344. [CrossRef]
4. Jang, H.; Park, Y.J.; Chen, X.; Das, T.; Kim, M.S.; Ahn, J.H. Graphene-based flexible and stretchable electronics. *Adv. Mater.* **2016**, *28*, 4184–4202. [CrossRef]
5. Tang, J. Carbon nanotube-based flexible electronics. In *Flexible, Wearable, and Stretchable Electronics*; CRC Press: Boca Raton, FL, USA, 2020; pp. 137–156.
6. Cai, L.; Wang, C. Carbon nanotube flexible and stretchable electronics. *Nanoscale Res. Lett.* **2015**, *10*, 320. [CrossRef] [PubMed]
7. Hu, Y.; Wang, F.; Ye, H.; Jiang, J.; Li, S.; Dai, B.; Li, J.; Yang, J.; Song, X.; Zhang, J. MXene-based flexible electronic materials for wound infection detection and treatment. *npj Flex. Electron.* **2024**, *8*, 30. [CrossRef]
8. Lyu, B.; Kim, M.; Jing, H.; Kang, J.; Qian, C.; Lee, S.; Cho, J.H. Large-area MXene electrode array for flexible electronics. *ACS Nano* **2019**, *13*, 11392–11400. [CrossRef]
9. Meena, J.S.; Choi, S.B.; Kim, J.-W. Review on Ti3C2-based MXene nanosheets for flexible electrodes. *Electron. Mater. Lett.* **2022**, *18*, 256–274. [CrossRef]
10. Das, P.; Marvi, P.K.; Ganguly, S.; Tang, X.; Wang, B.; Srinivasan, S.; Rajabzadeh, A.R.; Rosenkranz, A. MXene-based elastomer mimetic stretchable sensors: Design, properties, and applications. *Nano-Micro Lett.* **2024**, *16*, 135. [CrossRef] [PubMed]
11. Wong, W.S.; Salleo, A. *Flexible Electronics: Materials and Applications*; Springer Science & Business Media: Berlin/Heidelberg, Germany, 2009; Volume 11.
12. Chen, X.; Rogers, J.A.; Lacour, S.P.; Hu, W.; Kim, D.-H. Materials chemistry in flexible electronics. *Chem. Soc. Rev.* **2019**, *48*, 1431–1433. [CrossRef] [PubMed]
13. Cheng, I.-C.; Wagner, S. Overview of flexible electronics technology. In *Flexible Electronics: Materials and Applications*; Springer: Boston, MA, USA, 2009; pp. 1–28.
14. Kirchner, E.; Wallmersperger, T.; Gwosch, T.; Menning, J.D.; Peters, J.; Breimann, R.; Kraus, B.; Welzbacher, P.; Küchenhof, J.; Krause, D. A Review on Sensor-Integrating Machine Elements. *Adv. Sens. Res.* **2024**, *3*, 2300113. [CrossRef]
15. Gao, W.; Ota, H.; Kiriya, D.; Takei, K.; Javey, A. Flexible electronics toward wearable sensing. *Acc. Chem. Res.* **2019**, *52*, 523–533. [CrossRef] [PubMed]
16. Nag, A.; Mukhopadhyay, S.C.; Kosel, J. Wearable flexible sensors: A review. *IEEE Sens. J.* **2017**, *17*, 3949–3960. [CrossRef]
17. Liu, H.; Wang, L.; Lin, G.; Feng, Y. Recent progress in the fabrication of flexible materials for wearable sensors. *Biomater. Sci.* **2022**, *10*, 614–632. [CrossRef] [PubMed]
18. Nan, X.; Wang, X.; Kang, T.; Zhang, J.; Dong, L.; Dong, J.; Xia, P.; Wei, D. Review of flexible wearable sensor devices for biomedical application. *Micromachines* **2022**, *13*, 1395. [CrossRef]
19. Shi, Y.; Fu, X.; Wang, W.; Yu, D. Stretchable, adhesive and low impedance hydrogel prepared by one-pot method used as ECG electrodes. *Colloids Surf. A Physicochem. Eng. Asp.* **2023**, *662*, 130998. [CrossRef]
20. Lee, Y.; Yim, S.-G.; Lee, G.W.; Kim, S.; Kim, H.S.; Hwang, D.Y.; An, B.-S.; Lee, J.H.; Seo, S.; Yang, S.Y. Self-adherent biodegradable gelatin-based hydrogel electrodes for electrocardiography monitoring. *Sensors* **2020**, *20*, 5737. [CrossRef] [PubMed]
21. Kim, D.; Lee, J.; Park, M.K.; Ko, S.H. Recent developments in wearable breath sensors for healthcare monitoring. *Commun. Mater.* **2024**, *5*, 41. [CrossRef]
22. Chen, Z.; Wang, Q.; Bi, Y.; Lin, J.; Yang, W.; Deng, C.; Guo, S.; Liao, M. Analyzing human muscle state with flexible sensors. *J. Sens.* **2022**, *2022*, 5227955. [CrossRef]

23. Wang, P.; Liu, J.; Yu, W.; Li, G.; Meng, C.; Guo, S. Flexible, stretchable, breathable and sweatproof all-nanofiber iontronic tactile sensor for continuous and comfortable knee joint motion monitoring. *Nano Energy* **2022**, *103*, 107768. [CrossRef]
24. Han, S.; Li, S.; Fu, X.; Han, S.; Chen, H.; Zhang, L.; Wang, J.; Sun, G. Research Progress of Flexible Piezoresistive Sensors Based on Polymer Porous Materials. *ACS Sens.* **2024**, *9*, 3848–3863. [CrossRef]
25. Kim, S.; Truong, T.; Jang, J.; Kim, J. The programmable design of large-area piezoresistive textile sensors using manufacturing by jacquard processing. *Polymers* **2022**, *15*, 78. [CrossRef] [PubMed]
26. Kim, S.; Choi, S.; Oh, E.; Byun, J.; Kim, H.; Lee, B.; Lee, S.; Hong, Y. Revisit to three-dimensional percolation theory: Accurate analysis for highly stretchable conductive composite materials. *Sci. Rep.* **2016**, *6*, 34632. [CrossRef]
27. Jung, S.; Choi, H.W.; Mocanu, F.C.; Shin, D.-W.; Chowdhury, M.F.; Han, S.D.; Suh, Y.-H.; Cho, Y.; Lee, H.; Fan, X. Modeling electrical percolation to optimize the electromechanical properties of CNT/polymer composites in highly stretchable fiber strain sensors. *Sci. Rep.* **2019**, *9*, 20376. [CrossRef]
28. Ma, H.; Gao, X.-L. A three-dimensional Monte Carlo model for electrically conductive polymer matrix composites filled with curved fibers. *Polymer* **2008**, *49*, 4230–4238. [CrossRef]
29. Åkerfeldt, M.; Strååt, M.; Walkenström, P. Electrically conductive textile coating with a PEDOT-PSS dispersion and a polyurethane binder. *Text. Res. J.* **2013**, *83*, 618–627. [CrossRef]
30. Wang, Y.; Adam, M.L.; Zhao, Y.; Zheng, W.; Gao, L.; Yin, Z.; Zhao, H. Machine learning-enhanced flexible mechanical sensing. *Nano-Micro Lett.* **2023**, *15*, 55. [CrossRef] [PubMed]
31. Balberg, I.; Binenbaum, N.; Wagner, N. Percolation thresholds in the three-dimensional sticks system. *Phys. Rev. Lett.* **1984**, *52*, 1465. [CrossRef]
32. Hansen, G. High aspect ratio sub-micron and nano-scale metal filaments. *SAMPE J.* **2005**, *41*, 24–33.

materials

MDPI

Article

Qualitative Research of Composite Graphene Membranes Using the Electric Mode in SEM and AFM

Grzegorz Romaniak, Konrad Dybowski *, Łukasz Kołodziejczyk and Paulina Kowalczyk

Faculty of Mechanical Engineering, Institute of Materials Science and Engineering, Lodz University of Technology, 1/15 Stefanowskiego St., 90-924 Lodz, Poland
* Correspondence: konrad.dybowski@p.lodz.pl

Abstract: The development of new graphene-based materials necessitates the application of suitable material imaging techniques, especially for the identification of defects in the graphene structure and its continuity. For this purpose, it is natural to use one of the main properties of graphene—electrical conductivity. In this work, we prepare a 9 cm^2 large-area monolayer graphene membrane on porous scaffolding sealed with either GO or rGO. Then, we use electrostatic force microscopy (EFM) AFM mode along with SE and AEE SEM modes to characterize the as-prepared graphene membranes thoroughly. The combination of SEM-AEE and AFM-EFM techniques not only assesses the quality of graphene itself but also characterizes the selectivity and effectiveness of masking graphene layer defects by applying GO or rGO. This makes these methods valuable in optimizing the production of advanced graphene nanocomposites such as semipermeable membranes.

Keywords: graphene; membrane; filtration; AFM imaging; SEM imaging

1. Introduction

Two-dimensional-structured nanomaterials, such as graphene, hold great potential for the development of filtration membranes, including the separation of ion/molecule contaminants from aqueous solutions [1–3]. This potential arises from their unique structural characteristics and excellent physicochemical properties [4–6]. Graphene exhibits exceptional mechanical, chemical, electrical, and thermal properties [7–11]. Monolayer graphene is particularly important in the construction of filtration membranes, serving as a single-atom barrier to contaminant migration. Such a thin membrane can potentially significantly reduce the resistance to water molecule flow. Large-area monolayer graphene, used in membrane production, is synthesized via Chemical Vapour Deposition (CVD) on solid or liquid copper [12–20] and metallurgical methods on liquid copper (HSMG®) [21,22]. This graphene, used for water filtration membranes, cannot have a perfect structure; it must have appropriately sized structural defects that can trap ion/molecule contaminants while allowing water molecules to pass through [23–25]. In addition to sub-nanometer structural pores in graphene (ranging from 0.2 to 0.7 nm) capable of separating ions from water, there are larger defects that may be introduced during the graphene synthesis process, transfer procedure, or membrane preparation. Such defects should be detected and characterized, and then effectively eliminated by covering these areas with another carbon material, such as graphene derivatives—graphene oxide (GO); reduced graphene oxide (rGO); or a polymer [26–29]. Graphene oxide (GO) is primarily produced using the Hummers method. This method results in small graphene flakes with numerous oxygen-containing functional groups (hydroxyl, epoxy, carbonyl, and carboxyl) attached to their edges and surfaces. It can be reduced again through thermal, electrical, or chemical treatment [30–33] to produce reduced graphene oxide (rGO), which has a structure of small graphene-like flakes with trace amounts of oxygen groups. Depending on the reduction method used, the degree of reduction, hydrophobicity, number of oxygen groups, and electrical properties of rGO can vary, affecting its surface properties, including its ability to adsorb contaminants. Large-area graphene, GO, and rGO are used in the production of new filtration membranes. Here, large-area graphene serves as the filtering material, while GO/rGO is used to mask defects in the graphene layer. The effectiveness of selective defect masking in graphene by GO/rGO depends on the structural characterization methods used for these materials [27,29].

Graphene is primarily characterized by Raman spectroscopy (to determine the number of layers and degree of defects) [34], transmission electron microscopy (structural vacancies, dislocations, and grain boundaries) [35], and scanning tunnelling microscopy which allows for the determination of structural defects and electrical properties [36].

Light microscopy, scanning electron microscopy, and AFM allow for the assessment of microscopic defects on adequate substrates [37,38]. The continuity of the graphene layer after transfer to a polymer substrate can be evaluated using scanning electron microscopy (SEM). A technique that has found particular application in the study of conductive layers, such as graphene on non-conductive substrates (polymers, ceramics), utilizes the signal of electrons absorbed by the sample (AEE mode). The image contrast generated during the measurement results from differences in the currents flowing through the sample depending on whether the electron beam is in a conductive or non-conductive area. Dark areas on a bright (i.e., conductive) background illustrate structural discontinuities in the graphene on a polymer or ceramic substrate.

So far, AFM imaging has not been widely used for membrane characterization purposes. One of the advanced techniques employed for this purpose may be electrostatic force microscopy (EFM), a variant of atomic force microscopy (AFM).

EFM allows for the measurement of the electrostatic force between a probe tip and the sample surface. This is achieved by applying a voltage to the probe tip and measuring the resulting force. The data obtained from EFM can provide valuable information about the surface potential and charge distribution of the sample, which are critical parameters for understanding the behaviour of graphene-based membranes. In the context of water filtration, EFM can be used to investigate the distribution of nanoscale pores in the graphene sheet, which are responsible for the selective permeability of the membrane. By understanding how these pores are distributed and how they interact with water molecules and contaminants, it is possible to optimize the membrane design for maximum filtration efficiency.

In this paper, we present a method for the characterization of nanometric and micrometric defects in graphene using the electrical conductivity of graphene as the main property used for detection. Graphene filtration membranes were fabricated by transferring large-area graphene onto a non-conductive, porous polymer substrate. Then, selective coverage of graphene defects by GO or rGO was performed. Using scanning electron microscopy, with particular emphasis on the AEE mode and one of the atomic force microscope electrical modes, electrostatic force microscopy (EFM), the effectiveness of graphene defect masking was determined. It has been shown that both modes can be used complementary to effectively identify graphene defects and, consequently, serve as useful tools for characterizing new graphene-based semipermeable membranes.

2. Materials and Methods

2.1. Materials

High-Strength Metallurgical Graphene was synthesized according to the patent (method of producing graphene from liquid metal, US 9284640, USA) and the procedure described in the previous article [39] with minor modifications. As a matrix for graphene synthesis, a 0.2 mm nickel foil electroplated with copper (0.15 mm) in a saturated solution of copper sulphate pentahydrate ($CuSO_4 \cdot 5H_2O$) was used. Then, Cu/Ni substrates were placed in a SuperCarb Seco Warwick SA generator (Świebodzin, Poland) and heated up to 1060 °C at a pressure of 10 Pa and carburized in an acetylene (0.4 ln/min), ethylene (0.4 ln/min), and hydrogen (0.2 ln/min) mixture. Four cycles each consisting of gas dosing (5 s) and soaking (15 min) were used. Next, the samples were heated up to 1100 °C in a pure argon atmosphere (5 ln/min) and soaked for 5 min to melt Cu. The pressure during the graphene growth stage was 2 kPa. Finally, the samples were cooled down with the reactor to room temperature.

After synthesis graphene was characterized via Raman spectroscopy and was confirmed to be a monolayer with limited defects [39,40]. The graphene nanostructure was characterized by a TEM Talos F200X from FEI (Hillsboro, OR, USA). The previous studies identified and described nanostructural defects with sizes ranging from 0.2 to 0.5 nm [22,39].

MicroPES® 1F EL (3M, Jüchen, Germany) porous support substrate with a thickness of 110 μm, a pore diameter of 0.1 to 1.0 μm, and a pore density 1.05×10^6 mm^{-2} was used for graphene transfer.

Poly(methyl methacrylate) (PMMA) with a molecular weight of 996,000 g/mol from the Aldrich Chemistry Company (St. Louis, MO, USA), dissolved in chlorobenzene (Chempur, Piekary Śląskie, Poland) at a concentration of 0.10 mol dm^{-3} (9.2 g of PMMA per 100 mL of chlorobenzene) was used to cover the graphene during the transfer procedure.

The solvent used to dissolve and remove the PMMA film from the graphene during the transfer procedure was 2-propanol of technical grade 70% (Chempur, Poland).

Graphene oxide with 0.4% dispersion in water (4 mg mL^{-1}) was purchased from the Advanced Graphene Products Company, Zielona Góra, Poland.

For the reduction of graphene oxide, diluted commercial 80% hydrazine hydrate from POCh, Gliwice, Poland was used.

2.2. Graphene Transfer

The HSMG® graphene transfer process onto polysulfone membrane was conducted using a modified wet transfer method, in which the graphene growth substrate etching was replaced by graphene electrochemical delamination [40]. Firstly, graphene samples on the Cu/Ni substrate were cut, straightened, and then drop/blade coated with 10 μL cm^{-2} PMMA solution. Then, the polymer was dried at 50 °C for 40 min. The graphene electrochemical delamination from the growth substrate was conducted on a laboratory semi-automatic stand according to the parameters described elsewhere [40,41]. Then, graphene on PMMA was washed twice with deionized water. After that, it was placed on a water-wetted polysulfone membrane, rolled, and dried at 70 °C for 5 min. To remove the PMMA, the as-prepared stack was placed in a boiling isopropanol vapour for one hour and then washed twice in cold isopropanol.

2.3. GO and rGO Application

Reduced graphene oxide was prepared using a 1% solution of hydrazine in 10 mL of a 0.25% GO suspension in DI water and left for 1 h to reduce at room temperature (RT).

The solution was sonicated before the addition of the hydrazine. The properties of GO and rGO were described in a previous work [41].

Graphene membranes were drop-coated with 200 μL of GO or rGO (both 0.25% suspension in DI water), respectively. The suspensions were applied on the membrane surface, and after 3 min, the excess was poured off. The suspensions were sonicated for 1 min before being applied. The as-prepared membranes were baked at 80 °C for 30 min.

2.4. Optical Microscopy and SEM Examination

Graphene membrane analysis was conducted on a Nikon Eclipse MA200 light microscope (Nishioi, Japan) with inverted optics. Then, the same areas were identified and analyzed using the HITACHI S 3000 N scanning electron microscope (Tokyo, Japan). The tests were performed at an accelerating voltage of 5 kV, using SE (secondary electrons) and AEE (absorbed electron emission) detectors. The AEE imaging mode is better suited for this purpose, as it shows the contrast of the image resulting from the differences in the conductivity of the tested surface areas. The combination of the two images from the SE and AEE modes provides an effective assessment of the continuity of the graphene layer, even in areas where the substrate pores are present.

2.5. AFM Examination

Surface morphology, topography, and local electrical properties measurements were performed under ambient conditions using a Multimode 5 atomic force microscope equipped with a Nanoscope V controller (Bruker Corporation, Billerica, MA, USA) operating in one of its electrical modes—EFM. Commercial silicon cantilevers of type HQ: NSCPt15 (MicroMasch, Wetzlar, Germany) with a platinum-coated tip radius of <30 nm, a cantilever spring constant of 40 N/m, and a resonant frequency of 325 kHz were used. Image acquisition was performed using Nanoscope 7.3 software, and further image processing was carried out using Nanoscope Analysis 1.9 (Bruker Corporation). For the AFM/EFM imaging, the membrane substrates were attached to a magnetic puck using electrically conductive Cu tape along with the use of Ag conductive paste to connect the top of the membrane (with graphene structures) to the puck surface.

The EFM method allows primarily for the qualitative mapping of surface potential. Electrostatic force microscopy measurements were made in a two-pass (lift) mode and the size of the images was up to 100 × 100 μm. The electrostatic force was measured as a change in the cantilever resonant frequency (phase shift) due to the differences in charge between the sample and the tip. The lift height and DC bias were set to 100 nm and −4 V, respectively.

3. Results and Discussion

Optical Microscopy Characterization

Graphene membranes prepared according to Figure 1 were proven to be semipermeable in our previous studies [39,41]. The key components in these membranes are active monolayer graphene with subnanometric pores and a sealing material (GO or rGO). Using rGO as a sealing material enhances both solute rejection and water flow. Previous studies conducted for HSMG® graphene showed that the use of only a monolayer of quaimonocrystalline metallurgical graphene does not result in obtaining a semipermeable membrane, mainly due to the presence of transfer defects and defects from the synthesis process. Optimizing the graphene transfer method and effectively sealing these defects enables an increase in osmotic water flow (25 mL h^{-1} m^{-2} bar^{-1}) and rejection rate (up to 95%). Understanding the characteristics of graphene defects and the effectiveness of sealing are important steps in membrane preparation methodology. The use of electrical modes based on the electrical conductivity of the characterized material offers a natural approach to identifying graphene on a non-conductive substrate.

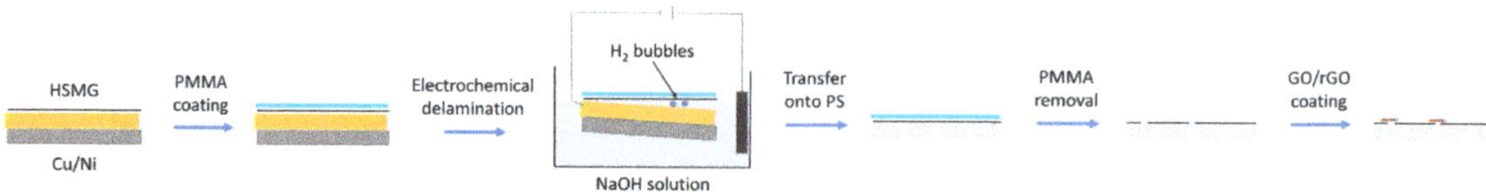

Figure 1. Graphene membrane preparation scheme.

The membranes on PS porous substrate with a graphene layer size of 9 cm^2 were fabricated (Figure 2a). Then, GO or rGO was applied as a defect-sealing material. The microscopic graphene quality was evaluated. The bare graphene layer on the porous substrate (Figure 2b) is mostly monolayer (bright grey area) with multilayers visible as darker, flat areas. Sealing material on graphene was observed under an optical microscope as a black, uneven layer (Figure 2c).

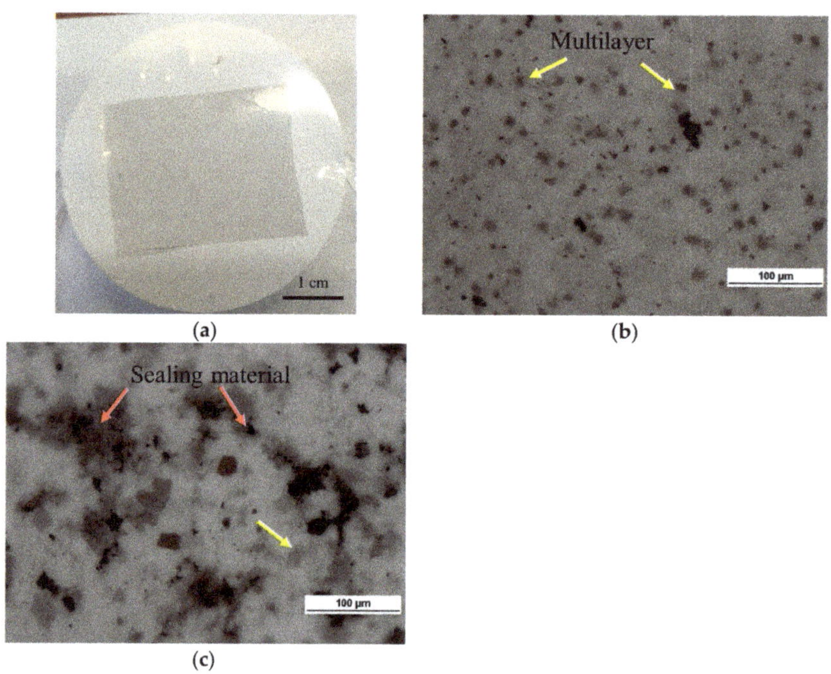

Figure 2. Graphene membranes. (**a**)—bare graphene on a porous substrate. (**b**)—an optical image of graphene on a porous substrate. (**c**)—rGO selectively sealed graphene membrane.

Figure 3a shows the graphene layer transferred onto PS with line defects in the graphene (marked with a red arrow). Despite these defects, graphene is stretched over the pores of the supporting substrate. After sealing with rGO (Figure 3b) platelets of rGO are arranged along a line, in this case, on a shape similar to a hexagon (the discontinuity occurring at the border of graphene crystallites has been covered). The size of the hexagons is of the order of tens of micrometres, which corresponds to the size of the hexagons in graphene after synthesis. rGO was also deposited in the pore region of the substrate, which is a crucial area to be sealed when it comes to filtration.

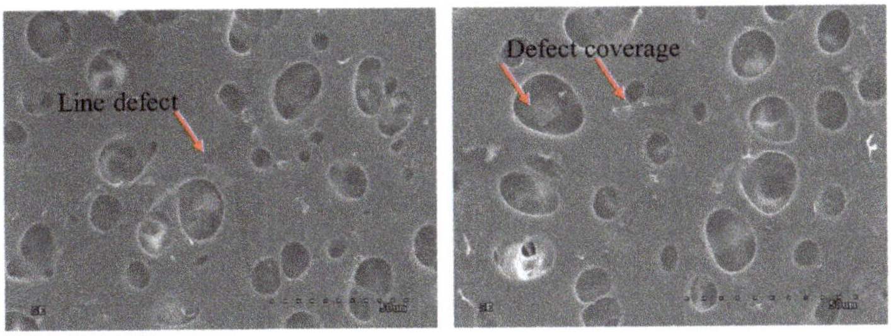

(a) (b)

Figure 3. (**a**)—graphene with line defects on a porous substrate. (**b**)—selectively sealed graphene membrane.

Figure 4a–d presents the boundary of rGO sealing applied on the graphene membrane (boundary marked with the red dotted line, rGO applied on the left side of the images). Most of the defects are in substrate unevenness that reflects copper grains (HSMG® graphene growth substrate). This may be due to the presence of defects in the graphene itself from the synthesis or a privileged place for the formation of transfer defects. Both SE and AEE modes in SEM show good masking of these discontinuities. Almost all point discontinuities were covered. A large part of the continuous, defect-free graphene has been preserved uncoated. Yellow arrows indicate rGO conglomerates, which are not common due to the sonication of rGO solution before application on the membrane. Effective sonication also has an impact on obtaining a thin masking layer, wherein the former grain boundary, despite the linear defect being sealed, and substrate pores are still visible in the SEM image (Figure 4c). This sealing method is not 100% efficient. In Figure 4c, there are still graphene clustered point defects that have not been sealed (bright areas, marked with green arrows).

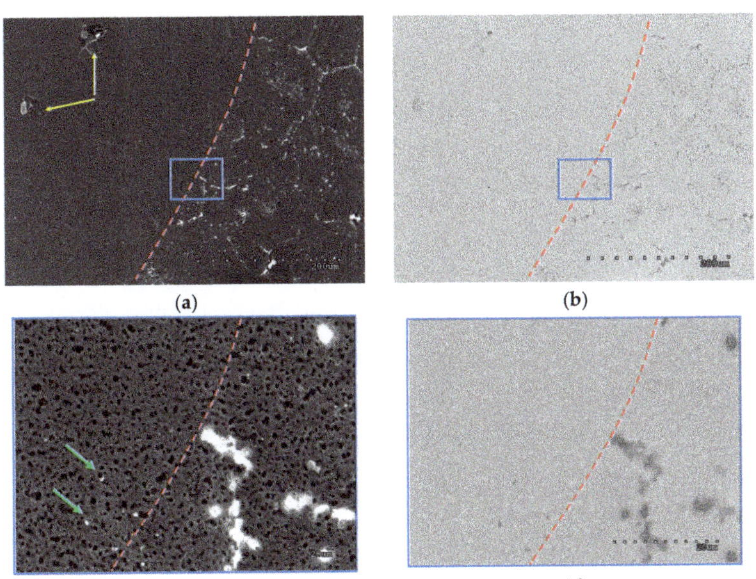

(a) (b) (c) (d)

Figure 4. (**a–d**) graphene membrane sealed with rGO. The left side of the images is covered with sealing material, and the right side is bare graphene on a porous substrate. SEM images, SE mode (**a–c**), and AEE mode (**b**,**d**). (**c**,**d**) are the enlarged areas marked with a blue frame.

Next, optical microscopy, SEM, and AFM analysis were conducted in selected areas with larger graphene discontinuities to better determine the possibility of imaging different membrane areas. The imaging was carried out in such a way that all the membrane components were present in each analyzed area. On the images from the optical microscope, for ease of use, the areas are marked according to Table 1.

Table 1. Signatures used to describe components of the graphene membrane on figures.

Signature on Figures	Membrane's Area
S	Bare substrate
G	Graphene layer
GO	Graphene oxide on PS
rGO	Reduced graphene oxide on PS
G + GO	Graphene oxide selectively coating graphene defects
G + rGO	Reduced graphene oxide selectively coating graphene defects

Figure 5 shows the selective coverage of graphene defects by graphene oxide. SEM images in SE and AEE modes (Figure 5b,c) allow us to observe both large areas of graphene discontinuity (right area in Figure 5b) and identify places completely covered by GO. On the left side is a part completely covered with GO, while in the upper left corner, there is a selectively deposited GO layer on graphene. In this area, substrate pores are visible in areas covered only with graphene. The defects in graphene have almost completely been masked by GO. Here, too, no substrate pores are visible. In SE mode, the graphene oxide morphology is visible (characteristic wrinkles). In AEE mode (Figure 5c), the image of the surface coverage with graphene and graphene oxide is similar to Figure 5b. Imaging in the conduction mode gives a contrasting image of the surface—the areas covered with graphene or graphene oxide are bright, while the areas not covered with graphene are grey. In the areas not covered with graphene, pores of the polysulfone substrate can be seen. However, there is no difference in imaging the areas covered with graphene and those covered with graphene oxide, although graphene oxide conducts electricity much worse. Then, atomic force microscopy imaging was performed for the area marked in purple. GO wrinkles are visible in AFM imaging in all three modes. The electric mode (EFM) imaging (Figure 5e) allowed us to obtain a contrast between the layer covered by large-area graphene and the bare substrate (bright areas). The boundary of graphene coverage and open pores are visible. Open pores are also visible in the amplitude method, but in this case, it is impossible to distinguish the graphene coverage boundary in the area between the opened pores. Thus, the EFM mode, due to the electrical properties of both large-area graphene and GO, turned out to be effective in characterizing the performance quality of composite graphene membranes.

Imaging of sealing graphene defects with rGO (Figure 6) was conducted the same way as GO sealing (Figure 5). High-quality graphene in the upper left corner, despite being exposed to the rGO aqueous suspension, was not covered with rGO platelets, thus the pores of the substrate under graphene are still visible in SEM images (Figure 6b). This phenomenon was confirmed on EFM imaging (Figure 6e) where large-area graphene is not covered by rGO. This is related to the hydrophobic properties of graphene. The rGO aqueous suspension only covers the places where the graphene layer is damaged, because the porous polysulfone substrate exhibits hydrophilic properties.

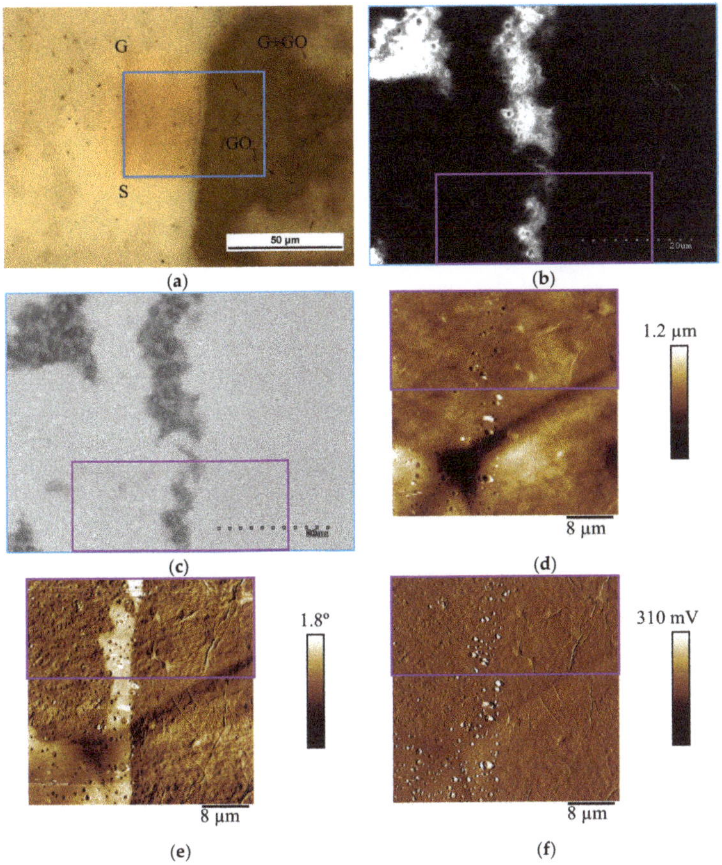

Figure 5. Imaging the effectiveness of covering graphene discontinuities by selectively applied GO. (**a**)—optical microscopy. (**b**)—SEM SE and (**c**)—SEM AEE images of the composite membrane. (**d–f**)—atomic force microscopy images in topography, EFM, and amplitude modes, respectively.

The images from the SEM microscope—SE mode (Figure 6b) and SEM—AEE mode (Figure 6c) compared with the images obtained from the AFM microscope show the full correlation of the obtained test results. The same areas not covered with graphene can be seen. In Figure 6b, the pores of the substrate can be seen in both graphene-coated and non-graphene-coated areas. Only the contrast of the image can indicate the lack of coverage in this place—the areas not covered with graphene are bright. In contrast, no substrate pores are visible in the rGO-coated areas due to the thickness of the rGO coating. The results obtained in Figure 6c can be interpreted similarly. The AEE mode indicates good conductivity of the substrate (bright areas) in the graphene and reduced graphene oxide-coated areas. Non-conductive areas where graphene or rGO is absent are grey. In these images, the substrate's pores are visible only in the non-graphene-coated areas. It is also difficult to distinguish which area is covered only with graphene and which with reduced graphene oxide. No characteristic wrinkles of the coating can be seen here for rGO. The results obtained in AEE mode are in full agreement with those obtained in SE mode. Analyzing the images captured with the AFM microscope (Figure 6d–f), it can be concluded that the results obtained in the topography mode do not allow for an accurate interpretation of the quality of the graphene coating on the membrane, and for an assessment of the effectiveness of the sealing of this layer of rGO. Visible here are the pores of the substrate, basically regardless of the area of the membrane, and point-like particles, probably small rGO conglomerates or graphene impurities. In the lower corner, the finely outlined wrinkle shape of the rGO coating can be observed. EFM mode (Figure 6e) gives a more complete result. The substrate's pores (dark dots) are clearly visible in areas not covered with graphene. The rGO-coated sites do not have visible pores, while the graphene-coated pores are also imaged, but in a different way than on bare polysulfone. Here, one can clearly distinguish between pores covered by graphene (they look like convex particles) and pores over which there is no stretched graphene (dark dots). In the amplitude mode (Figure 6f), porosity is also visible both on the substrate not covered with graphene and where it is present. Here, it is also possible to indicate pores covered by a graphene layer and those where graphene is damaged. As before, the areas where rGO is present are difficult to pinpoint unambiguously. These are areas where the pores are not well mapped. Similarly to the case of coating graphene membranes with graphene oxide, the EFM mode proved to be equally effective in imaging areas covered with graphene and sealed with rGO.

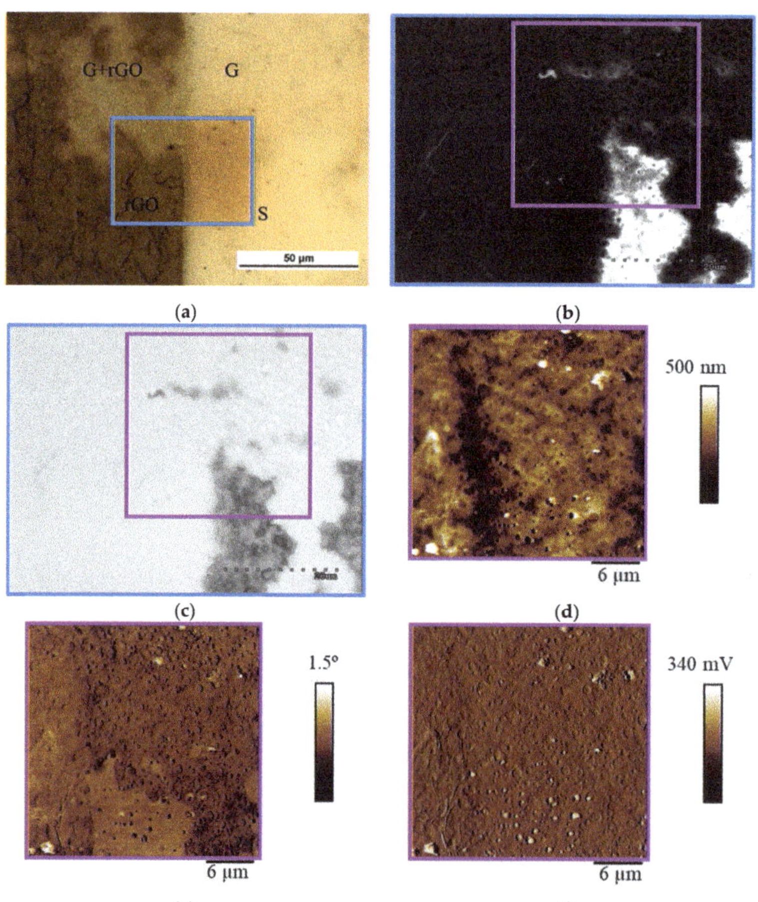

Figure 6. Imaging the effectiveness of covering graphene discontinuities by selectively applied rGO. (**a**)—optical microscopy. (**b**)—SEM SE and (**c**)—SEM AEE images of the composite membrane. (**d**–**f**)—atomic force microscopy images in topography, EFM, and amplitude modes, respectively.

4. Conclusions

The characterization of graphene defects and the evaluation of the effectiveness of their sealing are important steps in the methodology of preparing graphene membranes for water filtration. The selectivity and efficiency of such membranes ultimately depend on the accuracy and precision of these studies. The use of electrical modes based on the electrical conductivity of the characterized material is a natural approach to identifying graphene on a non-conductive substrate such as polysulfone. As demonstrated in this article, the use of such test methods as the SEM–AEE mode and AFM–EFM mode provides the opportunity for good characterization of the continuity of graphene layers and the effectiveness of their sealing with GO and rGO. Interestingly, although the difference in the conductivity of GO and rGO is significant, it did not fundamentally affect the imaging efficiency of the methods mentioned above. However, it should be noted that the use of only one test method may produce inconclusive results. The combination of SEM-AEE and AFM-EFM techniques yields more reliable results. Both areas are not covered by a graphene layer and the absence of graphene on individual pores of the polysulfone substrate can be seen. These techniques also make it possible to assess the selectivity and effectiveness of masking graphene layer defects by applying GO or rGO. In summary, imaging techniques based on differences in the electrical conductivity of the tested composite materials are highly effective for characterizing the quality of graphene filtration membranes.

Author Contributions: Conceptualization, G.R. and K.D.; methodology, G.R., K.D., Ł.K. and P.K.; validation, G.R., Ł.K. and P.K.; formal analysis, K.D.; investigation, G.R., Ł.K. and P.K.; resources, G.R. and Ł.K.; data curation, G.R. and Ł.K.; writing—original draft preparation, G.R., K.D. and Ł.K.; writing—review and editing, K.D., G.R. and Ł.K.; visualization, G.R. and Ł.K.; supervision, K.D.; project administration, G.R.; funding acquisition, K.D. All authors have read and agreed to the published version of the manuscript.

Funding: The research was performed with the use of the European Union funds within the framework of the project no. POIR.04.01.04–00-0089/15, Measure 4.1 "Scientific research and development works", Sub-measure 4.1.4 "Application projects" of the Smart Growth Operational Programme, 2014–2020, project title: "Graphene based composite materials for water purification".

Institutional Review Board Statement: Not applicable.

Informed Consent Statement: Not applicable.

Data Availability Statement: The original contributions presented in the study are included in the article, further inquiries can be directed to the corresponding author.

Acknowledgments: The authors would like to thank Radomir Atraszkiewicz and Adam Rzepkowski for their assistance with graphene fabrication processes.

Conflicts of Interest: The authors declare no conflicts of interest. The funders had no role in the design of the study; in the collection, analyses, or interpretation of data; in the writing of the manuscript; or in the decision to publish the results.

References

1. Jang, D.; Idrobo, J.C.; Laoui, T.; Karnik, R. Water and Solute Transport Governed by Tunable Pore Size Distributions in Nanoporous Graphene Membranes. *ACS Nano* **2017**, *11*, 10042–10052. [CrossRef] [PubMed]
2. Surwade, S.P.; Smirnov, S.N.; Vlassiouk, I.; Unocic, R.R.; Veith, G.M.; Dai, S.; Mahurin, S.M. Water Desalination Using Nanoporous Single-Layer Graphene. *Nat. Nanotechnol.* **2015**, *10*, 459–464. [CrossRef] [PubMed]
3. Jain, T.; Rasera, B.C.; Guerrero, R.J.S.; Boutilier, M.S.H.; O'Hern, S.C.; Idrobo, J.C.; Karnik, R. Heterogeneous Sub-Continuum Ionic Transport in Statistically Isolated Graphene Nanopores. *Nat. Nanotechnol.* **2015**, *10*, 1053–1057. [CrossRef]
4. Goh, P.S.; Matsuura, T.; Ismail, A.F.; Hilal, N. Recent Trends in Membranes and Membrane Processes for Desalination. *Desalination* **2016**, *391*, 43–60. [CrossRef]

5. Johnson, D.J.; Hilal, N. Can Graphene and Graphene Oxide Materials Revolutionise Desalination Processes? *Desalination* **2021**, *500*, 114852. [CrossRef]
6. Wu, W.; Shi, Y.; Liu, G.; Fan, X.; Yu, Y. Recent Development of Graphene Oxide Based Forward Osmosis Membrane for Water Treatment: A Critical Review. *Desalination* **2020**, *491*, 114452. [CrossRef]
7. Pop, E.; Varshney, V.; Roy, A. Thermal Properties of Graphene: Fundamentals and Applications. *MRS Bull.* **2013**, *37*, 1273–1281. [CrossRef]
8. Lee, C.; Wei, X.; Kysar, J.W.; Hone, J. Measurement of the Elastic Properties and Intrinsic Strength of Monolayer Graphene. *Science* **2008**, *321*, 385–388. [CrossRef]
9. Dash, G.N.; Pattanaik, S.R.; Behera, S. Graphene for Electron Devices: The Panorama of a Decade. *IEEE J. Electron Devices Soc.* **2014**, *2*, 77–104. [CrossRef]
10. Meyer, J.C.; Geim, A.K.; Katsnelson, M.I.; Novoselov, K.S.; Booth, T.J.; Roth, S. The Structure of Suspended Graphene Sheets. *Nature* **2007**, *446*, 60–63. [CrossRef] [PubMed]
11. Novoselov, K.S.; Geim, A.K.; Morozov, S.V.; Jiang, D.; Zhang, Y.; Dubonos, S.V.; Grigorieva, I.V.; Firsov, A.A. Electric Field Effect in Atomically Thin Carbon Films. *Science* **2004**, *306*, 666–669. [CrossRef] [PubMed]
12. Guo, W.; Xu, C.; Xu, K.; Deng, J.; Guo, W.; Yurgens, A.; Sun, J. Rapid Chemical Vapor Deposition of Graphene on Liquid Copper. *Synth. Met.* **2016**, *216*, 93–97. [CrossRef]
13. Cho, S.Y.; Kim, M.S.; Kim, M.; Kim, K.J.; Kim, H.M.; Lee, D.J.; Lee, S.H.; Kim, K.B. Self-Assembly and Continuous Growth of Hexagonal Graphene Flakes on Liquid Cu. *Nanoscale* **2015**, *7*, 12820–12827. [CrossRef] [PubMed]
14. Zheng, S.; Zeng, M.; Cao, H.; Zhang, T.; Gao, X.; Xiao, Y.; Fu, L. Insight into the Rapid Growth of Graphene Single Crystals on Liquid Metal via Chemical Vapor Deposition. *Sci. China Mater.* **2019**, *62*, 1087–1095. [CrossRef]
15. Kim, M.-S.; Cho, S.-Y.; Kim, M.; Kim, K.-J.; Lee, S.-H.; Kim, H.-M.; Kim, K.-B. Comparison of Growth Behavior and Electrical Properties of Graphene Grown on Solid and Liquid Copper by Chemical Vapor Deposition. *J. Nanosci. Nanotechnol.* **2019**, *20*, 316–323. [CrossRef] [PubMed]
16. Ding, G.; Zhu, Y.; Wang, S.; Gong, Q.; Sun, L.; Wu, T.; Xie, X.; Jiang, M. Chemical Vapor Deposition of Graphene on Liquid Metal Catalysts. *Carbon* **2013**, *53*, 321–326. [CrossRef]
17. Geng, D.; Wu, B.; Guo, Y.; Huang, L.; Xue, Y.; Chen, J.; Yu, G.; Jiang, L.; Hu, W.; Liu, Y. Uniform Hexagonal Graphene Flakes and Films Grown on Liquid Copper Surface. *Proc. Natl. Acad. Sci. USA* **2012**, *109*, 7992–7996. [CrossRef] [PubMed]
18. Liu, J.; Fu, L. Controllable Growth of Graphene on Liquid Surfaces. *Adv. Mater.* **2019**, *31*, e1800690. [CrossRef] [PubMed]
19. Chen, X.; Chen, S. Large Area CVD Growth of Graphene. *Synth. Met.* **2015**, *210*, 95–108. [CrossRef]
20. Li, X.; Cai, W.; Colombo, L.; Ruoff, R. Evolution of Graphene Growth on Ni and Cu by Carbon Isotope Labeling. *Nano Lett.* **2009**, *9*, 4268–4272. [CrossRef]
21. Kula, P.; Pietrasik, R.; Dybowski, K.; Atraszkiewicz, R.; Szymanski, W.; Kolodziejczyk, L.; Niedzielski, P.; Nowak, D. Single and Multilayer Growth of Graphene from the Liquid Phase. *Appl. Mech. Mater.* **2014**, *510*, 8–12. [CrossRef]
22. Romaniak, G.; Cheng, P.; Dybowski, K.; Kula, P.; Kidambi, P.R. Assessing the Quality of Large-Area Monolayer Graphene Grown on Liquid Copper for Size-Selective Ionic/Molecular Membrane Separations. *Mater. Res. Express* **2023**, *10*, 105101. [CrossRef]
23. Kafiah, F.M.; Khan, Z.; Ibrahim, A.; Karnik, R.; Atieh, M.; Laoui, T. Monolayer Graphene Transfer onto Polypropylene and Polyvinylidenedifluoride Microfiltration Membranes for Water Desalination. *Desalination* **2016**, *388*, 29–37. [CrossRef]
24. O'Hern, S.C.; Jang, D.; Bose, S.; Idrobo, J.-C.; Song, Y.; Laoui, T.; Kong, J.; Karnik, R. Nanofiltration across Defect-Sealed Nanoporous Monolayer Graphene. *Nano Lett.* **2015**, *15*, 3254–3260. [CrossRef] [PubMed]
25. Kazemi, A.S.; Abdi, Y.; Eslami, J.; Das, R. Support Based Novel Single Layer Nanoporous Graphene Membrane for Efficacious Water Desalination. *Desalination* **2019**, *451*, 148–159. [CrossRef]
26. Kafiah, F.; Khan, Z.; Ibrahim, A.; Atieh, M.; Laoui, T. Synthesis of Graphene Based Membranes: Effect of Substrate Surface Properties on Monolayer Graphene Transfer. *Materials* **2017**, *10*, 86. [CrossRef] [PubMed]
27. Boutilier, M.S.H.; Jang, D.; Idrobo, J.C.; Kidambi, P.R.; Hadjiconstantinou, N.G.; Karnik, R. Molecular Sieving Across Centimeter-Scale Single-Layer Nanoporous Graphene Membranes. *ACS Nano* **2017**, *11*, 5726–5736. [CrossRef] [PubMed]
28. O'Hern, S.C.; Stewart, C.A.; Boutilier, M.S.H.; Idrobo, J.-C.; Bhaviripudi, S.; Das, S.K.; Kong, J.; Laoui, T.; Atieh, M.; Karnik, R. Selective Molecular Transport through Intrinsic Defects in a Single Layer of CVD Graphene. *ACS Nano* **2012**, *6*, 10130–10138. [CrossRef]
29. Cheng, P.; Kelly, M.M.; Moehring, N.K.; Ko, W.; Li, A.P.; Idrobo, J.C.; Boutilier, M.S.H.; Kidambi, P.R. Facile Size-Selective Defect Sealing in Large-Area Atomically Thin Graphene Membranes for Sub-Nanometer Scale Separations. *Nano Lett.* **2020**, *20*, 5951–5959. [CrossRef]
30. Pei, S.; Cheng, H.-M. The Reduction of Graphene Oxide. *Carbon* **2012**, *50*, 3210–3228. [CrossRef]
31. Zhang, J.; Yang, H.; Shen, G.; Cheng, P.; Zhang, J.; Guo, S. Reduction of Graphene Oxide Vial-Ascorbic Acid. *Chem. Commun.* **2010**, *46*, 1112–1114. [CrossRef]

32. Fernández-Merino, M.J.; Guardia, L.; Paredes, J.I.; Villar-Rodil, S.; Solís-Fernández, P.; Martínez-Alonso, A.; Tascón, J.M.D. Vitamin C Is an Ideal Substitute for Hydrazine in the Reduction of Graphene Oxide Suspensions. *J. Phys. Chem. C* **2010**, *114*, 6426–6432. [CrossRef]
33. Yang, Z.; Zheng, Q.; Qiu, H.; Li, J.; Yang, J. A Simple Method for the Reduction of Graphene Oxide by Sodium Borohydride with $CaCl_2$ as a Catalyst. *Carbon* **2015**, *86*, 372. [CrossRef]
34. Grodecki, K. Spektroskopia Ramanowska Grafenu. *Electron. Mater.* **2013**, *41*, 47–53.
35. Gass, M.; Bangert, U.; Bleloch, A.; Wang, P.; Raveendran-Nair, R.; Geim, A.K. Free-Standing Graphene at Atomic Resolution. *Nat. Nanotechnol.* **2008**, *3*, 676–681. [CrossRef] [PubMed]
36. Xu, K.; Cao, P.; Heath, J.R. Scanning Tunneling Microscopy Characterization of the Electrical Properties of Wrinkles in Exfoliated Graphene Monolayers. *Nano Lett.* **2009**, *9*, 4446–4451. [CrossRef]
37. Abergel, D.S.L.; Russell, A.; Fal'ko, V.I. Visibility of Graphene Flakes on a Dielectric Substrate. *Appl. Phys. Lett.* **2007**, *91*, 63125. [CrossRef]
38. Hiura, H.; Miyazaki, H.; Tsukagoshi, K. Determination of the Number of Graphene Layers: Discrete Distribution of the Secondary Electron Intensity Stemming from Individual Graphene Layers. *Appl. Phys. Express* **2010**, *3*, 95101. [CrossRef]
39. Romaniak, G.; Dybowski, K.; Jeziorna, A.; Kula, P.; Kaźmierczak, T. Synthesis and Characterization of Semi-Permeable Graphene/Graphene Oxide Membranes for Water Desalination. *J. Mater. Sci.* **2020**, *55*, 9775–9786. [CrossRef]
40. Dybowski, K.; Romaniak, G.; Kula, P.; Jeziorna, A.; Kowalczyk, P.; Atraszkiewicz, R.; Kołodziejczyk, Ł.; Nowak, D.; Zawadzki, P.; Kucińska, M. Impact of the Method of Separating Graphene from the Growth Substrate on the Quality of the 2D Material Obtained. *Arch. Metall. Mater.* **2019**, *64*, 1321–1326. [CrossRef]
41. Romaniak, G.; Dybowski, K.; Jędrzejczak, A.; Sobczyk-Guzenda, A.; Januszewicz, B.; Szymański, W.; Kowalczyk, P.; Kaźmierczak, T.; Siniarski, J.; Kula, P. Impact of a Graphene Oxide Reducing Agent on a Semi-Permeable Graphene/Reduced Graphene Oxide Forward Osmosis Membrane Filtration Efficiency. *Membranes* **2021**, *11*, 679. [CrossRef]

nanomaterials

Article

Functionalization of Polypropylene by TiO_2 Photocatalytic Nanoparticles: On the Importance of the Surface Oxygen Plasma Treatment

Karolina Zajac [1,2,3], Joanna Macyk [2], Konrad Szajna [4], Franciszek Krok [4], Wojciech Macyk [1,2] and Andrzej Kotarba [1,*]

1 Faculty of Chemistry, Jagiellonian University in Kraków, Gronostajowa 2, 30-387 Krakow, Poland; karolina.zajac@doctoral.uj.edu.pl (K.Z.); wojciech.macyk@uj.edu.pl (W.M.)
2 InPhoCat—Innovative Photocatalytic Solutions Sp. z o. o., Brzask 49, 30-381 Krakow, Poland; jmacyk@inphocat.pl
3 Doctoral School of Exact and Natural Sciences, Jagiellonian University, Prof. St. Łojasiewicza 11, 30-348 Kraków, Poland
4 Marian Smoluchowski Institute of Physics, Faculty of Physics, Astronomy and Applied Computer Science, Jagiellonian University, Lojasiewicza 11, 30-348 Krakow, Poland; konrad.szajna@uj.edu.pl (K.S.); franciszek.krok@uj.edu.pl (F.K.)
* Correspondence: ak@uj.edu.pl

Abstract: A new two-step method for developing a nanocomposite of polypropylene (PP) decorated with photocatalytically active TiO_2 nanoparticles ($nTiO_2$) is proposed. This method involves the low-temperature plasma functionalization of polypropylene followed by the ultrasound-assisted anchoring of $nTiO_2$. The nanoparticles, polymeric substrate, and resultant nanocomposite were thoroughly characterized using nanoparticle tracking analysis (NTA), microscopic observations (SEM, TEM, and EDX), spectroscopic investigations (XPS and FTIR), thermogravimetric analysis (TG/DTA), and water contact angle (WCA) measurements. The photocatalytic activity of the nanocomposites was evaluated through the degradation of methyl orange. The individual TiO_2 nanoparticles ranged from 2 to 6 nm in size. The oxygen plasma treatment of PP generated surface functional groups (mainly -OH and -C=O), transforming the surface from hydrophobic to hydrophilic, which facilitated the efficient deposition of $nTiO_2$. Optimized plasma treatment and sonochemical deposition parameters resulted in an active photocatalytic $nTiO_2$/PP system, degrading 80% of the methyl orange under UVA irradiation in 200 min. The proposed approach is considered versatile for the functionalization of polymeric materials with photoactive nanoparticles and, in a broader perspective, can be utilized for the fabrication of self-cleaning surfaces.

Keywords: TiO_2; polypropylene; surface functionalization; plasma treatment; nanocomposite; photocatalyst

1. Introduction

Polypropylene (PP), due to its vast diversity of applications, is one of the most widespread polymeric materials in the world. It gained popularity almost instantly after its discovery, and the production of PP steadily increased [1]. Diverse types of industries, including biomedical, automobile, aerospace, textiles, and packaging, to mention a few, employ polypropylene using their unique and adjustable characteristics. The most crucial advantages of these materials are chemical resistance, mechanical properties, and cost-effectiveness [2]. Polypropylene properties could be additionally modified via chemical and physical methods, among which are copolymerization and structural modification using additives and forming composites. The resultant materials could be processed in many ways, like injection or extrusion, and form final products. From the practical point of view, surface functionalization is particularly interesting, where the enhancement and/or addition of new surface properties is achieved without changing the bulk characteristics [3,4].

Currently, one of the modern trends in polymeric surface functionalization concerns antibacterial and self-cleaning surfaces, which can be achieved through the introduction of photocatalytic nanoparticles on the surface [5]. When irradiated, the photocatalysts generate reactive oxygen species such as hydroxyl radical ($HO^\bullet$) and superoxide radical ($O_2^{\bullet-}$). The primarily formed $O_2^{\bullet-}$ species start a cascade of ROS formation, including H_2O_2, which, together with $HO^\bullet$, can destroy a broad range of contaminants. They typically include inanimate matter, such as inorganic/organic substances, and living organisms, such as bacteria, fungi, and algae. Inorganic contaminants such as dust could be washed away due to superhydrophilic properties. In this case, the water contact angle should be below 10° so the water droplets can spread over and clean the surface [6]. Organic impurities could be removed from the surface via total degradation to CO_2 and H_2O (mineralization) by photogenerated charges or reactive oxygen species. The main mechanism in destroying bacterial cell walls also involves damage to bacteria cell walls and operates for both Gram-negative and Gram-positive strains. However, it should be noted that the generated oxygen radicals are short-lived and are not found further than 1 μm from the photocatalyst surface [7]. As the distance between the bacteria and the photocatalyst is essential for the self-disinfection process of all surfaces, the preparation of a system requires a high dispersion of photoactive nanoparticles over the polymer surface.

Titanium dioxide in the form of nanoparticles is one of the best-known robust photocatalysts with superior properties, particularly photoactivity, nontoxicity, and commercial availability. TiO_2-based photocatalytic systems, including films, have been shown to effectively remove organic pollutants and destroy pathogens in water, as reviewed in [8]. Thus, the deposition of titanium dioxide nanoparticles on the polymers could effectively trigger their photocatalytic cleaning [9,10]. The most commonly used approach is to add nanoparticles to monomers before forming them into a final product. There are several methods reported for the successful preparation of PP-nTiO_2 composites like the direct blending of titania NPs with polymers, in situ formation of NPs within the polymeric matrix, copolymerization of surface-modified NPs, and grafting or self-assembly of NPs and polymers [11,12]. It is worth mentioning that in such a case, a majority of nanoparticles are located in the bulk and, due to limited light penetration, are photocatalytically inactive. The bulk embedment could also result in the undesired modifications of bulk properties of the commercial polymeric materials [13].

Considering the above considerations, optimal functionalization should be limited to the surface decoration of polymeric materials with photocatalytic nanoparticles. The polypropylene surface is, however, chemically inert, which is a huge advantage for its stability in harsh chemical environments. However, it also makes functionalization difficult, especially via coating or attachment of nanoparticles [14].

From a practical point of view, the best approach is to create surface chemical bonds between polypropylene and TiO_2 nanoparticles. The strategy involves the generation of abundant surface functional groups on the polypropylene to enhance nanoparticle anchoring [15]. For this purpose, the effective modification of the PP surface can be achieved by plasma treatment. The exposure of polymeric surfaces to ionized gas molecules, e.g., oxygen plasma, effectively changes their properties, like wettability and adhesion. Using low-temperature plasma, the surface properties can be changed effectively by generating functional groups while preserving the bulk properties of the polymeric materials [16,17]. Moreover, the surface modification could be precisely controlled by adjusting plasma parameters (power, oxygen partial pressure, and treatment time) [18,19]. Such functionalized surfaces can then be decorated with photoactive nanoparticles with the use of a sonochemical process [20,21]. The advantages of applying ultrasound for inorganic particle deposition over polymeric surface consist in employing ultrasound energy not only for the deposition process but also for softening the polymeric matrix, thus making the nanoparticle embedding effective. Consequently, the final product is expected to exhibit higher mechanical stability [22].

In this paper, oxygen plasma treatment on polypropylene material has been investigated. New functional groups generated on the polymer surface after plasma treatment were monitored by the water contact angle, while their chemical nature was characterized by employing FTIR and XPS analysis. The process of depositing nanoparticles on the surface was performed using the sonochemical method, modifying the sonication time and concentration of the $nTiO_2$ suspension. The influence of oxygen surface groups on the attachment and dispersion of the TiO_2 nanoparticles was documented by SEM images. The photocatalytic activity was tested in the methyl orange (as model contamination compound) degradation test using UV irradiation. The effect of plasma treatment on the properties of the obtained $nTiO_2/PP$ composites is discussed in terms of the nanoparticle dispersion and related photocatalytic activity.

2. Materials and Methods

2.1. Material Preparation

Titanium dioxide nanoparticles (CCA 100 BS) were purchased from Cinkarna Celje d. d. (Celje, Slovenia). The TiO_2 had an anatase crystal structure with a specific surface area of approximately 250 m^2/g. The nanoparticles were supplied in a 20% stabilized aqueous suspension with a neutral pH (7–9), which was diluted to 1% or 0.1% before the experiments.

In the experiments, 1 mm thick polypropylene sheets (Merck, Darmstadt, Germany) were used. The sheets were washed with isopropanol and dried in air to prepare the surface for further functionalization and measurement.

To functionalize the polymer's surface, oxygen plasma treatment (FEMTO system, Diener Electronics, Ebhausen, Germany) was used with a controlled oxygen partial pressure of 0.2 mbar, plasma generator power of 100 W, and a defined sample exposure time (0, 1, 2, 5, or 10 min).

A sonochemical method was used to anchor the nanoparticles to oxygen plasma-modified polypropylene. After the plasma treatment, the polymer was immediately immersed in 5 mL of titanium dioxide solution (0.1%/0.2%/0.5%/1%). The samples were irradiated at a frequency of 20 kHz, amplitude of 30%, and time from 1 min to 5 min using a homogenizer (Ti-horn, Q500, QSonica, Newtown, CT, USA). A cold-water bath was used to avoid overheating.

2.2. Nanoparticle Tracking Analysis (NTA)

The hydrodynamic diameter of the nanoparticles was investigated using the nanoparticle tracking analysis method with Nanosight LM10 (A.P. Instruments, Warsaw, Poland). The instrument's software calculates the particle diffusion coefficient and counts the particles' hydrodynamic diameter using the Stokes–Einstein equation. Every solution was investigated three times.

2.3. Attenuated Total Reflection–Fourier Transformation Infrared Spectroscopy (ATR-FTIR)

To examine changes in chemical groups on the surface after the plasma treatment, the ATR-FTIR method was used with the Nicolet Summit FTIR Spectrometer (Thermo Fisher Scientific, Waltham, MA, USA) equipment. The polypropylene samples were measured right after the plasma treatment. Each studied sample was scanned 64 times in the range of 600–4000 cm^{-1}.

2.4. X-ray Photoelectron Spectroscopy (XPS)

To identify the surface functional groups formed during the plasma treatment, the polypropylene samples were investigated by XPS. The measurements were conducted in an ultrahigh vacuum chamber (vacuum level > 5×10^{-9} mbar) using an SES R4000 analyzer (Gammadata Scienta, Seattle, WA, USA) with a monochromatic Al K_α X-ray source (1486.6 eV) at 250 W (pass energy: 100 eV) for the survey and narrow scans. The received XPS spectra were analyzed using the Casa-XPS 2.3.15 software. The electron binding energy of the C 1*s* peak was calibrated at 285 eV.

2.5. SEM

The fabricated nanocomposites were investigated using an FEI (Hillsboro, OR, USA) Quanta 3D FEG field-emission scanning electron microscope (SEM) at low vacuum conditions (10 mbar of H_2O pressure, evaporated from a distilled water reservoir) in the so-called Environmental SEM (ESEM) mode. This experimental setting was required due to the non-conductive nature of the polymeric substrate. To avoid disturbing the morphology of the nanocomposite, the samples were investigated in their native forms without any additional coatings. To verify the elemental composition of the nanocomposite, EDX measurements were also carried out for specific areas of interest.

2.6. TG/DTA

Thermogravimetric measurements with differential thermal analysis were performed with TGA 3+ equipment (Mettler Toledo, Warsaw, Poland). The measurements were carried out under airflow in the temperature range of 30–700 °C with a heating rate of 10 °C/min. The measured mass loss of samples (~7 mg) was normalized to 100% for the reliable comparison of the thermogravimetric profiles.

2.7. Photocatalytic Tests

The photocatalytic activity was tested using the methyl orange photodegradation reaction under UV-A light (365 nm) using a Sylvania blacklight bulb 20 W (Newhaven, UK). The prepared nanocomposites—polypropylene (35 mm × 8 mm) with different $nTiO_2$ coatings—were placed in a quartz cuvette with standard 3.5 mL of methyl orange aqueous solution (25 mmol/dm^3). Absorbance was measured periodically by UV-Vis spectrometer (model 1900i, Shimadzu, Kioto, Japan) to determine changes in the methyl orange concentration upon illumination. The characteristic absorbance maximum for methyl orange at 464 nm was used for kinetic measurements.

2.8. Diffusion Reflectance Spectroscopy

Ultraviolet-visible diffuse reflectance spectra (UV-vis DRS) were recorded for the nanocomposites ($nTiO_2$/PP) using a Perkin Elmer (Waltham, MA, USA) Lambda 365 instrument equipped with an integrating sphere. The results were calculated using the Kubelka–Munk equation.

2.9. XRD Measurements

The crystallinity of the photocatalyst (TiO_2 powder) was examined using XRD measurements before and after the photocatalytic processes. PXRD patterns were collected using a Bruker (Billerica, MA, USA) D8 Advance Eco diffractometer equipped with a Cu sealed tube radiation source and a capillary stage. Powder samples were loaded into glass capillaries (0.5 mm in diameter). The background in the obtained diffractograms was corrected using the DIFFRAC algorithm implemented in the DIFFRAC.EVA V5 program.

3. Results

3.1. Polypropylene Substrate

Before the preparation of the $nTiO_2$/PP nanocomposite, the main constituents were characterized in their pristine forms. The SEM and FTIR characterization of the polypropylene substrate is shown in Figure 1. The SEM image demonstrates that the polypropylene exhibits a fairly smooth surface morphology on the micrometer scale, with scratches or irregularities caused by the production process not typically observed for polymers. A surface examination using ATR-FTIR (Figure 1b) provides insight into the chemical structure of the polypropylene. All the characteristic bands present in the polypropylene fingerprint can be easily observed [23], whereas no maxima characteristic for any contaminants or impurities was identified.

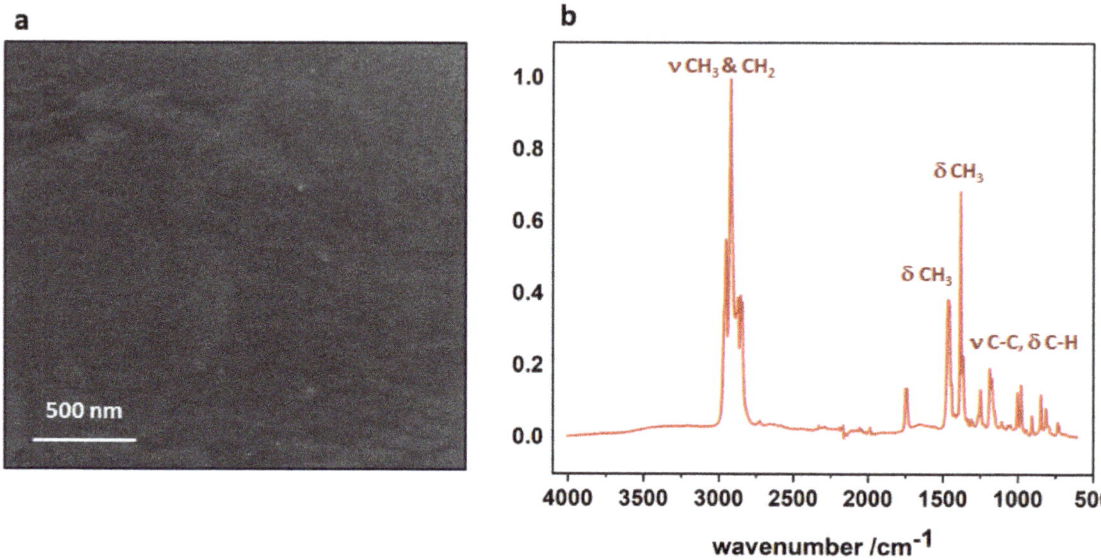

Figure 1. Pristine polypropylene substrate: (**a**) SEM image and (**b**) ATR-FTIR spectrum indicating the main functional groups in the polypropylene structure.

3.2. Characteristics of Titanium Dioxide Nanoparticles

As the photocatalytic reactions occur on the surface of the photocatalyst, the size of the photocatalyst particles is a crucial factor for the efficacy of photocatalytic processes. Therefore, the second constituent of the prepared nanocomposites—photoactive titania nanoparticles—was characterized in terms of their size. The results are presented in Figure 2a–f. The most abundant hydrodynamic diameter of the used nTiO_2 determined with the method of nanoparticle tracking analysis was found to be 35 nm (Figure 2a). It is worth noting that other peaks at 70 nm, 140 nm, and 210 nm represent the multiplicity of the smallest particle diameter and can be attributed to agglomerates, which are formed due to a strong interaction between nanoparticles. The size distribution of the nanoparticles was also examined by SEM, showing a particle size range between 20 and 55 nm (Figure 2b). The majority of the particles present are approximately 35 nm in size, as can be inferred from the particle size distribution presented in Figure 2c. However, bigger particles of around 200 nm were also occasionally observed. The results of the EDX analysis confirmed that all the particles contain only titanium and oxygen. To analyze the nanoparticles in more detail, the TEM analysis was conducted. Figure 2d shows a typical TEM image of the TiO_2 nanoparticles, revealing that the nanoparticles observed in the SEM are actually agglomerates of much smaller crystallites. An inset in Figure 2d provides an HR-TEM image, illustrating the individual nanoparticles at higher magnification. As shown in Figure 2e, the size distribution of the individual TiO_2 crystallites ranges from 2 to 6 nm. However, when the entire agglomerates are measured based on the TEM images (Figure 2f), their size matches the 35 nm diameter observed in SEM.

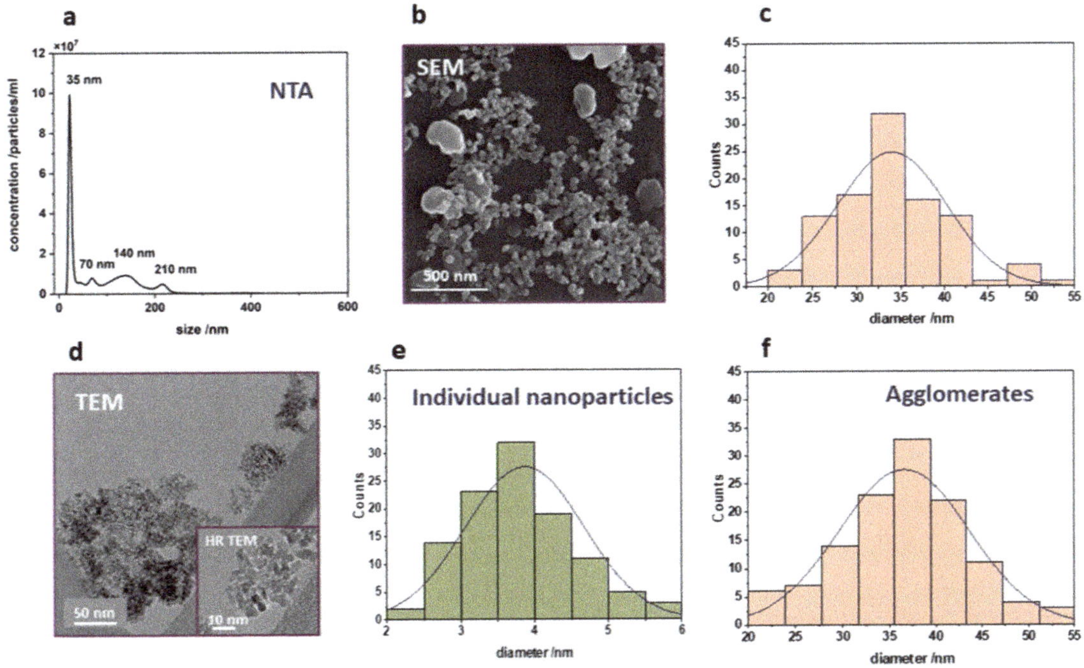

Figure 2. Titanium dioxide nanoparticles: (**a**) nanoparticle tracking analysis showing the hydrodynamic diameter; (**b**) SEM image of TiO_2; (**c**) particle size distribution based on the SEM images; (**d**) TEM image showing that the observed SEM nanoparticles are agglomerates of smaller TiO_2 crystallites, with the insert presenting a high resolution (HR-TEM) image; (**e**) size distribution of individual TiO_2 crystallites; and (**f**) size distribution of agglomerates from TEM images.

3.3. Surface Functional Groups

The oxygen plasma treatment was used to increase surface adhesion and facilitate the attachment of nanoparticles to polypropylene. The main changes introduced by low-temperature oxygen plasma on the polypropylene surface consist of generating surface functional groups. Modifying the polypropylene surface depends on the chemical nature and the surface concentration of the functional groups. Therefore, the surfaces before and after the plasma treatment were characterized by XPS. The results are shown in Figure 3. The XPS survey scans show only the main constituents of the polymeric substrate: oxygen O 1*s* at 533 eV and carbon C 1*s* at 285 eV. It can be observed that before the plasma treatment, the polypropylene contained surface oxygen at a level as low as 2.8%. Its concentration, however, dramatically changes upon the plasma surface modification, as can be seen comparing the ratio of C 1*s* to O 1*s*. The longer the plasma treatment was applied, the higher the surface oxygen concentration was observed on the polypropylene, with 21.3% reached after 10 min.

The XPS results also permit us to characterize the chemical nature of the generated functional groups. The narrow scan XPS spectra, including the analysis of C 1*s* peak, are shown in Figure 3b. As expected, a broadening of the C 1*s* peak was observed after the plasma treatment, which is related to its oxidation resulting from the generation of oxygen groups. After 1 min of oxygen plasma, the new component can be distinguished at 287 eV, corresponding to hydroxyl (-OH) and carbonyl groups (-C=O). After 5 min of the plasma treatment, a new peak at 288 eV can be distinguished and assigned to carboxyl bonds (O–C=O) [23]. Following the application of a plasma treatment for 10 min, the peak with the broadest distribution is observed, corresponding to the carbon–oxygen bonds. This clearly indicates the increase in the concentration of functional groups on the surface of the polypropylene with the time of exposure to oxygen plasma.

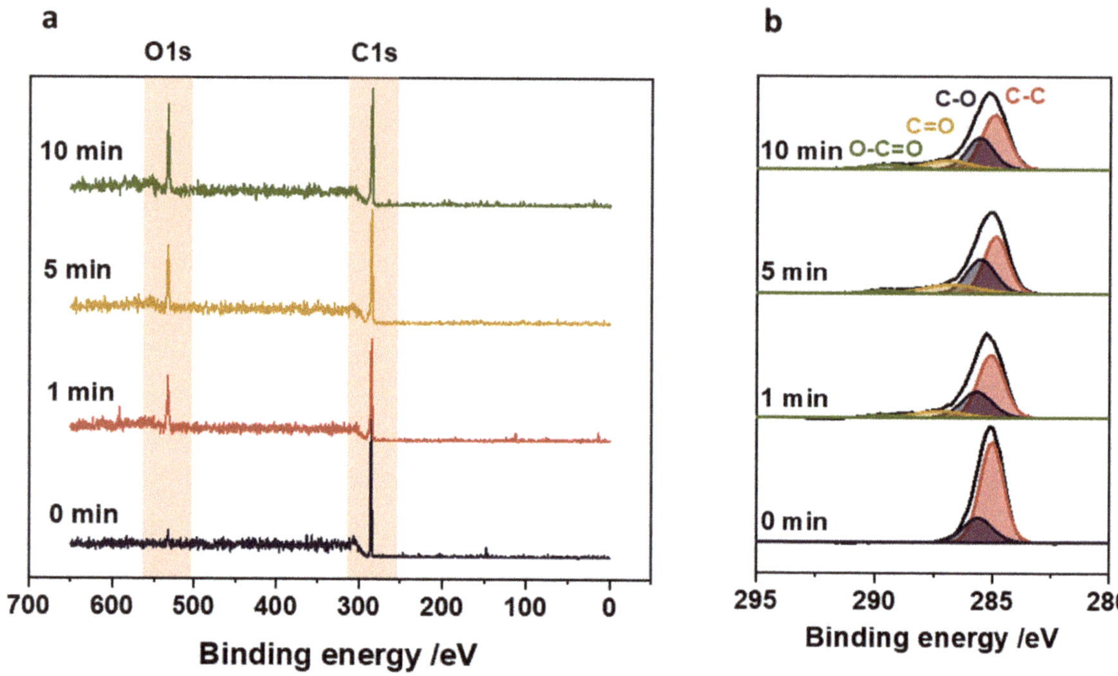

Figure 3. XPS results for polypropylene surface before and after oxygen plasma treatment. (**a**) XPS survey scans showing the increase in surface oxygen to carbon ratio with the plasma treatment time; (**b**) narrow XPS scan of the C 1s peak indicating the presence of hydroxyl (-OH) and carbonyl (C=O) groups, and the appearance of carboxyl bonds (O–C=O) upon plasma treatment.

The surface changes induced by plasma were also confirmed by the FTIR measurements (Figure S1). After the plasma treatment, two new bands appeared in the polypropylene spectra. A broad band at 3000–3500 cm^{-1} corresponds to the vibration of –OH bonds. A second band at 1730 cm^{-1} can be assigned to –C=O vibrations [23]. Both of these maxima indicate surface oxidation. Additionally, the intensity of the observed bands increased with the time of exposure to plasma, which aligns with the XPS results described above.

Introducing surface functional groups on the hydrophobic surface of polypropylene has very important practical implications. Since oxygen-containing functional groups are polar in nature, they stimulate stronger interactions with water molecules. Consequently, the surface becomes more hydrophilic as can be quantified experimentally by water contact angle values. They are presented as a function of the plasma treatment time in Figure 4. It can be noticed that after just 1 min of plasma, the water contact angle drops down from 95° for pristine PP to 30°. After that, a gentler water contact angle decrease is observed, reaching a stable level of about 20° after 5 min of plasma. The induced wettability of PP has a further impact on the deposition of $nTiO_2$, as discussed below.

During the plasma treatment, several processes occur on the polymer surface that can be distinguished, such as cleaning, functionalization, and etching. The optimal strategy of polymer functionalization via plasma treatment is to add functional groups to the surface without disturbing the optimized bulk properties of the polymer. The effect of plasma treatment on the structural properties of the polypropylene can be easily checked by the TG/DTA measurements. The results of such thermogravimetric analysis are presented in Figure 5. The TG profiles clearly illustrated that the polymeric material is stable up to 270 °C independently of the plasma treatment time. This result strongly supports the working hypothesis that bulk polymeric properties are preserved. Additionally, there is no change in the melting temperature after the plasma treatment for 1 to 5 min (the typical melting temperature of polypropylene is approximately 170 °C [2]). Even the rapid thermal degradation of the untreated and plasma-treated polypropylene, observed in the temperature range of 270–400 °C, is the same within the experimental error.

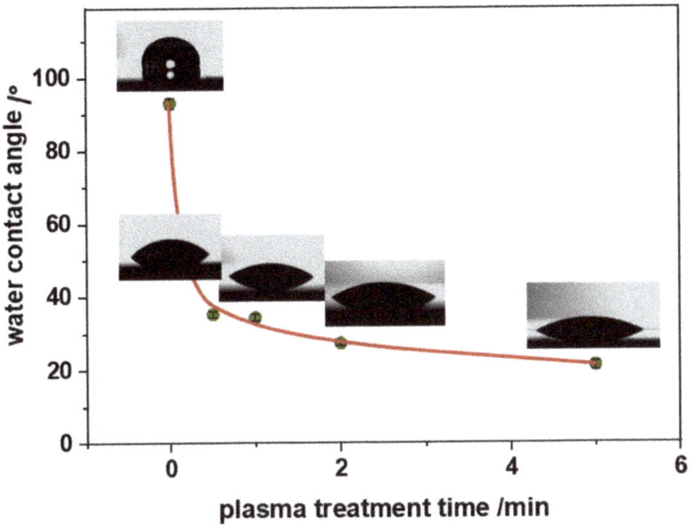

Figure 4. Changes in the water contact angle of the polypropylene surface as a function of oxygen plasma treatment time.

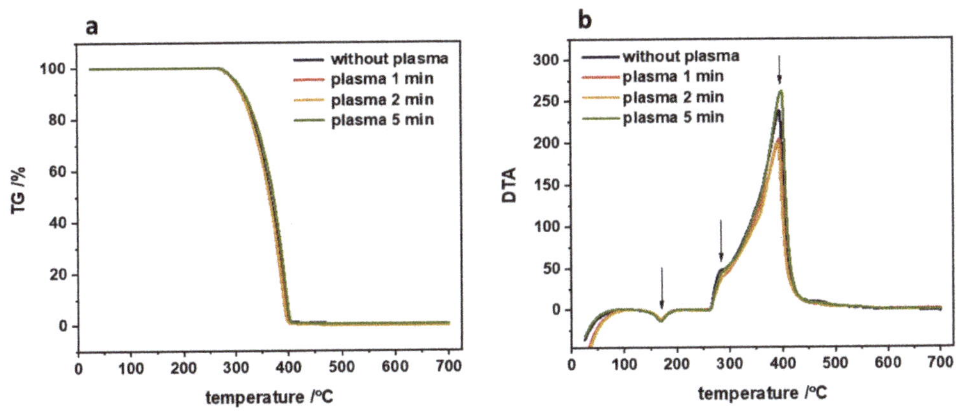

Figure 5. Thermogravimetric analysis of pristine polypropylene and PP after plasma treatment: (**a**) loss of mass profile and the corresponding (**b**) differential thermal analysis.

Summarizing the effect of oxygen plasma on the polypropylene substrate, it can be concluded that while the XPS and FTIR spectroscopic results, as well as the water contact angle measurement, confirm the generation of the oxygen functional groups on the polymeric surface, its bulk properties remain unchanged. Such surface-modified polypropylene was then used as the substrate for $nTiO_2$ deposition to prepare a functional nanocomposite, which is described in the next section.

3.4. Nanocomposite

The deposition of titania nanoparticles on the untreated or plasma-modified polypropylene substrates was performed via the sonochemical method. Figure 6 presents the SEM images comparing the effectiveness of the $nTiO_2$ deposition on these surfaces together with the pristine PP substrate. The images represent the surfaces of (a) polypropylene, (b) polypropylene untreated by plasma after the deposition of nanoparticles from a 1% titanium dioxide suspension, and (c) the plasma-treated polypropylene after the deposition of nanoparticles from a 1% titanium dioxide suspension. As can be inferred from the images in Figure 6b,c, the effect of oxygen plasms is tremendous. To assure a rational comparison during the deposition process on the untreated and plasma-treated PP surfaces, the identical sonication parameters, i.e., TiO_2 concentration (1%), deposition time, power, and amplitude, were employed. Single titania nanoparticles were deposited on the untreated polypropylene substrate by oxygen plasma, whereas the plasma-treated surface exhibited a completely covered surface with TiO_2 nanoparticles. It is evident that the efficiency of nanoparticle deposition on polymer substrates is influenced by the plasma treatment process and the introduction of functional groups on their surfaces.

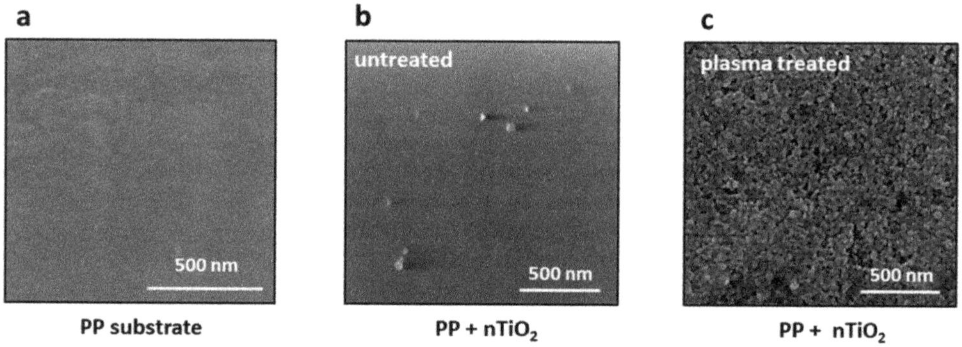

Figure 6. SEM images of the polypropylene substrate showing the effectiveness of TiO_2 nanoparticle deposition on untreated and plasma-treated substrates: (**a**) pristine polypropylene surface, (**b**) TiO_2 nanoparticles deposited on the untreated surface, and (**c**) TiO_2 nanoparticles deposited on the plasma-treated surface.

In order to ascertain the optimal degree of surface coverage and, consequently, the most developed photocatalytically active surface, the effect of nanoparticle suspension concentrations on the application process was also screened. Figure 7 illustrates the scanning electron microscopy (SEM) images of a polypropylene substrate following the application of TiO_2 from 0.1% and 1% suspensions (Figure 7b, Figure 7c, respectively). Figure 7a presents a polypropylene substrate as a reference; the surface of this polymer is not decorated with nanoparticles. As expected, reduced surface coverage is evident when employing a lower initial concentration (Figure 7b). Individual nanoparticles, as well as clusters and polymer uncovered surface regions, can be distinguished. Figure 7c illustrates the full coverage of polypropylene with nanoparticles. However, the SEM image also shows pores formed between the TiO_2 nanoparticles at high coverage. Although these pores may influence the photodegradation process, it is important to note that UV light penetration will be limited in comparison to the outermost surface of the photocatalyst. The optimal concentration of TiO_2 in the initial suspension and the most effective sonication time will yield the most favorable results. The obtained results showed that by optimizing both the plasma treatment and sonochemical deposition parameters, the surface concentration and the accessibility of the photocatalyst surface can be controlled.

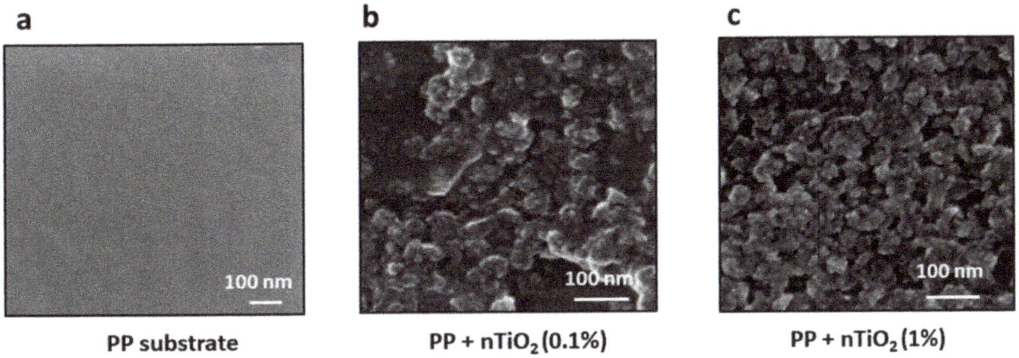

Figure 7. SEM images of polypropylene substrate after deposition of TiO_2 from suspension with varying nanoparticle concentrations: (**a**) reference polypropylene substrate before nanoparticle deposition, (**b**) polypropylene surface after deposition from TiO_2 0.1% suspension, and (**c**) polypropylene surface after deposition from 1% TiO_2 suspension.

3.5. Photocatalytic Tests

The photocatalytic activity was investigated in the degradation of methyl orange (as a model pollutant). The degradation of methyl orange in the presence of TiO_2/PP nanocomposite upon UVA light exposure was observed by monitoring the decrease in absorbance at the characteristic band of 464 nm in a time range of 0–200 min (Figure 8a). Based on these results, the degradation kinetics curves were determined and are presented in Figure 8b. The rate constants for the photocatalytic decomposition of methyl orange have been calculated based on the first-order reaction kinetics and the resulting values are presented beside the corresponding kinetic curves in Figure 8c. It can be observed that the polymer (PP) without nanoparticles (reference sample) does not exhibit photocatalytic activity. However, the deposition of the TiO_2 nanoparticles onto its surface strongly promotes its activity. A huge difference between the plasma-untreated (degradation of ~20% of methyl orange after 200 min) and plasma-treated PP (~80% degradation for 200 min) can also be noted. This results from the much more efficient attachment of nanoparticles onto the functionalized PP substrate (see Figure 6c).

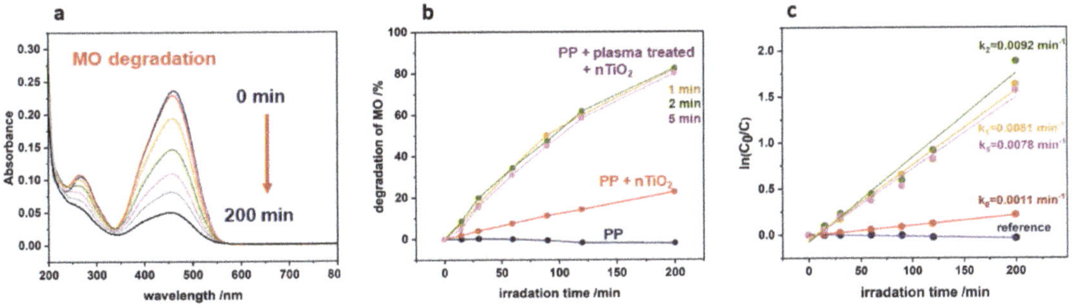

Figure 8. Photocatalytic activity of $nTiO_2$/PP nanocomposite in methyl orange degradation under UVA irradiation: (**a**) absorbance decrease at 464 nm upon irradiation, (**b**) kinetic curves illustrating enhanced activity for plasma-treated PP composite, and (**c**) the linearized kinetic data together with the rate constants calculated based on first-order reaction kinetics.

The effect of plasma treatment is schematically presented in Figure 9. It can be noted that a 1 min plasma treatment is enough to make a significant impact on the activity, and a further increase in the plasma treatment time does not result in any discernible changes To investigate the relationship between the photocatalytic activity and light absorption, UV light absorption measurements for the nanocomposites ($nTiO_2$/PP) were also conducted. As shown in Figure S2 (Supplementary Information), absorption increases significantly with TiO_2 loading. However, comparing these results with the data in Figure 9, where all the plasma-treated samples exhibit similar photocatalytic activity regardless of TiO_2 loading, suggests that UV light absorption may not be the critical factor for the dye degradation.

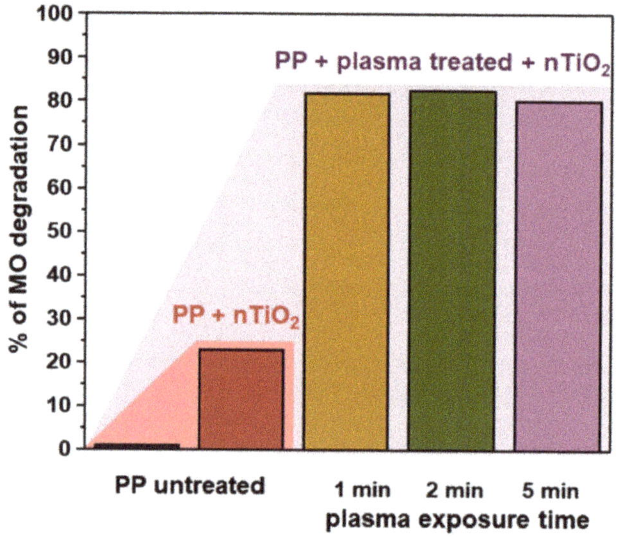

Figure 9. Schematic representation of the plasma treatment effect on the photocatalytic performance of the developed $nTiO_2$/PP composite.

To assess the stability and photocatalytic activity of the nanocomposite systems, the dye degradation process was conducted over five cycles. The results are shown in Figure 10. In each cycle, the model pollutant was degraded by 70% to 85%. Although there is a slight decrease in the degradation of methyl orange in the subsequent cycles, it remains at a reasonably high level. This stability suggests that the nanocomposite systems maintain their photocatalytic efficiency over multiple uses. Additionally, the crystallinity of the photocatalyst (TiO_2 powder) was examined using the XRD measurements before and after the photocatalytic processes to investigate any potential structural changes. The results show that there are no visible changes in the TiO_2 structure after the UV irradiation, indicating that the nanocomposite maintains its bulk integrity under photocatalytic conditions The corresponding diffractograms can be found in the Supplementary Information in Figure S3.

Figure 11a illustrates the effect of the initial concentration of the $nTiO_2$ suspension on the degradation of methyl orange (MO) with the irradiation time. As the concentration of the TiO_2 suspension used increases, the level of MO degradation increases in the whole-time range of irradiation. After 200 min of irradiation, for 0.1% TiO_2 and 0.2% TiO_2, approximately 40% and 60% degradation of the model pollutant was achieved, whereas for higher $nTiO_2$ concentrations of 0.5% and 1%, the MO degradation reached 80%. As can be observed, the degradation of the model pollutant did not increase linearly with the content of TiO_2. This may be attributed to the addition of the subsequent layers of nanoparticles, which lead to limitations in the accessibility of the TiO_2 surface for photocatalytic processes.

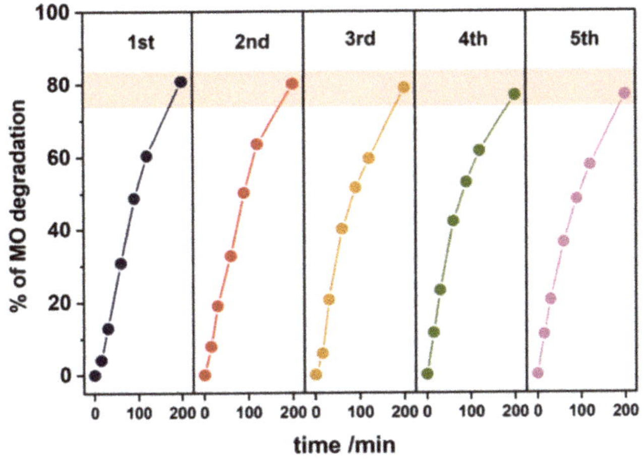

Figure 10. The dye degradation process over five cycles of irradiation in the presence of the $nTiO_2$/PP nanocomposite.

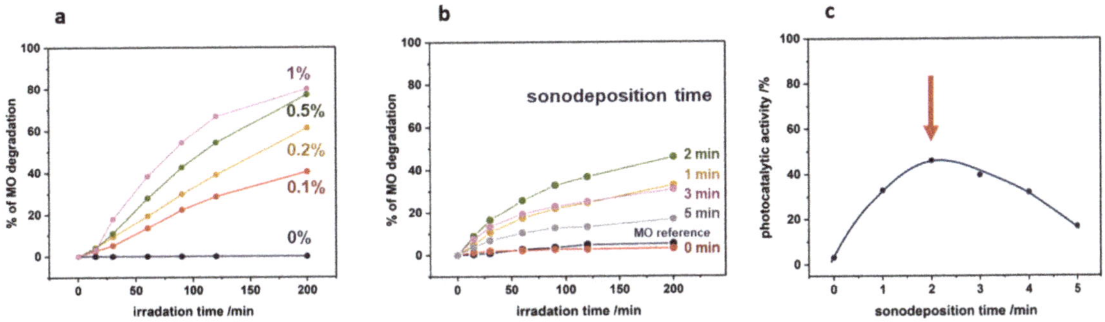

Figure 11. Optimization of sonochemical deposition parameters for enhanced photocatalytic activity: (**a**) effect of initial $nTiO_2$ suspension concentration, (**b**) effect of sonication time on methyl orange degradation using 0.1% TiO_2 suspension, (**c**) photocatalytic activity as a function of sonodeposition time, with 2 min identified as the optimal for maximum efficiency.

To achieve optimal photocatalytic effect, the optimization of the sonochemical deposition stage was essential. The degradation results of methyl orange upon light exposure, as a function of sonication time using a 0.1% TiO_2 suspension, are depicted in Figure 11b, while the relationship between photocatalytic activity and sonodeposition time is illustrated in Figure 11c. It becomes evident that insufficient and excessive sonication times can reduce the effective anchoring of the nanoparticles onto the polypropylene surface. The experimental data indicate that a sonication time of precisely two minutes is required to maximize the photocatalytic efficiency of the $nTiO_2$/PP nanocomposite system.

Furthermore, the analysis of the plasma treatment durations revealed that beyond one minute, there were no significant improvements in the photocatalytic properties of the $nTiO_2$/PP. This suggests that while initial plasma exposure is crucial for generating surface functional groups necessary for nanoparticle adhesion, extended treatment times do not proportionally enhance photocatalytic activity. Therefore, a balanced approach is necessary: ensuring adequate plasma treatment to functionalize the surface, coupled with optimized sonochemical deposition parameters, is essential for achieving high photocatalytic performance. The findings illustrate the importance of both surface functionalization and precise control of sonodeposition conditions in the development of effective photocatalytic nanocomposites. Finally, it is worth noting that the proposed approach combining oxygen plasma treatment and ultrasound shows significant potential and can be used for any polymer–nanoparticle system to develop functional composites for diverse specific applications.

4. Conclusions

The low-temperature oxygen plasma treatment effectively introduced polar surface functional groups (mainly -OH and -C=O) on the polypropylene, transforming its surface from hydrophobic to hydrophilic. This increase in wettability is evidenced by a significant reduction in the water contact angle from 95° to 20°. Moreover, the generated surface oxygen groups were found to be crucial as they enhance the interaction between the polymer and nanoparticles and thus significantly improve the anchoring efficiency of the TiO_2 nanoparticles. The optimized parameters for the plasma treatment (100 W power, 0.2 mbar O_2 pressure, and 1 min duration) ensured effective surface modification. The sonochemical method was found to be facile and effective for depositing the TiO_2 nanoparticles on the plasma-treated PP surface. Optimal nanoparticle dispersion and surface coverage were achieved with a 1% TiO_2 suspension and a sonication time of 2 min at 20 kHz, resulting in well-distributed photocatalytic TiO_2 nanoparticles on the PP substrate. The developed $nTiO_2$/PP nanocomposite demonstrated significant photocatalytic activity, degrading 80% of methyl orange under UVA irradiation within 200 min. Despite the superior surface modifications, the bulk properties of PP remained unchanged, as confirmed by the thermogravimetric analysis (TG/DTA). The polymer maintained its stability up to 270 °C and did not exhibit any changes in the melting temperature after the plasma treatment. The proposed two-step method (plasma treatment followed by sonochemical deposition) can be considered universal, versatile, and scalable, making it applicable for functionalizing various polymeric materials with photoactive nanoparticles. From a broader perspective, this approach shows high potential for designing and fabricating self-cleaning and antibacterial polymeric surfaces for diverse applications.

Supplementary Materials: The following supporting information can be downloaded at: https://www.mdpi.com/article/10.3390/nano14161372/s1, Figure S1. Introducing functional groups after plasma treatment on ATR-FTIR spectra. Figure S2. Kubelka–Munk function of $nTiO_2$/PP nanocomposites showing material absorbance Figure S3. XRD pattern of photocatalytic phase (TiO_2) before and after the photocatalytic processes indexing with characteristic peaks of the anatase. All authors have read and agreed to the published version of the manuscript..

Author Contributions: Conceptualization, K.Z., J.M., A.K. and W.M.; methodology, K.Z. and K.S.; validation, J.M., F.K., W.M. and A.K.; formal analysis, J.M., F.K., W.M. and A.K.; investigation, K.Z. and K.S.; data curation, K.Z. and K.S.; writing—original draft preparation, K.Z.; writing—review and editing, J.M., F.K., W.M. and A.K.; visualization, K.Z. and A.K.; supervision, J.M. and A.K.; project administration, J.M. and A.K.; funding acquisition, J.M. and A.K. All authors have read and agreed to the published version of the manuscript.

Funding: This research was funded by the Ministry of Education and Science, Implementation Doctorate within "Funkcjonalne powłoki nanokompozytowe" project (DWD/6/257/2022). The study was carried out using research infrastructure purchased with the funds of the European Union in the framework of the Smart Growth Operational Programme, Measure 4.2; Grant No. POIR.04.02.00-00-D001/20, "ATOMIN 2.0—ATOMic scale science for the INnovative economy".

Data Availability Statement: Data are contained within the article and Supplementary Materials.

Acknowledgments: The authors acknowledge D. Majda and M. Drozdek for performing the TG/DTA and XPS measurements.

Conflicts of Interest: Conflicts of Interest: Authors Karolina Zajac, Joanna Macyk, Wojciech Macyk were employed by the company InPhoCat—Innovative Photocatalytic Solutions Sp. z o. o. The remaining authors declare that the research was conducted in the absence of any commercial or financial relationships that could be construed as a potential conflict of interest.

References

1. Statista Research Department. Global Production of Plastics since 1950. 2024. Available online: https://www.statista.com/statistics/282732/global-production-of-plastics-since-1950/ (accessed on 1 July 2024).
2. Maddah, H.A. Polypropylene as a Promising Plastic: A Review. *Am. J. Polym. Sci.* **2016**, *6*, 1–11.
3. Hossain, M.T.; Shahid, M.A.; Mahmud, N.; Habib, A.; Rana, M.; Khan, S.A.; Hossain, D. Research and application of polypropylene: A review. *Discov. Nano* **2024**, *19*, 2. [CrossRef]
4. Harish, V.; Ansari, M.M.; Tewari, D.; Gaur, M.; Yadav, A.B.; García-Betancourt, M.-L.; Abdel-Haleem, F.M.; Bechelany, M.; Barhoum, A. Nanoparticle and Nanostructure Synthesis and Controlled Growth Methods. *Nanomaterials* **2022**, *12*, 3226. [CrossRef] [PubMed]
5. Ubaldi, F.; Valeriani, F.; Volpini, V.; Lofrano, G.; Romano Spica, V. Antimicrobial Activity of Photocatalytic Coatings on Surfaces: A Systematic Review and Meta Analysis. *Coatings* **2024**, *14*, 92. [CrossRef]
6. Wei, Y.; Wu, Q.; Meng, H.; Zhang, Y.; Cao, C. Recent advances in photocatalytic self-cleaning performances of TiO_2-based building materials. *RSC Adv.* **2023**, *13*, 20584–20597. [PubMed]
7. Howard, A. Foster Photocatalytic disinfection using titanium dioxide: Spectrum and mechanism of antimicrobial activity. *Appl. Microbiol. Biotechnol.* **2011**, *90*, 1847–1868. [CrossRef]
8. Pant, B.; Park, M.; Park, S.-J. Recent Advances in TiO_2 Films Prepared by Sol-Gel Methods for Photocatalytic Degradation of Organic Pollutants and Antibacterial Activities. *Coatings* **2019**, *9*, 613. [CrossRef]
9. Guo, Q.; Zhou, C.; Ma, Z.; Yang, X. Fundamentals of TiO_2 Photocatalysis: Concepts, Mechanisms, and Challenges. *Adv. Mater.* **2019**, *31*, 1901997. [CrossRef]
10. Iesalnieks, M.; Eglītis, R.; Juhna, T.; Šmits, K.; Šutka, A. Photocatalytic Activity of TiO_2 Coatings Obtained at Room Temperature on a Polymethyl Methacrylate Substrate. *Int. J. Mol. Sci.* **2022**, *23*, 12936. [CrossRef]
11. Sun, L.; Guan, J.; Xu, Q.; Yang, X.; Wang, J.; Hu, X. Synthesis and Applications of Molecularly Imprinted Polymers Modified TiO_2 Nanomaterials: A Review. *Polymers* **2018**, *10*, 1248. [CrossRef]
12. Zhang, H.; Han, J.; Yang, B. Structural Fabrication and Functional Modulation of Nanoparticle-Polymer Composites. *Adv. Funct. Mater.* **2010**, *20*, 1533–1550. [CrossRef]
13. Cazan, C.; Enesca, A.; Andronic, L. Synergic Effect of TiO_2 Filler on the Mechanical Properties of Polymer Nanocomposites. *Polymers* **2021**, *13*, 2017. [CrossRef]
14. Lai, J.; Sunderland, B.; Xue, J.; Yan, S.; Zhao, W.; Folkard, M.; Michael, B.D.; Wang, Y. Study on hydrophilicity of polymer surfaces improved by plasma treatment. *Appl. Surf. Sci.* **2006**, *252*, 3375–3379.
15. Nemani, S.K.; Annavarapu, R.K.; Mohammadian, B.; Raiyan, A.; Heil, J.; Haque, M.A.; Sojoudi, H. Surface Modification of Polymers: Methods and Application. *Adv. Mater. Interfaces* **2018**, *5*, 1801247. [CrossRef]
16. Sadowski, R.; Wach, A.; Buchalska, M.; Kuśtrowski, P.; Macyk, W. Photosensitized TiO_2 films on polymers—Titania-polymer interactions and visible light induced photoactivity. *Appl. Surf. Sci.* **2019**, *475*, 710–719. [CrossRef]
17. Chytrosz-Wróbel, P.; Golda-Cepa, M.; Stodolak-Zych, E.; Rysz, J.; Kotarba, A. Effect of oxygen plasma-treatment on surface functional groups, wettability, and nanotopography features of medically relevant polymers with various crystallinities. *Appl. Surf. Sci. Adv.* **2023**, *18*, 100497. [CrossRef]
18. Ricard, A. The production of active plasma species for surface treatments. *J. Phys. D Appl. Phys.* **1997**, *30*, 2261. [CrossRef]
19. Fleischer, M.; Kielar Tuceková, Z.; Galmiz, O.; Baková, E.; Plšek, T.; Kolárová, T.; Kovácik, D.; Kelar, J. Plasma Treatment of Large-Area Polymer Substrates for the Enhanced Adhesion of UV–Digital Printing. *Nanomaterials* **2024**, *14*, 426. [CrossRef] [PubMed]
20. Gedanken, A. Using sonochemistry for the fabrication of nanomaterials. *Ultrason. Sonochemistry* **2004**, *11*, 47–55. [CrossRef]
21. Ścibik, Ł.; Ochońska, D.; Gołda-Cępa, M.; Kwiecień, K.; Pamuła, E.; Kotarba, A.; Brzychczy-Włoch, M. Sonochemical Deposition of Gentamicin Nanoparticles at the PCV Tracheostomy Tube Surface Limiting Bacterial Biofilm Formation. *Materials* **2023**, *16*, 3765. [CrossRef]
22. Tanaka, Y.; Okawa, H.; Kato, T.; Sugawara, K. Effects of ultrasound irradiation on Au nanoparticles deposition on carbon-coated $LiNi_{0.5}Mn_{1.5}O_4$ and its performance as a cathode material for Li ion batteries. *Ultrason. Sonochemistry* **2022**, *82*, 105879. [CrossRef]
23. Morent, R.; De Geyter, N.; Leys, C.; Gengembre, L.; Payen, E. Comparison between XPS- and FTIR-analysis of plasma-treated polypropylene film surfaces. *Surf. Interface Anal.* **2008**, *40*, 597–600. [CrossRef]

molecules

MDPI

Article

Percolation-Triggered Negative Permittivity in Nano Carbon Powder/Polyvinylidene Fluoride Composites

Guangyue Shi [1,*], Xiaolei Sun [1] and Yao Liu [2,*]

[1] School of Materials Science and Engineering, Tianjin Key Laboratory for Rare Earth Materials and Applications, Center for Rare Earth and Inorganic Functional Materials, Nankai University, Tianjin 300350, China

[2] Key Laboratory for Liquid-Solid Structural Evolution and Processing of Materials (Ministry of Education), Shandong University, Jinan 250061, China

* Correspondence: sgy@mail.nankai.edu.cn (G.S.); liuyao@sdu.edu.cn (Y.L.)

Abstract: Percolating composites exhibiting negative permittivity have garnered considerable attention due to their promising applications in the realm of electromagnetic shielding, innovative capacitance devices, coil-less inductors, etc. Nano carbon powder/polyvinylidene fluoride (CP/PVDF) percolating composites were fabricated that exhibit Drude-type negative-permittivity behavior upon reaching the CP percolation threshold. This phenomenon is attributed to the formation of a plasmonic state within the interconnected CP network, enabling the delocalization of electrons under the alternating electric field. Furthermore, a significant (nearly two orders of magnitude) increase in the conductivity of sample is observed at a CP content of 12.5 wt%. This abrupt change coincides with the percolation phenomenon, suggesting a transition in the conduction mechanism. To elucidate this behavior, comprehensive analyses of the phase composition, microstructure, AC conductivity, and relative permittivity were performed. Additionally, the sample containing 5 wt% CP exhibits a remarkably high permittivity of 31.5, accompanied by a relatively low dielectric loss (tanδ < 0.2). The findings expand the potential applications of PVDF, while the fabricated percolating composites hold promise for electromagnetic shielding, antennas, and other electromagnetic devices.

Keywords: polyvinylidene fluoride; nano carbon powder; percolating composites; electrical properties; negative permittivity

1. Introduction

The negative permittivity ($\varepsilon_r' < 0$), commonly defined as the condition where the real part of the relative permittivity is negative, serves as a cornerstone of metamaterial research, holding immense promise for applications in meta-antennas, unconventional lenses, and wave absorption [1–3]. This exotic property arises from the incorporation of meticulously designed periodic structures within metamaterials. The resulting effective permittivity can be actively tuned by manipulating the size, shape, or geometrical arrangement of these constituent units. However, the fabrication of such intricate periodic architectures often presents significant challenges, including complexity and high cost. To address these limitations, researchers have explored alternative approaches utilizing random percolating composites. This strategy successfully achieves negative permittivity through the controlled distribution of conductive and dielectric components within the composite. Importantly, the resulting negative permittivity remains tunable [4,5]. By strategically modifying the functional phase content or chemical composition of these composites, researchers can achieve a high degree of design flexibility, enabling the creation of novel metamaterial functionalities. This newfound tunability paves the way for the development of next-generation metamaterial devices with an optimized performance [6–8].

Percolating composites, a class of composite materials, are characterized by a dispersed conductive phase embedded within a continuous insulating matrix phase. Conductive fillers, such as metals, carbon materials, and intrinsically conducting polymers, can be incorporated to achieve specific electrical properties [9,10]. Wei et al. explored this concept by fabricating copper/titanium dioxide composites. As the copper content exceeded the percolation threshold, the resulting composites demonstrated negative permittivity at low frequencies. This phenomenon can be ascribed to the delocalization of electrons within the percolating copper network [11]. Similarly, Song et al. synthesized percolating composites using ZrO_2/C nanoparticles, derived from metal–organic frameworks (MOFs), as the conductive phase and polyvinylidene fluoride (PVDF) as the matrix. With ZrO_2/C content approaching 30 wt%, a distinct transition from hopping to metallic conduction was observed. This transition, accompanied by negative permittivity, signifies the establishment of a conductive network throughout the composite, confirming the occurrence of electrical percolation [12]. Fan's group investigated the weak negative permittivity of graphene–carbon black/$CaCu_3Ti_4O_{12}$ percolating composites. With 8%~10% graphene–carbon black, the composites exhibited weak negative dielectric properties stemming from the electrical dipole resonance in the functional phase, corresponding to the Lorentz model. Upon further increasing the functional phase content to 14%~16%, the negative permittivity intensified with the increasing functional-phase content, originating from plasma resonance within the three-dimensional conductive network, corresponding to the Drude model [13]. Consequently, the negative-permittivity behavior of percolating composites strongly depends on the conductive functional fillers. This dependence can be attributed to the excitation of delocalized electrons within the conductive fillers, leading to a phenomenon known as plasma oscillation. This phenomenon is a key contributor to the emergence of negative permittivity in these materials. However, the use of metals with high intrinsic electron density as the conductive phase often results in large negative-permittivity values (up to the order of 10^5), thus hindering the precise control and practical application of negative-permittivity properties. Additionally, metals have high percolation thresholds and high intrinsic densities, posing challenges for the fabrication of percolation composites. Therefore, researchers are increasingly turning their attention to carbon materials, which possess advantages such as a light weight, chemical stability, and, more importantly, diverse morphologies, enabling lower percolation thresholds and negative-permittivity values, and ultimately allowing for finer control of negative-permittivity performance [14–16].

Our investigation leverages nano carbon powder (CP) as the conductive phase due to its high specific surface area. This zero-dimensional nanomaterial offers a unique opportunity to explore the influence of carbon morphology on the electromagnetic properties of percolating composites and their negative-permittivity response. Notably, the selection of nano carbon powder aligns with the growing interest in understanding the role of various carbon nanostructures in regulating percolation behavior. Polyvinylidene fluoride (PVDF) is chosen as the dielectric matrix owing to its exceptional chemical stability. Furthermore, the inherent piezoelectric properties and high dielectric constant of PVDF make it a well-established choice for electromagnetic composites [17]. This combination of materials offers a promising platform for investigating the interplay between nano carbon morphology, percolation network formation, and the resulting negative permittivity in these composites.

In this work, nano carbon powder/polyvinylidene fluoride (CP/PVDF) percolating composites were fabricated via a hot-pressing technique. The CP/PVDF composites exhibited negative permittivity once the CP content surpassed the percolation threshold. Notably, composites with CP content below the percolation threshold demonstrated a relatively high permittivity and low dielectric loss. The underlying mechanisms governing the dielectric and conductive behavior were subsequently elucidated in detail.

2. Results and Discussion

2.1. Phase Composition and Microstructure

The fabrication of CP/PVDF composites is described in the Experimental Section. In essence, a high-energy ball-milling process was employed to thoroughly mix nano carbon powder and PVDF powder, followed by hot pressing to consolidate the mixture into a form suitable for electrical measurement, using an LCR meter. Based on the mass fraction of nano carbon powder in the samples, we labeled the samples as follows: 0, 2.5, 5, 7.5, 10, 12.5, 15, and 17.5 wt%.

The FTIR spectrum in Figure 1 delineates the compositional fingerprint of PVDF and CP/PVDF composites, with distinct peaks attributed to the α and β phases of PVDF. The peak observed at 3400~3500 cm^{-1} originates from the stretching of O-H, which could be attributed to moisture in the PVDF powder or water absorbed by the sample during the preparation process for the FTIR test. The peak near 1627 cm^{-1} originates from the C=C bonds present in both PVDF and nano carbon powder. Additionally, the spectral region from 2850 to 3000 cm^{-1} is characterized by peaks arising from the vibrational modes of the C-H bonds within the PVDF matrix [18]. For the α phase, the 762 cm^{-1} absorption peak is indicative of the molecule's in-plane motion and wagging, the 1072 cm^{-1} peak signals the rocking vibration of the CF_2 group, and the 1185 cm^{-1} peak corresponds to the C-F bond's stretching. In the β phase, the 1402 cm^{-1} peak is associated with the CH_2 group's rocking vibration, the 1280 cm^{-1} peak is linked to the all-trans TT structure's vibration within PVDF, the 879 cm^{-1} peak is attributed to the symmetrical stretching vibration of the CF_2, and the 838 cm^{-1} peak denotes the asymmetrical stretching vibration of CF_2. Notably, the dual peaks at 879 cm^{-1} and 838 cm^{-1} are characteristic of the all-trans TT structure in PVDF, underscoring the spectrum's utility in elucidating the molecular dynamics and phase characteristics of the composites [19–21]. According to the FTIR test results, the main crystal structures of the PVDF powder used were the α phase and the β phase.

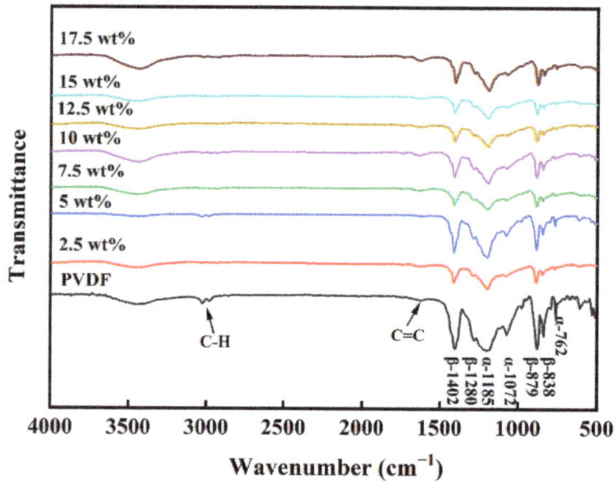

Figure 1. FTIR spectra of the PVDF and CP/PVDF composites.

The phase compositions of PVDF, nano carbon powder, and CP/PVDF composites were investigated using X-ray diffraction (XRD) and Raman spectroscopy. The XRD patterns in Figure 2a confirm the coexistence of α and β phases in the PVDF powder. Three prominent diffraction peaks are evident at $2\theta = 18.8°$, $2\theta = 20.2°$, and $2\theta = 26.6°$. These peaks correspond to the (020), (110), and (021) crystal planes of the α-phase, respectively. The diffraction peak at $2\theta = 20.6°$ represents a superposition of the (110) and (200) crystal planes of the β-phase [22]. The XRD pattern of nano carbon powder exhibits a broad hump in the range of 20°~25°, which is characteristic of amorphous carbon [23,24]. This hump overlaps with the diffraction peak of PVDF molecules at $2\theta = 26.6°$. It is noticeable that, with the increasing CP content, there is a slight shift in the peaks of the XRD. This shift may be attributed to the addition of nano carbon powder, which interacts with the polymer matrix, affecting the crystal structure of the polymer and consequently causing a slight shift in the peak positions [25,26].

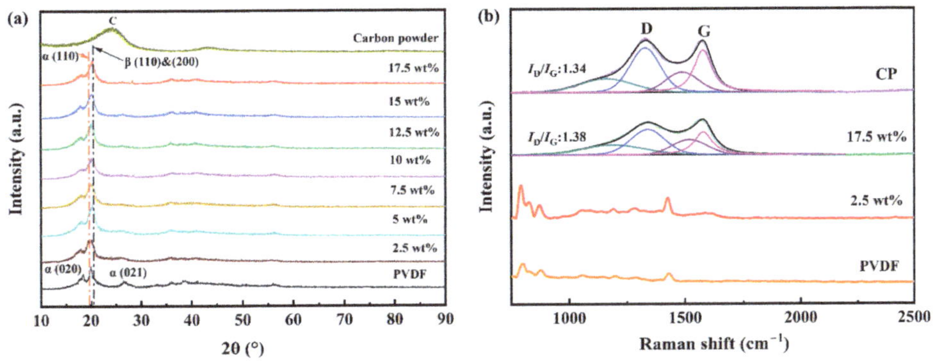

Figure 2. (**a**) XRD spectra. (**b**) Raman spectra of CP, PVDF, and CP/PVDF composites, with the spectra of CP and 17.5 wt% samples deconvoluted into four peaks: D (blue), TPA (green), A (purple), and G (pink).

Raman spectroscopy, as shown in Figure 2b, further elucidates the phase composition of the composites. When the nano carbon-powder content is low, the Raman spectrum of the sample resembles that of PVDF. The peak at 1433 cm^{-1} is still observable in the 2.5 wt% CP sample, corresponding to the C-H bond stretching vibration of PVDF molecules [27]. With increasing nano carbon-powder content, the Raman spectrum simplifies, exhibiting only two characteristic peaks: a D band at 1340 cm^{-1} and a G band at 1570 cm^{-1}, commonly associated with carbon materials. The D band originates from amorphous carbon and structural defects within nano carbon powder, while the G band arises from the C-C stretching vibrations of sp^2 hybridized carbon atoms in graphitic carbon [28]. For a more precise analysis of the D and G peaks, the Raman spectra of nano carbon powder and the 17.5 wt% sample are deconvoluted into four peaks, as shown in Figure 2b. For both the carbon powder and the sample, the I_D/I_G values are 1.34 and 1.38, respectively, revealing the presence of numerous defects [29]. These Raman spectroscopy results indicate that, as the nano carbon-powder content increases, the characteristic peaks of PVDF gradually become obscured, and only the characteristic peaks of nano-carbon black become apparent in the Raman spectrum of the sample.

Figure 3a–f present SEM images of nano carbon powder, PVDF matrix, and CP/PVDF composites with varying carbon-powder contents (2.5, 10, 12.5, and 17.5 wt%), respectively. As illustrated in Figure 3a, the nano carbon powder used in the experiment has a particle size ranging from 30 to 50 nm. It is also worth noting that the carbon powder exhibits significant agglomeration in its natural state, which necessitates the use of high-energy ball milling to uniformly mix the carbon powder with PVDF powder during the sample-preparation process. Figure 3b displays the cross-sectional morphology of the matrix, where the PVDF has formed a relatively dense structure following the hot-pressing method. At the 2.5 wt% content level (Figure 3c), the CP appears as isolated particles dispersed within the PVDF matrix. As the content increased to 10 wt% (Figure 3d), some agglomerates of carbon powder begin to form. However, these clusters remain largely isolated. The noticeable morphological changes in the nano carbon powder can be observed in Figure 3e. Here, the CP clusters begin to interconnect, forming a rudimentary three-dimensional network throughout the composites. This network becomes more extensive with a further increase in CP content to 17.5 wt% (Figure 3f). The observed microstructural evolution strongly suggests the occurrence of a percolation phenomenon for CP within the CP/PVDF composites at a critical loading between 10 and 12.5 wt%. Figure 3g offers a clearer depiction of the percolation dynamics in the CP/PVDF composite materials. When the functional-phase content is below the percolation threshold, the nanoparticles and their formed aggregates are isolated and dispersed within the matrix. As the functional-phase content increases and exceeds the percolation threshold, the nanoparticles become widely distributed and interconnected within the matrix, forming a three-dimensional network. The structural significance of conductive carbon path disconnection and connection, collectively termed percolation behavior, extends far beyond mere structural considerations. Percolation signifies the formation of interconnected pathways through the CP phase, enabling potential electrical conductivity throughout the composites.

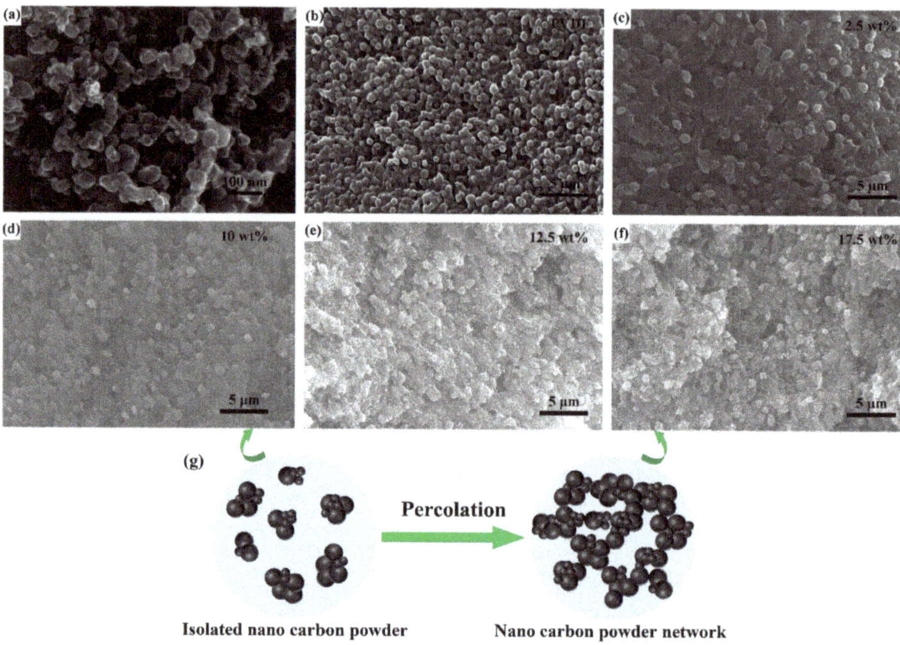

Figure 3. SEM images of the CP/PVDF composites, (**a**) nano carbon powder, (**b**) PVDF matrix, (**c**) 2.5 wt%, (**d**) 10 wt%, (**e**) 12.5 wt%, and (**f**) 17.5 wt%. (**g**) Schematic diagram of percolation behavior of the CP/PVDF composites.

2.2. AC Conductivity

Figure 4a reveals a positive correlation between AC conductivity and frequency for CP/PVDF composites containing less than 10 wt% nano carbon powder. However, the conductivity values remain relatively low within this range. This behavior aligns with the power-law relationship typically observed in samples with low functional-phase content (i.e., below 10 wt%). The power-law model accurately describes the frequency dependence of electrical conductivity in samples containing low amounts of nano carbon powder [30]. The expression for this relationship is given as follows:

$$\sigma_{ac} = \sigma_{dc} + A(2\pi f)^n \quad (1)$$

where σ_{dc} represents the DC conductivity, A is a constant coefficient, f denotes the frequency, and n signifies the exponent factor. The power-law model frequently serves as a valuable tool for elucidating the frequency dependence of AC conductivity in composite systems characterized by high degrees of disorder [31]. As depicted in Figure 4a, the excellent agreement between the power-law fitting curves (red solid lines) and the experimental data underscores the applicability of this model for composites with low functional-phase content (below 10 wt%). This alignment suggests a dominance of the hopping conduction mechanism in these sparsely filled composites, further implying their insulating behavior.

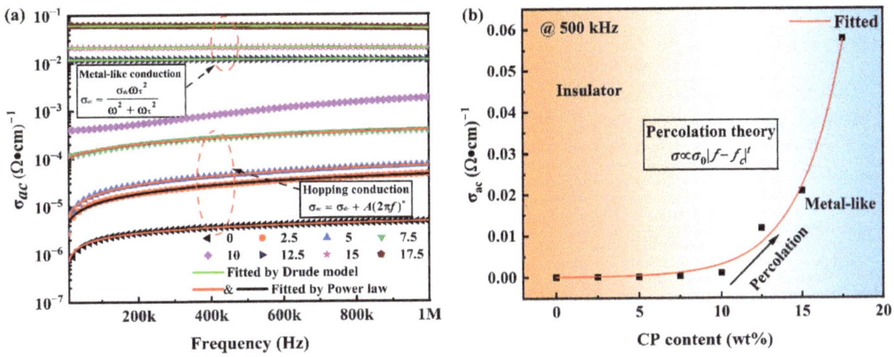

Figure 4. (**a**) Frequency dependences of ac conductivity of CP/PVDF composites. (**b**) Variation in AC conductivity with CP content at 500 kHz.

As the content exceeds 12.5 wt%, the AC conductivity significantly increases, while the conductivity decreases with the increasing frequency, obeying the free electron theory, also known as the Drude model [32].

$$\sigma_{ac} = \frac{\sigma_{dc}\omega_\tau^2}{\omega^2 + \omega_\tau^2} \tag{2}$$

where ω_τ is the collision frequency. As depicted in Figure 4a, the green solid lines illustrate the fitting results based on the Drude model. This agreement suggests a transition toward metal-like conductivity in CP/PVDF composites with carbon powder loadings exceeding 12.5 wt%. This behavior aligns with the concept of conductor behavior, signifying a substantial increase in electrical conductivity. The observed shift in the dominant conduction mechanism can be attributed to the well-established phenomenon of the percolation. Above this critical concentration, the formation of interconnected conductive networks within the composite facilitates efficient charge transport, leading to a conductor-like response.

The variation in AC conductivity with CP content at 500 kHz is shown in Figure 4b. In samples with carbon-powder contents of 12.5 wt% and higher, the AC conductivity increases significantly, with values from 2 to 3 orders of magnitude greater than those of samples with 10 wt% or less. In conjunction with the previous analysis, the percolation threshold is between 10 and 12.5 wt%. The observed increase in electrical conductivity suggests the establishment of a percolating network within the composites and can be described by percolation theory [33].

$$\sigma \propto |f - f_c|^t \tag{3}$$

where f represents the CP fraction, and f_c denotes the percolation threshold for the CP/PVDF composites. The red solid line in Figure 4b represents the fitting results obtained using percolation theory. The fitted data coincide with the experimental results, further corroborating the occurrence of the threshold-penetration phenomenon.

In brief, for the CP/PVDF composites, when the content exceeds the threshold, percolation occurs, and the nano carbon powder particles contact each other, forming a three-dimensional conductive network and expanding to the whole sample, thus leading to a change in the conductive mechanism and an increase in the AC conductivity.

2.3. Reactance Analysis

The impedance–frequency spectroscopy data presented in Figure 5a unveil a critical aspect of the CP/PVDF composites with functional phase content below the percolation threshold. These samples exhibit negative reactance values at lower frequencies. This observation signifies that the internal current phase leads to the applied voltage phase, a characteristic response of capacitive materials. Interestingly, the reactance behavior of the 7.5 wt% and 10 wt% CP samples deviates from this trend. Their reactance spectra initially exhibit an increase followed by a decrease with increasing frequency. This unique response suggests the coexistence of both capacitive and inductive components within the material. The emergence of these inductive contributions can be attributed to the formation of localized conductive networks arising from the agglomeration of carbon powder particles.

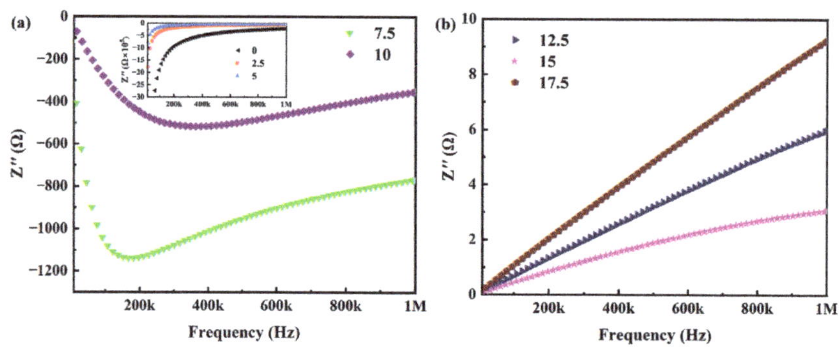

Figure 5. Reactance spectra of the CP/PVDF composites. (**a**) Samples with CP content below percolation threshold, (**b**) samples with CP content exceeding percolation threshold.

Conversely, Figure 5b shows that the impedance–frequency response of the composites surpasses the percolation threshold. In this case, the measured reactance values are positive across the entire frequency range. This positive reactance signifies that the internal voltage phase leads to the current phase, a behavior characteristic of inductive materials. It is noteworthy that the sample containing 15 wt% exhibits the lowest impedance value, which may stem from the composites' dielectric properties being sensitively dependent on the microstructure. Samples with varying contents of functional phases possess distinct internal microstructures; in the case of the 15 wt% sample, the lower impedance is likely due to the formation of a more interconnected percolation network within [34].

In essence, the combined observations from Figure 5a,b illustrate a percolating phenomenon. As the CP content within the CP/PVDF composites transitions beyond the percolation threshold, a distinct shift in the electrical response is observed. The composites exhibit a transition from capacitive to inductive character, a hallmark response associated with materials exhibiting negative permittivity [35].

2.4. Negative-Permittivity Analysis

As illustrated in Figure 6, the real part of relative permittivity exhibits a negative value when the nano carbon powder loading surpasses the percolation threshold. The absolute value of the negative permittivity increases with increasing CP content. In conjunction with the analysis of the reactance behavior of the material in Section 2.3, the negative-permittivity behavior is associated with the inductive characteristics exhibited by the composites. A further analysis of the negative-permittivity behavior exhibited by the composites revealed that the Drude model in free electron theory can be used to fit the trend of the negative permittivity with frequency. The Drude-model formula is as follows [36]:

$$\varepsilon_r'(\omega) = 1 - \frac{\omega_p^2}{\omega^2 + \omega_\tau^2} \tag{4}$$

$$\omega_p = \sqrt{\frac{n_{eff}e^2}{m_{eff}\varepsilon_0}} \tag{5}$$

where ω represents the angular frequency ($\omega = 2\pi f$), ω_p denotes the plasma frequency, and ω_τ signifies the collisional frequency, which is also the inverse of the relaxation time ($\omega_\tau = 1/\tau$). Additionally, n_{eff} refers to the effective concentration of free electrons, m_{eff} represents the effective mass of the electron, and ε_0 denotes the vacuum permittivity. Figure 6a presents the fitting results for the negative permittivity of the 17.5 wt% carbon-powder sample using the Drude model (green solid line). This fitting process yields plasma frequency (ω_p) and collision frequency (ω_τ) values of 2.58×10^8 rad/s and 6.8×10^5 rad/s, respectively. With a reliability factor as high as 0.998, free electron theory emerges as a powerful tool for understanding the negative permittivity in CP/PVDF composites. The observed negative permittivity can be understood within the framework of percolation theory and the formation of conductive networks. When the carbon-powder content surpasses the percolation threshold, the interconnectivity between nano carbon powder establishes a continuous three-dimensional conductive network. This network houses a significant population of free electrons. Under the influence of an external alternating electric field, these free electrons experience a restoring electrostatic force but also possess inertia, leading to reciprocating motion along the direction of the field. At a specific frequency, a resonance phenomenon known as plasma oscillation occurs within the free electron population. This collective oscillation manifests as a negative permittivity. Furthermore, Equations (4) and (5) elucidate a positive correlation between the absolute value of the real part of the relative dielectric constant and the concentration of free electrons. As demonstrated in Figure 6a, the absolute value of the real part increases with the rising content of nano carbon powder. This increase in the CP content is what leads to the augmentation in the concentration of free electrons. This relationship suggests that the negative permittivity of percolating composites can be precisely controlled by tailoring the content of the conductive phase in the composites.

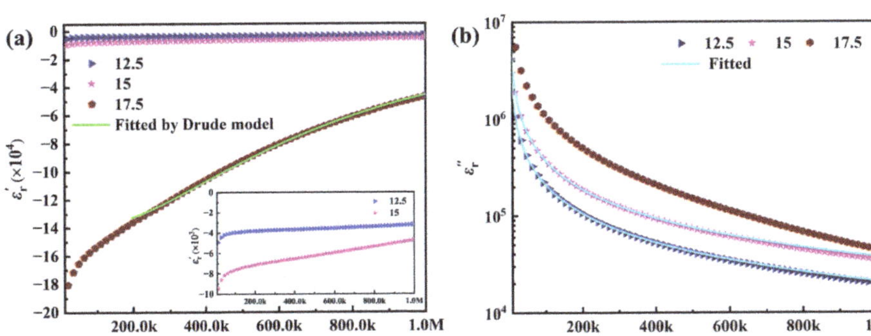

Figure 6. Frequency dependence of the complex permittivity for the CP/PVDF composites exceeding the *fc*: (**a**) real part and (**b**) imaginary part.

The imaginary part (ε_r'') of the relative permittivity is a key parameter for quantifying dielectric loss within materials. As illustrated in Figure 6b, for composites exceeding the percolation threshold for functional phase content, ε_r'' exhibits a dependence on frequency. This loss can be attributed to two primary mechanisms: polarization loss (ε_P''), arising from the interaction of the electric field with polarizable components; and conductive loss (ε_C'') associated with the movement of charges within the composite. The following equation expresses the relationship between these contributions and the total imaginary permittivity [37]:

$$\varepsilon_r'' = \varepsilon_C'' + \varepsilon_P'' \quad (6)$$

Furthermore, ε''_C can be expressed as follows:

$$\varepsilon''_C = \frac{\sigma_{dc}}{2\pi f \varepsilon_0} \tag{7}$$

where σ_{dc} denotes the DC conductivity. From Equation (7), the conductive loss of the composites is negatively correlated with the frequency, i.e., $\varepsilon''_C \propto 1/f$. As shown in Figure 6b, the imaginary component (ε''_r) of the permittivity for all three samples exhibits a decreasing trend with increasing frequency. This behavior is consistent with the dominance of conductive loss mechanisms within the composites. To quantify this behavior, Equation (7) was used to fit the ε''_r spectra of the samples containing 12.5 and 15 wt% carbon powder. The excellent agreement between the fitted curves (green solid lines) and the experimental data underscores the validity of this approach. This alignment suggests that internal conductive loss is the primary contributor to the overall dielectric loss in these composites with high carbon-powder loadings. As can be further observed from the figure, the imaginary part of the relative permittivity increases with increasing nano carbon-powder content, indicating that the loss of the composites increases with increasing functional phase content. This is because the increased content of the conductive functional phase leads to the expansion of conductive pathways, resulting in increased conductive loss.

Table 1 presents a comparison of the negative-permittivity properties of CP/PVDF composites with other polymer-based percolating composites using nano carbon materials as functional phases. The filler contents listed represent the amounts of functional phases just exceeding the percolation threshold, while the negative-permittivity values correspond to the dielectric constants measured at these concentrations. Research is mainly focused on the frequency range from kHz to MHz. Owing to the diverse morphologies of nano carbon materials, percolating composites utilizing different carbon materials exhibit considerable variations in their percolation threshold and negative permittivity. Notably, the percolation threshold in this study is higher compared to composites employing graphene and carbon nanotubes as functional phases. This phenomenon can be attributed to the morphology of the zero-dimensional nano carbon powder. Due to its particle-like morphology and small three-dimensional scale, a higher content of CP is required to form a conductive network in the PVDF matrix, resulting in a higher percolation threshold for the CP/PVDF composites. This indicates that precise control over negative-permittivity performance can be achieved by selecting different types of carbon materials or by managing the content of the functional phase.

Table 1. Comparison of negative-permittivity properties of the CP/PVDF composites with some polymer-based composites with nano carbon functional phase.

Composites	Filler Content	Test Frequency	Negative Permittivity	Ref.
Carbon nanotubes/epoxy	12 wt%	1–1000 MHz	~−100	[18]
Graphene/poly (vinyl alcohol)	10 wt%	10–1000 kHz	~−500	[15]
Graphene/polyaniline	2 wt%	100–1000 MHz	~−50	[34]
Graphene–carbon nanotubes/PVDF	14 wt%	100–1000 MHz	~−25	[5]
Acetylene black/polyimide	24 wt%	10–1000 kHz	~−350	[38]
Nano carbon powder/PVDF	12.5 wt%	10–1000 kHz	~−4000	This work

2.5. High Permittivity Analysis

PVDF exhibits a high dielectric constant ranging from 5 to 9, surpassing that of other polymer materials, making it a widely employed material for high-permittivity research [39]. Beyond the negative-permittivity properties, we also investigated the high-permittivity behavior of CP/PVDF composites with nano carbon-powder content below the percolation threshold. The dielectric properties of the CP/PVDF composites with carbon-powder contents of 10 wt% and lower are shown in Figure 7a. The dielectric constant of the PVDF material employed in this experiment is approximately 5. Upon the incorporation of nano carbon powder, the dielectric constant of the CP/PVDF composites is enhanced, demonstrating a continuous increase with increasing functional-phase content. This enhancement in the dielectric constant is attributed to the addition of nano carbon powder. Carbon powder particles and the PVDF matrix form micro capacitors, which in turn establish a micro capacitor network. The presence of micro capacitors and their network significantly elevate the dielectric constant of the CP/PVDF composites. This interfacial polarization phenomenon, often referred to as the Maxwell–Wagner effect, manifests distinct behaviors [40,41]. As the content increases, the distribution of nano carbon powder particles becomes more extensive, leading to a further augmentation of micro capacitors and the micro capacitor network. Consequently, the dielectric constant exhibits a continuous increase with the rising content of the functional phase.

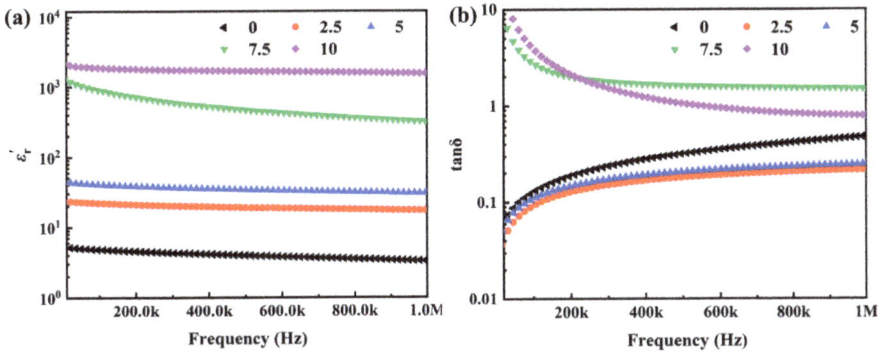

Figure 7. (**a**) Frequency dependence of the real part of the complex permittivity for CP/PVDF composites with CP content below percolation threshold. (**b**) The dielectric loss tangent (tanδ) of the CP/PVDF composites.

The CP/PVDF sample containing 10 wt% functional phase exhibits a remarkable permittivity of approximately 1589 at 800 kHz. This value represents a 450-fold increase compared to that of the pristine PVDF matrix (permittivity of 3.5). This significant increase in the permittivity can be attributed to the Maxwell–Wagner interfacial polarization effect. Due to their small dimensions and high specific surface area, the abundant nano carbon powder within the 10 wt% CP sample establish numerous interfacial capacitors with the surrounding PVDF matrix. These interfacial regions contribute significantly to the overall permittivity of the composites. However, this high permittivity comes at the expense of an elevated dielectric loss tangent of approximately 0.85. This loss likely originates from a combination of two mechanisms: the intrinsic polarization loss of the PVDF matrix and the conductive loss associated with the nano carbon powder [42]. It is noteworthy that the 5 wt% CP/PVDF composite displays a more favorable balance of permittivity and loss at 800 kHz. This sample exhibits a permittivity of 31.5, accompanied by a significantly lower dielectric loss tangent below 0.2. The reduced loss in this case can be ascribed to the lower content of the functional phase. With less nano carbon powder, the composite has a diminished interfacial area between the conductive phase and the PVDF matrix. This, in turn, leads to a decrease in conductive loss, allowing the intrinsic polarization loss of the PVDF substrate to become the dominant contributor to the overall dielectric loss.

3. Materials and Methods

3.1. Materials Preparation

Nano carbon powder (CP) with a purity greater than 99.5% and a diameter ranging from 30 to 40 nm was acquired from Aladdin Co. Ltd., Shanghai, China, with the product number C109965-5g. Polyvinylidene fluoride (PVDF) with a purity exceeding 99% was obtained from Shanghai Haoxi Nanotech, Co. Ltd., Shanghai, China. In this study, a hot-pressing technique was employed to fabricate CP/PVDF composites. Samples with seven different mass fractions of nano carbon powder were prepared, with the carbon-powder content increasing by 2.5% sequentially, labeled as 2.5, 5, 7.5, 10, 12.5, 15, and 17.5 wt%. The CP and PVDF powders were meticulously combined using a high-energy ball-milling apparatus to ensure homogenous dispersion of the CP within the PVDF matrix. The well-mixed powder blend was then transferred into a hot-pressing mold and subjected to a controlled heating and pressing cycle. The temperature was elevated to 160 °C to facilitate proper melting of the PVDF matrix, while a pressure of 15 MPa was applied for 15 min to promote densification of the composite and enhance interfacial bonding between the CP and PVDF phases. Following the hot-pressing cycle, the mold was cooled to room temperature at a controlled rate. The chosen mold has an inner diameter of 20 mm. Prior to hot pressing, an equal mass of mixed powder was weighed to ensure that the thickness of each sample remains essentially consistent. This process yielded disc-shaped samples with a diameter of approximately 20 mm and a thickness between 1 and 2 mm, thus meeting the test fixture requirements for a specimen diameter from 10 to 56 mm and a thickness of less than 10 mm.

3.2. Materials Characterization and Electrical Measurement

The Fourier-transform infrared (FTIR) measurements were conducted using a Nicolet 460 instrument from Thermo Fisher Scientific (Waltham, MA, USA). The Bruker D8 Advance diffractometer from Bruker Corporation (Billerica, MA, USA) was employed for X-ray diffraction (XRD) analysis to elucidate the phase composition of the composites. Raman spectroscopy was employed for further compositional analysis using a Renishaw RM2000 system which is from Renishaw plc (Gloucestershire, England, UK). Scanning electron microscopy (SEM) analysis was carried out on a Hitachi SU-70 instrument from Hitachi, Ltd. (Tokyo, Japan) to gain insights into the microstructure and morphology of the samples. The dielectric properties of the composites were evaluated using a high-precision Keysight E4980AL LCR meter equipped with a 16451B fixture (Santa Rosa, CA, USA). The fixture utilizes a parallel plate method to clamp the sample between two electrodes to form a capacitor. It is used in conjunction with the LCR meter to measure the capacitance; impedance; and other dielectric parameters, such as loss tangent. Measurements were conducted across a wide frequency range, spanning from 20 Hz to 1 MHz, employing a low-test voltage of 100 mV to minimize material distortions. This comprehensive approach enabled the direct acquisition of key electrical parameters, including capacitance (Cp), resistance (*R*p), impedance (Z), and loss tangent (tanδ). The dielectric loss angle (δ), also known as the dielectric phase angle, is the angle between the phase of the electric field and the phase of the dielectric displacement in a dielectric material subjected to an alternating electric field. It quantifies the ability of a material to convert electric field energy into heat. A larger dielectric loss angle indicates greater energy loss and a more lossy material. The dielectric loss tangent (tanδ) is the tangent of the dielectric loss angle and is a dimensionless quantity that also measures the material's loss. A higher dielectric loss tangent corresponds to a more lossy material. The dielectric loss tangent is calculated using the following formula:

$$\tan\delta = \frac{\varepsilon_r'}{\varepsilon_r''} \tag{8}$$

where ε_r' is the real part, and ε_r'' is the imaginary part of the relative dielectric constant.

Subsequently, the complex permittivity (the real part, ε_r'; and imaginary part, ε_r'') and AC conductivity (σ_{ac}) are calculated based on the obtained data. The calculation equations are as follows:

$$\varepsilon_r' = \frac{dC_p}{S\varepsilon_0} \tag{9}$$

$$\varepsilon_r'' = \frac{d}{2\pi f s R_p \varepsilon_0} \tag{10}$$

$$\sigma_{ac} = \frac{d}{R_p A} \quad (11)$$

The thickness of the sample is represented by the symbol d. S is the electrode area, f is the electric field frequency, and A is the electrode–sample contact area.

4. Conclusions

In this paper, CP/PVDF percolating composites were successfully prepared using the hot-pressing method, and a transition in the permittivity from positive to negative was observed. For carbon-powder contents of 12.5 wt% and higher, a transition to negative permittivity is observed in the composites. This behavior stems from the collective motion of free electrons within the established conductive network of the nano carbon powder, a phenomenon referred to as plasma oscillation. The AC conductivity analysis reveals that a percolation phenomenon occurred in the composites, which is also confirmed by microstructure characterization. The percolation threshold of the CP/PVDF composites is between 10 and 12.5 wt%. The negative permittivity in these composites can be effectively tailored by manipulating the carbon-powder content, highlighting the potential for controlling this property. Intriguingly, CP/PVDF composites with 5 wt% carbon powder achieve a remarkable balance between high permittivity and low dielectric loss. The development of CP/PVDF percolating composites with tunable negative permittivity represents a significant breakthrough. This paves the way for their use in diverse electromagnetic applications, including shielding, antennas, and beyond.

Author Contributions: Conceptualization, G.S.; investigation, G.S.; writing—original draft preparation, G.S.; writing—review and editing, X.S. and Y.L. All authors have read and agreed to the published version of the manuscript.

Funding: This study was funded by the National Natural Science Foundation of China (grant No. 52171141, No. 52272117, and No. 52207239), the Shandong Natural Science Foundation for Excellent Young Scholars (grant No. ZR2020YQ32), and the National Key Research and Development Program of China (2022YFB3505104 and 2022YFB3706604).

Institutional Review Board Statement: Not applicable.

Informed Consent Statement: Not applicable.

Data Availability Statement: Data presented in this study are available upon request from the corresponding author.

Conflicts of Interest: The authors declare no conflicts of interest.

References

1. Xiao, S.; Wang, T.; Liu, T.; Zhou, C.; Jiang, X.; Zhang, J. Active metamaterials and metadevices: A review. *J. Phys. D Appl. Phys.* **2020**, *53*, 503002. [CrossRef]
2. Veselago, V.G. Electrodynamics of substances with simultaneously negative electrical and magnetic permeabilities. *Sov. Phys. Usp.* **1968**, *10*, 504–509. [CrossRef]
3. Smith, D.R.; Pendry, J.B.; Wiltshire, M.C. Metamaterials and negative refractive index. *Science* **2004**, *305*, 788–792. [CrossRef]
4. Swetha, P.; Vidyadharan, D.M.; Sridharan, K.; Binesh, M.; Swaminathan, S. Investigation of negative permittivity in $BaTiO_3$/TiN metacomposites at the radio frequency region. *Mater. Chem. Phys.* **2024**, *317*, 129156.
5. Yuan, J.; Zhou, Y.; Manshaii, F.; Wang, S.; Yin, J.; Li, D.; Wang, S.; Qu, Y. Ultraweakly low-dispersion epsilon-negative response of GR-CNT/PVDF ternary metacomposites. *Compos. Commun.* **2024**, *47*, 101877. [CrossRef]
6. Luo, H.; Qiu, J. Carbon nanotube/polyolefin elastomer metacomposites with adjustable radio-frequency negative permittivity and negative permeability. *Adv. Electron. Mater.* **2019**, *5*, 1900011. [CrossRef]
7. Dou, R.; Shao, Y.; Li, S.; Yin, B.; Yang, M. Structuring tri-continuous structure multiphase composites with ultralow conductive percolation threshold and excellent electromagnetic shielding effectiveness using simple melt mixing. *Polymer* **2016**, *83*, 34–39. [CrossRef]
8. Kasagi, T.; Yamamoto, S. Effect of particle shape on electrical conductivity and negative permittivity spectra of Cu granular composite materials. *J. Mater. Sci. Mater. Electron.* **2022**, *33*, 4974–4983. [CrossRef]

9. Kono, K.; Tsutaoka, T.; Kasagi, T.; Yamamoto, S. Electromagnetic properties of Au/$Fe_{53}Ni_{47}$ hybrid granular composite materials. *IEEE Trans. Magn.* **2018**, *55*, 1–4. [CrossRef]
10. Jiang, T.; Wang, Y.; Wu, X.; Gao, Y.; Tang, B.; Yang, K.; Sun, K.; Zhao, Y.; Li, W.; Yu, J. A study of preparation and properties of epoxy resin/carbon fiber/phenolic residual carbon composites with adjustable negative permittivity behavior. *Fuller. Nanotub. Carbon Nanostruct.* **2022**, *30*, 675–682. [CrossRef]
11. Wei, Z.; Wang, Z.; Fan, G.; Xu, C.; Shi, G.; Zhang, G.; Liu, Y.; Fan, R. Low-frequency plasmonic state and negative permittivity in copper/titanium dioxide percolating composites. *Ceram. Int.* **2021**, *47*, 2208–2213. [CrossRef]
12. Song, X.; Fan, G.; Liu, Y.; Fan, R. MOF-derived ZrO_2/C-polyvinylidene fluoride composite towards negative permittivity regulation mechanism. *J. Mater. Sci. Technol.* **2024**, *179*, 50–56. [CrossRef]
13. Qu, Y.; Lin, J.; Wu, J.; Wang, Z.; Sun, K.; Chen, M.; Dong, B.; Guo, Z.; Fan, R. Graphene–carbon black/$CaCu_3Ti_4O_{12}$ ternary metacomposites toward a tunable and weakly ε-negative property at the radio-frequency region. *J. Phys. Chem. C* **2020**, *124*, 23361–23367. [CrossRef]
14. Tang, X.; Zhang, Z.; Xie, P.; Hou, Q.; Liang, G.; Ni, Z.; Huang, Q.; Wu, H. Theoretically designed epsilon-negative materials: Ni nanoparticles encapsulated CNTs and their magnetic driven properties. *Chem. Eng. J.* **2024**, *488*, 150899. [CrossRef]
15. Tang, X.; Zhang, Z.; Zheng, K.; Wu, Y.; Chen, Z.; Wang, C.; Shi, Z. Low dielectric loss in poly (vinyl alcohol)/graphene metacomposite films with negative permittivity prepared by spin coating. *Polymer* **2023**, *281*, 126092. [CrossRef]
16. Leng, Z.; Wu, H.; Tang, X.; Li, Y.; Xin, Y.; Xie, P.; Li, G.; Yan, K.; Liu, C. Carbon nanotube/epoxy composites with low percolation threshold and negative dielectric constant. *J. Mater. Sci. Mater. Electron.* **2022**, *33*, 26015–26024. [CrossRef]
17. Choudhary, H.K.; Kumar, R.; Pawar, S.P.; Sundararaj, U.; Sahoo, B. Enhancing absorption dominated microwave shielding in Co@C–PVDF nanocomposites through improved magnetization and graphitization of the Co@C-nanoparticles. *Phys. Chem. Chem. Phys.* **2019**, *21*, 15595–15608. [CrossRef] [PubMed]
18. Liu, M.; Wu, H.; Wu, Y.; Xie, P.; Pashameah, R.A.; Abo-Dief, H.M.; El-Bahy, S.M.; Wei, Y.; Li, G.; Li, W. The weakly negative permittivity with low-frequency-dispersion behavior in percolative carbon nanotubes/epoxy nanocomposites at radio-frequency range. *Adv. Compos. Hybrid Mater.* **2022**, *5*, 2021–2030. [CrossRef]
19. Lee, M.; Chen, C.Y.; Wang, S.; Cha, S.N.; Park, Y.J.; Kim, J.M.; Chou, L.J.; Wang, Z.L. A hybrid piezoelectric structure for wearable nanogenerators. *Adv. Mater.* **2012**, *24*, 1759–1764. [CrossRef]
20. Martins, P.; Lopes, A.; Lanceros-Mendez, S. Electroactive phases of poly (vinylidene fluoride): Determination, processing and applications. *Prog. Polym. Sci.* **2014**, *39*, 683–706. [CrossRef]
21. Salimi, A.; Yousefi, A. FTIR studies of β-phase crystal formation in stretched PVDF films. *Polym. Test.* **2003**, *22*, 699–704. [CrossRef]
22. Indolia, A.P.; Gaur, M. Investigation of structural and thermal characteristics of PVDF/ZnO nanocomposites. *J. Therm. Anal. Calorim.* **2013**, *113*, 821–830. [CrossRef]
23. Rajan, A.S.; Sampath, S.; Shukla, A.K. An in situ carbon-grafted alkaline iron electrode for iron-based accumulators. *Energy Environ. Sci.* **2014**, *7*, 1110–1116. [CrossRef]
24. Choudhary, H.K.; Kumar, R.; Pawar, S.P.; Sundararaj, U.; Sahoo, B. Superiority of graphite coated metallic-nanoparticles over graphite coated insulating-nanoparticles for enhancing EMI shielding. *New J. Chem.* **2021**, *45*, 4592–4600. [CrossRef]
25. Shen, J.; Liu, J.; Gao, Y.; Li, X.; Zhang, L. Elucidating and tuning the strain-induced non-linear behavior of polymer nanocomposites: A detailed molecular dynamics simulation study. *Soft Matter* **2014**, *10*, 5099–5113. [CrossRef] [PubMed]
26. Maity, N.; Mandal, A.; Nandi, A.K. Interface engineering of ionic liquid integrated graphene in poly (vinylidene fluoride) matrix yielding magnificent improvement in mechanical, electrical and dielectric properties. *Polymer* **2015**, *65*, 154–167. [CrossRef]
27. Kobayashi, M.; Tashiro, K.; Tadokoro, H. Molecular vibrations of three crystal forms of poly (vinylidene fluoride). *Macromolecules* **1975**, *8*, 158–171. [CrossRef]
28. Guo, Y.; Ruan, K.; Yang, X.; Ma, T.; Kong, J.; Wu, N.; Zhang, J.; Gu, J.; Guo, Z. Constructing fully carbon-based fillers with a hierarchical structure to fabricate highly thermally conductive polyimide nanocomposites. *J. Mater. Chem. C* **2019**, *7*, 7035–7044. [CrossRef]
29. Kumar, R.; Choudhary, H.K.; Pawar, S.P.; Bose, S.; Sahoo, B. Carbon encapsulated nanoscale iron/iron-carbide/graphite particles for EMI shielding and microwave absorption. *Phys. Chem. Chem. Phys.* **2017**, *19*, 23268–23279. [CrossRef]
30. Shi, Z.; Fan, R.; Yan, K.; Sun, K.; Zhang, M.; Wang, C.; Liu, X.; Zhang, X. Preparation of iron networks hosted in porous alumina with tunable negative permittivity and permeability. *Adv. Funct. Mater.* **2013**, *23*, 4123–4132. [CrossRef]
31. Dyre, J.C.; Schrøder, T.B. Universality of AC conduction in disordered solids. *Rev. Mod. Phys.* **2000**, *72*, 873. [CrossRef]
32. Shi, G.; Qu, Y.; Fan, G.; Liu, Y. Tunable and weakly negative permittivity in CNTs-CBs/polystyrene metacomposites. *ECS J. Solid State Sci. Technol.* **2019**, *8*, N141. [CrossRef]
33. Chen, M.; Sun, K.; Wang, X.; Wang, Y. Tunable negative permittivity of cobalt and epoxy composites at 3 kHz∼ 1 MHz frequency regions. *ECS J. Solid State Sci. Technol.* **2021**, *10*, 123007. [CrossRef]
34. Qu, Y.; Peng, Q.; Zhou, Y.; Manshaii, F.; Wang, S.; Wang, K.; Xie, P.; Qi, X.; Sun, K. Fine-tunable ε'-negative response derived from low-frequency plasma oscillation in graphene/polyaniline metacomposites. *Compos. Commun.* **2023**, *44*, 101750. [CrossRef]
35. Ma, R.; Cheng, C.; Liu, Y.; Wang, J.; Zhou, J.; Hu, Z.; Cui, H.; Li, J.; Fan, R. Temperature dependence of negative permittivity behavior in graphene/alumina ceramic metacomposites. *J. Eur. Ceram. Soc.* **2024**, *44*, 3012–3019. [CrossRef]
36. Band, Y.; Kuzmenko, I.; Trippenbach, M. Negative refraction in isotropic achiral and chiral materials. *Phys. Rev. A* **2024**, *109*, 063514. [CrossRef]

37. Tian, J.; Fan, R.; Zhang, Z.; Li, Y.; Wu, H.; Yang, P.; Xie, P.; Duan, W.; Lee, C.-S. Flexible and biocompatible poly (vinyl alcohol)/multi-walled carbon nanotubes hydrogels with epsilon-near-zero properties. *J. Mater. Sci. Technol.* **2022**, *131*, 91–99. [CrossRef]
38. Lu, Y.; Yang, P.; Wang, Z.; Liu, B.; Sun, K.; Pan, M.; Fan, R. Negative permittivity of polyimide-based percolative composites with zero/one dimensional functional fillers. *Compos. Commun.* **2024**, *47*, 101878. [CrossRef]
39. Li, W.; Song, Z.; Zhong, J.; Qian, J.; Tan, Z.; Wu, X.; Chu, H.; Nie, W.; Ran, X. Multilayer-structured transparent MXene/PVDF film with excellent dielectric and energy storage performance. *J. Mater. Chem. C* **2019**, *7*, 10371–10378. [CrossRef]
40. Nan, C.-W.; Shen, Y.; Ma, J. Physical properties of composites near percolation. *Annu. Rev. Mater. Res.* **2010**, *40*, 131–151. [CrossRef]
41. Bie, X.; Yin, P.; Xu, W.; Shi, Z. Significantly enhanced permittivity and suppressed loss of P(VDF-HFP)-based composites by introducing PPy@BT-OH fillers. *Mater. Lett.* **2024**, *364*, 136368. [CrossRef]
42. Li, H.; Jiang, M.; Dong, L.; Xie, H.; Xiong, C. Particle size dependence of the dielectric properties of polyvinyledene fluoride/silver composites. *J. Macromol. Sci. Part B* **2013**, *52*, 1073–1081. [CrossRef]

molecules

MDPI

Review

Thermal Conductive Polymer Composites: Recent Progress and Applications

Jianfeng Tan and Yuan Zhang *

College of Intelligent Systems Science and Engineering, Hubei Minzu University, Enshi 445000, China
* Correspondence: zhangy@hbmzu.edu.cn

Abstract: As microelectronics technology advances towards miniaturization and higher integration, the imperative for developing high-performance thermal management materials has escalated. Thermal conductive polymer composites (TCPCs), which leverage the benefits of polymer matrices and the unique effects of nano-enhancers, are gaining focus as solutions to overheating due to their low density, ease of processing, and cost-effectiveness. However, these materials often face challenges such as thermal conductivities that are lower than expected, limiting their application in high-performance electronic devices. Despite these issues, TCPCs continue to demonstrate broad potential across various industrial sectors. This review comprehensively presents the progress in this field, detailing the mechanisms of thermal conductivity (TC) in these composites and discussing factors that influence thermal performance, such as the intrinsic properties of polymers, interfacial thermal resistance, and the thermal properties of fillers. Additionally, it categorizes and summarizes methods to enhance the TC of polymer composites. The review also highlights the applications of these materials in emerging areas such as flexible electronic devices, personal thermal management, and aerospace. Ultimately, by analyzing current challenges and opportunities, this review provides clear directions for future research and development.

Keywords: thermal conductive; polymer nanocomposites; advanced application

1. Introduction

In the context of the ongoing trend towards miniaturization and high integration in microelectronics technology, improvements in device performance and lifespan are increasingly dependent on the effective implementation of advanced thermal management systems [1–5]. With the expansion of 5G technology and the prevalence of high-power devices, effective thermal conduction and dissipation have become key to ensuring stable device operation and extending their service lives, thus significantly highlighting the urgency of developing high-performance thermal management materials (TMMs) [6–12]. Moreover, thermal interface materials (TIMs), a critical subset of TMMs, are essential for managing heat in contemporary electronic devices by facilitating effective heat transfer and preserving electrical insulation [13–15]. The rapid development of technology has driven the demand for innovative high-performance materials, especially in high-temperature operating environments such as server rooms, electric vehicles, and aviation electronics, where higher demands are placed on TMMs [16]. Thermal management issues are not only crucial to the efficiency of device performance but also directly impact the efficiency of energy use and the sustainable development of the environment. For instance, in electric vehicles, the battery management system must effectively control and dissipate heat generated during the charging and discharging processes, failing which may decrease battery efficiency, shorten its lifespan, and even pose safety risks [17–21]. The development of new thermally conductive materials is thus not only a technical requirement for enhancing device performance and stability but also a response to the societal needs for environmental protection and efficient use of energy. Additionally, with the emergence of wearable devices and flexible electronics, there is an acute increase in demand for composite materials that possess good mechanical properties and can effectively manage thermal energy [22–25]. These devices often operate in environments sensitive to temperature changes, requiring materials that not only exhibit good thermal conductivity (TC) but also sufficient flexibility and stretchability to accommodate device bending and stretching.

Thermal conductive polymer composites (TCPCs) have secured an important position in modern thermal management systems due to their light weight, low cost, and excellent electrical insulation properties [26–30]. These materials combine the ease of processing of polymer matrices with the unique physicochemical properties of nanoscale enhancers, particularly excelling in enhancing TC and mechanical stability. However, these composites also face certain non-negligible limitations in practical applications, primarily their inherent TC often being lower than that of metallic materials, which limits their use in applications requiring very high TC [31–33]. Although the thermal performance of TCPCs can be enhanced by adding conductive fillers such as carbon nanotubes and graphene, this addition tends to increase the material's interfacial thermal resistance and decrease its overall mechanical strength [34,35]. To overcome these challenges, researchers have developed various strategies, including optimizing the type, morphology, distribution, and surface treatment of fillers, to improve the thermal performance of polymer composites [36–40]. The implementation of these strategies has led to certain advancements; for instance, improving the compatibility between nanofillers and polymer matrices can significantly reduce interfacial thermal resistance, thereby enhancing the overall TC of the composites [41–45]. Additionally, modern techniques also allow for precise control over the dispersion and directional alignment of nanofillers, further enhancing the composite's TC [46–48]. Despite ongoing challenges in enhancing thermal performance, the broad application prospects of TCPCs in sectors such as electronics, personal thermal management, and aerospace remain promising [49–52]. These materials not only meet the needs of modern electronic devices for efficient thermal management but are also considered key to enhancing the overall performance of devices due to their excellent electrical insulation and mechanical properties. Therefore, despite the challenges, the research and development of TCPCs continue to be a hot topic in the field of materials science.

This review article aims to comprehensively assess the latest advancements in TCPCs, particularly their potential applications in modern high-tech sectors. The article first details the thermal conduction mechanisms of TCPCs, exploring various factors that influence their thermal performance, including but not limited to the intrinsic properties of the polymers, interfacial thermal resistance, and the thermal properties of the fillers. Additionally, the article summarizes the main methods currently employed to enhance the TC of these materials and discusses specific application cases in emerging sectors such as personal thermal management and aerospace. While extensive research has been conducted on the preparation, performance enhancement, and mechanistic analysis of TCPCs, this paper additionally seeks to address the gaps related to their practical application details and requirements. Through an in-depth analysis of the current challenges and opportunities, this review seeks to provide clear directions and insights for future research and development of TCPCs. It is hoped that this systematic analysis will assist materials scientists and engineers in selecting and designing new thermal conductive materials, providing scientific foundations and innovative ideas, while also helping readers to understand the application potential and practical needs more fully in various fields.

2. Thermal Conductive Mechanism

2.1. Intrinsic Thermal Conductive Polymers

In the realm of solid materials, thermal conduction primarily operates via electrons and phonons, with the dominant mechanism depending on the material type. In metals, the high mobility of free electrons not only facilitates efficient electrical current transfer but also rapid heat dissipation. This phenomenon is explained by models such as the Drude model, where the movement and scattering of electrons determine the efficiency of thermal conduction [53–55]. Unlike metals, the thermal conduction mechanism in polymers predominantly relies on phonons, which are vibrations of the molecular structure (Figure 1) [56–58]. Due to the near absence of free electrons in polymers, heat transfer is significantly influenced by the tangled nature of molecular chains, the structure of long-chain molecules, and the presence of micro-defects. In amorphous polymers, phonon transport faces increased scattering due to the lack of long-range ordered structures, resulting in lower thermal conductivity. However, in crystalline polymers, while the ordered lattice structure can enhance phonon transport efficiency, the presence of irregular structures and defects within the molecular chains, such as chain folding, grain boundaries, and microvoids, continues to cause phonon scattering and affect thermal conduction [59–61].

Given the inherently low TC of polymers, material scientists have developed various methods to enhance their thermal properties, including the development of TCPCs and optimization of their molecular structures through design. These innovations not only enhance the thermal performance of polymers but also expand their applications in advanced manufacturing, energy transmission, and cooling of electronic devices [62–65]. Through a deeper understanding of the intrinsic thermal properties of polymers and technological innovation, the limitations in specific applications can be effectively addressed, thereby expanding their use in modern technology.

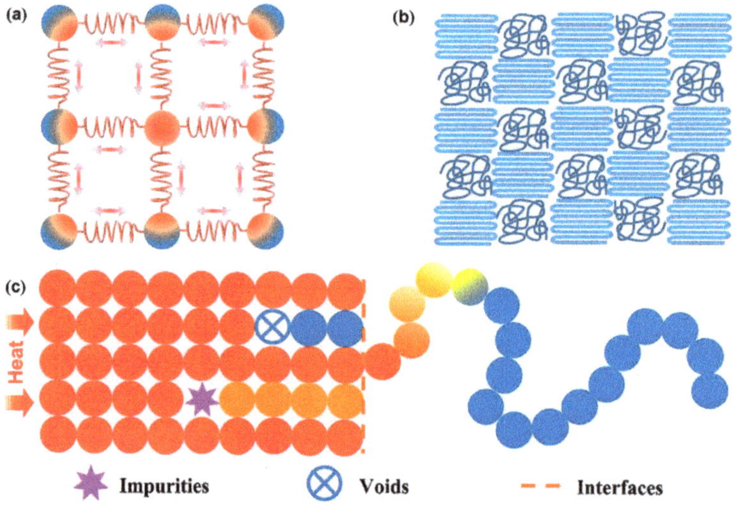

Figure 1. (**a**) Illustration of molecular chains or atoms vibration by mass and spring system; (**b**) crystalline and amorphous regions in polymers; (**c**) factors causing phonon scattering in polymers [66].

2.2. Filled Thermal Conductive Polymer Composites

2.2.1. Thermal Conductive Path Theory

In TCPCs enhanced with fillers, thermal conduction pathways are established through the incorporation of thermal conductive fillers into the polymer matrix [66–68]. The distribution of these fillers and their interfacial characteristics with the matrix are critical factors affecting the thermal performance of the composites. At low filler concentrations, as shown in Figure 2a, the conductive particles remain dispersed, isolated by the matrix, which contributes to considerable interfacial thermal resistance and keeps the composite's TC relatively low [69–71]. As the filler concentration increases, as depicted in Figure 2b, particles begin to make direct contact, gradually forming a continuous thermal network. the conductive particles remain dispersed, isolated by the matrix, which contributes to considerable interfacial thermal resistance and keeps the composite's TC relatively low [71,

72]. However, if the thermal network does not form along the direction of heat flow, high thermal resistance persists, limiting improvements in thermal conductivity.

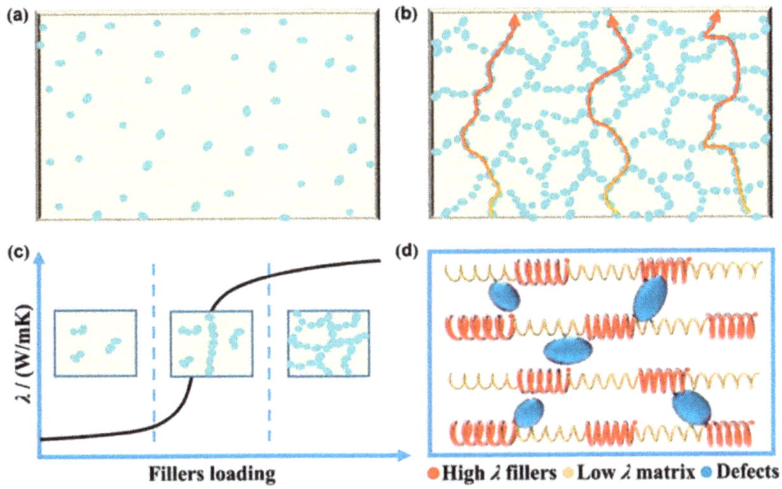

Figure 2. (**a**) "Sea-island" in low fillers loading; (**b**) thermal conduction paths in high fillers loading; (**c**) percolation phenomenon; (**d**) thermoelastic coefficient theory [66].

Additionally, at lower filler content, the common "sea–island" structure isolates the fillers within the polymer matrix, creating multiple isolated conductive islands. In such structures, due to the high interfacial thermal resistance, phonon transmission primarily occurs through these interfaces, resulting in limited enhancement of TC [73–75]. As the filler content increases, direct contact between filler particles becomes more frequent, gradually forming effective thermal conduction paths that allow heat to flow along paths of lower thermal resistance, significantly enhancing the composite's thermal conductivity. The development and refinement of these pathways are critical to advancing high-performance TCPCs. Consequently, optimizing filler distribution and minimizing interfacial thermal resistance are essential strategies for significantly boosting the thermal performance of polymer composites, addressing the growing needs for adept thermal management.

2.2.2. Thermal Conductive Percolation Theory

Percolation theory, originally used to explain the electrical conductivity phenomena in conductive composites, has been extended to analyze the thermal conduction behavior in TCPCs. This theory posits that when the filler content reaches a certain critical threshold, the properties of the composite undergo significant changes due to the formation of a continuous network by the filler particles, causing abrupt changes in thermal conductivity, as shown in Figure 2c [76–79]. At lower filler contents, with fillers dispersed uniformly within the polymer matrix and not forming a network, the "sea–island" structure is formed, and TC increases only gradually with filler content. However, surpassing the percolation threshold causes the fillers to configure into a "sea–sea" structure, triggering a marked enhancement in thermal conductivity. Nonetheless, in thermal conductive composites, the percolation point may not always be clearly discernible, and the variation in TC (λ value) might not exhibit a sudden spike. This phenomenon can be attributed to substantial phonon scattering at the interfaces between the fillers and the polymer matrix, which introduces notable interfacial thermal resistance.

Although percolation theory is somewhat applicable in explaining the thermal behavior of filled TCPCs, it has limitations in realistically interpreting the λ value changes of these materials. Near the percolation transition point, the increase in TC tends to be more gradual compared to the often-dramatic rise in electrical conductivity observed in these materials. Notably, significant thermal percolation effects are typically evident only with the incorporation of high-TC fillers. In such scenarios, the thermal resistance between the fillers is markedly lower than the interfacial thermal resistance between the fillers and the polymer matrix. Consequently, the formation of a continuous network of these fillers substantially boosts thermal conductivity. However, the complexities of phonon scattering and interfacial thermal resistance mean that percolation theory's explanatory power and predictive accuracy for practical applications remain limited. However, due to the complexities of phonon scattering and interfacial thermal resistance, the explanatory power and predictive accuracy of percolation theory in practical applications require further research and validation.

2.2.3. Thermoelastic Coefficient Theory

Thermoelastic coefficient theory views filled TCPCs as a macroscopic whole, emphasizing that their TC is influenced by macroscopic properties such as elastic coefficients and modulus changes, as depicted in Figure 2d. This theory posits that by enhancing the thermoelastic coefficient in the material, i.e., optimizing phonon transmission efficiency, TC can be significantly improved. Specifically, by incorporating high-TC fillers like carbon nanotubes or graphene, an effective thermal conduction network is formed. This setup significantly reduces phonon scattering and interfacial thermal resistance, thereby substantially enhancing the composite's thermal conductivity. However, increases in TC are usually not abrupt, as phonon transmission is still subject to the base properties and interfacial characteristics of the matrix. Consequently, enhancing the thermal performance of polymer composites hinges on optimizing the types, concentrations, and spatial distribution of fillers, alongside refining the interfacial interactions between the fillers and the matrix [80,81].

3. Key Factors Influencing the TC of Polymer Composite Materials

3.1. Key Factors Affecting the TC of Polymer Composites

3.1.1. Intrinsic TC of the Polymer

Understanding the crystalline and amorphous states in polymers is crucial for grasping their TC [82–84]. In polymers, macromolecular segments typically consist of many molecular chains that form through entanglement and folding, containing both crystalline and amorphous regions. The crystalline regions, where molecular chains are uniformly oriented and tightly packed, exhibit clear periodicity. This ordered structure allows for efficient phonon transmission, thus enhancing the material's thermal conductivity. Conversely, in the amorphous regions, the arrangement of molecular chains is more irregular, leading to structural defects and an increase in interfaces, which in turn increase phonon scattering and reduce the TC of the polymer bulk. The thermal performance of polymers is also significantly influenced by their molecular structure, including factors such as the degree of crystallinity, orientation, polarization of polar groups, number of atoms per unit volume, bond strength, molecular chain structure (including main chain bonds, side chains, and inter-chain coupling), and cross-linking density (Figure 3) [85,86]. For example, mechanical stretching can transform the amorphous structure of polymer chains into a crystalline structure with uniform orientation, which not only reduces phonon scattering but also effectively enhances TC through a straight, ordered polymer backbone and periodic structure of the molecules. In practical applications, research by Tu et al. employed a dual chemical–physical cross-linking and stretching orientation strategy to prepare highly thermally conductive, mechanically robust cellulose/hydroxylated boron nitride (BN) nanosheet composite films [87]. Furthermore, the number and polarization degree of polar groups in polymers directly impact TC [88]. The more polar groups present and the higher their degree of polarization, the higher the TC of the polymer.

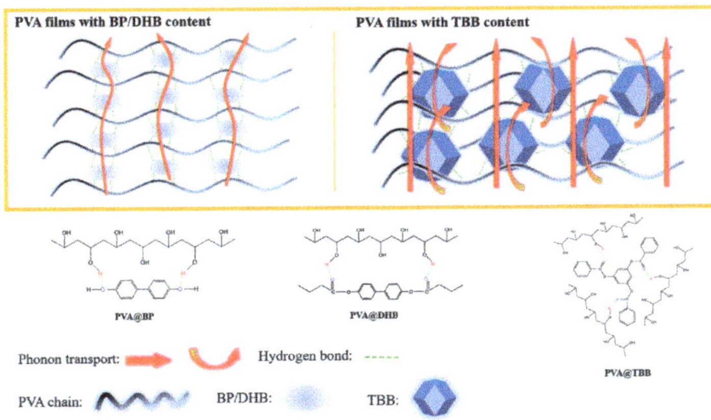

Figure 3. Microscopic ordered structures and the corresponding phonon transport model of polyvinyl alcohol (PVA) composite films [86].

In strategies to enhance the intrinsic TC of polymers, both the nanotemplating method and electrospinning have shown their unique effectiveness [89–91]. The nanotemplating method involves melting the polymer material and infiltrating it into a porous template, which orders the molecular chains as they flow into the nanoscale pores, thereby increasing the TC of the nanofibers. On the other hand, the electrospinning technique uses electrostatic forces to draw a polymer solution from the tip of an injection needle, forming polymer nanofibers where most molecular chains are aligned along the fiber axis. This orderly arrangement significantly enhances the TC along the fiber direction. By altering the structure of polymer molecules and chains, special physical structures can be obtained through synthesis and molding processes, thereby improving their thermal conductivity. Additionally, owing to their high aspect ratio, nanofibers provide a larger surface area, shorter thermal conduction paths, and increased interfacial heat transfer areas. These methods not only include the aforementioned high-stretching, electrospinning, and templating methods but also involve tuning the physical properties of polymers by optimizing molecular weight and the number of chains ends to achieve higher thermal conductivity. This series of research and development efforts continually pushes the application of polymer materials in the field of thermal conductivity, especially in the widespread use within electronic packaging domains.

3.1.2. Filler Properties

In the development of polymer composite materials, the selection, morphology, and spatial distribution of thermal conductive fillers within the matrix are pivotal factors that dictate the thermal performance of the composites [92–96]. These fillers are classified into carbon-based, inorganic, metallic, and specially functionalized hybrid fillers, each contributing uniquely to the composites due to their distinct physical and chemical properties. Carbon-based fillers are frequently used in high-TC composites that require electromagnetic shielding due to their excellent thermal and electrical properties. Inorganic fillers like BN and aluminum oxide are suited for applications requiring high thermal stability and good insulation properties. Metallic fillers, including silver (Ag), copper (Cu), and aluminum (Al), are extensively used in efficient thermal management systems for their extremely high thermal conductivity, although they are prone to oxidation and have higher thermal expansion coefficients at elevated temperatures.

The morphology of fillers—including zero-dimensional nanoparticles, one-dimensional nanowires, two-dimensional layers, and three-dimensional networks—significantly impacts thermal conductivity. Zero-dimensional fillers typically struggle to form effective thermal conduction paths due to their point-like structure. In contrast, one-dimensional and two-dimensional fillers, with their longer thermal transfer paths and larger surface contact areas, are more effective at constructing thermal networks and reducing thermal contact resistance. Three-dimensional fillers can form continuous thermal conduction channels in multiple directions, greatly reducing thermal resistance. The size of the fillers is also critically important. Although smaller-sized fillers can increase the interface area and phonon scattering sites, overly small sizes may lead to uneven distribution and aggregation within the polymer matrix, creating thermal bridging effects that impact overall thermal conductivity. Larger-sized fillers aid in forming continuous thermal conduction paths, but their distribution uniformity and interface quality within the polymer are equally crucial. More importantly, the thermal expansion coefficients and the impact on the conductivity due to the expansions in different fillers vary on the basis of the dimensionality.

Furthermore, the orientation of fillers significantly influences the TC of the composites. Fillers oriented by electric fields, magnetic fields, or mechanical forces can form efficient thermal conduction channels in specific directions. For example, Li et al. achieved high TC in epoxy composites by orienting carbon fibers under an applied stress field and curing them in situ in an epoxy resin. With a carbon fiber content of 46%, these composites achieved a thermal conductivity 171 times higher than that of pure epoxy resin [95]. Similarly, Shang et al. used an alternating vacuum filtration method to produce cellulose nanofiber/BN nanosheet/$Ti_3C_2T_x$ MXene films. With a 70 wt% BN nanosheet content, the films exhibited an in-plane TC of 19.97 $W{\cdot}m^{-1}{\cdot}K^{-1}$, underlining their potential for use in highly integrated electronic devices [96]. In the fields of microelectronics and high-performance materials, comprehensive consideration of the choice, morphology, size, surface treatment, and orientation of fillers is essential for designing and manufacturing polymer composites.

3.1.3. Interfacial Thermal Resistance

Interfacial thermal resistance significantly impedes the transfer of thermal energy. In polymer composites, the transfer of heat across interfaces encounters obstacles, often causing a sharp drop in temperature at these interfaces. This phenomenon is primarily due to the vibrational, acoustic, and modulus mismatch during phonon transmission, which causes severe scattering of phonons, significantly reducing their mean free path and thus decreasing the λ of the composites. Poor thermal contact between the filler and the matrix is a core reason for high interfacial thermal resistance; therefore, improving the thermal transfer performance of these microscopic interfaces is a key strategy [43].

3.2. Strategies to Enhance the TC of Polymer Composite Materials

In enhancing the thermal performance of composite materials, scientists have primarily adopted two strategies [97–108]. Firstly, researchers have functionalized the interfaces by adding functional groups to the filler surfaces through chemical or physical means. Secondly, by precisely controlling the microscopic structural parameters such as the size, shape, dispersity, and orientation of the fillers, more optimized thermal networks can be constructed. Research on strategies to enhance the TC of polymer composite materials is summarized in Table 1. These strategies not only enhance the functionality of the composites but also provide key technical support for designing efficient thermal management systems.

Table 1. Strategies to enhance the TC of polymer composite materials.

Strategies	Polymer Matrix	Fillers	Enhancement in k (%)	k ($W\ m^{-1}\ K^{-1}$)	Ref.
Interfacial Functionalization	PI	Graphene oxide	820	7.13	[99]
	PVA	BNNS	30.6	1.909	[100]
	Natural rubber	CF	71.3	0.218	[101]
	TPU	BN and CNT	210	4.52	[102]
	Epoxy natural rubber	Al_2O_3 and BN	370	0.5147	[103]

Table 1. *Cont.*

Strategies	Polymer Matrix	Fillers	Enhancement in *k* (%)	*k* ($W m^{-1} K^{-1}$)	Ref.
Microstructural Design	Cyclic butylene terephthalate	Graphene nanoplatelets	1518.8	8.094	[78]
	Cellulose	BNNS	3000	20.41	[87]
	PDMS	C@BNNS	1600	4.09	[89]
	PI	BNNS	4773	13.1	[91]
	Epoxy resin	CF	17100	32.6	[95]
	Cellulose	BNNS/$Ti_3C_2T_x$ MXene	/	19.97	[96]
	Polyurethane	BN	85.0	39.0	[105]
	Poly(p-phenylene-2,6-benzobisoxazole)	MXene	164	42.2	[107]

3.2.1. Interfacial Functionalization Reduces Interfacial Thermal Resistance

Interface functionalization involves the introduction of functional groups or layers onto the filler surface through chemical or physical methods, improving the compatibility of and interfacial bonding strength between the filler and the matrix [98–102]. This approach effectively reduces phonon scattering at the interface, enhancing phonon transmission and thereby lowering interfacial thermal resistance. From a microscopic mechanism perspective, interface functionalization typically includes covalent and non-covalent functionalization. Covalent functionalization involves forming stable chemical bonds on the filler surface, which also helps improve the dispersity of fillers and reduce agglomeration. However, it is important to note that excessive chemical modification may damage the surface structure of the fillers. Non-covalent functionalization utilizes physical adsorption or weaker chemical bonds, such as hydrogen bonds, to maintain the structural integrity of the fillers. Chen et al. employed one-dimensional thin-walled carbon nanotube (1D-TWCNT) and two-dimensional boron nitride nanosheet (2D-BNNS) fillers to enhance the thermal properties of PVA (Figure 4). A critical aspect in boosting PVA's thermal properties is the interfacial configuration strategy, which facilitates phonon transport and minimizes the intrinsic thermal property loss of the filler nanomaterials. The inclusion of borax enhanced the TC of the 1D-TWCNT/PVA nanocomposites by up to 14.5% with 4 wt.% borax and the 2D-BNNS/PVA nanocomposites by up to 30.6% with 2 wt.% borax [100].

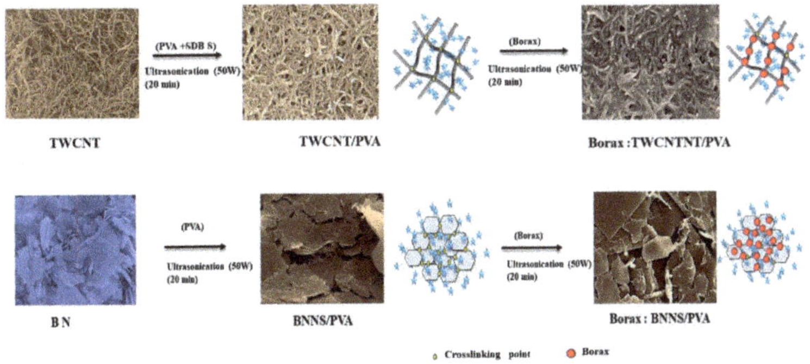

Figure 4. Enhancement mechanism through reinforced crosslinking and formation of thermal transport network by adding borax [100].

Chen et al. innovatively developed a composite material by incorporating crushed carbon fibers into natural rubber. The surface of the carbon fibers was first modified with dopamine, which self-polymerized to provide an initial coating. Subsequently, a silane coupling agent (SCA), specifically KH590, was grafted onto this platform to enhance interfacial bonding between the filler and the matrix. Testing revealed that the modifications with KH590 resulted in TC and tensile strength increases of 71.3% and 73.3%, respectively, when compared to the unmodified rubber, thus marking substantial enhancements in material performance [101]. Bashir et al. advanced composite materials by hydroxylating hexagonal BN (h-BN) and carbon nanotubes (CNTs), subsequently creating a polyurethane crosslink between the interfaces of the fillers and the thermoplastic polyurethane (TPU) matrix. The synergistic effect of the mixed fillers, comprising only 60 wt% h-BN and 2 wt% CNT, yielded a remarkable TC of 4.52 $W \cdot m^{-1} \cdot K^{-1}$ [102]. In summary, interface functionalization techniques, by optimizing the microstructure and interfacial performance within the composites, significantly reduce interfacial thermal resistance, greatly enhancing the TC of the composites. These techniques not only provide new avenues for the development of high-performance composite materials but also hold significant practical application value in demanding fields such as electronic devices and energy systems. As materials science continues to advance, these techniques are expected to demonstrate their unique advantages in a broader range of applications in the future.

3.2.2. Microstructural Design Governs the Formation of Thermal Conductive Pathways

Scientists have developed various microstructural control strategies to construct more optimized thermal networks [103–110]. The size and morphology of fillers are crucial for forming effective thermal pathways. Smaller-sized nanofillers can form denser networks within the polymer matrix, thus enhancing thermal conductivity. For instance, carbon nanotubes and graphene, with their unique one-dimensional and two-dimensional structures, can form continuous thermal networks across the composite material (Figure 5). Additionally, fillers of different morphologies, such as spherical, flake-like, or fibrous, also affect their orientation and dispersity within the matrix, further influencing the thermal conductivity.

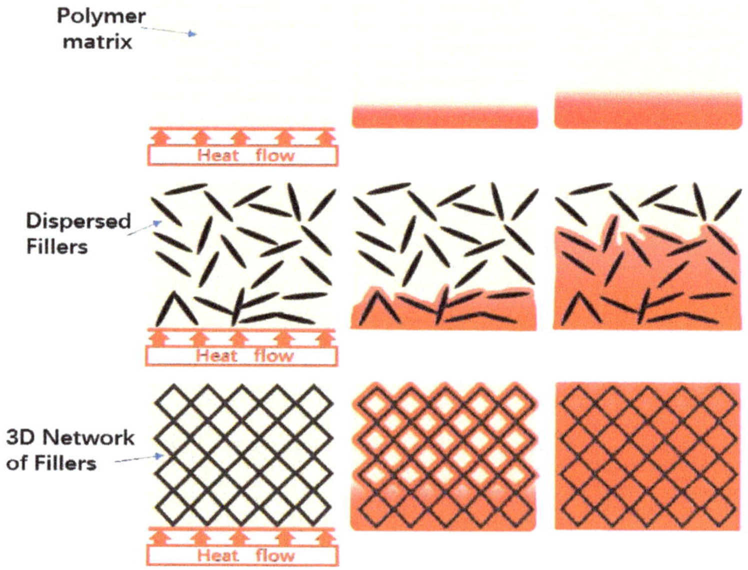

Figure 5. Schematic diagrams of heat transport in a pure polymer, a polymer composite filled with a traditional dispersed filler, and a composite filled with a 3D interconnected network of fillers [106].

Uniform dispersion of fillers is another key factor in achieving high thermal conductivity. Unevenly dispersed fillers can cause interruptions in the thermal conduction path, thereby reducing thermal efficiency. Techniques such as high shear mixing and ultrasonic dispersion can effectively improve the uniform distribution of fillers in the polymer matrix.

Moreover, the orientation of fillers significantly impacts TC performance. For example, aligning long fibers or flake-like fillers in a specific direction can create more effective thermal conduction paths. Additionally, mixed filler strategies employ various types, shapes, and sizes of thermal conductive fillers, orchestrated through tailored microstructural designs to establish effective thermal conduction pathways. Research shows that mixed filler systems can achieve high TC at lower total filler loadings, significantly outperforming traditional single-filler systems. At the microscale, by precisely controlling the size, morphology, dispersity, and orientation of fillers, a continuous and effective thermal network can be constructed. Liu et al. have synthesized a high-performance poly(p-phenylene-2,6-benzobisoxazole) (PBO)/MXene nanocomposite film using a sol–gel film conversion method (Figure 6). The resultant film benefits from an optimized brick and mortar structure, enhanced by the strong bridging and caging effects of the fine PBO nanofiber network on the MXene nanosheets. Critically, the synergistic orientation of the PBO nanofibers and MXene nanosheets imparts an in-plane TC of 42.2 $W \cdot m^{-1} \cdot K^{-1}$ to the film [107]. This requires meticulous design and optimization of the material's microstructure, such as predicting the impact of internal structures on TC through computer simulations and utilizing advanced microscopy techniques for precise measurement and analysis of material structures [108].

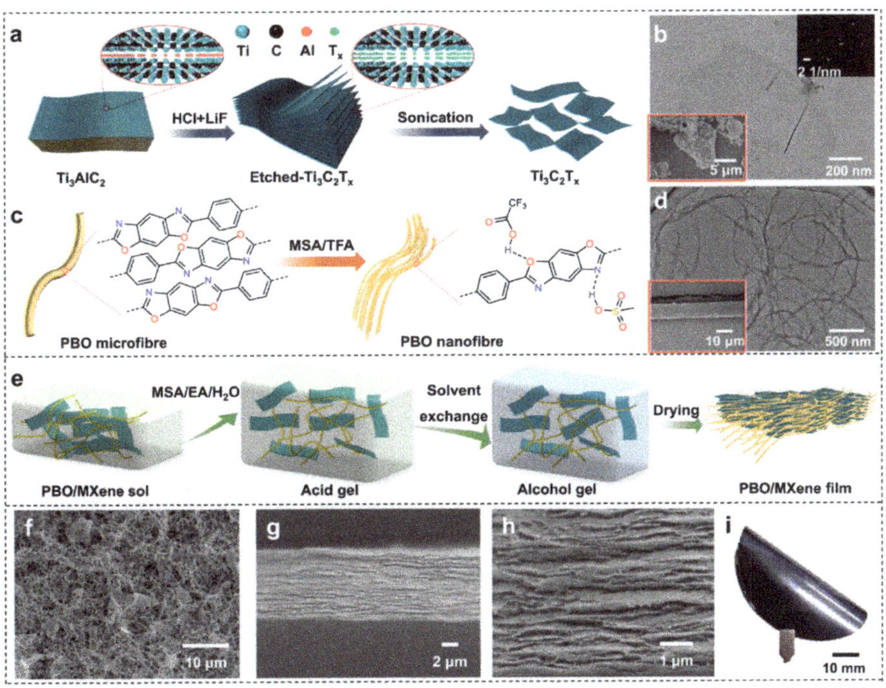

Figure 6. (**a**) Illustration for $Ti_3C_2T_x$ MXene prepared from Ti_3AlC_2 MAX precursor by HCl/LiF etching. (**b**) TEM image of MXene nanosheets (insets: electron diffraction pattern of MXene and SEM image of Ti_3AlC_2 MAX). (**c**) Illustration for PBO nanofibres prepared from commercial PBO fibres by MSA/TFA exfoliation. (**d**) TEM image of PBO nanofibres (inset: SEM image of PBO fibre). (**e**) Illustration for the nacre-inspired PBO/MXene films prepared by the sol-gel-film conversion approach including a proton-consumption homogeneous gelation process. (**f–h**) SEM images of (**f**) the freeze-dried PBO/MXene gel and (**g**,**h**) the cross-section of PBO/MXene film. (**i**) The optical photograph of PBO/MXene film [107].

In practical applications, the processing methods for TCPCs are also crucial. Common processing methods include templating, melt extrusion, injection molding, and hot pressing [109–111]. These methods need to consider the thermal stability and rheological properties of the materials and ensure uniform dispersion and appropriate orientation of fillers during processing to maintain the TC performance of the materials.

4. Applications of Thermal Conductive Polymer Composites

4.1. Electronic Devices (Wearable Electronics)

With technological advancements, the integration and power density of electronic devices, such as chips, are continually increasing, leading to serious issues with heat accumulation. TCPCs have become indispensable materials in modern electronic devices and systems due to their excellent performance in thermal management. TCPCs combine the high TC and electrical insulation properties of inorganic materials with the lightweight, easy processing, and low-cost advantages of organic polymers, making them widely used in underfill materials and TIMs [112–120]. Epoxy resin composites, due to their low cost and ease of processing, hold a significant position in electronic packaging. Xiao et al. proposed a simple, environmentally friendly method to prepare high-TC epoxy resin/BN nanosheet composites. Due to the interconnected network, this composite material exhibits excellent thermal performance in all directions, promising substantial applications in thermal management of electronic devices. Furthermore, the thermal stability of the composite material was significantly enhanced, with the T 20 increasing by 227.3 °C compared to pure epoxy resin [119]. Yan et al. prepared high-TC epoxy resin composites using 60 wt% of f-Al_2O_3, 3 wt% of MWCNTs, and 8 wt% of SiO_2 NPs. Through uniform dispersion and a synergistic mixing effect, various components were combined to construct an interconnected thermal network, achieving efficient thermal conduction. The composite material exhibited an out-of-plane TC of 1.73 $W{\cdot}m^{-1}{\cdot}K^{-1}$, suitable for thermal management components in electronic devices [120]. Liu et al., based on synthesized epoxy vitrimer and BN (BN) nanosheets, fabricated a fully recyclable high-TC thermal TIM. Simple hot pressing, with 40 wt% BN filling, increased the TC to 3.85 $W{\cdot}m^{-1}{\cdot}K^{-1}$, thirty times that of the original epoxy resin, significantly enhancing the heat dissipation of electronic devices. Epoxy/BN-40 wt% exhibited a high-onset thermal degradation temperature of approximately 280 °C, indicating excellent thermal stability [121].

Yan et al. proposed an efficient method for preparing ultra-high-aspect-ratio hexagonal BN nanosheets (BNNSs) based on microfluidic technology. Achieving a length-to-width ratio of approximately 1500 with minimal defects, BNNS composite films prepared with PVA exhibited an in-plane TC of 67.6 $W{\cdot}m^{-1}{\cdot}K^{-1}$ at 83 wt% BNNS content, a TC enhancement of about 35,500 times. Used as a heat sink for high-power LED modules, this composite film showed excellent heat dissipation efficiency [122]. Li et al. present an innovative interpenetrating architecture that integrates an electrocaloric polymer with highly thermally conductive pathways (Figure 7). This configuration achieves a 240% increase in electrocaloric performance and a 300% enhancement in the polymer's thermal conductivity. Utilizing this electrocaloric composite, we fabricated a scaled-up prototype device specifically designed for single heat spot cooling of a 5G chip, incorporating electromagnetic actuation [123].

Due to their compact designs, wearable devices are prone to heat accumulation under high-frequency operating conditions. Since these devices come into direct contact with human skin, their temperature should not significantly exceed that of the human body, necessitating the use of high-TC materials for packaging to ensure device cooling and safety [124–126]. Miniaturized, high-power-density wearable electronics demand advanced thermal management, excellent flexibility, and superior permeability. Balancing these requirements is challenging due to the inherent conflict between thermal conductivity and the properties of flexibility and permeability. Despite these challenges, significant progress has been made in managing the thermal properties of flexible wearable devices through the use of thermal conductive polymer nanocomposites. High-TC elastomer nanocomposites not only ensure rapid dissipation of heat generated within the device but also quickly respond to artificial thermal feedback, providing heating and cooling functions.

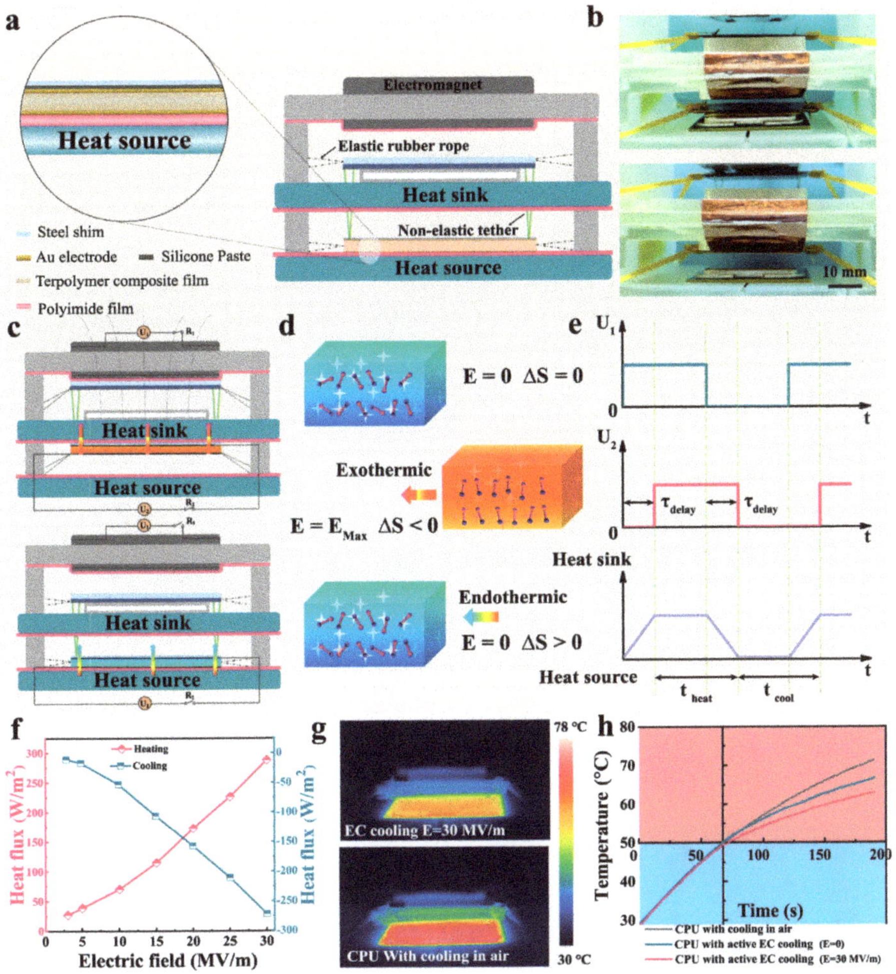

Figure 7. (**a**) Schematic illustration of the electrocaloric polymer stack and solid-state cooling device. (**b**) Photograph of the active EC device; the elaborated main framework was obtained by 3-D printing. (**c**) The schematic shows how an electromagnetic field drives an electrocaloric polymer stack to move heat from a heat source to a heat sink. The high-speed heat transfers from heat source to sink can be achieved by associating the active cooling of electrocaloric polymer stack with the heat dissipation cycle. (**d**) The working mechanism of the ECE based on the change in dipole entropy. (**e**) Time domain illustration of the cooling cycle. (**f**) The maximum heat flux of the electrocaloric stack on the heating and cooling side versus the applied electric field is measured by a heat flux sensor at an operation frequency of 0.1 Hz. (**g**) Infrared thermal images of CPU in active electrocaloric cooling. (**h**) Temperature versus time curves of CPU in air, by active electrocaloric cooling (U1 = 12 V at 1 Hz, U2/d2 = E = 0 MV m^{-1}; d2 is the thickness of the electrocaloric cooling stack) and active electrocaloric cooling (U1 = 12 V at 1 Hz, E = 30 MV m^{-1}) [123].

For example, Ma et al. employed emulsion electrospinning to create smart temperature-controlled nanofibers, incorporating lauric acid PCM, carbon nanotubes, and zinc oxide to endow them with excellent TC (0.665 W·m^{-1}·K^{-1}) and UV resistance. The polyvinyl alcohol base ensures flexibility, with a break elongation of 25%. A non-polluting polydimethylsiloxane coating provides hydrophobicity and self-cleaning properties, making it suitable for smart garments and protective fabrics [127]. Chen et al. reported a method for creating flexible, breathable composite materials by applying high-TCBN nanosheets BNNS onto a patterned electrospun TPU fiber mat grid. This composite material significantly enhanced TC while maintaining breathability. When integrated into flexible devices, its saturation operating temperature was significantly reduced, with surface temperature fluctuations less than 0.5 °C after more than 2000 bending–release cycles [128]. Tan et al. developed a stretchable strain sensor with exceptional thermal management performance. Its unique feature is a cast TPU-BN nanosheet layer, which not only enhances TC but also rapidly transfers heat to the environment, while the porous electrospun TPU film acts as a thermal insulator, reducing the real-time saturation temperature by 32%. [129]. Wu et al. developed high-TC regenerated cellulose/BNNS composite yarns through wet spinning and confinement-induced self-assembly, achieving a TC of 9.22 W·m^{-1}·K^{-1}, 4.5 times that of traditional materials. This composite yarn was woven into textiles and integrated with a thermoelectric generator (TEG) to effectively harvest body heat, continuously powering wearable devices [130]. Guo et al. introduced highly flexible composites consisting of aligned and overlapping interconnected boron nitride nanosheets (BNNSs) assembled in wrinkle structures (Figure 8). These composites exhibited a high in-plane thermal conductivity exceeding 26.58 W·m^{-1}·K^{-1} and enhanced through-plane conduction, improving with increased pre-strain. During an extensive bending test exceeding 3000 cycles, the maximum temperature fluctuation in the flexible device using a 100%-prestrained composite remained within 0.9 °C, which is less than one-third of that observed with a commercial thermal pad [131].

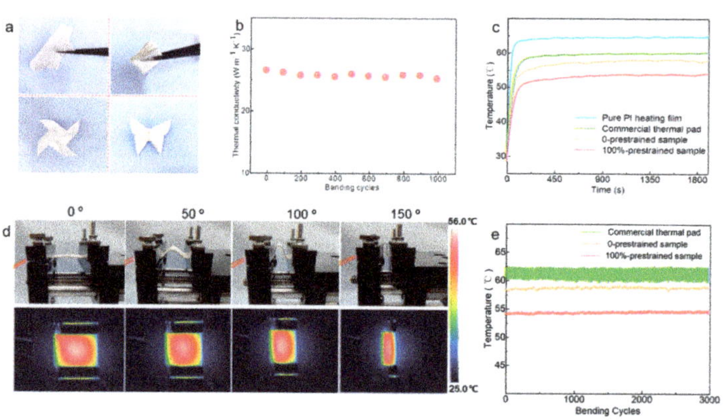

Figure 8. (**a**) The flexibility of the composite. (**b**) In-plane thermal conductivity of the 100%-prestrained composite in response to 1000 bending cycles. (**c**) Surface temperature variations of the flexible PI heating film against time. (**d**) Temperature fluctuation of the flexible PI heating film with the 100%-prestrained composite under a bending angle of 0−150°. (**e**) Temperature fluctuation of the flexible PI heating film with different TIMs under more than 3000 bending cycles [131].

Polyimide (PI) materials are recognized for their exceptional heat resistance and stability, making them highly suitable for applications in the electronics industry. Employing in situ polymerization, wet spinning, and subsequent heat drawing techniques, Fang et al. achieved uniform dispersion of nanofillers and covalent bonding between polydopamine (PDA)@BNNS and the PI matrix, with high orientation of the fillers. This process resulted in notable enhancements in thermal diffusion, while a 10 wt% nanofiller addition achieved a TC of 3.44 W·m^{-1}·K^{-1}. The filament containing 0.5 wt% PDA@BNNS exhibits a maximum tensile strength of 2.9 GPa. The good dispersion of the filler and the strong interfacial interactions between PDA@BNNS and the PI matrix, induced by chemical bonds, are expected to play a significant role in enhancing the mechanical properties and thermal conductivity of the composite fibers [132]. In another study, Yoon et al. focused on improving the TC of PI by enhancing chain orientation and intermolecular interactions. Their research demonstrated that the thermal conductivities of a single electrospun PI nanofiber (INF-PI) and a hot-pressed nanofiber PI film (HP-PI) were 0.44 and 0.98 $W \cdot m^{-1} \cdot K^{-1}$ at room temperature, respectively. The electrospinning process was found to improve chain orientation, while the high pressure and heat treatment during hot pressing facilitated the development of π–π interactions, significantly increasing the TC of the HP-PI. Even after 3000 cycles, no cracks or delamination between the alternately stacked layers were observed in the film, demonstrating the exceptional flexibility and durability of the synthesized material [133]. Additionally, Ruan et al. developed a novel high-TC liquid crystal PI (LC-PI) film by optimizing the liquid crystal range to align with the curing temperature. The in-plane and through-plane thermal conductivities of the LC-PI IV film at room temperature were 2.11 $W \cdot m^{-1} \cdot K^{-1}$ and 0.32 $W \cdot m^{-1} \cdot K^{-1}$, respectively, which are significantly higher than those of films cured outside the liquid crystal range. Simultaneously, the LC-PI film exhibited excellent mechanical and thermal properties, indicating its potential for application in high-temperature flexible electronics [134].

Room temperature fast-self-healing structures and functions of composite materials have broad application prospects; however, given the complexity of composite material structures and compositions, their self-healing faces challenges. Yu et al. developed a high-TC self-healing composite material, 2-[[(butylamino) carbonyl] oxy] ethyl ester (PBA)-polydimethylsiloxane (PDMS)/folded graphene film (FGf), capable of fully self-healing within 10 min at room temperature. This material exhibited high fracture tensile strength (2.23 ± 0.15 MPa) and TC (13 ± 0.2 $W \cdot m^{-1} \cdot K^{-1}$), with self-healing efficiencies reaching 100% and 98.65%, achieved based on supramolecular interactions between polymers and graphene [135]. Through this research and its applications, TCPCs are demonstrating significant value in the field of flexible electronic devices, with the potential for widespread adoption in many practical applications in the future. Whether in electronic packaging or wearable devices, the application prospects of TCPCs are very broad and are expected to have a profound impact on the development of electronic technology [136].

4.2. Energy Storage Devices

With the rapid advancement of electric vehicles and renewable energy, the demand for energy storage devices has surged, making lithium-ion batteries and supercapacitors focal points of attention. These storage devices generate substantial heat during charging and discharging cycles, and ineffective heat management can lead to performance degradation or even safety hazards. TCPCs, known for their excellent thermal conductivity, have become key materials in the thermal management of energy storage devices [137,138]. During charging and discharging, if the heat generated by batteries is not efficiently dissipated, it can lead to excessive temperatures that adversely affect their lifespan and performance. TCPCs are widely used in the TIMs and separators of batteries, aiding rapid heat dissipation and maintaining battery operation within safe temperature limits. For instance, Huang et al. developed a novel flexible composite material, styrene butadiene styrene@paraffin/expanded graphite (SBS@PA/EG). In battery application tests, the battery modules using this material maintained a maximum temperature below 46 °C during 5C discharges, with a temperature difference controlled within 4 °C, proving its efficiency and practicality in battery thermal management [139]. Yin's research team developed an electrolyte with an improved thermal response by adding two-dimensional BN nanosheets to a polyethylene oxide-based electrolyte (Figure 9). The addition of BN not only enhanced the thermal stability of the electrolyte but also improved its ionic conduction and mechanical properties. This discovery provides a vital pathway for improving the safety and efficiency of solid-state batteries [140].

Guo and his research team successfully developed graphene-filled TPU composites during the 3D printing process, achieving an asymmetric alignment of graphene. This vertically aligned 3D printed structure exhibited an in-plane TC of 12 W·m^{-1}·K^{-1} at a graphene content of 45 wt%, eight times higher than that of horizontally printed structures and significantly surpassing traditional particle-reinforced polymer composites. This substantial increase in thermal conductivity was primarily due to the preferred orientation of graphene and finely controlled print parameters of multi-scale dense structures [141]. Cheng et al. developed a ternary composite PCM battery thermal management structure based on expanded graphite/polymer/paraffin, featuring high TC and electrical conductivity. This structure utilizes a dual-layer design to preheat batteries using Joule heat, rapidly increasing the temperature (20.5 °C/min) at low temperatures and improving discharge efficiency by 35.5%. Additionally, at high temperatures, it acts as a cooling medium, keeping the battery temperature at 42.2 °C, optimizing battery performance and reliability [142]. Pei et al. developed ultra-light, double-sided BNNS/poly(acrylic acid) composite-modified separators employing a layer-by-layer (LBL) self-assembly method to enhance the TC of these composite separators. These separators demonstrated excellent PS shuttling and lithium dendrite suppression capabilities under high temperatures and temperature gradients, significantly enhancing the cycle stability and rate performance of Li-S batteries [143].

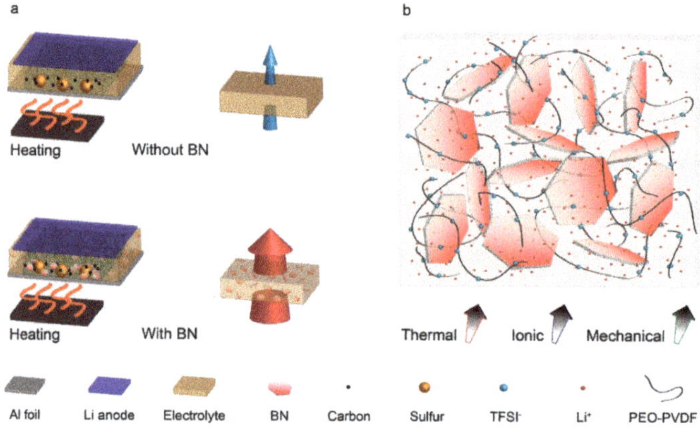

Figure 9. Rational design of solid-state polymer electrolyte with BN additive. (**a**) Illustration of solid-state Li-S batteries with polymer-based electrolytes and schematic comparison of heat transport through electrolytes with and without BN additives and (**b**) sketch of a composite electrolyte consisting of 2D BN flakes and a blended PEO-PVDF polymer with LiTFSI [140].

In supercapacitors, which are valued for their high power density, rapid charge–discharge rates, and long cycle life, TCPCs play a crucial role. These materials significantly enhance the electrochemical performance and electrical conductivity of supercapacitor electrodes [144,145]. Li et al. developed a cross-linked polymer nanocomposite incorporating BN nanosheets, exhibiting exceptional dielectric properties across a broad range of temperatures and frequencies. The inclusion of BN nanosheets notably boosted the TC of the material, enhancing its operational stability at elevated temperatures. Moreover, this nanocomposite is lightweight, photolithographically patternable, and exhibits excellent mechanical flexibility [146]. Chen's research on ladderphane copolymers, materials with inherently low electrical conductivity under high electric fields and temperatures, demonstrated a unique self-assembling mechanism through π–π stacking interactions that form highly ordered arrays, resulting in an intrinsic vertical TC of 1.96 $\pm$ 0.06 W·m^{-1}·K^{-1}. These copolymer films effectively dissipate heat, thus exhibiting superb cyclic stability under challenging conditions [147]. TCPCs are revolutionizing energy storage device applications by effectively managing heat, which enhances device performance and safety and drives innovation in energy storage technology. As advancements continue, TCPCs are expected to further propel the development of energy storage technologies [148–152].

4.3. Transmission Systems

In the realm of power transmission systems, the thermal management capabilities of insulation materials are crucial for enhancing overall performance and extending the lifespan of equipment. Traditional low-conductivity insulation materials often lead to heat accumulation during operation, increasing the risk of failures and reducing efficiency. Recent advances in high-TC polymer composites have shown great potential to address these issues [153–156]. Improving the TC of cable insulation can effectively reduce internal cable temperatures, thereby enhancing the cable's current carrying capacity and operational stability. This improvement not only reduces maintenance costs but also extends the cable's lifespan. To address dielectric material distortions in cable accessories and thermal management issues, Chi et al. grew silver nanoparticles (Ag NPs) on hexagonal BN and then incorporated them into ethylene propylene diene monomer (EPDM) to form a Ag NPs/BN/EPDM composite material. Their studies showed that increasing the Ag NPs content enhanced both the nonlinear electrical conductivity and TC of the composite [157].

Enhancing the TC of the main insulation materials can significantly lower the operating temperatures inside generators, thereby improving their output capacity and efficiency. Lin and others developed a phase change nanocomposite (PCN) using coaxial electrospinning, electrostatic spraying, and hot-pressing techniques (Figure 10). This advanced PCN film, containing 32 wt% of aligned and interconnected BN nanosheets, exhibited an extraordinary TC of 28.3 $W \cdot m^{-1} \cdot K^{-1}$. The successful implementation of this core-sheath strategy has positioned PCNs as ideal cooling solutions for high-power-density devices like 5G base station chips and thermoelectric generators, demonstrating significant potential in thermal management applications [158].

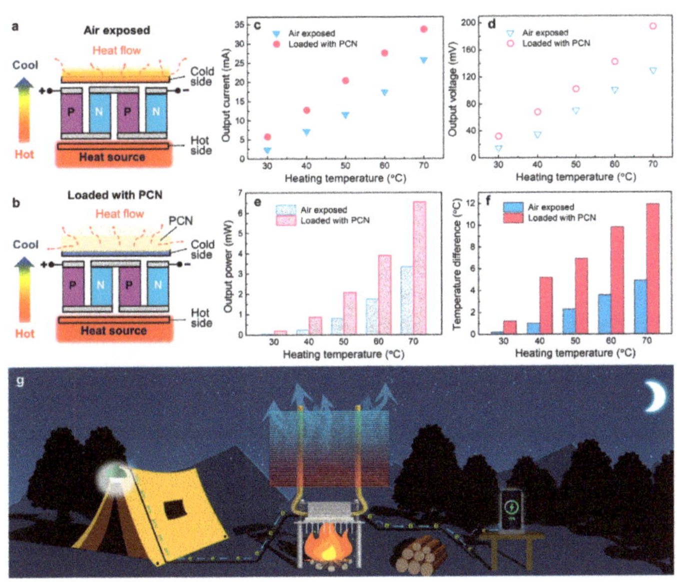

Figure 10. Thermal management application of PEG@TPU/BNNS-es film (with 32 wt% BNNSs content) in thermoelectric generator (TEG). (**a**) Schematic mechanism of TEG exposed to air (top) and (**b**) loaded with PEG@TPU/BNNS-es (PCN) (bottom). (**c**–**e**) Output current, output voltage and output power of TEGs with/without PCC at different heating temperature, respectively. (**f**) Temperature difference between the hot and cold sides versus heating temperature of TEGs. (**g**) Conception diagram of potential application of TEGs integrated with the PEG@TPU/BNNS-es film for outdoor activities [158].

Transformers and other power devices, such as solid insulation switchgear and saturable reactors, also face overheating issues [159,160]. Wei and his team developed a thermofluidic coupled cooling model for saturable reactors to address thermal losses in high-voltage direct current systems. Their research indicated that increasing the TC of the epoxy resin insulation layer (from 0.8 to 1.2 $W \cdot m^{-1} \cdot K^{-1}$) significantly enhanced the cooling effect. This improvement effectively reduced core temperatures, validating the strategy of enhancing thermal dissipation efficiency by optimizing TC [161]. Overall, the development and application of high-TC polymer composites have brought revolutionary improvements to power transmission systems. These materials not only effectively resolve thermal management challenges but also propel the technology of power devices towards greater efficiency and reliability. As materials science continues to advance, these new materials are expected to play an increasingly critical role in future power systems.

4.4. Solar Energy

Solar thermal storage is crucial for enhancing the efficiency of solar energy utilization. PCMs are ideal for solar thermal storage due to their ability to absorb or release substantial amounts of heat during phase transitions. However, the low TC of PCM limits their application under high-temperature concentrated solar beams. Incorporating high-TC nano-fillers into PCM can significantly improve their thermal performance, thereby enhancing the efficiency of solar thermal storage systems [162–164]. In solar collectors and storage systems, PCMs are widely used for thermal energy storage and release. These materials provide latent heat by changing the state of the material without altering the temperature, effectively addressing the discontinuity of solar supply. However, the low TC of PCMs (about 0.2 $W \cdot m^{-1} \cdot K^{-1}$) can lead to hotspots under concentrated solar radiation, preventing uniform heat conduction throughout the receiver and leading to suboptimal energy conversion efficiencies and potential material structural damage due to overheating. Therefore, improving the TC of PCMs is key to optimizing solar thermal storage technology.

To enhance the thermal performance of PCMs, researchers have experimented with incorporating various high-TC nano-fillers. Peng et al. developed an anisotropic graphene aerogel phase change composite with an optimized structure through directional freezing and high-temperature processing, achieving high-transverse (2.68 $W \cdot m^{-1} \cdot K^{-1}$) and longitudinal (8.87 $W \cdot m^{-1} \cdot K^{-1}$) thermal conductivities and a latent heat retention rate of 98.7% [165]. Mo's research team developed a novel polyethylene glycol (PEG 2000)-based PCM composite using a $Ti_3C_2T_x$@PVA foam skeleton as a carrier. By vacuum impregnating and freeze-drying techniques, PEG was introduced into the 3D interconnected skeleton, significantly increasing the TC to 0.428 $W \cdot m^{-1} \cdot K^{-1}$, which is 4.2 times that of pure PEG 2000. This structure not only enhanced TC but also exhibited excellent thermal and chemical reliability [166]. Aftab et al. developed a highly effective solar thermal storage solution by embedding phosphorene nanoflakes (PNFs) derived from high-quality bulk black phosphorus (BP) single crystals into a polyurethane (PU)-based solid–solid PCM matrix using a precipitation/crystallization-induced encapsulation strategy (Figure 11). This method ensured uniform distribution of PNFs and enhanced the composite's structural stability and thermal performance, providing an effective technological path for efficient solar thermal storage [167].

Gao et al. developed a 3D thermal conductive framework combining aramid nanofibers (ANF) and graphene nanoplates (GNP) through ball milling techniques to enhance interfacial interactions and utilized unidirectional freeze-casting to produce a highly oriented honeycomb porous structure. With a GNP volume fraction of only 4.26%, the PCM composite material achieved a TC of 3.9 $W \cdot m^{-1} \cdot K^{-1}$, greatly enhancing the rate and stability of thermal charging and discharging [168]. Moreover, the application of TCPCs in photovoltaic modules has also garnered wide attention. Photovoltaic cells, while converting solar energy, lose a significant portion of solar radiation as heat, reducing the efficiency of the cells. Incorporating high TC and electrically insulating fillers in the encapsulating materials can effectively enhance the thermal dissipation of solar cells [169]. Huang et al. utilized graphene nanoplates (GNs) as fillers and successfully fabricated a polyvinyl butyral (PVB) composite material with significantly enhanced TC through a solution blending method. The study found that the PVB composite material containing 30 wt% GN achieved a TC of 4.521 W·m^{-1}·K^{-1}, which is 20.55 times that of pure PVB [170]. In summary, the application of TCPCs in the solar sector not only enhances the efficiency of solar thermal storage systems but also optimizes the thermal management of photovoltaic cells, improving their conversion efficiency and lifespan. As materials science and nanotechnology continue to advance, these new materials are set to play an increasingly significant role in solar technology, driving the utilization and development of renewable energy.

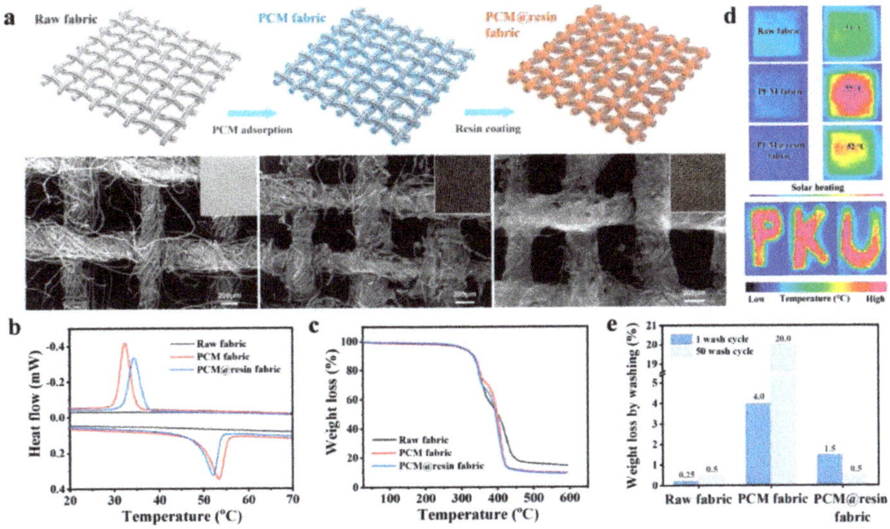

Figure 11. Solar heat storage fabric. (**a**) Schematic illustration of fabric treatment and corresponding SEM images. The inset of SEM images shows real photographs of corresponding fabric. (**b**) DSC thermograms of raw fabric, PCM fabrics and PCM@resin fabric. (**c**) TGA thermograms of raw fabric, PCM fabrics and PCM@resin fabric. (**d**) IR images of raw fabric, PCM fabrics and PCM@resin fabric before and after 2 min of solar irradiation at 80 mW/cm^2. The PKU letter pattern is IR images of developed fabrics after 2 min of solar irradiation. (**e**) Bar graph showing fabric weight loss after washing [167].

4.5. Personal Thermal Management

The application of TCPCs in personal thermal management is gaining increased attention, as their integration into textiles and garments effectively facilitates heat transfer between the human body and the environment, aiming to enhance and regulate warming and cooling effects. These composites show great potential in this context. The porous structures in textiles and the combination of fibers and air significantly impact thermal conductivity. In cold environments, composites with low TC help retain warmth, whereas in hot settings, materials with high TC aid in dissipating heat, preventing thermal stress. This versatile application of material properties lays the foundation for developing functional garments tailored to various environmental and activity demands [171–173]. Soong et al. have developed a multilayered BN-GNP/TPU composite film by mechanically blending GNP and BN fillers into a TPU base film. This composite film exhibits a high TC of 6.86 W·m^{-1}·K^{-1} with 20 wt% of BN and GNP, nearly 30 times that of pure TPU. After 5000 bending cycles, its TC remains at 6.25 W·m^{-1}·K^{-1}, and it maintains 6.85 W·m^{-1}·K^{-1} after 10 wash cycles, demonstrating excellent mechanical stability and hydrophobicity, making it highly suitable for smart cooling garments. Additionally, the composite film exhibits good resistance to bending fatigue and washing, maintaining a TC of 6.25 $W{\cdot}m^{-1}{\cdot}K^{-1}$ after 5000 bending fatigue tests and 6.85 $W{\cdot}m^{-1}{\cdot}K^{-1}$ after 10 wash cycles. Furthermore, enhanced thermal stability, cooling, and hydrophobicity of the multilayer BN-GNP/TPU composite film were observed in the resulting composite film [174]. Huang et al. proposed a direct strategy for fabricating dual-functional wearable Janus fabrics by electrospinning TPU elastomers onto cellulose cotton fibers while loading thermal conductive BN nanoplates. The resulting fabric exhibited excellent thermal management and humidity control characteristics, with a TC of 0.307 $W{\cdot}m^{-1}{\cdot}K^{-1}$, reducing the total thermal resistance to 10.62 K cm^2/W. Indoor and outdoor winter experiments showed that the fabric surface in contact with human skin was 2.9 °C and 3.1 °C warmer than regular fabrics, respectively, demonstrating effective heat dissipation capabilities [175]. Xi et al. developed a simple, scalable method to produce breathable nanofiber membranes with high TC and super-hydrophobic properties, aimed at enhancing the thermal management of personal cooling textiles. Specifically, the in-plane TC reached 17.9 $W{\cdot}m^{-1}{\cdot}K^{-1}$, the transverse TC was 0.29 $W{\cdot}m^{-1}{\cdot}K^{-1}$, the water vapor transmission rate stood at 11.6 $kg/m^2{\cdot}day$, the water contact angle was 153°, and the hydrostatic pressure was 32 kPa. These properties make it highly practical and effective for next-generation cooling fabrics, offering a mix of active and passive cooling capabilities [176]. Jing et al. present a chemical cross-linking strategy to develop ultra-flexible polymer-based phase change composites (Figure 12). This method involves a dual 3D crosslinked network of olefin block copolymers (OBCs) and styrene–ethylene–butylene–styrene (SEBS) within paraffin wax (PW). The enhanced C–C bond within the OBC–SEBS networks synergistically improves the mechanical, thermal, and leakage-proof properties of PW@OBC-SEBS. Importantly, the peroxide-initiated chemical cross-linking technique surpasses the limitations of conventional physical blending methods and is applicable across various polymer matrices [177].

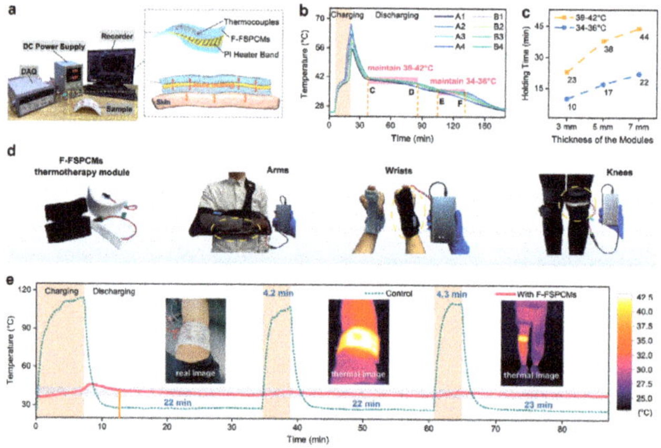

Figure 12. (**a**) Experimental setup for the evaluation of heat storage/release of the F-FSPCMs module. (**b**) Surface temperature of the F-FSPCMs module during a heat storage/discharge cycle (7 mm thickness). (**c**) Temperature holding time of the F-FSPCMs module with the same heat supply for 3 mm, 5 mm and 7 mm thickness. (**d**) Several types of integrated and portable F-FSPCMs modules for personal thermotherapy. (**e**) Temperature evolution of insulating device based on F-FSPCMs module (5 mm thickness) acting on a knee during charging and discharging [177].

In summary, the application of TCPCs in personal thermal management not only enhances thermal comfort but also reduces energy consumption through intelligent design, leading to personalized, energy-efficient thermal management solutions [178,179]. The development of these technologies marks a step toward a more intelligent and efficient future in personal health and comfort management through functional textiles and garments.

4.6. Aerospace

The application of TCPCs in aerospace is a testament to the advances in modern technology. These materials play a crucial role in the design and operation of spacecraft and aircraft, particularly in managing heat within extreme temperature conditions. Ensuring efficient thermal management is a significant engineering challenge, crucial for maintaining performance and extending the lifespan of aerospace equipment. In this context, TCPCs, due to their light weight, high strength, and superior thermal conductivity, are increasingly preferred over traditional metal materials [180–182]. These composites have been enhanced by the incorporation of inorganic fillers like carbon nanotubes, graphene, and metal oxides, facilitated by advances in nanotechnology. This not only optimizes heat dissipation but also enhances electromagnetic shielding and corrosion resistance, making these materials viable for use in advanced aerospace applications. For instance, they are widely used in manufacturing the external shells, wings, and other critical structural components of spacecraft, enhancing the overall mechanical properties while significantly reducing weight.

In spacecraft thermal management systems, these efficient thermal materials ad-dress the enormous temperature differentials created by one side being exposed to sunlight and the other facing the cold void of space. Adequate TC ensures temperature equilibrium, preventing overheating that could lead to electronic equipment failure. Additionally, advancements in micro- and nanoscale technologies have led to the development of new microfluidic cooling systems and encapsulation structures that further enhance thermal management efficiency. Wang et al. enhanced the in-plane orientation of pitch-based carbon fibers within TPU composite films through solution blending and hot-pressing techniques, successfully creating films with excellent anisotropic TC and high EMI shielding performance. The studies show that with 30 vol% filler, the TPU composite films at 0.2 mm thickness achieved an in-plane TC of 15.11 $W{\cdot}m^{-1}{\cdot}K^{-1}$ while providing more than 29 dB of EMI shielding. Through subsequent lamination and hot pressing, the in-plane TC of the laminated composite film increased with the number of layers, reaching up to 17.69 $W{\cdot}m^{-1}{\cdot}K^{-1}$ for a ten-layer laminate. These laminates also demonstrated excellent EMI shielding effectiveness and good mechanical properties across the 8.2–12.5 GHz frequency range [183]. Ryu et al. used lyotropic liquid crystal and wet-spinning processes to develop composite fibers of aromatic polyamide and BN nanotubes (BNNTs). These uniaxially oriented one-dimensional composite fibers are continuous and flexible, enabling excellent thermal stability (up to 479 °C) and significant thermal neutron shielding performance (0.73 mm^{-1}). Due to the dense packing and uniaxial orientation of the fibers, their TC reached 7.88 $W{\cdot}m^{-1}{\cdot}K^{-1}$. These characteristics make this composite textile a promising material for space applications, providing effective radiation and thermal protection for astronauts and electronic equipment [184]. In the future, as nanotechnology and materials science continue to advance, the prospects for TCPCs will broaden further. These materials not only meet the demands of spacecraft for lightweight, high strength, and high TC but will also drive innovation and development in spacecraft thermal management systems [185–189]. In summary, the development of TCPCs has not only enhanced the performance and reliability of spacecraft but has also pushed the entire aerospace field towards higher technological levels, demonstrating the significant potential of modern materials science in applications within extreme environments.

4.7. Others

TCPCs are displaying extensive application potential across various emerging fields, notably within crucial technologies such as building materials, transportation, and medical devices, due to their exceptional TC and mechanical properties.

In the realm of construction, TCPCs play a pivotal role as green and smart buildings rise in prominence. These materials are key in maintaining stable indoor temperatures and enhancing energy efficiency. They are utilized in the production of efficient insulative and conductive building materials, which regulate the internal distribution of heat within structures. This not only increases energy utilization efficiency but also reduces the energy consumption of heating, ventilation, and air conditioning systems, aiding in energy conservation and emission reduction efforts [190,191].

In the transportation sector, TCPCs show immense potential. Modern automotive and railway systems require efficient thermal management systems to ensure stability and safety. Many critical components in these systems benefit from the advanced thermal management properties of these composites, helping to keep vital parts functioning within optimal temperature ranges, thus ensuring the safety and efficiency of the transportation infrastructure [192–194].

The application of TCPCs in medical devices is also noteworthy. Many medical devices generate significant amounts of heat during operation, necessitating efficient cooling solutions to ensure the devices operate correctly and safely. These composites are used to create effective thermal management components, such as conductive pads and cooling plates, which help maintain a stable operating temperature for medical equipment. This not only enhances the reliability and lifespan of the devices but also ensures patient safety [195–197].

As advancements in nanotechnology and materials science continue, TCPCs are set to broaden their impact further across various domains. These materials not only meet modern technology's demands for high TC and mechanical strength but also enhance the performance and reliability of various devices through intelligent design.

5. Conclusions and Outlook

5.1. Conclusions

In this comprehensive review, we have thoroughly examined the development of TCPCs, detailing the advancements in this field, elucidating the thermal conduction mechanisms of these materials, and discussing factors that influence their thermal properties, such as the inherent characteristics of the polymers, interface thermal resistance, and thermal performance of fillers. Additionally, methods to enhance the TC of polymer composites have been categorized and summarized. Lastly, this article highlights the applications of these materials in emerging areas such as flexible electronics, personal thermal management, and aerospace.

5.2. Outlook

Despite the progress made with TCPCs, several challenges persist. Future developments should be application driven, focusing on the following breakthroughs:

Development and Application of Novel Fillers: With continual advancements in technology, the development of new, highly efficient thermal fillers, such as novel two-dimensional materials and three-dimensional network structure fillers including black phosphorus and MXenes, is expected. These materials are becoming research focal points due to their unique thermal conduction mechanisms and structural properties. Through the further development of functionalization and surface treatment techniques, it is possible to optimize the interface bonding between fillers and the polymer matrix, achieving higher thermal conduction efficiencies, particularly crucial for electronic packaging and energy storage systems.

Multi-scale Composite Design: Future designs of TCPCs will likely emphasize optimization of multi-scale structures. From the nano to the macro level, fine control over the dispersion, orientation, and network structure of fillers can effectively construct thermal conduction pathways. Additionally, structural design and interface engineering of materials will be key to enhancing overall thermal conductivity, necessitating a combination of physical and chemical methods along with the use of computational modeling and simulation technologies to predict and optimize material designs.

Efficient Manufacturing Processes and Scalable Production: As the demand for TCPCs continues to grow, developing efficient and cost-effective manufacturing processes will become a research focus. New technologies, such as 3D printing and electrospinning, are expected to facilitate the efficient fabrication and scalable production of these materials.

Moreover, optimizing process parameters during manufacturing and improving production equipment are crucial for enhancing production efficiency and material consistency.

Multifunctionality and Performance Enhancement: Future TCPCs will not only require excellent TC but also need to possess electrical insulation, mechanical strength, and environmental stability (such as thermal stability). This multifunctionality is particularly prominent in the electronics and automotive industries. Through functional design and comprehensive performance optimization, high-performance composite materials that meet specific application needs can be developed. The high loading of fillers not only leads to agglomeration but also negatively impacts the mechanical properties and thermal conductivity of the composites. Although constructing thermal conductive networks through modification, hybridization, or macroscopic construction can enhance thermal conductivity, the filler content in composites still needs to exceed 20 wt% to achieve high thermal conductivity. Generally, the mechanical threshold for polymer composites is within 10 wt%, and the addition of fillers continues to degrade the mechanical properties of the polymer. Therefore, achieving an effective combination of properties at low filler content remains a significant challenge that needs to be addressed.

In-depth Study of Thermal Mechanisms: Although significant progress has been made, the thermal mechanisms of TCPCs still require deeper investigation. In basic theoretical research of composite materials, the forms of thermal conduction are simplistic, and theories of thermal conduction lack depth. Further exploration of thermal mechanisms, multi-scale phonon conduction, and phonon transmission at interfaces is needed. Additionally, the precise mathematical modeling of interface thermal resistance, effective testing methods, and related measurement equipment are lacking. By utilizing advanced characterization techniques and computational simulations, a better understanding of the interactions at the interface between fillers and the polymer matrix and their impact on thermal conduction can be achieved. Further research into the specific impacts of the distribution and orientation of fillers within the polymer matrix on TC will provide theoretical support and experimental basis for designing more efficient TCPCs.

Overall, research on TCPCs is moving towards more efficient, smarter, and more environmentally friendly directions. With the discovery of new materials and application of new technologies, these materials are expected to find broader applications in industries such as energy, electronics, and automotive, providing effective solutions for thermal management challenges.

Author Contributions: J.T. drafted the manuscript. J.T. and Y.Z. participated in discussion and manuscript editing. All authors have read and agreed to the published version of the manuscript.

Funding: This research received no external funding.

Conflicts of Interest: The authors declare no conflicts of interest.

References

1. Mathew, J.; Krishnan, S. A review on transient thermal management of electronic devices. *J. Electron. Packag.* **2022**, *144*, 010801. [CrossRef]
2. Afaynou, I.; Faraji, H.; Choukairy, K.; Arshad, A.; Arıcı, M. Heat transfer enhancement of phase-change materials (PCMs) based thermal management systems for electronic components: A review of recent advances. *Int. Commun. Heat Mass Transf.* **2023**, *143*, 106690. [CrossRef]
3. Deng, C.; Zhang, X.; Peng, J.; Yang, X.; Shang, B.; Luo, X. In51Bi32.5Sn16.5@SiO_2 microcapsules-based composite phase change materials with high thermal conductivity and heat storage density for electronics thermal management. *J. Energy Storage* **2024**, *86*, 111432. [CrossRef]
4. Naghibi, S.; Kargar, F.; Wright, D.; Huang, C.Y.T.; Mohammadzadeh, A.; Barani, Z.; Salgado, R.; Balandin, A.A. Noncuring graphene thermal interface materials for advanced electronics. *Adv. Electron. Mater.* **2020**, *6*, 1901303. [CrossRef]
5. He, Z.; Yan, Y.; Zhang, Z. Thermal management and temperature uniformity enhancement of electronic devices by micro heat sinks: A review. *Energy* **2021**, *216*, 119223. [CrossRef]
6. Lohrasbi, S.; Hammer, R.; Essl, W.; Reiss, G.; Defregger, S.; Sanz, W. A comprehensive review on the core thermal management improvement concepts in power electronics. *IEEE Access* **2020**, *8*, 166880–166906. [CrossRef]

7. Chen, W.; Shi, X.; Zou, J.; Chen, Z. Thermoelectric coolers for on-chip thermal management: Materials, design, and optimization. *Mater. Sci. Eng. R Rep.* **2022**, *151*, 100700. [CrossRef]
8. Luo, J.; Zou, D.; Wang, Y.; Wang, S.; Huang, L. Battery thermal management systems (BTMs) based on phase change material (PCM): A comprehensive review. *Chem. Eng. J.* **2022**, *430*, 132741. [CrossRef]
9. Feng, C.P.; Yang, L.Y.; Yang, J.; Bai, L.; Bao, R.Y.; Liu, Z.Y.; Yang, M.; Yang, W. Recent advances in polymer-based thermal interface materials for thermal management: A mini-review. *Compos. Commun.* **2020**, *22*, 100528. [CrossRef]
10. Yang, G.; Zhou, L.; Zhang, X.; Pan, D.; Wang, M.; Luo, S.; Su, F.; Ji, Y.; Liu, C. Highly thermally conductive, electrically insulating, thermal resistance polyimide-base composites with fibrillated carbon networks for thermal management applications. *Compos. Commun.* **2023**, *37*, 101450. [CrossRef]
11. Jiang, K.; Liao, G.; Jiaqiang, E.; Zhang, F.; Chen, J.; Leng, E. Thermal management technology of power lithium-ion batteries based on the phase transition of materials: A review. *J. Energy Storage* **2020**, *32*, 101816. [CrossRef]
12. Cui, Y.; Li, M.; Hu, Y. Emerging interface materials for electronics thermal management: Experiments, modeling, and new opportunities. *J. Mater. Chem. C* **2020**, *8*, 10568–10586. [CrossRef]
13. Cao, J.; Meng, T.L.; Zhang, X.; Gong, N.; Karyappa, R.; Tan, C.K.I.; Suwardi, A.; Zhu, Q.; Liu, H. Recent advances in thermal interface materials. *ES Mater. Manuf.* **2020**, *7*, 4–24. [CrossRef]
14. Guo, X.; Cheng, S.; Cai, W.; Zhang, Y.; Zhang, X.A. A review of carbon-based thermal interface materials: Mechanism, thermal measurements and thermal properties. *Mater. Des.* **2021**, *209*, 109936. [CrossRef]
15. Nayak, S.K.; Mohanty, S.; Nayak, S.K. A new way synthesis of expanded graphite as a thermal filler to enhance the thermal conductivity of DGEBA resin as thermal interface material. *High Perform. Polym.* **2020**, *32*, 506–523. [CrossRef]
16. Shin, S.; Wang, Q.; Luo, J.; Chen, R. Advanced materials for high-temperature thermal transport. *Adv. Funct. Mater.* **2020**, *30*, 1904815. [CrossRef]
17. Gupta, P.; Toksha, B.; Patel, B.; Rushiya, Y.; Das, P.; Rahaman, M. Recent developments and research avenues for polymers in electric vehicles. *Chem. Rec.* **2022**, *22*, e202200186. [CrossRef] [PubMed]
18. Previati, G.; Mastinu, G.; Gobbi, M. Thermal management of electrified vehicles—A review. *Energies* **2022**, *15*, 1326. [CrossRef]
19. Widyantara, R.D.; Zulaikah, S.; Juangsa, F.B.; Budiman, B.A.; Aziz, M. Review on battery packing design strategies for superior thermal management in electric vehicles. *Batteries* **2022**, *8*, 287. [CrossRef]
20. Allen, J. Review of polymers in the prevention of thermal runaway in lithium-ion batteries. *Energy Rep.* **2020**, *6*, 217–224. [CrossRef]
21. Wang, Z.; Changqing, D. A comprehensive review on thermal management systems for power lithium-ion batteries. *Renew. Sustain. Energy Rev.* **2021**, *139*, 110685.
22. Ju, Y.S. Thermal management and control of wearable devices. *iScience* **2022**, *25*, 104587. [CrossRef]
23. Yun, J.; Yoo, Y.J.; Kim, H.R.; Song, Y.M. Recent progress in thermal management for flexible/wearable devices. *Soft Sci.* **2023**, *3*, 12. [CrossRef]
24. Nie, S.; Hao, N.; Zhang, K.; Xing, C.; Wang, S. Cellulose nanofibrils-based thermally conductive composites for flexible electronics: A mini review. *Cellulose* **2020**, *27*, 4173–4187. [CrossRef]
25. Niu, Z.; Yuan, W. Smart nanocomposite nonwoven wearable fabrics embedding phase change materials for highly efficient energy conversion–storage and use as a stretchable conductor. *ACS Appl. Mater. Interfaces* **2021**, *13*, 4508–4518. [CrossRef]
26. Ghaffari-Mosanenzadeh, S.; Aghababaei Tafreshi, O.; Dammen-Brower, E.; Rad, E.; Meysami, M.; Naguib, H.E. A review on high thermally conductive polymeric composites. *Polym. Compos.* **2022**, *43*, 692–711. [CrossRef]
27. Yu, H.; Guo, P.; Qin, M.; Han, G.; Chen, L.; Feng, Y.; Feng, W. Highly thermally conductive polymer composite enhanced by two-level adjustable boron nitride network with leaf venation structure. *Compos. Sci. Technol.* **2022**, *222*, 109406. [CrossRef]
28. Meng, X.; Yu, H.; Wang, L.; Wu, X.; Amin, B.U. Recent progress on fabrication and performance of polymer composites with highly thermal conductivity. *Macromol. Mater. Eng.* **2021**, *306*, 2100434. [CrossRef]
29. Tan, X.; Yuan, Q.; Qiu, M.; Yu, J.; Jiang, N.; Lin, C.; Dai, W. Rational design of graphene/polymer composites with excellent electromagnetic interference shielding effectiveness and high thermal conductivity: A mini review. *J. Mater. Sci. Technol.* **2022**, *117*, 238–250. [CrossRef]
30. Du, Y.K.; Shi, Z.X.; Dong, S.; Jin, H.; Ke, X.; Zhao, P.; Jiang, B.; You, F. Recent progress in fabrication and structural design of thermal conductive polymer composites. *Chin. J. Polym. Sci.* **2024**, *42*, 277–291. [CrossRef]
31. Amraei, J.; Katunin, A. Recent advances in limiting fatigue damage accumulation induced by self-heating in polymer–matrix composites. *Polymers* **2022**, *14*, 5384. [CrossRef] [PubMed]
32. Huang, X.; Zhi, C.; Lin, Y.; Bao, H.; Wu, G.; Jiang, P.; Mai, Y.W. Thermal conductivity of graphene-based polymer nanocomposites. *Mater. Sci. Eng. R Rep.* **2020**, *142*, 100577. [CrossRef]
33. Pathumudy, R.D.; Prabhu, K.N. Thermal interface materials for cooling microelectronic systems: Present status and future challenges. *J. Mater. Sci. Mater. Electron.* **2021**, *32*, 11339–11366. [CrossRef]
34. Liu, Y.; Zhou, J.; Ning, Z.; Huang, H.; Cheng, Z.; Duan, L.; Wang, Y.; Tao, X.; Liu, P.; Ma, Y.; et al. Multi-Interfacial Heterostructure Design of Carbon Fiber/Silicone Rubber-Oriented Composites for Microwave Absorption and Thermal Management. *Adv. Funct. Mater.* **2024**, 2411573. [CrossRef]
35. Guo, Y.; Lyu, Z.; Yang, X.; Lu, Y.; Ruan, K.; Wu, Y.; Kong, J.; Gu, J. Enhanced thermal conductivities and decreased thermal resistances of functionalized boron nitride/polyimide composites. *Compos. Part B Eng.* **2019**, *164*, 732–739. [CrossRef]

36. Gu, J.; Ruan, K. Breaking through bottlenecks for thermally conductive polymer composites: A perspective for intrinsic thermal conductivity, interfacial thermal resistance and theoretics. *Nano-Micro Lett.* **2021**, *13*, 110. [CrossRef] [PubMed]
37. Ma, H.; Gao, B.; Wang, M.; Yuan, Z.; Shen, J.; Zhao, J.; Feng, Y. Strategies for enhancing thermal conductivity of polymer-based thermal interface materials: A review. *J. Mater. Sci.* **2021**, *56*, 1064–1086. [CrossRef]
38. Ghosh, S.; Nath, K.; Chowdhury, S.; Paul, S.; Ghosh, T.; Katheria, A.; Das, P.; Das, N. Combination effect of functionalized high aspect ratio carbonaceous nanofillers and carbon black on electrical, thermal conductivity, dielectric and EMI shielding behavior of co-continuous thermoplastic elastomeric blend composite films. *Chem. Eng. J Adv.* **2023**, *15*, 100505. [CrossRef]
39. Lee, D.K.; Yoo, J.; Kim, H.; Kang, B.H.; Park, S.H. Electrical and thermal properties of carbon nanotube polymer composites with various aspect ratios. *Materials* **2022**, *15*, 1356. [CrossRef]
40. Yang, Y.; Zhu, J.; Yang, D.; Zhang, J.; Guo, Y.; Zhong, X.; Kong, J.; Gu, J. High-efficiency improvement of thermal conductivities for epoxy composites from synthesized liquid crystal epoxy followed by doping BN fillers. *Compos. Part B Eng.* **2020**, *185*, 107784. [CrossRef]
41. Tarannum, F.; Muthaiah, R.; Annam, R.S.; Gu, T.; Garg, J. Effect of alignment on enhancement of thermal conductivity of polyethylene–graphene nanocomposites and comparison with effective medium theory. *Nanomaterials* **2020**, *10*, 1291. [CrossRef] [PubMed]
42. Ruan, K.; Shi, X.; Guo, Y.; Gu, J. Interfacial thermal resistance in thermally conductive polymer composites: A review. *Compos. Commun.* **2020**, *22*, 100518. [CrossRef]
43. Idumah, C.I.; Obele, C.M. Understanding interfacial influence on properties of polymer nanocomposites. *Surf. Interfaces* **2021**, *22*, 100879. [CrossRef]
44. Saminathan, R.; Hadidi, H.; Fageehi, Y.A.; Manoj Kumar, P.; Venkatasudhahar, M.; Ankit; Ram, S.; Gebreyohannes, D.T. Experimental analysis of mechanical and thermal characteristics of luffa/epoxy polymer composite under the influence of nanosilica. *Adv. Mater. Sci. Eng.* **2022**, *2022*, 6040629. [CrossRef]
45. Zhang, B.; Dou, Z.L.; Zhang, Y.Z.; Fu, Q.; Wu, K. Exploring trade-offs in thermal interface materials: The impact of polymer-filler interfaces on thermal conductivity and thixotropy. *Chin. J. Polym. Sci.* **2024**, *42*, 916–925. [CrossRef]
46. Wu, W.; Ren, T.; Liu, X.; Davis, R.; Huai, K.; Cui, X.; Wei, H.; Hu, J.; Xia, Y.; Huang, S.; et al. Creating thermal conductive pathways in polymer matrix by directional assembly of synergistic fillers assisted by electric fields. *Compos. Commun.* **2022**, *35*, 101309. [CrossRef]
47. Hao, X.; Wan, S.; Zhao, Z.; Zhu, L.; Peng, D.; Yue, M.; Kuang, J.; Cao, W.; Liu, G.; Wang, Q. Enhanced thermal conductivity of epoxy composites by introducing 1D AlN whiskers and constructing directionally aligned 3D AlN filler skeletons. *ACS Appl. Mater. Interfaces* **2022**, *15*, 2124–2133. [CrossRef] [PubMed]
48. Griffin, A.; Guo, Y.; Hu, Z.; Zhang, J.; Chen, Y.; Qiang, Z. Scalable methods for directional assembly of fillers in polymer composites: Creating pathways for improving material properties. *Polym. Compos.* **2022**, *43*, 5747–5766. [CrossRef]
49. Luo, Z.; Yang, D.; Liu, J.; Zhao, H.; Zhao, T.; Li, B.; Yang, W.; Yu, Z.Z. Nature-Inspired Solar-Thermal Gradient Reduced Graphene Oxide Aerogel-based Bilayer Phase Change Composites for Self-Adaptive Personal Thermal Management. *Adv. Funct. Mater.* **2023**, *33*, 2212032. [CrossRef]
50. Zhou, M.; Tan, S.; Wang, J.; Wu, Y.; Liang, L.; Ji, G. "Three-in-One" multi-scale structural design of carbon fiber-based composites for personal electromagnetic protection and thermal management. *Nano-Micro Lett.* **2023**, *15*, 176. [CrossRef]
51. Wang, Z.; Wu, Z.; Weng, L.; Ge, S.; Jiang, D.; Huang, M.; Mulvihill, D.M.; Chen, Q.; Guo, Z.; Jazzar, A.; et al. A roadmap review of thermally conductive polymer composites: Critical factors, progress, and prospects. *Adv. Funct. Mater.* **2023**, *33*, 2301549. [CrossRef]
52. Kang, J.H.; Gordon, K.L.; Ward, D.; Belancik, G.; Jagtap, P.; Sauti, G. Highly Thermally Conductive Hybrid Carbon Fiber Polymer Composite for Radiator Application. In Proceedings of the 52nd International Conference on Environmental Systems (ICES 2023), Calgary, AB, Canada, 16–20 July 2023; ICES-2023-123.
53. Storm, M.L. Heat conduction in simple metals. *J. Appl. Phys.* **1951**, *22*, 940–951. [CrossRef]
54. Tong, Z.; Li, S.; Ruan, X.; Bao, H. Comprehensive first-principles analysis of phonon thermal conductivity and electron-phonon coupling in different metals. *Phys. Rev. B* **2019**, *100*, 144306. [CrossRef]
55. Inogamov, N.A.; Petrov, Y.V. Thermal conductivity of metals with hot electrons. *J. Exp. Theor. Phys.* **2010**, *110*, 446–468. [CrossRef]
56. Das, S.; Muthukumar, M. Thermal conduction and phonon transport in folded polyethylene chains. *Macromolecules* **2023**, *56*, 393–403. [CrossRef]
57. Choy, C.L. Thermal conductivity of polymers. *Polymer* **1977**, *18*, 984–1004. [CrossRef]
58. Wei, X.; Wang, Z.; Tian, Z.; Luo, T. Thermal transport in polymers: A review. *J. Heat Transf.* **2021**, *143*, 072101. [CrossRef]
59. Guo, Y.; Ruan, K.; Wang, G.; Gu, J. Advances and mechanisms in polymer composites toward thermal conduction and electromagnetic wave absorption. *Sci. Bull.* **2023**, *68*, 1195–1212. [CrossRef] [PubMed]
60. Qian, X.; Zhou, J.; Chen, G. Phonon-engineered extreme thermal conductivity materials. *Nat. Mater.* **2021**, *20*, 1188–1202. [CrossRef]
61. Zhang, Z.; Ouyang, Y.; Guo, Y.; Nakayama, T.; Nomura, M.; Volz, S.; Chen, J. Hydrodynamic phonon transport in bulk crystalline polymers. *Phys. Rev. B* **2020**, *102*, 195302. [CrossRef]
62. Cheng, P.; Chen, X.; Gao, H. Different dimensional nanoadditives for thermal conductivity enhancement of phase change materials: Fundamentals and applications. *Nano Energy* **2021**, *85*, 105948. [CrossRef]

63. Kim, T.; Drakopoulos, S.X.; Ronca, S.; Minnich, A.J. Origin of high thermal conductivity in disentangled ultra-high molecular weight polyethylene films: Ballistic phonons within enlarged crystals. *Nat. Commun.* **2022**, *13*, 2452. [CrossRef] [PubMed]
64. Dinpajooh, M.; Nitzan, A. Heat conduction in polymer chains: Effect of substrate on the thermal conductance. *J. Chem. Phys.* **2022**, *156*, 144506. [CrossRef] [PubMed]
65. Chen, A.; Wu, Y.; Zhou, S.; Xu, W.; Jiang, W.; Lv, Y.; Guo, W.; Chi, K.; Sun, Q.; Fu, T.; et al. High thermal conductivity polymer chains with reactive groups: A step towards true application. *Mater. Adv.* **2020**, *1*, 1996–2002. [CrossRef]
66. Guo, Y.; Ruan, K.; Shi, X.; Yang, X.; Gu, J. Factors affecting thermal conductivities of the polymers and polymer composites: A review. *Compos. Sci. Technol.* **2020**, *193*, 108134. [CrossRef]
67. Hands, D. The thermal transport properties of polymers. *Rubber Chem. Technol.* **1977**, *50*, 480–522. [CrossRef]
68. Mehra, N.; Mu, L.; Ji, T.; Yang, X.; Kong, J.; Gu, J.; Zhu, J. Thermal transport in polymeric materials and across composite interfaces. *Appl. Mater. Today* **2018**, *12*, 92–130. [CrossRef]
69. Stastna, J. *Transport Properties in Polymers*; CRC Press: Boca Raton, FL, USA, 1995.
70. Li, A.; Zhang, C.; Zhang, Y.F. Thermal conductivity of graphene-polymer composites: Mechanisms, properties, and applications. *Polymers* **2017**, *9*, 437. [CrossRef] [PubMed]
71. Burger, N.; Laachachi, A.; Ferriol, M.; Lutz, M.; Toniazzo, V.; Ruch, D. Review of thermal conductivity in composites: Mechanisms, parameters and theory. *Prog. Polym. Sci.* **2016**, *61*, 1–28. [CrossRef]
72. Kochetov, R.; Korobko, A.V.; Andritsch, T.; Morshuis, P.H.F.; Picken, S.J.; Smit, J.J. Modelling of the thermal conductivity in polymer nanocomposites and the impact of the interface between filler and matrix. *J. Phys. D Appl. Phys.* **2011**, *44*, 395401. [CrossRef]
73. Jasmee, S.; Omar, G.; Othaman, S.S.C.; Masripan, N.A.; Hamid, H.A. Interface thermal resistance and thermal conductivity of polymer composites at different types, shapes, and sizes of fillers: A review. *Polym. Compos.* **2021**, *42*, 2629–2652. [CrossRef]
74. You, H.; Zhao, Q.; Mei, T.; Li, X.; You, R.; Wang, D. Self-Reinforced Polymer Nanofiber Aerogels for Multifunctional Applications. *Macromol. Mater. Eng.* **2022**, *307*, 2100971. [CrossRef]
75. Xiao, H.; Huang, Z.X.; Zhang, Z.P.; Rong, M.Z.; Zhang, M.Q. Highly thermally conductive flexible copper clad laminates based on sea-island structured boron nitride/polyimide composites. *Compos. Sci. Technol.* **2022**, *230*, 109087. [CrossRef]
76. Kargar, F.; Barani, Z.; Salgado, R.; Debnath, B.; Lewis, J.S.; Aytan, E.; Balandin, A.A. Thermal percolation threshold and thermal properties of composites with high loading of graphene and boron nitride fillers. *ACS Appl. Mater. Interfaces* **2018**, *10*, 37555–37565. [CrossRef] [PubMed]
77. Fu, Y.; Hansson, J.; Liu, Y.; Chen, S.; Zehri, A.; Samani, M.K.; Wang, N.; Ni, Y.; Zhang, Y.; Zhang, Z.; et al. Graphene related materials for thermal management. *2D Mater.* **2020**, *7*, 012001. [CrossRef]
78. Jang, J.U.; Nam, H.E.; So, S.O.; Lee, H.; Kim, G.S.; Kim, S.Y.; Kim, S.H. Thermal percolation behavior in thermal conductivity of polymer nanocomposite with lateral size of graphene nanoplatelet. *Polymers* **2022**, *14*, 323. [CrossRef] [PubMed]
79. Li, R.; Yang, X.; Li, J.; Shen, Y.; Zhang, L.; Lu, R.; Wang, C.; Zheng, X.; Chen, H.; Zhang, T. Review on polymer composites with high thermal conductivity and low dielectric properties for electronic packaging. *Mater. Today Phys.* **2022**, *22*, 100594. [CrossRef]
80. Ma, T.; Ruan, K.; Gu, J. Theories and mechanisms of thermal conduction. *Therm. Conduct. Polym. Compos.* **2023**, *57*, 1–19.
81. Abbas, I.; Hobiny, A.; Alshehri, H.; Vlase, S.; Marin, M. Analysis of thermoelastic interaction in a polymeric orthotropic medium using the finite element method. *Polymers* **2022**, *14*, 2112. [CrossRef]
82. Kiessling, A.; Simavilla, D.N.; Vogiatzis, G.G.; Venerus, D.C. Thermal conductivity of amorphous polymers and its dependence on molecular weight. *Polymer* **2021**, *228*, 123881. [CrossRef]
83. Xi, Q.; Zhong, J.; He, J.; Xu, X.; Nakayama, T.; Wang, Y.; Liu, J.; Zhou, J.; Li, B. A ubiquitous thermal conductivity formula for liquids, polymer glass, and amorphous solids. *Chin. Phys. Lett.* **2020**, *37*, 104401. [CrossRef]
84. Xu, X.; Chen, J.; Zhou, J.; Li, B. Thermal conductivity of polymers and their nanocomposites. *Adv. Mater.* **2018**, *30*, 1705544. [CrossRef] [PubMed]
85. Bai, L.; Zhao, X.; Bao, R.Y.; Liu, Z.Y.; Yang, M.B.; Yang, W. Effect of temperature, crystallinity and molecular chain orientation on the thermal conductivity of polymers: A case study of PLLA. *J. Mater. Sci.* **2018**, *53*, 10543–10553. [CrossRef]
86. Zheng, H.; Wu, K.; Zhan, Y.; Wang, K.; Shi, J. High intrinsic thermal conductive polymer films by engineered interchain hydrogen bond interactions. *J. Polym. Sci.* **2023**, *61*, 1622–1633. [CrossRef]
87. Tu, H.; Xie, K.; Lin, X.; Zhang, R.; Chen, F.; Fu, Q.; Duan, B.; Zhang, L. Superior strength and highly thermoconductive cellulose/boron nitride film by stretch-induced alignment. *J. Mater. Chem. A* **2021**, *9*, 10304–10315. [CrossRef]
88. Zhao, L.; Yan, L.; Wei, C.; Li, Q.; Huang, X.; Wang, Z.; Fu, M.; Ren, J. Synergistic enhanced thermal conductivity of epoxy composites with boron nitride nanosheets and microspheres. *J Phys. Chem. C* **2020**, *124*, 12723–12733. [CrossRef]
89. Wang, Z.; Fan, L.; Li, R.; Xu, Y.; Fu, Q. Preparation of polymer composites with high thermal conductivity by constructing a "double thermal conductive network" via electrostatic spinning. *Compos. Commun.* **2022**, *36*, 101371. [CrossRef]
90. Wu, X.; Zhang, X.; Yan, X.; Zhang, C.; Zhang, Y.; Li, P.; Li, N.; Liu, H.; Wang, Z. A review: From the whole process of making thermal conductive polymer, the effective method of improving thermal conductivity. *J. Polym. Sci.* **2024**, *62*, 2410–2442. [CrossRef]
91. Wang, J.; Li, Q.; Liu, D.; Chen, C.; Chen, Z.; Hao, J.; Li, Y.; Zhang, J.; Naebe, M.; Lei, W. High temperature thermally conductive nanocomposite textile by "green" electrospinning. *Nanoscale* **2018**, *10*, 16868–16872. [CrossRef]

92. Ouyang, Y.; Bai, L.; Tian, H.; Li, X.; Yuan, F. Recent progress of thermal conductive polymer composites: Al2O3 fillers, properties and applications. *Compos. Part A Appl. Sci. Manuf.* **2022**, *152*, 106685.
93. Huang, Y.; Ellingford, C.; Bowen, C.; McNally, T.; Wu, D.; Wan, C. Tailoring the electrical and thermal conductivity of multi-component and multi-phase polymer composites. *Int. Mater. Rev.* **2020**, *65*, 129–163. [CrossRef]
94. Nilagiri Balasubramanian, K.B.; Ramesh, T. Role, effect, and influences of micro and nano-fillers on various properties of polymer matrix composites for microelectronics: A review. *Polym. Adv. Technol.* **2018**, *29*, 1568–1585. [CrossRef]
95. Li, M.; Ali, Z.; Wei, X.; Li, L.; Song, G.; Hou, X.; Do, H.; Greer, J.C.; Pan, Z.; Lin, C.; et al. Stress induced carbon fiber orientation for enhanced thermal conductivity of epoxy composites. *Compos. Part B Eng.* **2021**, *208*, 108599. [CrossRef]
96. Shang, Y.; Ji, Y.; Dong, J.; Yang, G.; Zhang, X.; Su, F.; Feng, Y.; Liu, C. Sandwiched cellulose nanofiber/boron nitride nanosheet/Ti3C2Tx MXene composite film with high electromagnetic shielding and thermal conductivity yet insulation performance. *Compos. Sci. Technol.* **2021**, *214*, 108974. [CrossRef]
97. Chandrashekar, A.; Hegde, M.; Krishna, S.; Gopi, J.A.; Kotresh, T.M.; Prabhu, T.N. Non-covalent surface functionalization of nanofillers towards the enhancement of thermal conductivity of polymer nanocomposites: A mini review. *Eur. Polym. J.* **2023**, *198*, 112379. [CrossRef]
98. Liang, J.; Cui, R.; Zhang, X.; Koumoto, K.; Wan, C. Polymer/Carbon Composites with Versatile Interfacial Interactions for High Performance Carbon-Based Thermoelectrics: Principles and Applications. *Adv. Funct. Mater.* **2023**, *33*, 2208813. [CrossRef]
99. Ruan, K.; Guo, Y.; Lu, C.; Shi, X.; Ma, T.; Zhang, Y.; Kong, J.; Gu, J. Significant reduction of interfacial thermal resistance and phonon scattering in graphene/polyimide thermally conductive composite films for thermal management. *Research* **2021**, *2021*, 8438614. [CrossRef]
100. Chen, G.; Yadav, A.A.; Jung, I.W.; Lee, J.; Choi, K.; Kang, S.W. Crosslinking effect of borax additive on the thermal properties of polymer-based 1D and 2D nanocomposites used as thermal interface materials. *Sci. Rep.* **2022**, *12*, 16029. [CrossRef] [PubMed]
101. Chen, Z.; Tu, Q.; Shen, X.; Fang, Z.; Bi, S.; Yin, Q.; Zhang, X. Enhancing the thermal and mechanical properties of carbon fiber/natural rubber composites by co-modification of dopamine and silane coupling agents. *Polym. Test.* **2023**, *108*, 108164. [CrossRef]
102. Bashir, A.; Maqbool, M.; Lv, R.; Usman, A.; Aftab, W.; Niu, H.; Kang, L.; Bai, S.L. Engineering of interfacial interactions among BN and CNT hybrid towards higher heat conduction within TPU composites. *Compos. Part A Appl. Sci. Manuf.* **2023**, *167*, 107428. [CrossRef]
103. Xie, X.; Yang, D. Construction of thermal conduction networks and decrease of interfacial thermal resistance for improving thermal conductivity of epoxy natural rubber composites. *Ceram. Int.* **2023**, *49*, 17650–17659. [CrossRef]
104. Zhao, H.Y.; Yu, M.Y.; Liu, J.; Li, X.; Min, P.; Yu, Z.Z. Efficient preconstruction of three-dimensional graphene networks for thermally conductive polymer composites. *Nano-Micro Lett.* **2022**, *14*, 129. [CrossRef] [PubMed]
105. Zhao, N.; Li, J.; Wang, W.; Gao, W.; Bai, H. Isotropically ultrahigh thermal conductive polymer composites by assembling anisotropic boron nitride nanosheets into a biaxially oriented network. *ACS Nano* **2022**, *16*, 18959–18967. [CrossRef]
106. Zhang, F.; Feng, Y.; Feng, W. Three-dimensional interconnected networks for thermally conductive polymer composites: Design, preparation, properties, and mechanisms. *Mater. Sci. Eng. R Rep.* **2020**, *142*, 100580. [CrossRef]
107. Liu, Y.; Zou, W.; Zhao, N.; Xu, J. Electrically insulating PBO/MXene film with superior thermal conductivity, mechanical properties, thermal stability, and flame retardancy. *Nat. Commun.* **2023**, *14*, 5342. [CrossRef]
108. Wu, S.; Kondo, Y.; Kakimoto, M.A.; Yang, B.; Yamada, H.; Kuwajima, I.; Lambard, G.; Hongo, K.; Xu, Y.; Shiomi, J.; et al. Machine-learning-assisted discovery of polymers with high thermal conductivity using a molecular design algorithm. *NPJ Comput. Mater.* **2019**, *5*, 66. [CrossRef]
109. Song, J.; Zhang, Y. Vertically aligned silicon carbide nanowires/reduced graphene oxide networks for enhancing the thermal conductivity of silicone rubber composites. *Compos. Part A Appl. Sci. Manuf.* **2020**, *133*, 105873. [CrossRef]
110. Zhao, C.; Li, Y.; Liu, Y.; Xie, H.; Yu, W. A critical review of the preparation strategies of thermally conductive and electrically insulating polymeric materials and their applications in heat dissipation of electronic devices. *Adv. Compos. Hybrid Mater.* **2023**, *6*, 27. [CrossRef]
111. Liu, Z.; Yin, X.; Zhang, H.; Wang, M.; Feng, Y. Efficient preparation of bn/uhmwpe composites with oriented thermal conductivity by powder solid-state extrusion. *Compos. Part A Appl. Sci. Manuf.* **2023**, *172*, 107598. [CrossRef]
112. Wen, Y.; Chen, C.; Ye, Y.; Xue, Z.; Liu, H.; Zhou, X.; Zhang, Y.; Li, D.; Xie, X.; Mai, Y. Advances on thermally conductive epoxy-based composites as electronic packaging underfill materials-a review. *Adv. Mater.* **2022**, *34*, 2201023. [CrossRef]
113. Qian, Y.; Luo, Y.; Haruna, A.Y.; Xiao, B.; Li, W.; Li, Y.; Xiong, T.; Jiang, Q.; Yang, J. Multifunctional epoxy-based electronic packaging material MDCF@LDH/EP for electromagnetic wave absorption, thermal management, and flame retardancy. *Small* **2022**, *18*, 2204303. [CrossRef] [PubMed]
114. Hu, Y.; Chen, C.; Wen, Y.; Xue, Z.; Zhou, X.; Shi, D.; Hu, G.; Xie, X. Novel micro-nano epoxy composites for electronic packaging application: Balance of thermal conductivity and processability. *Compos. Sci. Technol.* **2021**, *209*, 108760. [CrossRef]
115. Wang, D.; Ren, S.; Chen, J.; Li, Y.; Wang, Z.; Xu, J.; Jia, X.; Fu, J. Healable, highly thermal conductive, flexible polymer composite with excellent mechanical properties and multiple functionalities. *Chem. Eng. J.* **2022**, *430*, 133163. [CrossRef]
116. Jia, L.; Jin, Y.; Ren, J.; Zhao, L.; Yan, D.; Li, Z. Highly thermally conductive liquid metal-based composites with superior thermostability for thermal management. *J. Mater. Chem. C* **2021**, *9*, 2904. [CrossRef]

117. Liu, Z.; Li, J.; Liu, X. Novel functionalized BN nanosheets/epoxy composites with advanced thermal conductivity and mechanical properties. *ACS Appl. Mater. Interfaces* **2020**, *12*, 6503. [CrossRef] [PubMed]
118. Yan, H.; Dai, X.; Ruan, K.; Zhang, S.; Shi, X.; Guo, Y.; Cai, H.; Gu, J. Flexible thermally conductive and electrically insulating silicone rubber composite films with BNNS@Al_2O_3 fillers. *Adv. Compos. Hybrid Mater.* **2021**, *4*, 36. [CrossRef]
119. Xiao, C.; Tang, Y.; Chen, L.; Zhang, X.; Zheng, K.; Tian, X. Preparation of highly thermally conductive epoxy resin composites via hollow boron nitride microbeads with segregated structure. *Compos. Part A Appl. Sci. Manuf.* **2019**, *121*, 330–340. [CrossRef]
120. Yan, R.; Su, F.; Zhang, L.; Li, C. Highly enhanced thermal conductivity of epoxy composites by constructing dense thermal conductive network with combination of alumina and carbon nanotubes. *Compos. Part A Appl. Sci. Manuf.* **2019**, *125*, 105496. [CrossRef]
121. Liu, J.; Feng, H.; Dai, J.; Yang, K.; Chen, G.; Wang, S.; Jin, D.; Liu, X. A Full-component recyclable Epoxy/BN thermal interface material with anisotropy high thermal conductivity and interface adaptability. *Chem. Eng. J.* **2023**, *469*, 143963. [CrossRef]
122. Yan, Q.; Dai, W.; Gao, J.; Tan, X.; Lv, L.; Ying, J.; Lu, X.; Lu, J.; Yao, Y.; Wei, Q.; et al. Ultrahigh-aspect-ratio boron nitride nanosheets leading to superhigh in-plane thermal conductivity of foldable heat spreader. *ACS Nano* **2021**, *15*, 6489–6498. [CrossRef]
123. Li, M.; Shen, X.; Chen, X.; Gan, J.; Wang, F.; Li, J.; Wang, X.; Shen, Q. Thermal management of chips by a device prototype using synergistic effects of 3-D heat-conductive network and electrocaloric refrigeration. *Nat. Commun.* **2022**, *13*, 5849. [CrossRef] [PubMed]
124. Prunet, G.; Pawula, F.; Fleury, G.; Cloutet, E.; Robinson, A.J.; Hadziioannou, G.; Pakdel, A. A review on conductive polymers and their hybrids for flexible and wearable thermoelectric applications. *Mater. Today Phys.* **2021**, *18*, 100402. [CrossRef]
125. Faruk, M.O.; Ahmed, A.; Jalil, M.A.; Islam, M.T.; Adak, B.; Hossain, M.M.; Mukhopadhyay, S. Functional textiles and composite based wearable thermal devices for Joule heating: Progress and perspectives. *Appl. Mater. Today* **2021**, *23*, 101025. [CrossRef]
126. Sun, N.; Li, X. A flexible composite phase change material with ultrahigh stretchability for thermal management in wearable electronics. *J. Mater. Sci.* **2021**, *56*, 15937–15949. [CrossRef]
127. Ma, Q.; Wang, X.; Chen, Y.; Chen, L.; Zhang, L.; Zhao, X.; Bing, N.; Yu, W.; Xie, H. Poly (vinyl alcohol)-based nanofibers with improved thermal conductivity and efficient photothermal response for wearable thermal management. *ACS Appl. Nano Mater.* **2023**, *6*, 14733–14744. [CrossRef]
128. Chen, H.; Ding, Y.; Zhu, G.; Liu, Y.; Fang, Q.; Bai, X.; Zhao, Y.; Li, X.; Huang, X.; Zhang, T.; et al. A new route to fabricate flexible, breathable composites with advanced thermal management capability for wearable electronics. *NPJ Flex. Electron.* **2023**, *7*, 24. [CrossRef]
129. Tan, C.; Dong, Z.; Li, Y.; Zhao, H.; Huang, X.; Zhou, Z.; Jiang, J.; Long, Y.; Jiang, P.; Zhang, T.; et al. A high performance wearable strain sensor with advanced thermal management for motion monitoring. *Nat. Commun.* **2020**, *11*, 3530. [CrossRef]
130. Wu, K.; Zhang, Y.; Gong, F.; Liu, D.; Lei, C.; Fu, Q. Highly thermo-conductive but electrically insulating filament via a volume-confinement self-assembled strategy for thermoelectric wearables. *Chem. Eng. J.* **2021**, *421*, 127764. [CrossRef]
131. Guo, G.; Liu, Y.; Ding, Y.; Liu, W.; Zhu, G.; Hao, X.; Huang, X.; Xia, J.; Li, B.; Zhang, T.; et al. Flexible yet impermeable composites with wrinkle structured BNNSs assembling for high-performance thermal management. *NPJ Flex. Electron.* **2024**, *8*, 34. [CrossRef]
132. Fang, Y.; Dong, J.; Zhao, X.; Chen, T.; Xiang, L.; Xie, Y.; Zhang, Q. Covalently linked polydopamine-modified boron nitride nanosheets/polyimide composite fibers with enhanced heat diffusion and mechanical behaviors. *Compos. Part B Eng.* **2020**, *199*, 108281. [CrossRef]
133. Yoon, D.; Lee, H.; Kim, T.; Song, Y.; Lee, T.; Lee, J.; Seol, J.H. Enhancing the thermal conductivity of amorphous polyimide by molecular-scale manipulation. *Eur. Polym. J.* **2023**, *184*, 111775. [CrossRef]
134. Ruan, K.; Guo, Y.; Gu, J. Liquid crystalline polyimide films with high intrinsic thermal conductivities and robust toughness. *Macromolecules* **2021**, *54*, 4934–4944. [CrossRef]
135. Yu, H.; Chen, C.; Sun, J.; Zhang, H.; Feng, Y.; Qin, M.; Feng, W. Highly thermally conductive polymer/graphene composites with rapid room-temperature self-healing capacity. *Nano-Micro Lett.* **2022**, *14*, 135. [CrossRef] [PubMed]
136. Peng, S.; Yu, Y.; Wu, S.; Wang, C.H. Conductive polymer nanocomposites for stretchable electronics: Material selection, design, and applications. *ACS Appl. Mater. Interfaces* **2021**, *13*, 43831–43854. [CrossRef] [PubMed]
137. Agobi, A.U.; Louis, H.; Magu, T.O.; Dass, P.M. A review on conducting polymers-based composites for energy storage application. *J. Chem. Rev.* **2019**, *1*, 19–34.
138. Lin, X.; Zhang, X.; Liu, L.; Liang, J.; Liu, W. Polymer/expanded graphite-based flexible phase change material with high thermal conductivity for battery thermal management. *J. Clean. Prod.* **2022**, *331*, 130014. [CrossRef]
139. Huang, Q.; Li, X.; Zhang, G.; Deng, J.; Wang, C. Thermal management of lithium-ion battery pack through the application of flexible form-stable composite phase change materials. *Appl. Therm. Eng.* **2021**, *183*, 116151. [CrossRef]
140. Yin, X.; Wang, L.; Kim, Y.; Ding, N.; Kong, J.; Safanama, D.; Zheng, Y.; Xu, J.; Repaka, D.V.M.; Hippaigaonkar, K.; et al. Thermal Conductive 2D boron nitride for high-performance all-solid-state lithium–sulfur batteries. *Adv. Sci.* **2020**, *7*, 2001303. [CrossRef] [PubMed]
141. Guo, H.; Zhao, H.; Niu, H.; Ren, Y.; Fang, H.; Fang, X.; Lv, R.; Maqbool, M.; Bai, S. Highly thermally conductive 3D printed graphene filled polymer composites for scalable thermal management applications. *ACS Nano* **2021**, *15*, 6917–6928. [CrossRef]
142. Cheng, G.; Wang, Z.; Wang, X.; He, Y. All-climate thermal management structure for batteries based on expanded graphite/polymer composite phase change material with a high thermal and electrical conductivity. *Appl. Energy* **2022**, *322*, 119509. [CrossRef]

143. Pei, H.; Yang, C.; Wang, P.; Lin, J.; Yin, L.; Zhou, X.; Xie, X.; Ye, Y. Efficient thermal management of lithium-sulfur batteries by highly thermally conductive LBL-assembled composite separators. *Electrochim. Acta* **2022**, *407*, 139807.
144. Li, H.; Ren, L.; Zhou, Y.; Yao, B.; Wang, Q. Recent progress in polymer dielectrics containing boron nitride nanosheets for high energy density capacitors. *High Volt.* **2020**, *5*, 365–376. [CrossRef]
145. Li, Q.; Cheng, S. Polymer nanocomposites for high-energy-density capacitor dielectrics: Fundamentals and recent progress. *IEEE Electr. Insul. Mag.* **2020**, *36*, 7–28. [CrossRef]
146. Li, Q.; Chen, L.; Gadinski, M.R.; Zhang, S.; Zhang, G.; Li, H.U.; Lagodkine, E.; Haque, A.; Chen, L.; Jackson, T.N.; et al. Flexible high-temperature dielectric materials from polymer nanocomposites. *Nature* **2015**, *523*, 576–579. [CrossRef] [PubMed]
147. Chen, J.; Zhou, Y.; Huang, X.; Yu, C.; Han, D.; Wang, A.; Zhu, Y.; Shi, K.; Li, P.; Jiang, P.; et al. Ladderphane copolymers for high-temperature capacitive energy storage. *Nature* **2023**, *615*, 62–66. [CrossRef]
148. Liu, X.J.; Zheng, M.S.; Chen, G.; Dang, Z.M.; Zha, J.W. High-temperature polyimide dielectric materials for energy storage: Theory, design, preparation and properties. *Energy Environ. Sci.* **2022**, *15*, 56–81. [CrossRef]
149. Murali, G.; Sravya, G.S.N.; Jaya, J.; Naga Vamsi, V. A review on hybrid thermal management of battery packs and it's cooling performance by enhanced PCM. *Renew. Sustain. Energy Rev.* **2021**, *150*, 111513. [CrossRef]
150. Althobiti, R.A.; Morsi, M.A.; Alzahrani, E.; Al-Muntaser, A.A. Enhancing the performance of PVC/PMMA polymer blend through hybrid nanofiller of TiO2 NPs/GNPs for capacitive energy storage applications. *Ceram. Int.* **2024**, *50*, 19039–19047. [CrossRef]
151. Sumdani, M.G.; Islam, M.R.; Yahaya, A.N.A.; Safie, S.I. Recent advancements in synthesis, properties, and applications of conductive polymers for electrochemical energy storage devices: A review. *Polym. Eng. Sci.* **2022**, *62*, 269–303. [CrossRef]
152. Sanker, S.B.; Baby, R. Phase change material based thermal management of lithium ion batteries: A review on thermal performance of various thermal conductivity enhancers. *J. Energy Storage* **2022**, *50*, 104606. [CrossRef]
153. Hwang, F.S.; Confrey, T.; Reidy, C.; Picovici, D.; Callaghan, D.; Culliton, D.; Nolan, C. Review of battery thermal management systems in electric vehicles. *Renew. Sust. Energy Rev.* **2024**, *192*, 114171. [CrossRef]
154. Nilsson, F.; Karlsson, M.; Gedde, U.W.; Kadar, R.; Gaska, K.; Mueller, C.; Hagstrand, P.O.; Olsson, B.T.; Hedenqvist, M.S.; Gkourmpis, T. Nanocomposites and polyethylene blends: Two potentially synergistic strategies for HVDC insulation materials with ultra-low electrical conductivity. *Compos. Part B Eng.* **2021**, *204*, 108498. [CrossRef]
155. Paramane, A.; Chen, X.; Dai, C.; Guan, H.; Yu, L.; Tan, Y. Electrical insulation performance of cross-linked polyethylene/MgO nanocomposite material for ±320 kV high-voltage direct-current cables. *Polym. Compos.* **2020**, *41*, 1936–1949. [CrossRef]
156. Ren, H.; Zhong, L.; Yang, X.; Li, W.; Gao, J.; Yu, Q.; Chen, X.; Li, Z. Electric field distribution based on radial nonuniform conductivity in HVDC XLPE cable insulation. *IEEE Trans. Dielectr. Electr. Insul.* **2020**, *27*, 121–127. [CrossRef]
157. Chi, Q.G.; Yang, M.; Zhang, C.H.; Zhang, T.D.; Feng, Y.; Chen, Q.G. Nonlinear electrical conductivity and thermal properties of AgNPs/BN/EPDM composites for cable accessory. *IEEE Trans. Dielectr. Electr. Insul.* **2019**, *26*, 1081–1088. [CrossRef]
158. Lin, Y.; Kang, Q.; Liu, Y.; Zhu, Y.; Jiang, P.; Mai, Y.W.; Huang, X. Flexible, highly thermally conductive and electrically insulating phase change materials for advanced thermal management of 5G base stations and thermoelectric generators. *Nano-Micro Lett.* **2023**, *15*, 31. [CrossRef] [PubMed]
159. Imai, T. Polymer Composites for Switchgears. In *Polymer Composites for Electrical Engineering*; John Wiley & Sons: Hoboken, NJ, USA, 2021; pp. 339–375.
160. Li, P.; Huang, X. Thermally conductive polymer composites for high voltage insulation. In *Polymer Composites for Electrical Engineering*; John Wiley & Sons: Hoboken, NJ, USA, 2021.
161. Wei, Y.; Chen, W.; Yuan, Z.; Zhang, C.; Zhuo, Z.; Chen, X.; Zhao, Y.B. Heat dissipation characteristics of anode saturable reactors with high thermal conductivity epoxy resin used for ultra-high-voltage direct current converter valves. *High Volt.* **2020**, *5*, 598–604. [CrossRef]
162. Nazari, M.A.; Maleki, A.; Assad, M.E.H.; Rosen, M.A.; Haghighi, A.; Sharabaty, H.; Chen, L. A review of nanomaterial incorporated phase change materials for solar thermal energy storage. *Sol. Energy* **2021**, *228*, 725–743. [CrossRef]
163. Shanmugam, M.; Augustin, A.; Mohan, S.; Honnappa, B.; Chuaicham, C.; Rajendran, S.; Hoang, T.K.A.; Sasaki, K.; Sekar, K. Conducting polymeric nanocomposites: A review in solar fuel applications. *Fuel* **2022**, *325*, 124899. [CrossRef]
164. Wang, Q.; Yang, L.; Song, J. Preparation, thermal conductivity, and applications of nano-enhanced phase change materials (NEPCMs) in solar heat collection: A review. *J. Energy Storage* **2023**, *63*, 107047. [CrossRef]
165. Min, P.; Liu, J.; Li, X.; An, F.; Liu, P.; Shen, Y.; Koratkar, N.; Yu, Z.Z. Thermally conductive phase change composites featuring anisotropic graphene aerogels for real-time and fast-charging solar-thermal energy conversion. *Adv. Funct. Mater.* **2018**, *28*, 1805365. [CrossRef]
166. Mo, Z.; Mo, P.; Yi, M.; Hu, Z.; Tan, G.; Selim, M.S.; Chen, Y.; Chen, X.; Hao, Z.; Wei, X. Ti_3C_2Tx@polyvinyl alcohol foam-supported phase change materials with simultaneous enhanced thermal conductivity and solar-thermal conversion performance. *Sol. Energy Mater. Sol. Cells* **2021**, *219*, 110813. [CrossRef]
167. Aftab, W.; Khurram, M.; Jinming, S.; Tabassum, H.; Liang, Z.; Usman, A.; Guo, W.; Huang, X.; Wu, W.; Yao, R.; et al. Highly efficient solar-thermal storage coating based on phosphorene encapsulated phase change materials. *Energy Storage Mater.* **2020**, *32*, 199–207. [CrossRef]

168. Gao, J.; Han, G.; Song, J.; He, C.; Hu, J.; Wang, W.; Feng, Y.; Liu, C. Customizing 3D thermally conductive skeleton by 1D aramid nanofiber/2D graphene for high-performance phase change composites with excellent solar-to-thermal conversion ability. *Mater. Today Phys.* **2022**, *27*, 100811. [CrossRef]
169. Gaddam, S.K.; Pothu, R.; Boddula, R. Advanced polymer encapsulates for photovoltaic devices-a review. *J. Mater.* **2021**, *7*, 920–928. [CrossRef]
170. Huang, X.; Lin, Y.; Fang, G. Thermal properties of polyvinyl butyral/graphene composites as encapsulation materials for solar cells. *Sol. Energy* **2018**, *161*, 187–193. [CrossRef]
171. Zhang, X.; Chao, X.; Lou, L.; Fan, J.; Chen, Q.; Li, B.; Ye, L.; Shou, D. Personal thermal management by thermally conductive composites: A review. *Compos. Commun.* **2021**, *23*, 100595. [CrossRef]
172. Li, L.; Liu, W.D.; Liu, Q.; Chen, Z.G. Multifunctional wearable thermoelectrics for personal thermal management. *Adv. Funct. Mater.* **2022**, *32*, 2200548. [CrossRef]
173. Kwon, Y.J.; Park, J.B.; Jeon, Y.P.; Hong, J.Y.; Park, H.S.; Lee, J.U. A review of polymer composites based on carbon fillers for thermal management applications: Design, preparation, and properties. *Polymers* **2021**, *13*, 1312. [CrossRef]
174. Soong, Y.C.; Chiu, C.W. Multilayered graphene/boron nitride/thermoplastic polyurethane composite films with high thermal conductivity, stretchability, and washability for adjustable-cooling smart clothes. *J. Colloid Interface Sci.* **2021**, *599*, 611–619. [CrossRef]
175. Huang, X.; Li, Z.; Li, Y.; Wu, X.; Liu, C.; Xie, H.; Yu, W. Thermally conductive boron nitride nanosheets on electrospun thermoplastic polyurethane for wearable janus-type fabrics with simultaneous thermal and moisture management. *ACS Appl. Nano Mater.* **2024**, *7*, 8229–8237. [CrossRef]
176. Yu, X.; Li, Y.; Wang, X.; Si, Y.; Yu, J.; Ding, B. Thermoconductive, moisture-permeable, and superhydrophobic nanofibrous membranes with interpenetrated boron nitride network for personal cooling fabrics. *ACS Appl. Mater. Interfaces* **2020**, *12*, 32078–32089. [CrossRef] [PubMed]
177. Jing, Y.; Zhao, Z.; Cao, X.; Sun, Q.; Yuan, Y.; Li, T. Ultraflexible, cost-effective and scalable polymer-based phase change composites via chemical cross-linking for wearable thermal management. *Nat. Commun.* **2023**, *14*, 8060. [CrossRef]
178. Zahid, M.; Rathore, H.A.; Tayyab, H.; Rehan, Z.A.; Rashid, I.A.; Lodhi, M.; Zubair, U.; Shahid, I. Recent developments in textile based polymeric smart sensor for human health monitoring: A review. *Arab. J. Chem.* **2022**, *15*, 103480. [CrossRef]
179. Zhang, X.; Sun, K.; Liu, H.; Chen, J.; Yan, X.; Kou, Y.; Shi, Q. Flexible insulating phase change composite film with improved thermal conductivity for wearable thermal management. *Nano Energy* **2024**, *121*, 109256. [CrossRef]
180. Lv, Y.G.; Lv, Y.Y.G.; Wang, Y.T.; Meng, T.; Wang, Q.W.; Chu, W.X. Review on thermal management technologies for electronics in spacecraft environment. *Energy Storage Sav.* **2024**, *3*, 153–189. [CrossRef]
181. Khan, J.; Singh, P. Review on phase change materials for spacecraft avionics thermal management. *J. Energy Storage* **2024**, *87*, 111369. [CrossRef]
182. Ryan, E.A.; Seibers, Z.D.; Reynolds, J.R.; Shofner, M.L. Electrically conducting polymers and composites for applications in space exploration. *J. Appl. Polym. Sci.* **2024**, *141*, e55225. [CrossRef]
183. Wang, Y.; Zhu, L.; Wang, J.; Shangguan, Y.; Zheng, Q. Flexible, high thermal conductivity, and EMI shielding performance thermoplastic polyurethane composites through reinforced-orientation carbon fiber. *ACS Appl. Polym. Mater.* **2024**, *6*, 6057–6067. [CrossRef]
184. Ryu, K.H.; Kang, M.; Kim, J.; You, N.H.; Jang, S.G.; Jeong, K.U.; Ahn, S.; Kim, D.Y. Spacesuit textiles from extreme fabric materials: Aromatic amide polymer and boron nitride nanotube composite fiber for neutron shielding and thermal management. *Adv. Fiber Mater.* **2024**, 1–12. [CrossRef]
185. Almeida, C.M.; Ghica, M.E.; Ramalho, A.L.; Durães, L. Silica-based aerogel composites reinforced with different aramid fibres for thermal insulation in space environments. *J. Mater. Sci.* **2021**, *56*, 13604–13619. [CrossRef]
186. Ince, J.C.; Peerzada, M.; Mathews, L.D.; Pai, A.R.; Al-Qatatsheh, A.; Abbasi, S.; Yin, Y.; Hameed, N.; Lau, A.K.; Salim, N.V. Overview of emerging hybrid and composite materials for space applications. *Adv. Compos. Hybrid Mater.* **2023**, *6*, 130. [CrossRef]
187. Öztürkmen, M.B.; Demirel, M.Ö.; Ağaç, Ö.; Ece, R.E.; Öz, Y. Tailored multifunctional nanocomposites obtained by integration of carbonaceous fillers in an aerospace grade epoxy resin curing at high temperatures. *Diam. Relat. Mater.* **2023**, *135*, 109840. [CrossRef]
188. Barra, G.; Guadagno, L.; Raimondo, M.; Santonicola, M.G.; Toto, E.; Vecchio Ciprioti, S. A comprehensive review on the thermal stability assessment of polymers and composites for aeronautics and space applications. *Polymers* **2023**, *15*, 3786. [CrossRef] [PubMed]
189. Ghotekar, Y.; Vartak, D.; Deshpande, N.; Makwana, B.A.; Bhatt, P.; Upadhyay, M.; Lal, A.K. Optimization of nano fillers content to fabricate electrically conductive carbon fiber reinforced polymer for space use. *Compos. Mech. Comput. Appl. Int. J.* **2023**, *14*, 79–88. [CrossRef]
190. Shen, J.; Liang, J.; Lin, X.; Lin, H.; Yu, J.; Yang, Z. Recent progress in polymer-based building materials. *Int. J. Polym. Sci.* **2020**, *2020*, 8838160. [CrossRef]
191. Rupal, A.; Meda, S.R.; Gupta, A.; Tank, I.; Kapoor, A.; Sharma, S.K.; Sathish, T.; Murugan, P. Utilization of polymer composite for development of sustainable construction material. *Adv. Mater. Sci. Eng.* **2022**, *2022*, 1240738. [CrossRef]
192. Yuan, S.; Zhou, L.; Chen, T.; Wang, D.; Wang, L. Thermally conductive h-BN/EHTPB/epoxy composites with enhanced toughness for on-board traction transformers. *e-Polymers* **2022**, *22*, 821–833. [CrossRef]

193. Verma, A. A perspective on the potential material candidate for railway sector applications: PVA based functionalized graphene reinforced composite. *Appl. Sci. Eng. Prog.* **2022**, *15*, 5727–5737. [CrossRef]
194. Kapahi, A.; McKinnon, M.; Lattimer, B. Evaluation of standard and real fire exposures on thermal response of rail car floor assembly. *Fire Mater.* **2020**, *44*, 396–408. [CrossRef]
195. Zhang, C.; Li, Y.; Kang, W.; Liu, X.; Wang, Q. Current advances and future perspectives of additive manufacturing for functional polymeric materials and devices. *SusMat* **2021**, *1*, 127–147. [CrossRef]
196. Bahru, R.; Hamzah, A.A.; Mohamed, M.A. Thermal management of wearable and implantable electronic healthcare devices: Perspective and measurement approach. *Int. J. Energy Res.* **2021**, *45*, 1517–1534. [CrossRef]
197. Dong, M.; Zhang, J.; Liu, L.; Hou, G.; Yu, Y.; Yuan, C.; Wang, X. New gutta percha composite with high thermal conductivity and low shear viscosity contributed by the bridging fillers containing ZnO and CNTs. *Compos. Part B Eng.* **2019**, *173*, 106903. [CrossRef]

solids

MDPI

Article

Establishing the Link across the Synthesis Reaction Kinetics, Structural Changes, and Photocatalytic Efficiency of an Enhanced Chitosan–Clay (1:3) Nanocomposite

Wael Albouchi †, Malek Lahbib †, Chadha Mejri, Sana Jebali, Mahdi Meftah and Walid Oueslati *

LR19ES20: Resources, Materials and Ecosystems (RME), Faculty of Sciences of Bizerte, University of Carthage, Bizerte 7021, Tunisia; meftahmahdi@yahoo.fr (M.M.)
* Correspondence: walidoueslati@ymail.com
† These authors contributed equally to this work.

Abstract: This research investigates the influence of synthesis kinetics on the structural and photocatalytic properties of chitosan–clay nanocomposites (Cs/MMT) and chitosan–hectorite nanocomposites (Cs/HET), employing an optimized initial stoichiometry of 1:3. Utilizing a variety of analytical techniques, including X-ray diffraction (XRD), transmission electron microscopy (TEM), and Fourier-transform infrared spectroscopy (FTIR), the study explores the structural evolution of the nanocomposites and their photocatalytic performance using semiconductor catalysts TiO_2 and ZnO. The findings emphasize the significant impact of reaction kinetics, particularly after 3 h of reaction time, on the structural features of the nanocomposites. Notably, Cs/MMT demonstrates greater crystalline stability compared to Cs/HET due to variations in octahedral cavity occupancy in the initial clays. FTIR and TEM analyses depict the progressive evolution of the nanocomposites during the reaction, shedding light on how reaction kinetics drive the formation of specific bonds within the nanocomposites. In terms of photocatalytic activity, this study provides insights into the complex dynamics of photocatalytic degradation, with a specific focus on the performance of TiO_2 and ZnO under diverse experimental conditions. The superior efficacy of TiO_2 as a catalyst, particularly when integrated with Cs/MMT nanocomposites, is unequivocally demonstrated, with degradation rates exceeding 80%. This preference stems from TiO_2 consistently exhibiting higher degradation rates compared to ZnO, attributed to structural disparities between montmorillonite and hectorite, influencing catalyst–support interactions. The findings underscore the critical importance of selecting suitable catalyst and support matrix combinations for optimizing performance in specific applications.

Keywords: hybrid materials; chitosan–clay nanocomposite; structural features; photocatalytic activity; reaction time; Cibacron Brilliant Yellow 3GP dye

1. Introduction

The rapid development of the textile sector has led to a significant challenge in many countries: the accumulation of organic pollutants, particularly dyes, in wastewater. Due to their chemical stability and slow biodegradation in the environment, these organic compounds pose a potential threat to the local ecosystem. To prevent the discharge of untreated wastewater containing these toxins into aquatic environments, efficient and cost-effective treatment methods are crucial.

Advanced oxidation processes (AOPs) are a promising technology for wastewater treatment. AOPs are being increasingly used to reduce the amount of organic pollutants in various wastewater streams from different industrial sites [1–4]. Among these AOPs, heterogeneous photocatalysis has demonstrated high effectiveness in breaking down resistant organic molecules and transforming them into readily biodegradable compounds [5–8].

Heterogeneous photocatalysis is a powerful and versatile environmental remediation technique that leverages semiconductor materials to accelerate chemical reactions when exposed to light. This process excels at breaking down a broad spectrum of resistant organic molecules.

One of the key advantages of heterogeneous photocatalysis is its ability to tackle persistent organic pollutants (POPs) and other stubborn contaminants present in water and air. By harnessing the energy of photons (light particles), photocatalysts generate electron–hole pairs that trigger redox reactions. These reactions degrade complex organic compounds into simpler, less harmful substances. This process is particularly effective for treating challenging pollutants like pesticides, pharmaceutical residues, and industrial dyes, which often prove difficult to remove using conventional methods.

Despite its strengths, implementing heterogeneous photocatalysis comes with some challenges. The effectiveness of this process hinges heavily on choosing the right semiconductor material, designing the catalyst effectively, and optimizing reaction conditions [9–11].

Among semiconductor metal oxides, titanium dioxide (TiO_2) reigns supreme as the most widely used photocatalyst for organic molecule degradation. However, it possesses a relatively high bandgap energy (3.2 eV), limiting its light absorption to ultraviolet light. Similarly, zinc oxide (ZnO) has become a popular and cost-effective alternative due to its decent photoactivity and a bandgap energy comparable to TiO_2 [5,12,13].

A groundbreaking development in photocatalysis research is the emergence of nanocomposite materials, which offer a new approach to environmental cleanup and pollutant degradation. These innovative materials integrate nanoscale structures, often combining multiple materials or phases, to enhance catalytic activity and absorb a wider range of light. The unique properties of nanocomposites allow them to outperform their individual components. This breakthrough opens doors for utilizing renewable energy sources and eliminating persistent organic pollutants more effectively. By meticulously designing and synthesizing nanocomposite photocatalysts, researchers see immense potential for significantly increasing the efficiency and practicality of photocatalytic processes [14–25].

Natural biopolymer chitosan and clay minerals have been innovatively combined in chitosan–clay nanocomposites. This novel substance possesses exceptional qualities as a result of its effective synthesis. Due to the presence of amino groups (-NH_2) and hydroxyl groups (-OH) acting as adsorption sites, chitosan is widely used in food preservation, cosmetic products, drug delivery systems, and wastewater treatment [5–7].

Clay adds increased mechanical strength and structural stability, while chitosan, which is made from chitin, offers biodegradability and biocompatibility. The resulting nanocomposite has enormous potential for use in a variety of fields, including biomedicine, environmental remediation, and other fields [15,26–33].

The enhanced surface contact between each component strengthens the interaction between the nanofillers and the polymer matrix. The nature of the reinforcement (nanofillers) and the type of polymer fraction establish different types of nanocomposites.

The effects of reactant ratio and type of nanofiller on the microstructural characteristics, porosity, and capability of heavy metal removal of chitosan–clay nanocomposites were explored in an earlier study by the same group of authors [32]. Dioctahedral and trioctahedral smectite clays are the two different nonofiller types used [34–37]. The outcomes show that the synthesis of the nanocomposite was effective. According to the research, a stoichiometry ratio of 1:3 promotes intercalation and increases crystallite size and crystallinity. Atomic absorption spectroscopy tests the nanocomposites' ability to adsorb Pb^{2+} and Cd^{2+} cations in solutions of slightly saturated soil, with extremely encouraging results for dioctahedral smectite.

In continuation of the earlier study [34], the focus has shifted towards investigating and optimizing the synthesis kinetics of the 1:3 sample stoichiometry. This endeavor aims to enhance the photocatalytic activity of the resulting nanocomposites. This strategic approach represents a crucial step forward in unlocking the full potential of these materials for advanced applications in environmental remediation. The goal is to delve into the underlying kinetics and subsequently elevate the photocatalytic capabilities of the synthesized nanocomposites. Indeed, while numerous research endeavors have explored chitosan–clay nanocomposites [38–40], a critical aspect has remained largely unaddressed: the influence of octahedral cavity occupancy within the base sheet. Additionally, the synthesis kinetics and the potential for optimization towards an exceptionally efficient nanocomposite have seldom been examined, particularly within the context of serious consideration for its application as a photocatalyst. This represents an important gap in the existing body of study, as understanding and fine-tuning these parameters hold significant promise for advancing the field of nanocomposite technology.

2. Materials and Methods

2.1. Starting Reagents: Formula and Composition

2.1.1. Nanofillers

This study looks at how two different types of starting clays affect nanofiller structure. One clay, hectorite, has a tri-octahedral charge, while the other, montmorillonite, has a di-octahedral charge. Montmorillonite is called SWy-2 and comes from Wyoming, USA. Hectorite, known as SHCa-1, comes from San Bernardino, CA, USA. Its chemical composition (%): SiO_2: 34.7, Al_2O_3: 0.69, TiO_2: 0.038, Fe_2O_3: 0.02, FeO: 0.25, MnO: 0.008, MgO: 15.3, CaO: 23.4, Na_2O: 1.26, K_2O: 0.13, Li_2O: 2.18, F: 2.60, P_2O_5: 0.014, S: 0.01 [38]. Both clays mainly have Mg ions in their structure, but hectorite also has Li^+ ions. The formulas of montmorillonite and hectorite show their unique properties. Montmorillonite has layers and can swap ions, while hectorite has a trioctahedral structure. Its chemical composition (%): SiO_2: 62.9, Al_2O_3: 19.6, TiO_2: 0.090, Fe_2O_3: 3.35, FeO: 0.32, MnO: 0.006, MgO: 3.05, CaO: 1.68, Na_2O: 1.53, K_2O: 0.53, F: 0.111, P_2O_5: 0.049, S: 0.05 [38]. Before use, natural samples are treated with a NaCl solution to help them disperse evenly. The process for making montmorillonite or hectorite suspension follows a set method explained in references [34,39,40] and shown in Figure 1. The samples are named MMT-Na (montmorillonite) and HET-Na (hectorite).

2.1.2. Chitosan (Cs)

This study incorporates a class of linear polysaccharides, commercially known as Cs, which are produced by partially deacetylating chitin. This compound is derived from either the exoskeletons of crustaceans or the cell walls of fungi [41,42]. Cs manifests as copolymers consisting of -(1-4)-linked glucosamine and N-acetylglucosamine. The chitosan used in this study was provided by Glentham Life Sciences, based in Wiltshire, UK, with the CAS number 9012-76-4. Detailed specifications from the supplier are outlined in Table S1. Additionally, chemical reagents, including NaCl, NaOH, acetic acid, $PbCl_2$, and $CdCl_2$, were sourced from Merck in Germany. All supplementary chemicals, meeting analytical-grade standards, were also procured from Merck in Germany. Notably, all solutions were prepared using deionized water, ensuring high-quality standards for the experimental procedures.

2.2. Experimental Synthesis Process

The synthesis of Cs/HET and Cs/MMT nanocomposites followed a well-defined protocol. Precisely weighed amounts (0.25 g chitosan, 0.75 g each of MMT-Na and HET-Na) were used. The chitosan was dissolved in a 2% acetic acid solution and centrifuged. Meanwhile, the clay fraction was dispersed in 50 mL of distilled water. Subsequently, the clay suspension was added to the chitosan solution, and the mixture was stirred at 60 °C for 6 h. The resulting composite was then oven-dried at 60 °C for 48 h. Following this, the composite underwent a 5 h soaking step in a 1 M NaOH solution, followed by multiple rinses with distilled water. Finally, the composite was dried again in an oven at 60 °C for 48 h. This meticulous procedure ensured the successful synthesis of Cs/HET and Cs/MMT nanocomposites, paving the way for subsequent analyses and characterizations.

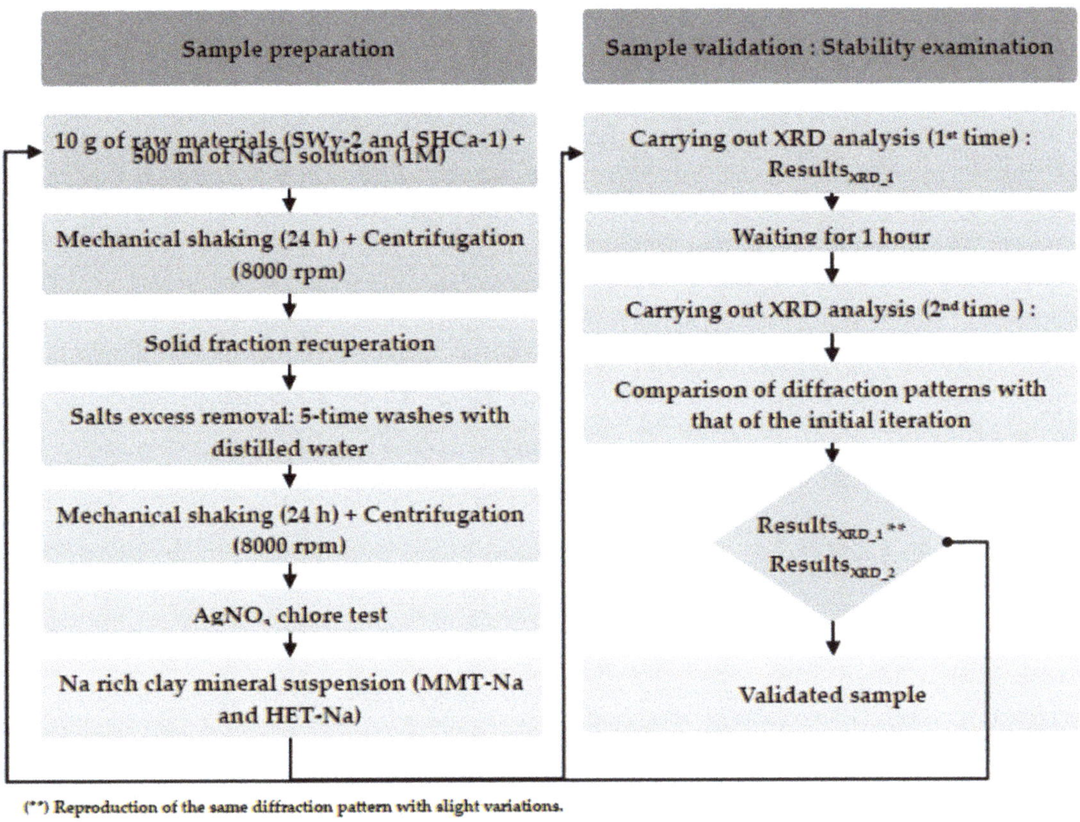

Figure 1. Pretreatment of the cationic clays MMT-Na and HET-Na [43–47].

2.3. Controlling Reaction Kinetics for Enhanced Nanocomposite Synthesis

Understanding the reaction kinetics during the synthesis of Cs/MMT and Cs/HET nanocomposites is crucial for optimizing production methods [48]. The final structure and functionality of these hybrid nanocomposites are heavily influenced by the reaction time. To monitor the kinetic behavior, samples are extracted from the reaction mixture at hourly intervals. Five samples are collected at each time point: 1, 2, 3, 4, and 5 h after the start of the reaction, for each initial mixture. It is important to note that both starting mixtures maintain a constant 1:3 ratio of chitosan to clay (Cs/MMT or Cs/HET). This thorough methodology allows for an in-depth examination of how different reaction times affect the entire synthesis process. Ultimately, by understanding these details, we can tailor the production process to achieve the desired properties in the final nanocomposite.

2.4. Characterization Techniques

2.4.1. X-ray Diffraction Analysis (XRD)

A Bruker D8 ADVANCE X-ray diffractometer equipped with monochromatic Cu-Kα radiation (λ = 1.5406 nm) was used to examine the crystalline structures of the samples. The analysis was conducted at 40 kV and 20 mA. Standard scanning parameters included a step size of 0.01° in 2θ and a counting time of 6 s per step across the investigated angular range. XRD profiles were systematically recorded for all specimens within a 2θ range of 3° to 40° [28,37,39,49–52]. The Bragg relation was employed to determine the basal spacing value, while the Scherrer equation was used to calculate the average crystallite size. It is important to note that all measurements were carried out at room temperature under atmospheric pressure [15,53–55].

2.4.2. Fourier Transform Infrared Spectroscopy (FTIR)

Thermo Nicolet's Avatar 370 FTIR spectrometer (Thermo Fisher Scientific., Schwerte, Germany), which has a KBr beam splitter and a DTGS detector, was used to obtain the FTIR spectrum. Data were gathered at a resolution of 4 cm^{-1} and over the wavelength range of 4000 to 400 cm^{-1} for each spectra using 56 scans.

2.4.3. TEM Scanning and Energy-Dispersive X-ray Spectroscopy (EDX)

Transmission electron microscopy (TEM) was employed to analyze the nanostructure and morphology of the powdered materials using a FEI Tecnai G2 (FEI, Eindhoven, Netherlands) operating at an acceleration voltage of 200 kV. To prepare TEM specimens, samples were first fractured and dispersed in ethanol. A drop of the resulting suspension was then deposited onto a holey carbon-coated copper grid. Local chemical analysis was performed using an energy-dispersive X-ray spectroscopy (EDX) system in conjunction with a Carl Zeiss Ultra 55 field-emission scanning electron microscope (SEM) (Carl Zeiss, Jena, Germany).

2.4.4. UV-Visible Spectroscopy

The Shimadzu UV-2600 spectrophotometer (Shimadzu Corporation, Kyoto, Japan) played a critical role in this investigation. This advanced equipment was employed to meticulously record both UV-visible diffuse reflectance and absorbance spectra for all samples across a broad spectral range of 200–400 nm. The high precision of the instrument ensured the acquisition of comprehensive data, facilitating a thorough analysis of the materials' optical properties within this crucial wavelength region. These insights proved invaluable in understanding the behavior of the materials under study.

2.4.5. Photocatalytic Application

The photocatalytic degradation method offers a promising approach for eliminating organic pollutants from wastewater. This technique utilizes semiconductor materials like titanium dioxide (TiO_2) and zinc oxide (ZnO) as supports for semiconductor particles with catalytic properties. In this study, we aim to improve our understanding of the Cibacron Brilliant Yellow 3GP dye degradation process by leveraging the photocatalytic properties of Cs/MMT and Cs/HET nanocomposites. These nanocomposites act as supports for the semiconductor particles, increasing the available surface area for dye adsorption and potentially leading to a synergistic effect and enhanced dispersion of the photocatalyst.

Cibacron Brilliant Yellow 3G-P ($C_{25}H_{15}Cl_3N_9Na_3O_{10}S_3$), obtained from Sigma-Aldrich (Merck KGaA, Darmstadt, Germany), possesses three sulfonate groups that become negatively charged in water. It exhibits an absorption peak at 404 nm, a color index value of 18,972, and a molecular weight of 831.02 g/mol. The stability of Cibacron Brilliant Yellow 3G-P, a synthetic product, is influenced by several factors. These factors encompass the light source, with UV light posing a greater risk of photodegradation compared to visible light. Furthermore, specific compounds like TiO and ZnO can exacerbate this degradation process, particularly in the presence of oxygen. Additionally, the pH level of the surrounding environment is pivotal, as acidic conditions can expedite the deterioration of the dye. Double-distilled water was used in all tests to eliminate impurities. NaOH (Sigma-Aldrich, Merck KGaA-Darmstadt, Germany) and nitric acid (TEDIA, Fairfield, OH, USA) were employed to adjust the pH to 5. All other chemicals, unless otherwise specified, were of analytical grade. The dye was utilized in its unpurified, as-received state [56–58].

For the experiment, a 2066 W, 230 V, 50–60 Hz UV lamp with a cooling water circuit served as the light source. The UV light was positioned 6.0 cm from the photoreactor. Figure 2 illustrates the photocatalytic reaction setup.

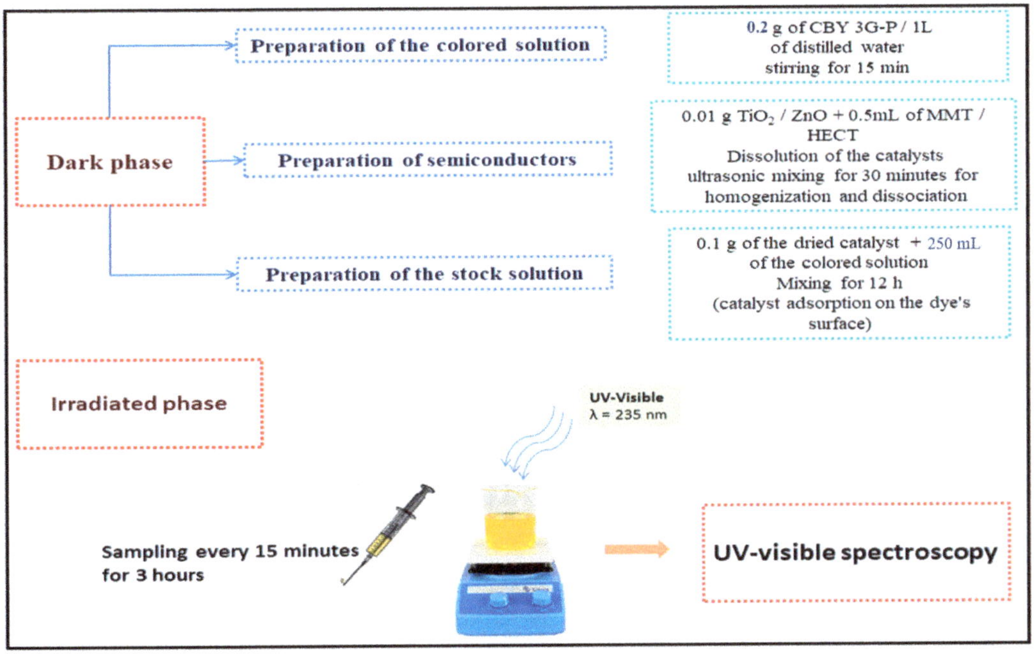

Figure 2. Experimental process for photocatalytic application.

The photocatalytic process is initiated by simultaneous exposure to ultraviolet (UV) light with a focused wavelength of 235 nm. To assess the method's efficacy, samples were collected at consistent 15 min intervals over a total duration of 180 min. This time series allows for the analysis of the degradation reaction's progress over time. Various factors are considered in this evaluation, including reaction kinetics, the type of semiconductor employed (TiO_2 or ZnO), and the octahedral occupation type of the initial nanofiller.

3. Results

3.1. XRD Analysis

XRD Study of the Structural Evolution of the Cs/MMT and Cs/HET Samples (1:3)

X-ray diffraction (XRD) analysis reveals intriguing differences in the structural evolution of Cs/MMT (1:3) and Cs/HET (1:3) samples.

Cs/MMT (1:3): After one hour of reaction (Figure 3), Cs/MMT exhibits a broad reflection at 21.82 Å, indicating an expansion of the interlamellar space due to chitosan intercalation. This is contrasted by Cs/HET, where the initial diffractogram lacks characteristic peaks, suggesting a near-complete exfoliation of hectorite layers. This difference highlights the varying reactivity of the clays—montmorillonite responds readily, while hectorite undergoes a slower, more progressive transformation.

Over time, Cs/MMT displays a dynamic interplay of reflections. The 15 Å and 12 Å peaks corresponding to hydrated interstratification fluctuate, likely due to variations in hydration and the formation of distinct layer populations. Conversely, Cs/HET exhibits a distinct reaction pattern (Figure 4). A single reflection at 16.76 Å emerges after two hours, indicative of an intermediate hydration state or the presence of chitosan within the layers. This slower reaction likely stems from differences in the clays' octahedral cavity occupancy.

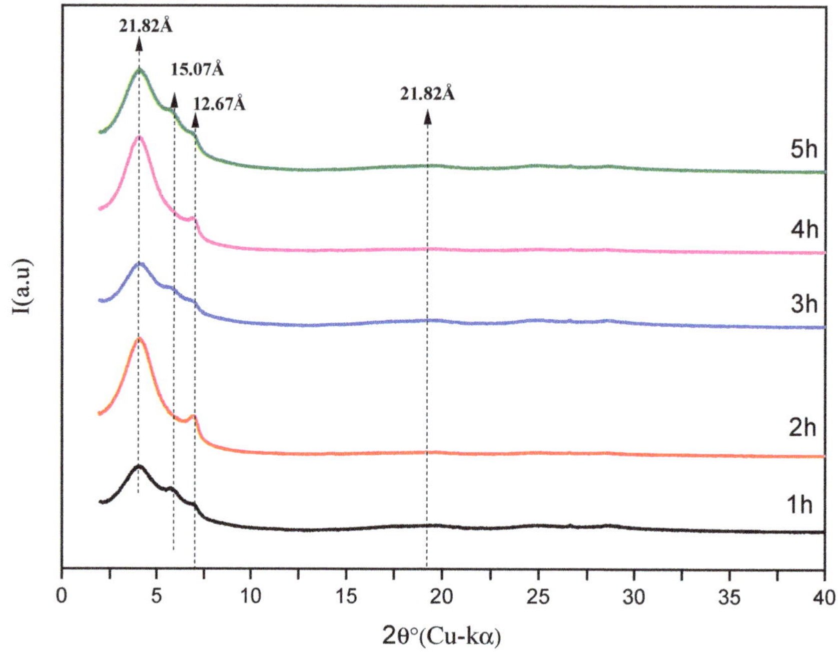

Figure 3. Evolution of experimental XRD profiles as a function of the reaction kinetics for the Cs/MMT (1:3) sample.

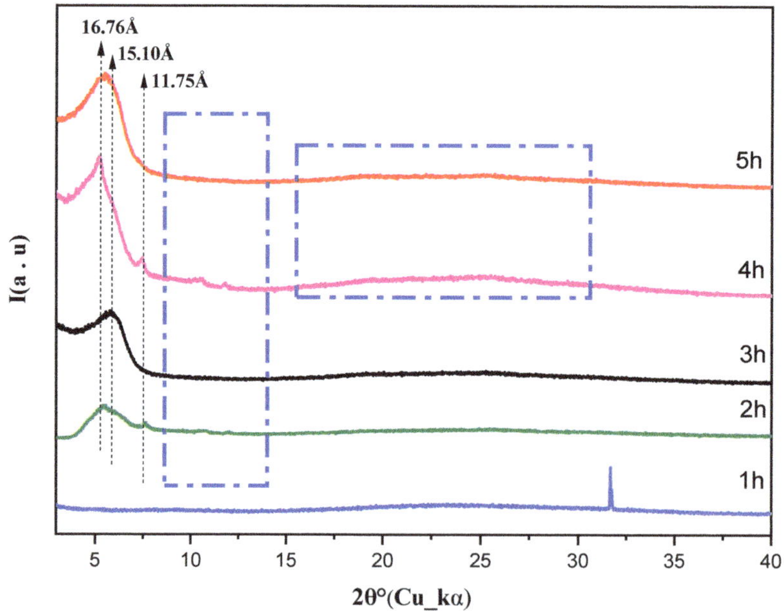

Figure 4. Evolution of experimental XRD profiles as a function of the reaction kinetics for the Cs/HET (1:3) sample.

Both samples ultimately reach a heterogeneous state, as evidenced by the emergence of chitosan reflections and the broadening of peaks (FWHM) in later stages. However, the contrasting reaction kinetics are undeniable. Cs/MMT undergoes a swifter structural reorganization due to its inherent reactivity, while Cs/HET exhibits a more gradual transformation. These findings emphasize the importance of clay type in dictating the final structure and properties of the synthesized nanocomposite [15,34,59–62].

Determining crystallite size is a pivotal procedure in characterizing the crystalline structure of materials, particularly in materials science. This parameter can be deduced from empirical data acquired through techniques like X-ray diffraction (XRD) or electron diffraction (TEM). The Scherrer method, a widely adopted approach, estimates crystallite size based on X-ray diffraction peaks. This estimation relies on Equation (1):

$$D = k\lambda/\beta\ \cos(\theta) \quad (1)$$

Indeed, the Scherrer method follows a systematic approach. First, it entails accurately measuring the full width at half maximum (FWHM) of the diffraction peak, denoted as β, from the experimental dataset. Subsequently, one needs to consider the wavelength of the X-rays (λ) used for the diffraction. Additionally, it is crucial to identify the specific diffraction angle (θ) corresponding to the peak under analysis. Once armed with these essential values, they are then inputted into the formula, culminating in the calculation of the crystallite size, represented as D. This methodological sequence is fundamental to extracting crucial insights about the crystalline structure of materials in the field of materials science. All calculated crystallite sizes are presented in Table 1.

Table 1. Calculation of crystallite size.

Samples	Time (h)	X-ray Wavelength λ (Å)	Diffraction Angle θ (°)	FWHM (rad)	Crystallite Size D (nm)
Cs/MMT	1	1.5406 Å	2.015	0.133	3.830
	2		2.045	0.116	2.129
	3		2.150	0.375	1.948
	4		2.112	0.373	1.698
	5		2.112	0.374	1.696
Cs/HET	1		*	*	**
	2		2.789	0.127	4.140
	3		2.634	0.593	3.847
	4		2.949	0.592	3.824
	5		2.789	0.591	3.559

Note: (*) denotes the absence of a peak, and (**) denotes complete exfoliation.

The crucial importance of crystallite size in optimized crafting and functional structures (intercalated or exfoliated) can be explained as follows:

When clay crystallites are of reduced dimensions, it significantly facilitates the polymer's intercalation between the clay layers. A smaller crystallite size provides more space and flexibility for the polymeric chains to fit between the layers, establishing tighter and more uniform bonds. This configuration promotes an optimal dispersion of the polymer within the clay matrix, creating an intercalated structure. Thanks to this specific arrangement, the properties and performance of the composite material are significantly enhanced. With increasing crystallite size, the challenge of polymer intercalation between the clay layers intensifies significantly. This dimensional increase restricts the available space for intercalation, making the process less straightforward. In this scenario, the clay layers are forced apart, resulting in an exfoliated structure. This transformation, offering a larger and more conducive space for polymer insertion, has notable implications for the properties and performance of the composite material.

3.2. Microstructural Study Using TEM and FTIR

3.2.1. TEM Study of the Cs/MMT (1:3) and Cs/HET (1:3) Samples

Transmission electron microscopy (TEM) unveils contrasting interaction dynamics between Cs/MMT (1:3) and Cs/HET (1:3) samples throughout the synthesis process.

Cs/MMT (1:3): After one hour, TEM images of Cs/MMT reveal a uniform structure with well-distributed chitosan (yellow circle) and montmorillonite crystallites (red circle), indicating immediate interaction (Figure 5). This contrasts sharply with Cs/HET, where TEM images show distinct phases with no interaction between hectorite and chitosan after one hour, aligning with the XRD data suggesting minimal exfoliation in Cs/HET (Figure 6).

Over time, Cs/MMT undergoes a continuous transformation. By the two-hour mark, Cs/MMT exhibits clear signs of exfoliation, with layer separation increasing around the organic component. This progressive interaction between chitosan and montmorillonite intensifies at three hours, leading to maximum exfoliation by four to five hours. TEM images effectively capture this transformation, showcasing the formation of a well-established and stable nanocomposite. In contrast, Cs/HET displays a more gradual interaction process. Initially segregated, the TEM images reveal a gradual shift towards interaction between chitosan and hectorite crystallites, starting at the two-hour mark. Similar to Cs/MMT, Cs/HET achieves maximum exfoliation in four to five hours, resulting in a stable nanocomposite. However, the slower kinetic interaction between chitosan and hectorite compared to montmorillonite is evident throughout the process. In conclusion, TEM analysis highlights the significant influence of clay type on the interaction kinetics and overall transformation process. Cs/MMT exhibits a swifter and more uniform interaction with chitosan, while Cs/HET undergoes a slower but progressive interaction, ultimately achieving a stable nanocomposite structure [15,34,61].

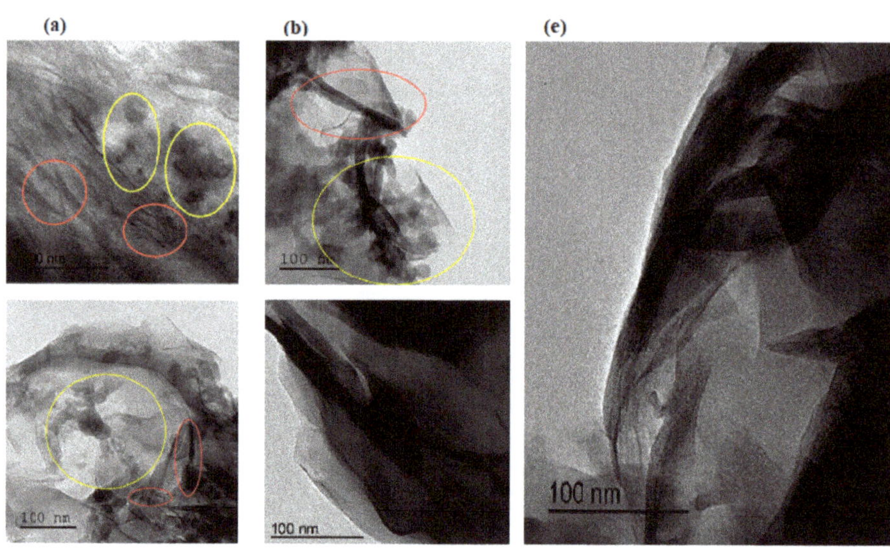

Figure 5. TEM images related to the kinetics of the preparation of the Cs/MMT sample (1:3); (**a**) after 1 h, (**b**) after 2 h, (**c**) after 3 h, (**d**) after 4 h, and (**e**) after 5 h.

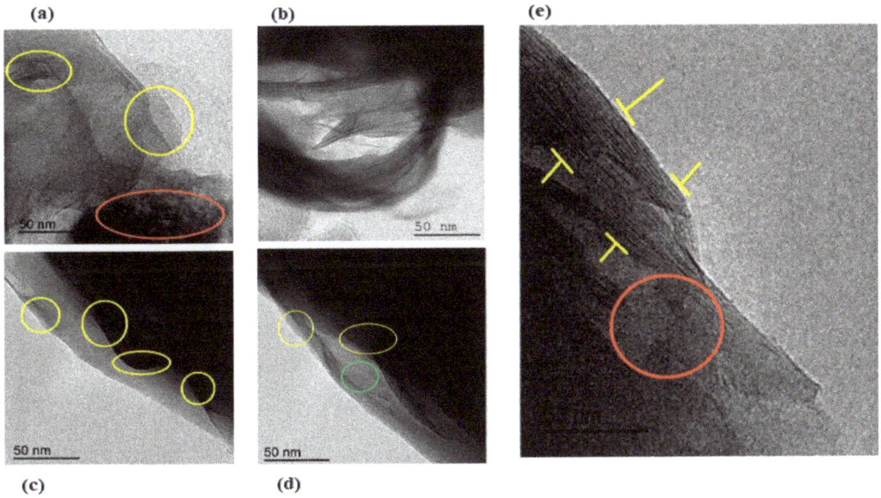

Figure 6. TEM images related to the kinetics of the preparation of the Cs/HET sample (1:3); (**a**) after 1 h, (**b**) after 2 h, (**c**) after 3 h, (**d**) after 4 h, and (**e**) after 5 h.

The comparative analysis of EDX (energy-dispersive X-ray spectroscopy) results (Table 2 and Figure 7) confirms the observations made by electron microscopy. For the CS/MMT sample, during the initial hour of synthesis, the absence of characteristic nitrogen peaks and the low presence of carbon from the polymer are consistent. This observation aligns with the predominance of the mineral fraction in the initial stoichiometry and suggests a slow reaction rate. However, after one hour of reaction, the emergence of a nitrogen peak, albeit of low intensity, likely indicates the exfoliation of clay layers, leading to a reduction in crystallite size and the appearance of a clear nitrogen fraction. This evolution is consistent with TEM observations, which demonstrated progressive exfoliation and increased interaction between chitosan and montmorillonite over time. This interpretation is also supported by the XRD results for the same sample. Additionally, the decrease in the mass fraction of exchangeable sodium and chlorine present in the interlayer space of the clay after pretreatment confirms this.

Table 2. Elemental analysis of the CS/MMT and CS/HET obtained by energy-dispersive spectrometry (EDX) coupled to TEM at 1 h and 5 h.

Element Sample	C	N	O	Na	Mg	Al	Si	Cl	Ca	Fe	Total (wt%)
CS/MMT_1h	5.87	0.10	36.13	5.75	2.43	12.24	31.52	2.74	n.d	3.22	100
CS/MMT_5h	7.12	1.27	36.38	4.89	2.08	11.92	31.04	2.43	n.d	2.87	100
CS/HET_1h	3.81	0.88	41.02	3.61	3.11	10.71	28.49	1.58	3.18	3.61	100
CS/HET_5h	3.92	2.57	39.46	4.36	1.78	11.02	27.01	1.92	4.14	3.82	100

n.d. indicates that the element was not detected or could not be determined within the limitations of the analysis.

For the CS/HET sample under the same synthesis and EDX characterization conditions, the presence of characteristic peaks of the polymer phase is observed, with intensities reflecting the initial 1:3 stoichiometry. This differs from the CS/MMT sample and is likely due to the nature of occupancy in the octahedral cavity of hectorite. The nitrogen peak is observable after one hour of reaction, indicating a faster reaction rate compared to montmorillonite.

Furthermore, the exfoliation of layers is confirmed by the appearance of a nitrogen-associated peak after one hour of synthesis. This observation corresponds with TEM results, which showed a progressive interaction between chitosan and hectorite, ultimately leading to a stable nanocomposite structure. In summary, the comparative analysis between electron microscopy and EDX results highlights the correlation between the evolution of interactions observed at the nanoscale and changes in the chemical composition of the samples, thereby confirming the mechanisms of formation of CS/MMT and CS/HET nanocomposites during the synthesis process.

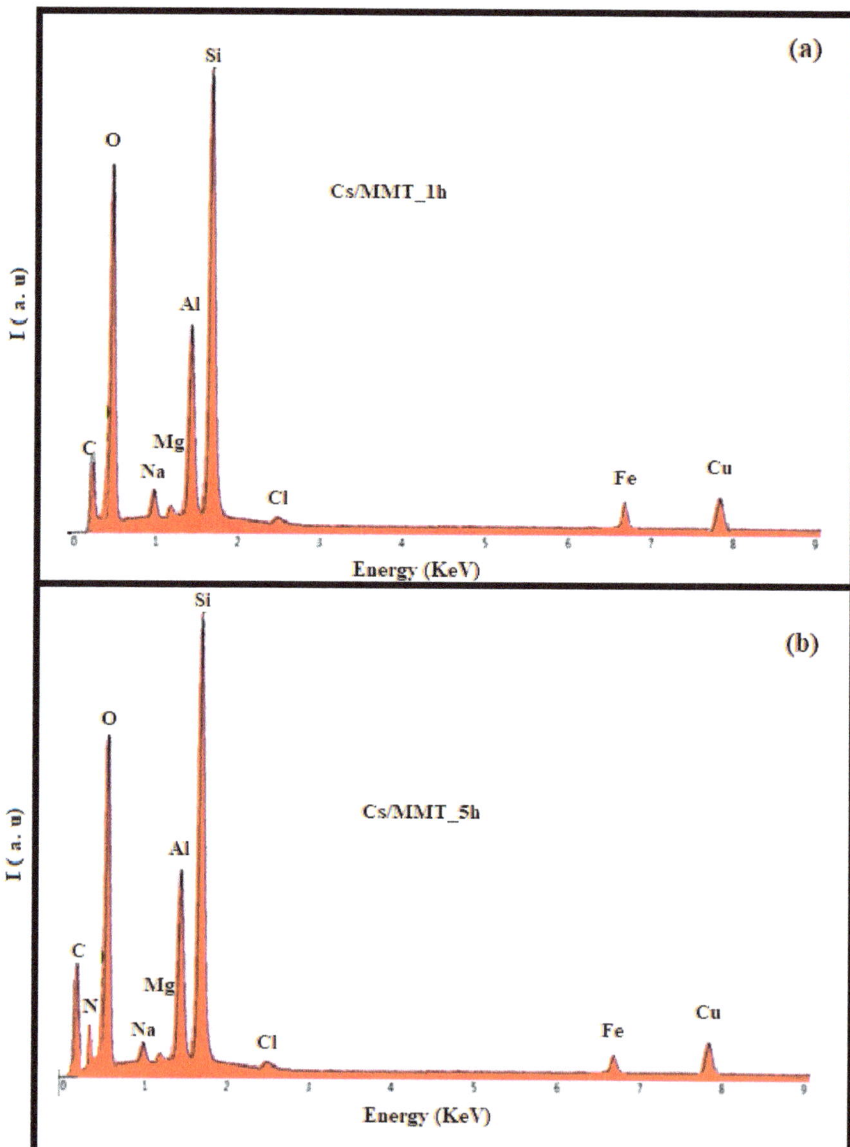

Figure 7. *Cont.*

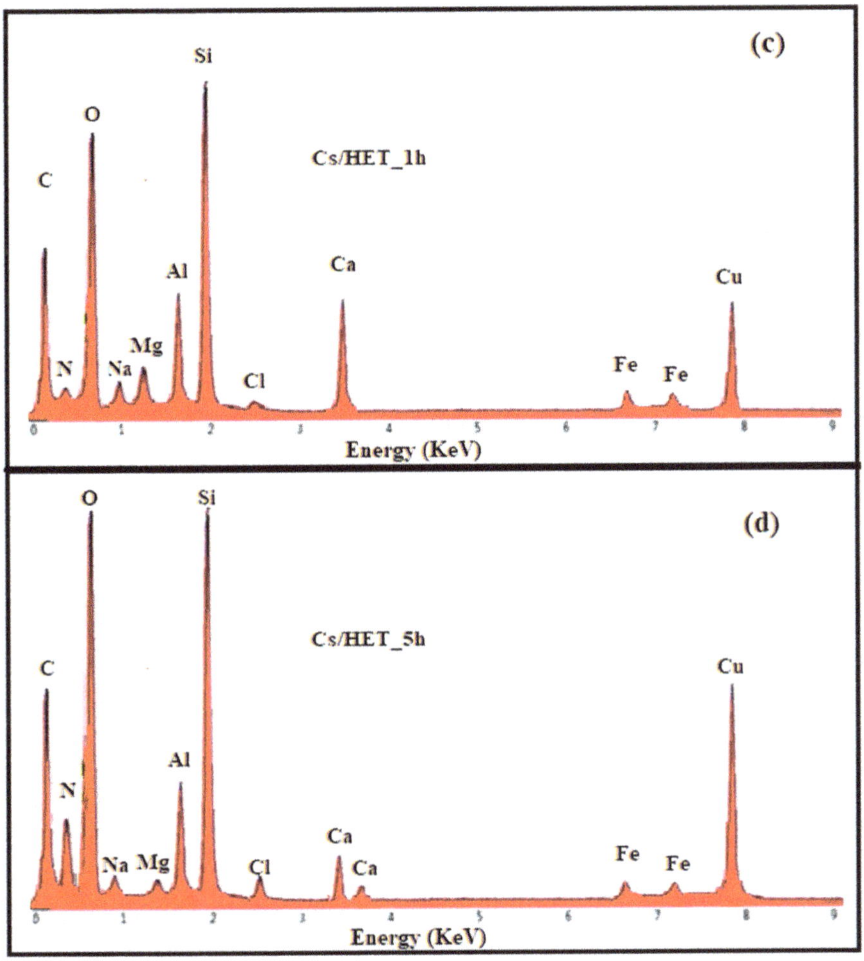

Figure 7. The EDX spectra of synthesized nanocomposite (**a**) Cs/MMT (1:3) after 1 h reaction time, (**b**) Cs/MMT (1:3) after 5 h reaction time, (**c**) Cs/HET (1:3) after 1 h reaction time, and (**d**) Cs/HET (1:3) after 5 h reaction time.

3.2.2. Characterization through FTIR of the Effect of Kinetics Reaction

Case of the Cs/MMT (1:3) and Cs/HET (1:3) Sample

FTIR analysis of Cs/MMT (1:3) and Cs/HET (1:3) nanocomposites (Figures 8 and 9) reveals both commonalities and contrasting interaction trends. While both exhibit characteristic bands for chitosan and clay components, the Cs/MMT spectra (Figure 7) show increasing band intensity with reaction time, suggesting ongoing interaction and potential new bond formation. Conversely, the Cs/HET spectra (Figure 8) display a decrease in band intensity, particularly for chitosan's amide II bond, implying potential disruption or deformation during synthesis. Additionally, the presence of a Si-O-Al bond (~500 cm^{-1}) solely in Cs/MMT confirms the distinct structure of montmorillonite compared to hectorite. Overall, FTIR analysis highlights the following contrasting interaction dynamics: Cs/MMT demonstrates progressive interaction, while Cs/HET exhibits a weakening interaction with reaction time, influencing the final structure and properties of the synthesized nanocomposites [15,34,63,64].

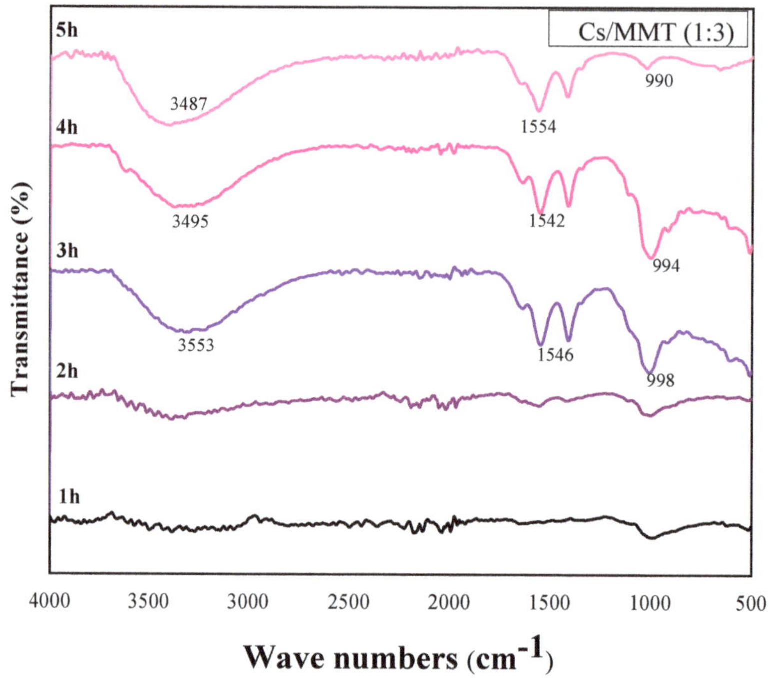

Figure 8. IR transmission spectra of Cs/MMT (1:3)_1h, Cs/MMT (1:3)_2h, Cs/MMT (1:3)_3h, Cs/MMT (1:3)_4h, and Cs/MMT (1:3)_5h.

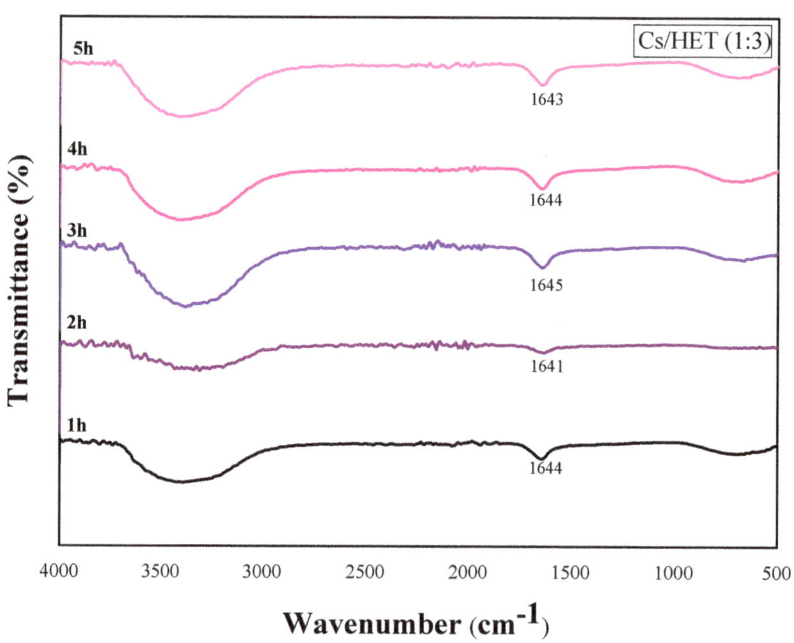

Figure 9. Transmission IR spectra of Cs/HET (1:3)_1h, Cs/HET (1:3)_2h, Cs/HET (1:3)_3h, Cs/HET (1:3)_4h, and Cs/HET (1:3)_5h.

3.3. *Photocatalytic Activity*

3.3.1. Effect of Reaction Kinetics on Dye Degradation

The kinetic data strikingly unveil the evolving dynamics orchestrated by the degradation of Brilliant Yellow Cibacron 3GP dye under irradiation. By examining samples subjected to various time intervals (1 h, 2 h, 3 h, 4 h, and 5 h), it is evident that their absorbance gradually decreases, a telling sign of the dye's progressive deterioration. This reduction in intensity visually illustrates the ongoing degradation process. Particularly noteworthy is that the intensification of this destructive phenomenon becomes markedly more pronounced as the exposure duration increases, highlighting a direct correlation between the reaction kinetics and the degradation efficiency. This significantly underscores that both the rate and extent of the degradation process are profoundly influenced by two key factors: the exposure duration and the degree of agitation. These interdependent elements are captured and depicted in Figure 10, Figure 11 and Figures S1–S9, offering a visual perspective on how the reaction kinetics shape the evolution of the dye degradation process.

Cs/MMT Sample

In the context of organic contaminants, TiO_2 demonstrated a notable 32% degradation efficiency within the chitosan–montmorillonite nanocomposites (Figure 10a). This observation signifies that TiO_2 possesses the capability to effectively degrade a portion of the organic contaminants present in the samples when employed as a photocatalyst. Depending on the specific application scenario, a 32% degradation rate may hold significant value, highlighting TiO_2's capacity to generate electron–hole pairs that actively contribute to the decomposition of organic molecules. In contrast, ZnO (Figure 10b) exhibited a slightly lower degradation efficiency of 23%. This disparity suggests that, within the given context of chitosan–montmorillonite nanocomposites, ZnO is comparatively less proficient in degrading organic pollutants. Nonetheless, ZnO retains the capacity to catalyze photocatalytic degradation processes, albeit with a reduced degradation rate compared to TiO_2 (Figure 10).

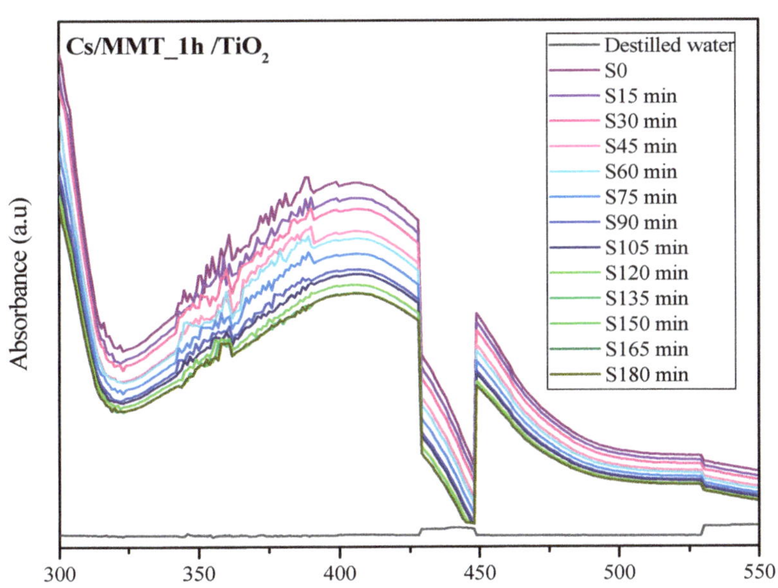

(**a**)

Figure 10. *Cont.*

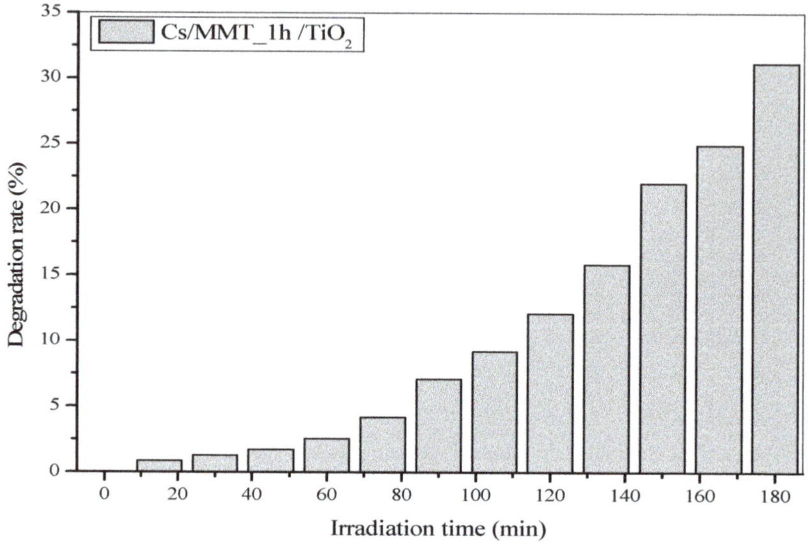

(b)

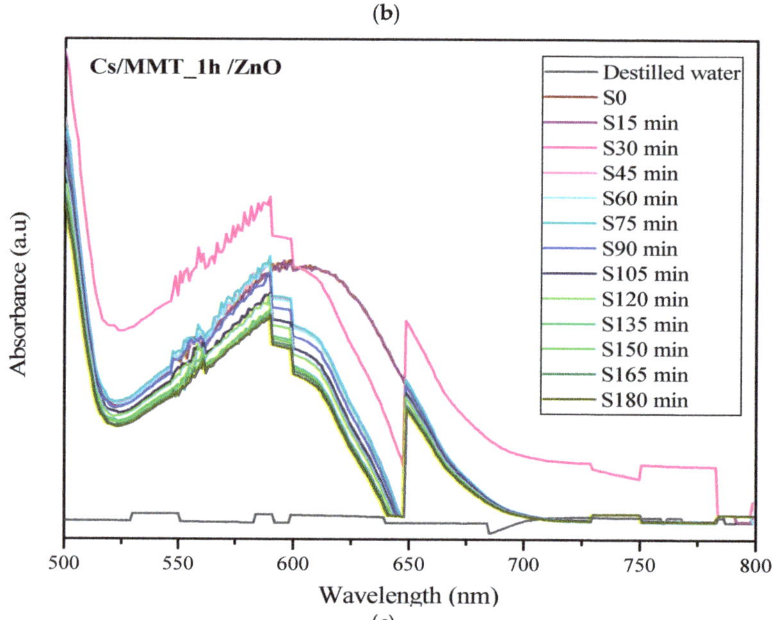

(c)

Figure 10. *Cont.*

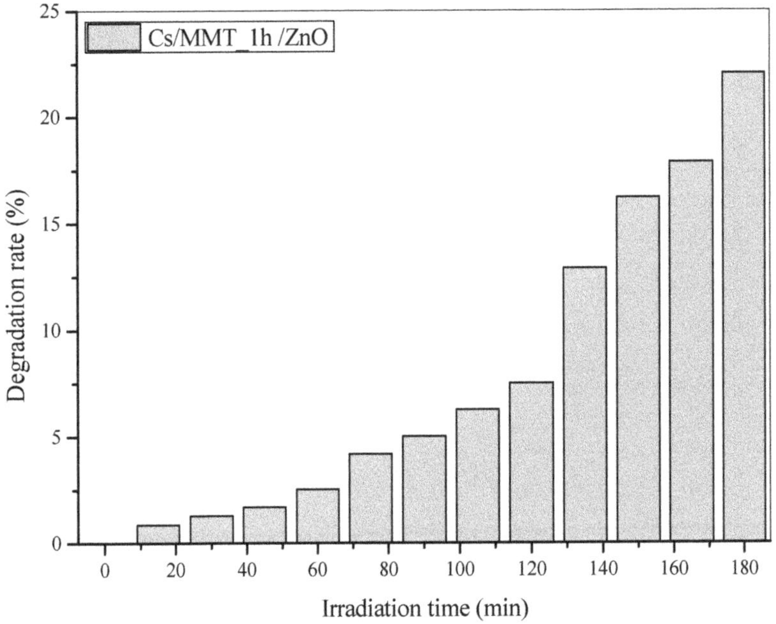

(d)

Figure 10. (**a**) Catalytic degradation of CBY3G-P by Cs/MMT nanocomposites for 1 h in the presence of TiO_2, (**b**) photocatalytic degradation rate of CBY3G-P dye on Cs/MMT, TiO_2 for 1 h under 180 min of irradiation, (**c**) catalytic degradation of CBY3G-P by Cs/MMT nanocomposites for 1 h in the presence of ZnO, and (**d**) photocatalytic degradation rate of CBY3G-P dye on Cs/MMT, ZnO for 1 h under 180 min of irradiation.

By extending the agitation time during the synthesis of chitosan–montmorillonite nanocomposites from 1 to 2 h (Figure S1), there was a marginal uptick in the degradation rate for ZnO, from 23% to 25%, and a more substantial rise for TiO_2, from 32% to 37%. Regarding ZnO, the increased agitation time had a modest impact on its efficacy in degrading organic pollutants. Although the difference in degradation rates between 23% and 25% is relatively small, it suggests a positive response of ZnO to prolonged reaction time. This could be attributed to the extended duration allowing for improved interaction between ZnO and organic pollutants, thereby facilitating degradation. In contrast, TiO_2 exhibited a more pronounced enhancement in efficiency with the extended agitation time, escalating from 32% to 37%. This more substantial increase indicates that TiO_2 is particularly sensitive to prolonged reaction time. It likely had more opportunity to generate electron–hole pairs under UV irradiation, resulting in a more efficient degradation of organic pollutants.

Increasing the agitation time from 2 h to 3 h (Figure S2) exerted a remarkable impact on the photocatalytic degradation rate of chitosan–montmorillonite nanocomposites, yielding impressive results. The degradation rate surged from 23% to 45% for ZnO and from 37% to an impressive 60% for TiO_2 (Figure 10). For ZnO, the extension in agitation time from 2 h to 3 h resulted in a noteworthy enhancement in its efficacy in degrading organic pollutants. The degradation rate leaped from 23% to 45%, signifying a substantial 22% increase. This enhancement can be attributed to the additional time that allowed ZnO to engage with organic pollutants over an extended period, leading to a more potent degradation process. This positive response to prolonged reaction times underscores ZnO's suitability for applications necessitating extended reaction durations. In comparison, TiO_2 exhibited an even more significant surge in efficiency with the extended agitation time, elevating from 37% to an impressive 60%, marking a notable 23% increase. This substantial improvement underscores the particular sensitivity of TiO_2 to extended reaction durations. The supplementary time afforded TiO_2 the opportunity to generate a greater number of electron–hole pairs, consequently enhancing the efficiency of organic pollutant degradation. TiO_2 emerged as an exceptional choice for applications demanding swift and effective degradation, especially in situations where extended reaction times are feasible.

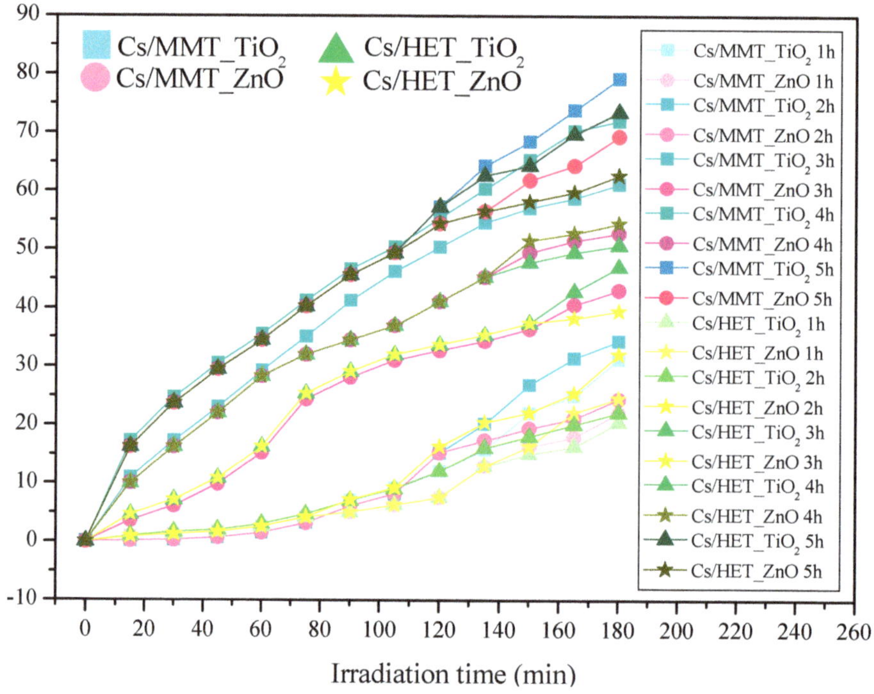

Figure 11. Photocatalytic degradation rate of CBY3G-P dye in the case of Cs/MMT and Cs/HET (catalysts TiO_2 and ZnO) versus irradiation time.

When the agitation time was extended to 4 h (Figure S3), a notable upswing in the photocatalytic degradation rate was observed for both semiconductors, ZnO and TiO_2, reaching approximately 52% for ZnO and 73% for TiO_2. For ZnO, prolonging the agitation time to 4 h brought about a substantial enhancement in its efficiency in degrading organic pollutants, achieving a degradation rate of 52%. This improvement stands out significantly when compared to the earlier rates. This marked increase underscores ZnO responsiveness to extended reaction durations. With more time for interaction with organic pollutants, ZnO demonstrated more effective degradation. ZnO proves to be a fitting choice for applications necessitating prolonged reaction times, such as continuous water or air treatment. Similarly, TiO_2 exhibited a noteworthy surge in efficiency with the extended agitation time, attaining an impressive degradation rate of 73%. Once again, this significant enhancement underscores TiO_2's particular sensitivity to extended reaction durations. The supplementary time enabled TiO_2 to generate a larger number of electron–hole pairs, thus facilitating highly efficient degradation of organic pollutants. TiO_2 emerges as an excellent option for applications where swift and efficient degradation of organic pollutants is imperative, especially when extended reaction times are viable. Further extension of the agitation time to 5 h resulted in a substantial improvement in the photocatalytic degradation rate for both semiconductors, ZnO and TiO_2, achieving a degradation rate of 69% for ZnO and 78% for TiO_2.

Extending the agitation time to 5 h (Figure S4) resulted in a remarkable enhancement in the efficiency of ZnO in degrading organic pollutants, achieving a degradation rate of 69%. This improvement is notably significant compared to the earlier rates. The substantial surge in the degradation rate highlights ZnO's remarkable sensitivity to extended reaction durations. With ample time for thorough interaction with organic pollutants, ZnO facilitated highly efficient degradation. It establishes ZnO as a highly promising option for applications necessitating in-depth and continuous degradation of organic pollutants, indicating its high material efficiency. Similarly, TiO_2 exhibited a significant boost in its efficiency with the extended agitation time, achieving an impressive degradation rate of 78%. This substantial enhancement once again underscores TiO_2's extreme sensitivity to extended reaction durations. The additional time provided TiO_2 with the opportunity to generate a large number of electron–hole pairs, leading to highly efficient degradation of organic pollutants. TiO_2 emerges as an exceptional choice for applications where rapid and comprehensive degradation of organic pollutants is essential, especially in situations where extended reaction times are feasible.

Cs/HET Sample

In the initial one-hour synthesis of the chitosan–hectorite nanocomposite (Figure S5), both ZnO and TiO_2 semiconductors exhibited low photocatalytic degradation, with a rate of 20% for ZnO and 25% for TiO_2. These initial degradation rates indicate a relatively modest level of organic pollutant degradation in the medium during the first hour of agitation. This suggests that the initial hour of reaction was insufficient to achieve significant pollutant degradation. It is important to note that the low initial degradation rates do not necessarily imply the ineffectiveness of the photocatalytic degradation process. Optimization of results can potentially be achieved by extending the agitation time.

Increasing the agitation time in the synthesis of chitosan–hectorite nanocomposites from 1 h to 2 h (Figure S6) resulted in a notable enhancement in the photocatalytic degradation rate for both semiconductors, ZnO and TiO_2. Specifically, the degradation rate increased from 20% to 22% for ZnO and from 25% to 32% for TiO_2. These results can be interpreted as follows:

The extension of agitation time from 1 h to 2 h led to a subtle yet discernible increase in the degradation rate for ZnO, progressing from 20% to 22%. While this augmentation is relatively modest, it signifies that ZnO exhibits a favorable response to a slightly prolonged reaction duration. This increment may be attributed to the additional time facilitating improved interaction between ZnO and the organic pollutants, resulting in a marginal enhancement in degradation efficiency. It can be inferred that ZnO may find suitability in applications where a moderately extended reaction time is viable, though it is plausible that it approaches its peak efficiency at this juncture.

In contrast, TiO_2 displayed a more pronounced surge in efficiency with the escalation of agitation time, escalating from 25% to 32%. This advancement is notably more substantial than what was observed for ZnO. The considerable improvement strongly suggests that TiO_2 is particularly responsive to an extended reaction time. The supplementary time afforded TiO_2 the opportunity to generate a greater number of electron–hole pairs under UV irradiation, consequently promoting a more efficient degradation of organic pollutants. This highlights TiO_2 as an excellent choice for applications necessitating rapid and thorough degradation, especially under conditions of extended reaction times.

The variation in agitation time from 2 h to 3 h during the synthesis process of chitosan–hectorite nanocomposites (Figure S7) resulted in a noteworthy upswing in the photocatalytic degradation rate for both semiconductors, ZnO and TiO_2. Specifically, the degradation rate increased from 22% to 40% for ZnO and from 32% to 46% for TiO_2. The extension of agitation time from 2 h to 3 h led to a substantial surge in the degradation rate for ZnO, from 22% to 40%. This augmentation underscores the high sensitivity of ZnO to an extended reaction period. The heightened degradation rate can be attributed to the additional time allowing for more extensive and effective interaction between ZnO and organic pollutants, culminating in a more thorough degradation. These results signify that ZnO proves to be an effective choice for applications requiring a 3 h reaction time, demonstrating a notable efficiency improvement at this stage. Likewise, TiO_2 exhibited a substantial increase in efficiency with the prolongation of agitation time, surging from 32% to 46%. This escalation reaffirms that TiO_2 is highly responsive to an extended reaction time. Upon further extending the agitation time to 4 h in the synthesis of chitosan–hectorite nanocomposites, a significant elevation in the photocatalytic degradation rate was observed for both semiconductors, ZnO and TiO_2. Specifically, the degradation rate increased from 40% to 50% for ZnO and from 46% to 54% for TiO_2 (Figure 10).

The extension of agitation time from 3 to 4 h notably boosted the degradation rate for ZnO, from 40% to 50% (Figure S8). This enhancement is substantial, highlighting ZnO's positive response to prolonged reaction periods. The considerable surge in degradation rate implies that the extra time facilitated extended and more effective interaction between ZnO and organic pollutants, leading to more comprehensive degradation. These findings strongly advocate ZnO as an exceptionally efficient option for applications necessitating a 4 h reaction time, demonstrating remarkable proficiency at this stage. In parallel, TiO_2 exhibited a noteworthy uptick in efficiency with the lengthening of agitation time, advancing from 46% to 54%. This increase underscores TiO_2's high sensitivity to extended reaction periods.

With the extension of agitation time to 5 h in the synthesis of chitosan–hectorite nanocomposites (Figure S9), both semiconductors, ZnO and TiO_2, achieved their peak rates of photocatalytic degradation. At this juncture, ZnO attained a degradation rate of approximately 62%, while TiO_2 reached a remarkable rate of 73%. Extending the agitation time to 5 h proved highly advantageous for ZnO, resulting in its highest degradation rate of 62%. This outcome underscores ZnO's extreme sensitivity to prolonged reaction periods. The noteworthy degradation rate of ZnO at 5 h suggests that the additional time facilitated extensive and effective interaction between ZnO and organic pollutants, leading to thorough degradation. ZnO emerges as an exceptionally efficient option for applications where a 5 h reaction time is viable, demonstrating outstanding proficiency at this stage. Similarly, TiO_2 exhibited a substantial increase in efficiency with prolonged agitation time, culminating in its highest degradation rate of 73% at 5 h. This high degradation rate of TiO_2 at 5 h indicates that the extra time permitted the generation of a substantial number of electron–hole pairs, promoting exceedingly effective degradation of organic pollutants. TiO_2 distinguishes itself as an exceptional choice for applications necessitating rapid and thorough degradation of organic pollutants, particularly when extended reaction times are feasible.

The photocatalytic performance of a TiO_2-based nanocomposite depends on several factors, such as particle size, material interface, crystalline structure, presence of co-catalysts, stability, and synthesis conditions. The lamellar structure of montmorillonite (MMT) and its higher TiO_2 content compared to hectorite (HET) make it more susceptible to degradation. This increased degradation can affect its photocatalytic performance in the long term. In summary, the photocatalytic properties of a nanocomposite depend not only on the presence of TiO_2 but also on a synergy of intrinsic and extrinsic factors. The material's structure, chemical composition, synthesis conditions, the presence of co-catalysts, and material stability play a crucial role (Table 3).

Table 3. Comparison between samples.

Samples	Experimental Conditions	Decolorization (%)	Refs.
TiO_2/MMT	Light = 250 W mercury lamp; MMT = 40 mg L^{-1}; V = 0.5 L; t = 135 min	100.00	[65]
TiO_2/MMT	Light = 30 W mercury lamp; MMT = 20 mg L^{-1}; V = 0.01 L; t = 50 min	100.00	[66]
TiO_2/Cs	Light = 6 W mercury lamp; MMT = 10 mg L^{-1}; V = 0.15 L; t = 75 min	70.00	[67]
TiO_2/Cs	Light = solar simulator; MMT = 16.4 mg L^{-1}; V = 1 L; t = 120 min	063.60	[68]
Cs/MMT (1:3)_1h	Light = 230 W mercury lamp; MMT = 75 mg L^{-1}; V = 0.05 L; t = 180 min	024.12	Our work
Cs/MMT (1:3)_5h	Light = 230 W mercury lamp; MMT = 75 mg L^{-1}; V = 0.05 L; t = 180 min	069.27	
Cs/HET (1:3)_1h	Light = 230 W mercury lamp; MMT = 75 mg L^{-1}; V = 0.05 L; t = 180 min	024.16	
Cs/HET (1:3)_5h	Light = 230 W mercury lamp; MMT = 75 mg L^{-1}; V = 0.05 L; t = 180 min	062.51	

3.3.2. Photocatalytic Efficiency and Catalyst Type

The apparent differences in performance between the two semiconductor photocatalysts, specifically TiO_2 and ZnO, are depicted graphically in Figure 11. Particularly, samples with TiO_2 treatment degrade Cibacron Brilliant Yellow 3GP dye much more quickly than samples with ZnO. This discrepancy provides unmistakable proof of TiO_2 supremacy in terms of photocatalytic activity within the particular parameters of this reaction. Furthermore, Figure 11 highlights the differences between the two types of nanocomposites, Cs/MMT and Cs/HET, in terms of how well they degraded the Cibacron Brilliant Yellow 3GP dye. In contrast to ZnO, TiO_2 consistently outperforms ZnO in both versions of nanocomposites, suggesting that the choice of support has less of an effect on TiO_2 performance. This finding shows that, unlike ZnO, which exhibits more variable efficiency depending on the supporting material, the degradation process accelerated by TiO_2 appears to be less dependent on the type of support used. In conclusion, our thorough research convincingly demonstrates the significant impact of the catalyst's inherent properties on photocatalytic efficiency while also illuminating the differences in interactions between the catalytic materials and their distinct support matrices.

3.3.3. Relationship between Photocatalytic Performance and Octahedral Cavity Occupation of the Starting Clay Fraction

Figure 12 encapsulates the overarching trends discerned from our results. A prominent conclusion is evident: TiO_2 emerges as the catalyst of choice for the degradation process, particularly when coupled with the dioctahedral charge nanocomposites Cs/MMT. This graphical representation consolidates and provides a coherent overview of the patterns observed throughout this study. Upon meticulous examination, Figure 12 enables the deciphering and emphasis of significant relationships and comparisons across various experimental conditions.

One salient deduction is the exceptional performance of TiO_2 as the favored catalyst for the degradation of the dye Cibacron Brilliant Yellow 3GP. This conclusion stems from the markedly higher degradation rates achieved with TiO_2 in contrast to ZnO across diverse experimental setups. This divergence in photocatalytic efficacy can be attributed to a fundamental dissimilarity in the crystalline structures of the employed clays. Montmorillonite, a dioctahedral charge clay, contrasts with hectorite, which exhibits a trioctahedral nature. This incongruity in the occupancy rate of the octahedral cavity by ions directly impacts behavior and photocatalytic efficiency. The observed discrepancies in performance can be explained by the major influence that this unique structural attribute has on the interactions between catalytic materials and their respective supports. In summary, Figure 12 visually elucidates this intricate dynamic, pinpointing TiO_2 as the preferred catalyst owing to its superior efficiency while underscoring the pivotal role played by the nature of the support in the performance of photocatalytic catalysts.

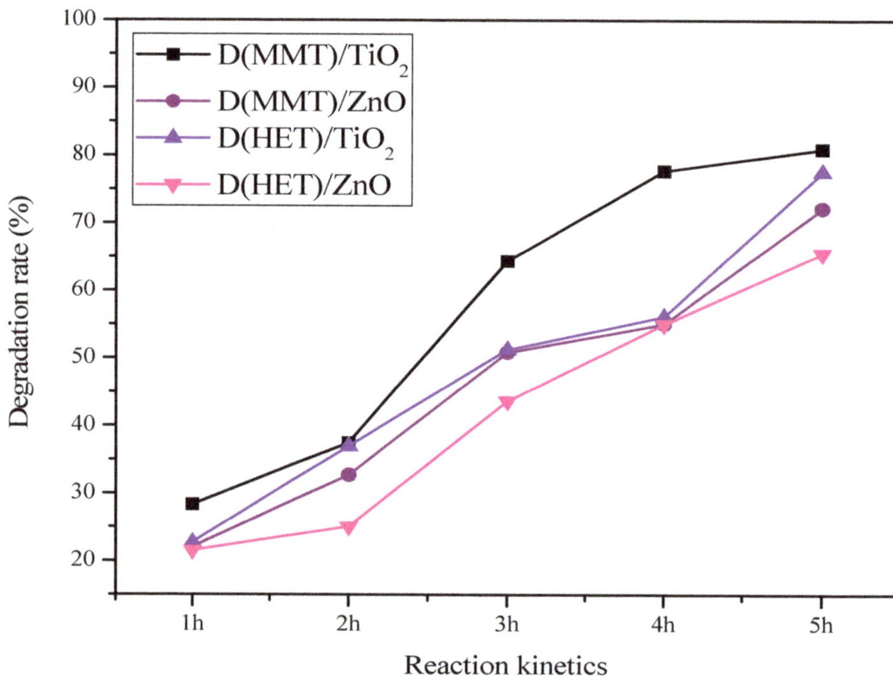

Figure 12. Correlation of results comparing the photocatalytic activity of different samples after 180 min of irradiation.

In this study, the stability of the samples is a crucial element, and to validate it, we have implemented a series of measures at various stages of the reaction. First, each manipulation was carried out with methodological rigor to ensure the reproducibility of the results. Furthermore, all manipulations were conducted under consistent atmospheric conditions for all samples, including humidity, temperature, and pressure, to minimize external variations that could influence the results. Additionally, a systematic blank test was performed before each radiographic recording, involving two passes of each sample under the diffractometer, spaced one hour apart. If any structural anomaly or difference in diffraction profiles was observed, it called into question the integrity of the entire manipulation, typically resulting in the experimental procedure being restarted from the beginning, regardless of the reaction kinetics (Figure 1).

4. Discussion

4.1. Utilizing Reaction Kinetics Assessment Tools to Optimize Nanocomposite Efficiency

Over the past 10 years, there has been a great deal of success with the use of chitosan/clay nanocomposites in environmental applications. The focus of research has been on agri-food and antibacterial uses [69]. However, the ability to integrate structural and physicochemical features from both organic and inorganic materials with nanoclays opens the door to improving numerous functionalities (such as mechanical, barrier, and antioxidant capabilities) of chitosan-based films and coatings. To achieve good adsorptive or degradative activity in the presence of organic molecules, the production of these nanocomposites has been optimized [34]. The success of the synthesized product is directly related to the decision to use the synthesis process and the method by which the application is carried out. The most commonly employed method for producing chitosan–clay nanocomposite films is solution casting, which is both cost-effective and straightforward to prepare. The approach utilized in this article aligns with this technique, underscoring its efficiency and applicability in nanocomposite film production [34,70,71]. A versatile method for creating thin coatings with mixed functional properties of various polymers is layer-by-layer (LbL) self-assembly [72]. Hydrophobic interaction, hydrogen bonding, van der Waals forces, and electrostatic interactions are among the mechanisms of deposition that depend on mutual attraction between layers [73]. The cationic and hydrophilic properties of chitosan in acidic environments enable high miscibility with negatively charged nanoclays, including Montmorillonite (MMT) platelets, in chitosan–clay nanocomposites [74]. Solutions can be sprayed over the substrate or submerged to accomplish experimental deposition [75]. Although studies employing LbL deposition for chitosan–clay nanocomposite films are limited, the approach exhibits potential for film creation [74]. The mechanical, barrier, antioxidant, and antibacterial properties of chitosan-based films can all be enhanced by incorporating nanoclays at an optimized initial stoichiometry. This is demonstrated in this study by establishing an optimized starting stoichiometry and delving further into quantifying the kinetics of the reaction, enabling the tracking of structural evolution and precise identification of the link between structure and photocatalytic effectiveness. The three fundamental structures—tactoid, intercalation, and exfoliation—have a strong correlation with the physicochemical characteristics of chitosan–clay nanocomposites, which is pertinent to note [76–78]. In accordance with the cited references, the same group of authors has previously worked on the optimization of the initial stoichiometry (from a structural standpoint) [34]. A highly detailed process that allowed reactant concentrations to be adjusted in accordance with possible applications made this synthesis possible [34]. The results of these experiments form the basis of this research, which attempts to improve the control and optimize the synthesis. Specifically, a starting stoichiometry of 1:3 was employed. After the synthesis reaction time was optimized by spaced-time samplings, we were able to fully comprehend the changes that affected the structure and even the photocatalytic performance in the presence of Cibacron Brilliant Yellow 3GP dye. Indeed, characteristic features of a nanocomposite emerge as early as the first hour of synthesis, demonstrating photocatalytic activity in the presence of ZnO or TiO_2. The nanocomposite efficiency increases regardless of the semiconductor used as a function of reaction time. This adaptation has made it possible to objectively define the synthesis time, which was, to the best of our knowledge, frequently addressed in qualitative terms in all earlier research.

4.2. Synchronized Influence of Clay Nature, Octahedral Cavity Occupancy, Initial Stoichiometry, Metal Oxide Type, and Polymer Concentration on Photocatalytic Efficiency in Nanocomposites

Previous studies have made significant contributions to understanding the photocatalytic activity of clay and clay–polymer nanocomposites combined with metal oxide nanoparticles. For instance, [79] highlighted the formation of a robust photocatalyst resulting from the heterojunction between ZnO/TiO_2 semiconductor nanoparticles and delaminated layers of natural clay. The obtained results indicate that the nanocomposite exhibits intense adsorption and a notable degradation rate, allowing for the complete decolorization of Methylene Blue in just 10 min. The acquired results corroborate those found during this study. Indeed, apart from the clay and metallic oxide (ZnO or TiO_2), a fraction of the polymer, with an optimally calibrated initial concentration, was employed to underscore the degradative potential of the nanocomposite in the presence of Cibacron Brilliant Yellow 3GP dye.

Similarly, the study by [80] emphasizes the significance of clay-based nanocomposites incorporating TiO_2 and ZnO in pioneering water treatment processes, highlighting the potential of nanostructured filters for water treatment applications. In this context, our contribution is characterized by the optimization of the initial stoichiometry between chitosan and clay (1:3) and an in-depth investigation into the effect and efficacy of the starting clay's nature, particularly concerning the occupancy of the octahedral cavity in two distinct smectites. This expansion of knowledge broadens the scope of potential applications in the field.

The study conducted by [81] on the adsorption and photocatalysis of Methylene Blue using a sodium carbonate-activated bentonite adsorbent and a ZnO/bentonite photocatalyst demonstrates an impressive 99.54% degradation efficiency with a photocatalyst of an average size of 30.06 nm. However, it is important to note that this study overlooks the potential influence of the polymer fraction in the nanocomposite synthesis process. This leads to results that are highly specific and limited to localized effects. While clay proves to be an effective porous medium for adsorption and degradation, the properties of the nanocomposite remain somewhat restricted due to the oversight of the enhanced mechanical characteristics provided by the presence of the polymer fraction. In contrast, our study, building upon the foundation laid by the same authors in their previous studies [34], introduces a novel aspect. We achieve highly encouraging degradation rates through the optimization of a complex structure, offering a more comprehensive understanding of the nanocomposite's capabilities.

In recent investigations, ref. [82] explored the suppression of photocatalytic activity by encapsulating TiO_2 nanoparticles with chitosan using a spray-drying method, targeting potential applications in sunblocking. The study revealed a significant reduction in photoactivity for chitosan-encapsulated nanoparticles compared to pristine TiO_2, with degradation rates dropping from 95% to 39.5% after 120 min. This highlights the potential of the straightforward coating process and chitosan material as an inactive protective layer for sunblocking applications. Additionally, ref. [83] delved into the adsorption and photocatalytic-conjugated activity of a chitosan-functionalized titanate coating for clonazepam removal from drinking water. Their research evaluated $H_2Ti_2O_7$ nanotubes functionalized with varying percentages of chitosan, demonstrating a substantial 80.79% removal of clonazepam after 240 min, surpassing both TiNT and commercial TiO_2 powders by 4.38% and 49.64%, respectively. Moreover, the 2TiCN coating exhibited a remarkable 6.88% increase in efficiency after four reuses. The Ref. [84] study focused on the synthesis of chitosan/SnO_2 nanocomposites through chemical precipitation for enhanced visible light photocatalytic degradation of Congo red and rhodamine B dye molecules. Their findings showcased exceptional degradation rates, with 95% degradation of rhodamine B dye within 60 min, significantly outperforming SnO_2 nanoparticles. Furthermore, the chitosan/SnO_2 nanocomposites achieved a remarkable 98% degradation of Congo red dye within 40 min of visible light exposure, underscoring their superior photodegradation capabilities. Ref. [8] provided a comprehensive examination of the physicochemical and photocatalytic properties of the ZnO-bentonite/chitosan hybrid-biocomposite for water remediation under visible light conditions. Their research demonstrated a substantial increase in methyl orange (MO) removal (86.1%) in the presence of a ZnO-bentonite/chitosan hybrid biocomposite, surpassing bentonite (75.8%) and chitosan (65.4%). Under visible light irradiation, the MO removal rate was an impressive 268 times higher compared to ZnO nanoparticles, showcasing the hybrid biocomposite's potential for highly effective water remediation applications. These recent studies have primarily focused on the development of nanocomposites using chitosan and metallic oxides, often overlooking the contribution of the clay fraction. The results obtained, however, unequivocally demonstrate the potential for dye degradation in the presence of these materials, offering valuable insights. They serve as pivotal starting points for our study, which seeks to further investigate the degradation phenomenon of a broader spectrum of organic molecules, incorporating various combinations of polymer, clay, and metallic oxide components to enhance the understanding of organic dye degradation processes.

Incorporating these prior studies within the context of our research underscores the exceptional nature of our contribution. Our investigation has not only optimized the initial stoichiometry between chitosan and clay but has also provided a comprehensive understanding of how the nature of clay profoundly influences the photocatalytic performance. Notably, our exploration of hectorite-based nanocomposites with such precision in initial concentration is unprecedented in the field. Additionally, we have ventured into uncharted territory by exploiting the effect of octahedral cavity occupancy on the photocatalytic properties of the developed material. Furthermore, our study stands out for its rigorous optimization of the synthesis protocol and nanocomposite architecture, which have been validated across various types of starting clays. This versatility adds a significant dimension to our study, enhancing its applicability and impact. Additionally, we conducted a meticulous comparative analysis of the efficacy of degrading the Cibacron Brilliant Yellow 3GP dye using both ZnO and TiO_2. This comparative approach provides a nuanced understanding of the distinct contributions of these metal oxides to the degradation process. Moreover, the meticulous monitoring of the reaction kinetics has allowed us to fine-tune the synthesis time while maintaining a fixed initial stoichiometry for all components. This optimization process has led to the development of a nanocomposite with superior catalytic potential, setting a new standard in the field. Altogether, our research represents a significant advancement, pushing the boundaries of knowledge in the realm of clay-based nanocomposites and their applications in photocatalysis.

5. Conclusions

This research explores the impact of synthesis kinetics on the structural and photocatalytic properties of chitosan–clay nanocomposites (Cs/MMT) and chitosan–hectorite nanocomposites (Cs/HET), employing an optimized initial stoichiometry of 1:3. A range of analytical techniques, including X-ray diffraction (XRD), transmission electron microscopy (TEM), and Fourier-transform infrared spectroscopy (FTIR), were utilized. Additionally, the photocatalytic performance of the nanocomposites was assessed using semiconductor catalysts TiO_2 and ZnO, focusing on reaction kinetics, catalyst characteristics, and octahedral cavity occupancy of the clay affecting photocatalytic efficiency. The results highlight the significant influence of reaction kinetics on the structural features of the nanocomposites. Notably, the structural evolution of Cs/MMT and Cs/HET was notably affected by reaction duration, with Cs/MMT exhibiting higher crystalline stability compared to Cs/HET due to differences in octahedral cavity occupancy in the initial clays. FTIR and TEM analyses facilitated the observation of component interactions, revealing the progressive evolution of the nanocomposites during the reaction. These findings elucidate how reaction kinetics drive the formation of specific bonds within the nanocomposites.

Regarding photocatalytic activity, this comprehensive study provides insights into the complex dynamics of photocatalytic degradation, with a specific focus on the performance of TiO_2 and ZnO under varied experimental conditions. The results unequivocally demonstrate the superior performance of TiO_2 as a catalyst, particularly when integrated with the dioctahedral charge nanocomposites Cs/MMT. This preference arises from TiO_2 consistently exhibiting higher degradation rates compared to ZnO across various experimental setups. A crucial factor influencing this difference in photocatalytic efficacy lies in the distinct crystalline structures of the utilized clays. The contrasting nature of montmorillonite, a dioctahedral charge clay, with hectorite, exhibiting a trioctahedral nature, significantly impacts the occupancy rate of the octahedral cavity by ions. This structural disparity plays a pivotal role in dictating the interactions between catalytic materials and their respective supports, thereby accounting for the observed variations in performance. The findings underscore the critical importance of selecting an appropriate catalyst and support matrix for photocatalytic processes, with TiO_2 emerging as the preferred catalyst due to its exceptional efficiency. This research emphasizes the need for tailored catalyst-material combinations to optimize performance in specific applications. Overall, this study offers valuable insights into the intricate interplay of catalyst properties and support matrices in photocatalytic processes, paving the way for more efficient and targeted applications in environmental remediation and other fields requiring effective pollutant degradation.

Supplementary Materials: The following supporting information can be downloaded at: https://www.mdpi.com/article/10.3390/solids5020015/s1, Table S1: The commercialized CS specifications given by the supplier (Glentham life sciences (Wiltshire, UK)); Figure S1: (a) Catalytic degradation of CBY3G-P by Cs/MMT nanocomposites for 2 h in the presence of TiO_2, (b) Photocatalytic degradation rate of CBY3G-P dye on Cs/MMT, TiO_2 for 2 h under 180 min of irradiation, (c) Catalytic degradation of CBY3G-P by Cs/MMT nanocomposites for 2 h in the presence of ZnO, and (d) Photocatalytic degradation rate of CBY3G-P dye on Cs/MMT, ZnO for 2 h under 180 min of irradiation; Figure S2: (a) Catalytic degradation of CBY3G-P by Cs/MMT nanocomposites for 3 h in the presence of TiO_2, (b) Photocatalytic degradation rate of CBY3G-P dye on Cs/MMT, TiO_2 for 3 h under 180 min of irradiation, (c) Catalytic degradation of CBY3G-P by Cs/MMT nanocomposites for 3 h in the presence of ZnO, and (d) Photocatalytic degradation rate of CBY3G-P dye on Cs/MMT, ZnO for 3 h under 180 min of irradiation; Figure S3: (a) Catalytic degradation of CBY3G-P by Cs/MMT nanocomposites for 4 h in the presence of TiO_2, (b) Photocatalytic degradation rate of CBY3G-P dye on Cs/MMT, TiO_2 for 4 h under 180 min of irradiation, (c) Catalytic degradation of CBY3G-P by Cs/MMT nanocomposites for 4 h in the presence of ZnO, and (d) Photocatalytic degradation rate of CBY3G-P dye on Cs/MMT, ZnO for 4 h under 180 min of irradiation; Figure S4: (a) Catalytic degradation of CBY3G-P by Cs/MMT nanocomposites for 5 h in the presence of TiO_2, (b) Photocatalytic degradation rate of CBY3G-P dye on Cs/MMT, TiO_2 for 5 h under 180 min of irradiation, (c) Catalytic degradation of CBY3G-P by Cs/MMT nanocomposites for 5 h in the presence of ZnO, and (d) Photocatalytic degradation rate of CBY3G-P dye on Cs/MMT, ZnO for 5 h under 180 min of irradiation; Figure S5: (a) Catalytic degradation of CBY3G-P by Cs/HET nanocomposites for 1 h in the presence of TiO_2, (b) Photocatalytic degradation rate of CBY3G-P dye on Cs/HET, TiO_2 for 1 h under 180 min of irradiation, (c) Catalytic degradation of CBY3G-P by Cs/HET nanocomposites for 1 h in the presence of ZnO, and (d) Photocatalytic degradation rate of CBY3G-P dye on Cs/HET, ZnO for 1 h under 180 min of irradiation; Figure S6: (a) Catalytic degradation of CBY3G-P by Cs/HET nanocomposites for 2 h in the presence of TiO_2, (b) Photocatalytic degradation rate of CBY3G-P dye on Cs/HET, TiO_2 for 2 h under 180 min of irradiation, (c) Catalytic degradation of CBY3G-P by Cs/HET nanocomposites for 2 h in the presence of ZnO, and (d) Photocatalytic degradation rate of CBY3G-P dye on Cs/HET, ZnO for 2 h under 180 min of irradiation; Figure S7: (a) Catalytic degradation of CBY3G-P by Cs/HET nanocomposites for 3 h in the presence of TiO_2, (b) Photocatalytic degradation rate of CBY3G-P dye on Cs/HET, TiO_2 for 3 h under 180 min of irradiation, (c) Catalytic degradation of CBY3G-P by Cs/HET nanocomposites for 3 h in the presence of ZnO, and (d) Photocatalytic degradation rate of CBY3G-P dye on Cs/HET, ZnO for 3 h under 180 min of irradiation; Figure S8: (a) Catalytic degradation of CBY3G-P by Cs/HET nanocomposites for 4 h in the presence of TiO_2, (b) Photocatalytic degradation rate of CBY3G-P dye on Cs/HET, TiO_2 for 4 h under 180 min of irradiation, (c) Catalytic degradation of CBY3G-P by Cs/HET nanocomposites for 4 h in the presence of ZnO, and (d) Photocatalytic degradation rate of CBY3G-P dye on Cs/HET, ZnO4 h under 180 min of irradiation; Figure S9: (a) Catalytic degradation of CBY3G-P by Cs/HET nanocomposites for 5 h in the presence of TiO_2, (b) Photocatalytic degradation rate of CBY3G-P dye on Cs/HET, TiO_2 for 5 h under 180 min of irradiation, (c) Catalytic degradation of CBY3G-P by Cs/HET nanocomposites for 5 h in the presence of ZnO, and (d) Photocatalytic degradation rate of CBY3G-P dye on Cs/HET, ZnO for 5 h under 180 min of irradiation.

Author Contributions: W.A.: formal analysis, methodology, funding acquisition. M.L.: conceptualization, methodology, funding acquisition. C.M.: conceptualization, visualization, funding acquisition, writing—original draft. S.J.: formal analysis, methodology, funding acquisition. M.M.: formal analysis, methodology, funding acquisition. W.O.: data curation, conceptualization, formal analysis, funding acquisition, supervision, writing—original draft, writing—review and editing, project administration. All authors have read and agreed to the published version of the manuscript.

Funding: This research received no external funding.

Data Availability Statement: Data will be made available on request.

Acknowledgments: Malek Lahbib (M.L.) and Wael Albouchi (W.A.) acknowledge Walid Oueslati (W.O.) (University of Carthage-Faculty of Science of Bizerte) for his outstanding contributions to this research endeavor. The outcome of this scientific endeavor has been significantly influenced by his (W.O.) originality in addressing the subject, his constant dedication, and his tremendous support in interpreting and developing ideas.

Conflicts of Interest: The authors declare no conflicts of interest.

References

1. Fatimah, I.; Wang, S.; Wulandari, D. ZnO/montmorillonite for photocatalytic and photochemical degradation of methylene blue. *Appl. Clay Sci.* **2011**, *53*, 553–560. [CrossRef]
2. Bahnemann, D. Photocatalytic water treatment: Solar energy applications. *Sol. Energy* **2004**, *77*, 445–459. [CrossRef]
3. Comninellis, C.; Kapalka, A.; Malato, S.; Parsons, S.A.; Poulios, I.; Mantzavinos, D. Advanced oxidation processes for water treatment: Advances and trends for R&D. *J. Chem. Technol. Biotechnol.* **2008**, *83*, 769–776. [CrossRef]
4. Goswami, D.; Vijayaraghavan, S.; Lu, S.; Tamm, G. New and emerging developments in solar energy. *Sol. Energy* **2004**, *76*, 33–43. [CrossRef]
5. Das, N.C.; Rahman, M.M.; Kabir, S.F. Preparation of novel clay/chitosan/ZnO bio-composite as an efficient adsorbent for tannery wastewater treatment. *Int. J. Biol. Macromol.* **2023**, *249*, 126136. [CrossRef]
6. Gupta, V.K.; Saravanan, R.; Agarwal, S.; Gracia, F.; Khan, M.M.; Qin, J.; Mangalaraja, R.V. Degradation of azo dyes under different wavelengths of UV light with chitosan-SnO_2 nanocomposites. *J. Mol. Liquids* **2017**, *232*, 423–430. [CrossRef]
7. Pomogailo, A.D.; Kestelman, V.N. Chemical methods of metal-polymer nanocomposite production. In *Metallopolymer Nanocomposites*; Springer: Berlin, Heidelberg, 2005; pp. 135–236. [CrossRef]
8. Aadnan, I.; Zegaoui, O.; Daou, I.; da Silva, J.C.E. Synthesis and physicochemical characterization of a ZnO-Chitosan hybrid-biocomposite used as an environmentally friendly photocatalyst under UV-A and visible light irradiations. *J. Environ. Chem. Eng.* **2020**, *8*, 104260. [CrossRef]
9. Ibhadon, A.O.; Fitzpatrick, P. Heterogeneous Photocatalysis: Recent Advances and Applications. *Catalysts* **2013**, *3*, 189–218. [CrossRef]
10. Li, H.; Wang, X.; Li, X.; Zhao, C.; Song, X.; Zhang, P.; Huo, P. A review on heterogeneous photocatalysis for environmental remediation: From semiconductors to modification strategies. *Chin. J. Catal.* **2022**, *43*, 178–214.
11. Wang, H.; Wan, Y.; Li, B.; Ye, J.; Gan, J.; Liu, J.; Huo, P. Rational design of Ce-doped CdS/N-rGO photocatalyst enhanced interfacial charges transfer for high effective degradation of tetracycline. *J. Mater. Sci. Technol.* **2024**, *173*, 137–148. [CrossRef]
12. Wang, B.; Zhang, R.; Li, Z.; Shi, C.; Liu, E.; Zheng, Z.; Zhou, B.; Ji, M.; Chen, H. Enhancing photocatalytic degradation efficiency via the construction of organic/inorganic S-scheme supramolecular hybrid heterostructures. *Appl. Surf. Sci.* **2023**, *641*, 158481. [CrossRef]
13. Aal, A.A.; Mahmoud, S.A.; Aboul-Gheit, A.K. Sol–Gel and Thermally Evaporated Nanostructured Thin ZnO Films for Photocatalytic Degradation of Trichlorophenol. *Nanoscale Res. Lett.* **2009**, *4*, 627–634. [CrossRef]
14. Schubnell, M.; Kamber, I.; Beaud, P. Photochemistry at high temperatures—Potential of ZnO as a high temperature photocatalyst. *Appl. Phys. A* **1997**, *64*, 109–113. [CrossRef]
15. Lahbib, M.; Meftah, M.; Mejri, C.; Oueslati, W.; Amara, A.B.H. The starting stoichiometry, keys parameter to enhance the intrinsic microstructural and functional properties of synthesized hybrid nanocomposites chitosan/Na-montmorillonite/ZnO nanoparticlest type. *Appl. Surf. Sci. Adv.* **2023**, *13*, 100369. [CrossRef]
16. Kango, S.; Kalia, S.; Celli, A.; Njuguna, J.; Habibi, Y.; Kumar, R. Surface modification of inorganic nanoparticles for development of organic–inorganic nanocomposites—A review. *Prog. Polym. Sci.* **2013**, *38*, 1232–1261. [CrossRef]
17. Wu, L.; Liao, L.; Lv, G. Influence of interlayer cations on organic intercalation of montmorillonite. *J. Colloid Interface Sci.* **2015**, *454*, 1–7. [CrossRef]
18. Hochella, M.F., Jr.; Mogk, D.W.; Ranville, J.; Allen, I.C.; Luther, G.W.; Marr, L.C.; McGrail, B.P.; Murayama, M.; Qafoku, N.P.; Rosso, K.M.; et al. Natural, incidental, and engineered nanomaterials and their impacts on the Earth system. *Science* **2019**, *363*, eaau8299. [CrossRef]
19. Putri, G.E.; Rilda, Y.; Syukri, S.; Labanni, A.; Arief, S. Enhancing morphological and optical properties of montmorillonite/chitosan-modified cerium oxide nanoparticles for antimicrobial applications. *Surf. Interfaces* **2022**, *32*, 102166. [CrossRef]
20. Giannakas, A.E.; Salmas, C.E.; Moschovas, D.; Baikousi, M.; Kollia, E.; Tsigkou, V.; Karakassides, A.; Leontiou, A.; Kehayias, G.; Avgeropoulos, A.; et al. Nanocomposite film development based on chitosan/polyvinyl alcohol using ZnO@ montmorillonite and ZnO@ halloysite hybrid nanostructures for active food packaging applications. *Nanomaterials* **2022**, *12*, 1843. [CrossRef]
21. Li, D.; Huang, X.; Hao, F.; Lv, Y.; Chen, H.; Wu, S.; Xiong, W.; Liu, P.; Luo, H. Preparation of organic-inorganic composites with high antibacterial activity based on sepiolite, chitosan and zinc: The study of the active antibacterial sites of chitosan-zinc oxide structure. *Appl. Clay Sci.* **2022**, *216*, 106325. [CrossRef]
22. Lee, D.-E.; Kim, M.-K.; Danish, M.; Jo, W.-K. State-of-the-art review on photocatalysis for efficient wastewater treatment: Attractive approach in photocatalyst design and parameters affecting the photocatalytic degradation. *Catal. Commun.* **2023**, *183*, 106764. [CrossRef]
23. Bagade, A.V.; Pund, S.N.; Nagwade, P.A.; Kumar, B.; Deshmukh, S.U.; Kanagare, A.B. Ni-doped Mg-Zn nano-ferrites: Fabrication, characterization, and visible-light-driven photocatalytic degradation of model textile dyes. *Catal. Commun.* **2023**, *181*, 106719. [CrossRef]
24. Ramirez, L.A.; Dennehy, M.; Alvarez, M. Metal oxide-biochar supported recyclable catalysts: A feasible solution for the reduction of 4-nitrophenol in water. *Catal. Commun.* **2023**, *181*, 106723. [CrossRef]

25. Bian, X.; Li, F.; Zhang, J.; Zhong, M.; Yang, Y.; Khan, S. Photocatalytic degradation of tetracycline antibiotics in swine wastewater using Fe^{3+}-loaded $NaBiO_3$ coupled with sodium persulfate. *Catal. Commun.* **2023**, *174*, 106579. [CrossRef]
26. Zehra, A.; Wani, S.M.; Bhat, T.A.; Jan, N.; Hussain, S.Z.; Naik, H.R. Preparation of a biodegradable chitosan packaging film based on zinc oxide, calcium chloride, nano clay and poly ethylene glycol incorporated with thyme oil for shelf-life prolongation of sweet cherry. *Int. J. Biol. Macromol.* **2022**, *217*, 572–582. [CrossRef]
27. Brahmi, M.; Essifi, K.; Bakirhan, N.K.; El Bachiri, A.; Ouldriane, S.D.; Tahani, A. New insights into physicochemical aspects involved in the formation of chitosan@ alginate biobased polyelectrolyte complexes on natural montmorillonite clay surface. *J. Mol. Liq.* **2023**, *387*, 122. [CrossRef]
28. Yahya, R.; Elshaarawy, R.F. Highly sulfonated chitosan-polyethersulfone mixed matrix membrane as an effective catalytic reactor for esterification of acetic acid. *Catal. Commun.* **2023**, *173*, 106557. [CrossRef]
29. Dutta, S.; Kumar, P.; Yadav, S.; Sharma, R.D.; Shivaprasad, P.; Vimaleswaran, K.S.; Srivastava, A.; Sharma, R.K. Accelerating innovations in CH activation/functionalization through intricately designed magnetic nanomaterials: From genesis to applicability in liquid/regio/photo catalysis. *Catal. Commun.* **2023**, *175*, 106615. [CrossRef]
30. Altıntıg, E.; Ates, A.; Angın, D.; Topal, Z.; Aydemir, Z. Kinetic, equilibrium, adsorption mechanisms of RBBR and MG dyes on chitosan-coated montmorillonite with an ecofriendly approach. *Chem. Eng. Res. Des.* **2022**, *188*, 287–300. [CrossRef]
31. Eltabey, R.M.; Abdelwahed, F.T.; Eldefrawy, M.M.; Elnagar, M.M. Fabrication of poly(maleic acid)-grafted cross-linked chitosan/montmorillonite nanospheres for ultra-high adsorption of anionic acid yellow-17 and cationic brilliant green dyes in single and binary systems. *J. Hazard. Mater.* **2022**, *439*, 129589. [CrossRef] [PubMed]
32. Tahari, N.; de Hoyos-Martinez, P.L.; Izaguirre, N.; Houwaida, N.; Abderrabba, M.; Ayadi, S.; Labidi, J. Preparation of chitosan/tannin and montmorillonite films as adsorbents for Methyl Orange dye removal. *Int. J. Biol. Macromol.* **2022**, *210*, 94–106. [CrossRef] [PubMed]
33. Wang, B.; Luo, C.-Y.; Zhu, P.; Liu, Y.; Xu, Y.-J. Facile construction of H_3PO_3-modified chitosan/montmorillonite coatings for highly efficient flame retardation of polyester–cotton fabrics. *Prog. Org. Coat.* **2023**, *184*, 107864. [CrossRef]
34. Albouchi, W.; Meftah, M.; Amara, A.B.H.; Oueslati, W. Effect of reactant ratio and nanofillers type on the microstructural properties, porosity fluctuations and heavy metal removal ability of chitosan-clay hybrid materials. *Appl. Surf. Sci. Adv.* **2023**, *13*, 100387. [CrossRef]
35. Franco, F.; Pozo, M.; Cecilia, J.A.; Benítez-Guerrero, M.; Lorente, M. Effectiveness of microwave assisted acid treatment on dioctahedral and trioctahedral smectites. The influence of octahedral composition. *Appl. Clay Sci.* **2016**, *120*, 70–80. [CrossRef]
36. Pálková, H.; Barlog, M.; Madejová, J.; Hronský, V.; Petra, L.; Šimon, E.; Billik, P.; Zimowska, M. Structural changes in smectites subjected to mechanochemical activation: The effect of the occupancy of the octahedral sites. *Appl. Clay Sci.* **2021**, *213*, 106214. [CrossRef]
37. Douillard, J.; Lantenois, S.; Prelot, B.; Zajac, J.; Henry, M. Study of the influence of location of substitutions on the surface energy of dioctahedral smectites. *J. Colloid Interface Sci.* **2008**, *325*, 275–281. [CrossRef] [PubMed]
38. Available online: https://www.clays.org/sourceclays_data/ (accessed on 29 March 2024).
39. Cheira, M.F.; Kouraim, M.N.; Zidan, I.H.; Mohamed, W.S.; Hassanein, T.F. Adsorption of U(VI) from sulfate solution using montmorillonite/polyamide and nano-titanium oxide/polyamide nanocomposites. *J. Environ. Chem. Eng.* **2020**, *8*, 104427. [CrossRef]
40. Mejri, C.; Oueslati, W.; Amara, A.B.H. Structure and reactivity assessment of dioctahedral montmorillonite during provoked variable sequential cation exchange process via XRD modelling approach. *Appl. Surf. Sci. Adv.* **2023**, *15*, 100403. [CrossRef]
41. Jiang, A.; Patel, R.; Padhan, B.; Palimkar, S.; Galgali, P.; Adhikari, A.; Varga, I.; Patel, M. Chitosan Based Biodegradable Composite for Antibacterial Food Packaging Application. *Polymers* **2023**, *15*, 2235. [CrossRef]
42. Biswal, A.; Sethy, P.K.; Swain, S.K. Change in orientation of polyacrylic acid and chitosan networks by imprintment of gold nanoparticles. *Polym. Technol. Mater.* **2020**, *60*, 182–194. [CrossRef]
43. Moll, W.F. Baseline studies of the clay minerals society source clays: Geological origin. *Clays Clay Miner.* **2001**, *49*, 374–380. [CrossRef]
44. Mermut, A.R.; Lagaly, G. Baseline studies of the clay minerals society source clays: Layer-charge determination and characteristics of those minerals containing 2:1 layers. *Clays Clay Miner.* **2001**, *49*, 393–397. [CrossRef]
45. Oueslati, W.; Meftah, M. Discretization of the water uptake process of Na-montmorillonite undergoing atmospheric stress: XRD modeling approach. *Adv. Mater. Sci. Eng.* **2018**, *2018*, 1–17. [CrossRef]
46. Ammar, M.; Oueslati, W.; Ben Rhaiem, H.; Amara, A.B.H. Effect of the hydration sequence orientation on the structural properties of Hg exchanged montmorillonite: Quantitative XRD analysis. *J. Environ. Chem. Eng.* **2014**, *2*, 1604–1611. [CrossRef]
47. Vinci, D.; Dazas, B.; Ferrage, E.; Lanson, M.; Magnin, V.; Findling, N.; Lanson, B. Influence of layer charge on hydration properties of synthetic octahedrally-charged Na-saturated trioctahedral swelling phyllosilicates. *Appl. Clay Sci.* **2019**, *184*, 105404. [CrossRef]
48. Fu, J.; Yang, F.; Guo, Z. The chitosan hydrogels: From structure to function. *New J. Chem.* **2018**, *42*, 17162–17180. [CrossRef]
49. Chipera, S.J.; Bish, D.L. Baseline studies of the clay minerals society source clays: Powder X-ray diffraction analyses. *Clays Clay Miner.* **2001**, *49*, 398–409. [CrossRef]
50. Borden, D.; Giese, R.F. Baseline studies of the clay minerals society source clays: Cation exchange capacity measurements by the ammonia-electrode method. *Clays Clay Miner.* **2001**, *49*, 444–445. [CrossRef]

51. Oueslati, W.; Ben Rhaiem, H.; Amara, A.B.H. Effect of relative humidity constraint on the metal exchanged montmorillonite performance: An XRD profile modeling approach. *Appl. Surf. Sci.* **2012**, *261*, 396–404. [CrossRef]
52. Oueslati, W.; Mejri, C.; Amara, A.B.H. Impact of uniaxial mechanical perturbation on structural properties and smectite porosity features: Ion exchanger efficiency and adsorption performance fate. *Adv. Civ. Eng.* **2022**, *2022*, 1–16. [CrossRef]
53. Jebali, S.; Meftah, M.; Mejri, C.; Ben Haj Amara, A.; Oueslati, W. Enhancement of Photocatalytic Activity and Microstructural Growth of Cobalt-Substituted Ba1−xCoxTiO$_3$ {x = 0,. . ., 1} Heterostructure. *ChemEngineering* **2023**, *7*, 43. [CrossRef]
54. Mejri, C.; Oueslati, W.; Amara, A.B.H. Structural alteration, hydration stability, heavy metal removal efficiency, and montmorillonite porosity fate by coupling the soil solution ph and a thermal gradient. *Adsorpt. Sci. Technol.* **2022**, *2022*, 1–28. [CrossRef]
55. Hussain, M.; Ceccarelli, R.; Marchisio, D.; Fino, D.; Russo, N.; Geobaldo, F. Synthesis, characterization, and photocatalytic application of novel TiO_2 nanoparticles. *Chem. Eng. J.* **2010**, *157*, 45–51. [CrossRef]
56. Shume, W.M.; Murthy, H.C.A.; Zereffa, E.A. A review on synthesis and characterization of Ag_2O nanoparticles for photocatalytic applications. *J. Chem.* **2020**, *2020*, 1–15. [CrossRef]
57. Kefeni, K.K.; Mamba, B.B. Photocatalytic application of spinel ferrite nanoparticles and nanocomposites in wastewater treatment: Review. *Sustain. Mater. Technol.* **2020**, *23*, e00140. [CrossRef]
58. Bhattacharjee, M.; Goswami, S.; Pramanik, N.B.; Barman, J.; Saikia, M.; Hulle, N.R.S.; Haloi, D.J. Chitosan/Clay biocomposite films with enhanced physico-chemical, mechanical and antimicrobial properties. *Anal. Chem. Lett.* **2023**, *13*, 609–625. [CrossRef]
59. Vijayalekshmi, V. UV-Visible, mechanical and anti-microbial studies of chitosan-montmorillonite clay/TiO_2 nanocomposites. *Res. J. Recent Sci.* **2015**, *4*, 131–135.
60. Bahrudin, N.N.; Nawi, M.A.; Zainal, Z. Insight into the synergistic photocatalytic-adsorptive removal of methyl orange dye using TiO_2/chitosan based photocatalyst. *Int. J. Biol. Macromol.* **2020**, *165*, 2462–2474. [CrossRef]
61. Gao, X.; Yin, H.; Li, M.; Xin, L.; Zhang, H.; Long, H. Photocatalytic degradation of methyl orange by a diethylenetriamine modified chitosan/bentonite composite. *React. Chem. Eng.* **2023**, *8*, 2505–2521. [CrossRef]
62. Bahrudin, N.N.; Nawi, M.A.; Sabar, S. Immobilized chitosan-montmorillonite composite adsorbent and its photocatalytic regeneration for the removal of methyl orange. *React. Kinet. Catal. Lett.* **2019**, *126*, 1135–1153. [CrossRef]
63. Aadnan, I.; Zegaoui, O.; El Mragui, A.; Esteves da Silva, J.C.G. Physicochemical and photocatalytic properties under visible light of ZnO-Bentonite/Chitosan hybrid-biocompositefor water remediation. *Nanomaterials* **2021**, *12*, 102. [CrossRef]
64. Bai, Q.; Yangchao, L. A review on the preparation and characterization of chitosan-clay nanocomposite films and coatings for food packaging applications. *Carbohydr. Polym. Technol. Appl.* **2021**, *2*, 100102.
65. Butman, M.F.; Ovchinnikov, N.L.; Karasev, N.S.; Kochkina, N.E.; Agafonov, A.V.; Vinogradov, A.V. Photocatalytic and adsorption properties of TiO_2-pillared montmorillonite obtained by hydrothermally activated intercalation of titanium polyhydroxo complexes. *Beilstein J. Nanotechnol.* **2018**, *9*, 364–378. [CrossRef]
66. Huo, M.; Guo, H.; Jiang, Y.; Ju, H.; Xue, B.; Li, F. A facile method of preparing sandwich layered TiO_2 in between montmorillonite sheets and its enhanced UV-light photocatalytic activity. *J. Photochem. Photobiol. A Chem.* **2018**, *358*, 121–129. [CrossRef]
67. Amir, M.N.I.; Julkapli, N.M.; Hamid, S.B.A. Incorporation of chitosan and glass substrate for improvement in adsorption, separation, and stability of TiO_2 photodegradation. *Int. J. Environ. Sci. Technol.* **2016**, *13*, 865–874. [CrossRef]
68. Saravanan, R.; Aviles, J.; Gracia, F.; Mosquera, E.; Gupta, V.K. Crystallinity and lowering band gap induced visible light photocatalytic activity of TiO_2/CS (Chitosan) nanocomposites. *Int. J. Biol. Macromol.* **2018**, *109*, 1239–1245. [CrossRef]
69. Cesur, S.; Köroğlu, C.; Yalçın, H.T. Antimicrobial and biodegradable food packaging applications of polycaprolactone/organo nanoclay/chitosan polymeric composite films. *J. Vinyl Addit. Technol.* **2018**, *24*, 376–387. [CrossRef]
70. Mohammadzadeh-Vazifeh, M.; Hosseini, S.M.; Mohammadi, A.; Jahanfar, M.; Maleki, H. Investigation of the antimicrobial properties of nanoclay and chitosan based nanocomposite on the microbial characteristics of Gouda cheese. *Iran. J. Microbiol.* **2020**, *12*, 121–126. [CrossRef]
71. Kusmono; Abdurrahim, I. Water sorption, antimicrobial activity, and thermal and mechanical properties of chitosan/clay/glycerol nanocomposite films. *Heliyon* **2019**, *5*, e02342. [CrossRef]
72. Khubiev, O.M.; Egorov, A.R.; Kirichuk, A.A.; Khrustalev, V.N.; Tskhovrebov, A.G.; Kritchenkov, A.S. Chitosan-Based Antibacterial Films for Biomedical and Food Applications. *Int. J. Mol. Sci.* **2023**, *24*, 10738. [CrossRef]
73. Priyadarshi, R.; Rhim, J.-W. Chitosan-based biodegradable functional films for food packaging applications. *Innov. Food Sci. Emerg. Technol.* **2020**, *62*, 102346. [CrossRef]
74. Ben Dhieb, F.; Dil, E.J.; Tabatabaei, S.H.; Mighri, F.; Ajji, A. Effect of nanoclay orientation on oxygen barrier properties of LbL nanocomposite coated films. *RSC Adv.* **2019**, *9*, 1632–1641. [CrossRef]
75. Kumar, S.; Mukherjee, A.; Dutta, J. Chitosan based nanocomposite films and coatings: Emerging antimicrobial food packaging alternatives. *Trends Food Sci. Technol.* **2020**, *97*, 196–209. [CrossRef]
76. Cankaya, N.; Sahin, R. Chitosan/clay bionanocomposites: Structural, antibacterial, thermal and swelling properties. *Cellul. Chem. Technol.* **2019**, *53*, 537–549. [CrossRef]
77. Bandyopadhyay, J.; Ray, S.S. Are nanoclay-containing polymer composites safe for food packaging applications?—An overview. *J. Appl. Polym. Sci.* **2019**, *136*, 47214. [CrossRef]

78. Rodrigues, C.; de Mello, J.M.M.; Dalcanton, F.; Macuvele, D.L.P.; Padoin, N.; Fiori, M.A.; Soares, C.; Riella, H.G. Mechanical, thermal and antimicrobial properties of chitosan-based-nanocomposite with potential applications for food packaging. *J. Polym. Environ.* **2020**, *28*, 1216–1236. [CrossRef]
79. El-Naggar, E.; Wassel, M.A.; Shoueir, R.K. Visible-light driven photocatalytic effectiveness for solid-state synthesis of ZnO/natural clay/TiO_2 nanoarchitectures towards complete decolorization of methylene blue from aqueous solution. *Environ. Nanotechnol. Monit. Manag.* **2021**, *15*, 100425. [CrossRef]
80. Mustapha, S.; Ndamitso, M.M.; Abdulkareem, A.S.; Tijani, J.O.; Shuaib, D.T.; Ajala, A.O.; Mohammed, A.K. Application of TiO_2 and ZnO nanoparticles immobilized on clay in wastewater treatment: A review. *Appl. Water Sci.* **2020**, *10*, 49. [CrossRef]
81. Jida, S.M.; Zerefa, E.A. Preparation and photocatalysis of ZnO/bentonite based on adsorption and photocatalytic activity. *Mater. Res. Express* **2023**, *10*, 035502. [CrossRef]
82. Morlando, A.; Sencadas, V.; Cardillo, D.; Konstantinov, K. Suppression of the photocatalytic activity of TiO_2 nanoparticles encapsulated by chitosan through a spray-drying method with potential for use in sunblocking applications. *Powder Technol.* **2018**, *329*, 252–259. [CrossRef]
83. Castillo, P.C.H.-D.; Castro-Velázquez, V.; Rodríguez-González, V. Adsorption and photocatalytic-conjugated activity of a chitosan-functionalized titanate coating for the removal of the drug clonazepam from drinking water. *Environ. Sci. Pollut. Res.* **2023**, 1–16. [CrossRef]
84. Maruthupandy, M.; Muneeswaran, T.; Chackaravarthi, G.; Vennila, T.; Anand, M.; Cho, W.-S.; Quero, F. Synthesis of chitosan/SnO_2 nanocomposites by chemical precipitation for enhanced visible light photocatalytic degradation efficiency of congo red and rhodamine-B dye molecules. *J. Photochem. Photobiol. A Chem.* **2022**, *430*, 113972. [CrossRef]

nanomaterials

MDPI

Article

Electrosprayed Nanoparticles Containing Mangiferin-Rich Extract from Mango Leaves for Cosmeceutical Application

Vissuta Sirirungsee [1], Pawitrabhorn Samutrtai [2], Padchanee Sangthong [3], Phakorn Papan [3], Pimporn Leelapornpisid [2,4], Chalermpong Saenjum [2,5,*] and Busaban Sirithunyalug [2,5,*]

1 Master's Degree Program in Cosmetic Science, Faculty of Pharmacy, Chiang Mai University, Chiang Mai 50200, Thailand; vissuta_s@cmu.ac.th

2 Department of Pharmaceutical Sciences, Faculty of Pharmacy, Chiang Mai University, Chiang Mai 50200, Thailand; pawitrabhorn.s@cmu.ac.th (P.S.); pim_leela@hotmail.com (P.L.)

3 Department of Chemistry, Faculty of Science, Chiang Mai University, Chiang Mai 50200, Thailand; padchanee.sangthong@cmu.ac.th (P.S.); phakorn_papan@cmu.ac.th (P.P.)

4 Innovation Center for Holistic Health, Nutraceuticals, and Cosmeceuticals, Faculty of Pharmacy, Chiang Mai University, Chiang Mai 50200, Thailand

5 Research Center for Innovation in Analytical Science and Technology for Biodiversity-Based Economic and Society, Chiang Mai University, Chiang Mai 50200, Thailand

* Correspondence: chalermpong.saenjum@gmail.com (C.S.); busaban.s@cmu.ac.th (B.S.)

Abstract: Mango (*Mangifera indica* L.) is one of the most economically important fruits in Thailand. Mango has been used as a traditional medicine because it possesses many biological activities, such as antioxidant properties, anti-inflammatory properties, microorganism-growth inhibition, etc. Among its natural pharmacologically active compounds, mangiferin is the main active component found in mango leaves. Mangiferin has the potential to treat a variety of diseases due to its multifunctional activities. This study aims to prepare a mangiferin-rich extract (MRE) from mango leaves and develop nanoparticles containing the MRE using an electrospraying technique to apply it in a cosmeceutical formulation. The potential cosmeceutical mechanisms of the MRE were investigated using proteomic analysis. The MRE is involved in actin-filament organization, the positive regulation of cytoskeleton organization, etc. Moreover, the related mechanism to its cosmeceutical activity is metalloenzyme-activity regulation. Nanoparticles were prepared from 0.8% *w*/*v* MRE and 2% *w*/*v* Eudragit® L100 solution using an electrospraying process. The mean size of the MRE-loaded nanoparticles (MNPs) received was 247.8 nm, with a PDI 0.271. The MRE entrapment by the process was quantified as 84.9%, indicating a high encapsulation efficiency. For the skin-retention study, the mangiferin content in the MNP-containing emulsion-gel membranes was examined and found to be greater than in the membranes of the MRE solution, illustrating that the MNPs produced by the electrospraying technique help transdermal delivery for cosmetic applications.

Keywords: mangiferin; aging; mango leaves; nanoparticles; electrospraying; cosmeceuticals

1. Introduction

Mango (*Mangifera indica* L.) is one of the most economically important fruits in Thailand. According to the information from the Trade Policy and Strategy Office of Thailand, Thailand exported around 113,806 tons of mangos, valued at THB 2.9 billion, in 2021. The expanding rate jumped to 50.25% higher than the previous month [1]. Mango is a stone fruit found in tropical and subtropical regions. It constitutes one of the most commercialized fruits in the world, belonging to the Anacardiaceae family [2]. It has long been used in Ayurvedic medicine, and different parts of the mango are used as an antiseptic, dentifrice, and diuretic, and to treat diarrhea, dysentery, asthma, hypertension, and toothache [3]. Mango possesses many biological activities, such as antioxidant, anti-inflammatory, antiobesity, anticancer activity, microbial-growth inhibition, etc. [4]. One related study reported that mango is composed of carotenoids, anthocyanins, dietary fiber, and phenolic compounds [5]. Phenolic compounds in mango include mangiferin, gallic acid, caffeic acid, and tannin [6].

One of the most bioactive compositions found in the leaves, bark, root, peel, and flesh of mango is mangiferin. Mangiferin, generally called C-glycosyl xanthone, is a xanthone connected to a C-glycoside (structure shown in Figure 1), displaying numerous biological actions contributing to therapeutic, pharmaceutical, and cosmetic applications [7]. Its biological activities include antioxidant effects, which are thought to have a greater potential than vitamins C and E [8], antiviral, anticancer, antidiabetic, antiaging, immunomodulatory, hepatoprotective, and analgesic effects [9]. Regarding its antioxidant activity, methanolic extract, which consists of mangiferin, exhibited a potent antioxidant activity (EC_{50} = 5.8 ± 0.96 μg/mL) that was comparable to rutin (EC_{50} = 5.56 ± 0.33 μg/mL) [10]. Mangiferin could reduce hydrogen peroxide-induced lipid peroxidation in human peripheral blood lymphocytes in a concentration-dependent manner, and inhibited the induction of $OH^{\bullet}$, O_2^{-}, DPPH, and ABTS radicals in a dose-dependent manner in cell-free studies. It demonstrated that mangiferin has radioprotective properties by suppressing the effects of free radicals [11]. The antioxidant activity of mangiferin prevented UVB-induced wrinkle formation by reducing the UVB-induced matrix metalloproteinase (MMP-9) protein expression and enzyme activity, and attenuating the UVB-induced phosphorylation of mitogen-activated protein kinase 1 (MEK) and extracellular-signal-regulated kinase (ERK). In one in vivo study, mangiferin inhibited UVB-induced skin wrinkles, epidermal thickening, and damage to collagen, suggesting mangiferin exerts antiphotoaging [12]. Basically, the skin dry weight is mainly composed of 70–80% collagen, which provides strength to the skin. Elastin is involved in skin elasticity, accounting for 2–4% of the extracellular matrix (ECM), while glycosaminoglycans (GAGs), such as hyaluronic acid, are responsible for the hydration of the skin. Collagen, elastin, and GAGs are proteins produced by fibroblasts, which can be affected by photoaging, leading to wrinkles, hyperpigmentation, and skin thickness.

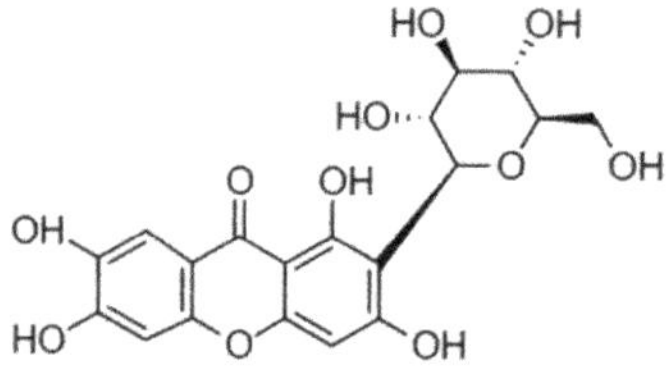

Figure 1. Chemical structure of mangiferin.

Electrospraying is a technique producing nano/microparticles by applying an electric field that has been widely used in food as well as the nutraceutical industry, etc. The electrospraying setup consists of a high-voltage supplier, a syringe-loading polymer solution, a needle equipped with a syringe, a syringe pump, and a collector [13]. When applying an electric field, the high voltage causes the polymer solution to become charged, and the charged solution is ejected from the needle. The density of charged droplets increases, so that repulsive force of the charge surpasses the surface tension, resulting in a "Taylor cone". After evaporating the solvent, the generated nanoparticles are sprayed onto a collector. The electrospraying technique is a simple and inexpensive technique to set up. It provides a high encapsulation efficiency (EE) and can be used with high-molecular-weight polymers [14]. According to its simplicity and versatility, the electrospraying technique is an effective option for drug delivery.

Proteomics is the study of the interactions, functions, composition of the organism, and structures of proteins and their cellular activities. It provides a better understanding of the structure and function of the organism than genomics. The study of proteomics has many applications, such as medicine, oncology, food microbiology, and agriculture. Proteomics is a fast, sensitive technology providing high proteome coverage [15]. It helps discover new biomarkers in different diseases, the mechanisms of drugs, as well as the cosmeceutical mechanisms of the MRE.

After harvesting, mango leaves are discarded as agricultural waste. Thus, this study aims to prepare the mangiferin-rich extract (MRE) from mango "Talapnak"-variety leaves and investigate its antioxidant, antiaging, and antityrosinase properties. The electrospraying technique was used to produce nanoparticles. Nanoparticles containing the MRE were then formulated into a cosmetic product. Additionally, proteomics will provide more information of a possible mechanism between the MRE and proteins associated to antiaging activity.

2. Material and Methods

2.1. Materials

Mango "Talapnak"-variety leaves were sourced from Banpan District, Lamphun Province, Thailand. DPPH (1,1-diphenyl-2-picryhydrazyl), mangiferin, Trolox, linoleic acid, N-succinyl-AlaAlaAla-p-nitroanilide (AAAPVN), L-tyrosine, collagenase from *Histidium histolyticum*, elastase from porcine pancreas, hyaluronidase from bovine, tyrosinase from mushroom, EGCG (epigallocatechin gallate), tannic acid, hyaluronic acid sodium salt, activated charcoal, and collagen were obtained from Sigma-Aldrich, St. Louis, MO, USA. Ascorbic acid and N,N-dimethylacetamide were purchased from Lobachemie Pvt. Ltd., Mumbai, India. Eudragit® L100 was obtained from Evonik, Darmstadt, Germany. Isopropanol, ethanol, H_2SO_4, and n-hexane were bought from RCI Labscan, Bangkok, Thailand. All solvents were of analytical grade.

2.2. Mango Leaf Extraction

Mango "Talapnak"-variety leaves were cleaned and dried in a hot-air oven at 50 °C for 24 h. The dried mango leaves were pulverized into powder. Mango leaf powder was then extracted according to the method of Vo et al. (2017) with modifications [16]. In all, 200 g of the mango leaf powder was extracted using 1600 mL of 50%, 60%, or 70% isopropanol (IPA) and using reflux extraction for 1 h. Then, the solution was cooled and filtered. After evaporating to 400 mL, the solution was partitioned with 400 mL of n-hexane three times, adjusted to a pH of 2 with 1 M H_2SO_4, and boiled in a reflux condenser for 1 h. The solution was evaporated to 80 mL. Exactly 1 g of charcoal and 80 mL of dioxane were added to the solution. The solution was then filtered before storing in a refrigerator for 24 h; the precipitate was filtered and washed with 60 mL of cold ethanol three times and dried in a vacuum oven at 70 °C for 4 h. The obtained precipitate is called MRE. The MRE extraction was performed ten times to obtain an adequate amount of the MRE. To purify the mangiferin, the MRE was dissolved in 50% IPA, then boiled while stirring for 5 min. The solution was then filtered and stored in a refrigerator for 24 h. The precipitate was filtered and washed with cold ethanol and dried in a vacuum oven, thus obtaining purified MRE (pMRE).

2.3. Characterization of the MRE and pMRE

2.3.1. UV–Vis Spectrophotometer

The MRE and pMRE were characterized for mangiferin using a UV–Vis spectrophotometer (Shimadzu UV2600i, Tokyo, Japan) by preparing 0.001% of the extract in methanol. The scanning wavelength was from 200 to 800 nm.

2.3.2. Fourier Transform Infrared Spectrometer (FT-IR)

An amount of 1 mg of the pMRE was measured using the potassium–bromide (KBr) pellet method in an FI-IR spectrometer. The resolution was 4.0 cm^{-1}. The spectral region was between 4000 and 400 cm^{-1}, and the result was compared with the reference standard mangiferin.

2.3.3. Differential Scanning Calorimetry (DSC)

The MRE was placed in a standard aluminum pan and covered. Measurement was performed from 25 to 300 °C at a constant heating rate of 10 °C/min in a dynamic nitrogen atmosphere (flow rate = 30 mL/min).

2.4. Standardization of the MRE

The mangiferin amount was determined using high-performance liquid chromatography (HPLC). Briefly, the HPLC system (Agilent, Santa Clara, CA, USA) with a C18 analytical column was used with a detection wavelength of 258 nm. The mobile phase was composed of 2% acetic acid (A) and 0.5% acetic acid: acetonitrile (1:1 *v*/*v*) (B). The gradient elution was performed as follows: 0 to 2 min, 5% of B; 2 to 10 min, 5 to 25% of B; 10 to 40 min, 25 to 55% of B; 40 to 45 min, 55 to 90% of B; 45 to 50 min, 90 to 55% of B, 50 to 55 min, 55 to 5% of B; 55 to 60 min, 5% of B. The injection volume, the flow rate, and the column temperature were 20 µL, 1.0 mL/min, and 25 °C, respectively. The mangiferin amount was quantified by comparing the peak area of the sample to the standard curve. The mangiferin stock solution of 500 ppm was diluted with 50% IPA to concentrations of 25, 50, 75, 100, and 125 ppm.

2.5. Bioactivities Assessment of the MRE

2.5.1. Antioxidant Activity

DPPH Radical Scavenging Assay

Antioxidant activity was determined by preparing various concentrations of the MRE. In a 96-well plate, 20 µL of the samples were mixed with 180 µL of the methanolic DPPH solution. The mixture was incubated for 30 min at RT in the dark. The absorbance was read at 520 nm using a microplate reader (BMG Labtech, Ortenberg, Germany), and ascorbic acid was used as a positive control. The % inhibition was calculated using the equation below:

$$\% \text{ Inhibition} = \frac{A_c - A_s}{A_c} \times 100$$

where A_c and A_s were the absorbances of the control and sample, respectively. The IC_{50} value was calculated from the polynomial equation of the plot between the % inhibition and concentration. The samples were performed in triplicate, and the average IC_{50} value was reported as mean ± SD.

Lipid Peroxidation Inhibition Assay

The linoleic acid peroxidation method was used to determine the lipid peroxidation inhibition. Various concentrations of the MRE were prepared. In Eppendorf tubes, 75 µL of the samples/solvent, 350 µL of 1.3% *w*/*v* of linoleic acid in methanol, 175 µL of DI water, 350 µL of 20 mM phosphate buffer with a pH of 7.0, and 50 µL of 46.35 mM 2,2′-Azobis(2-aminopropane) dihydrochloride (AAPH) were mixed. The mixtures were incubated at 50 °C for 4 h. Then, 5 µL of the mixture was added to a 96-well plate. After adding 5 µL of 10% NH4SCN, 5 µL of 20 mM FeCl2, and 185 µL of 75% methanol, the 96-well plate was incubated at RT for 5 min; then, the absorbance was measured at 500 nm using a microplate reader (BMG Labtech, Ortenberg, Germany). Trolox was used as a positive control.

2.5.2. Antiaging Activity

Anticollagenase Assay

Anticollagenase activity was investigated by mixing the samples with collagenase enzyme, 5 mM $CaCl_2$ in borate buffer with a pH of 7.5, and DI water. After incubating at 37 °C for 10 min, 1% collagen solution was added. The mixture was incubated at 37 °C for 1 h. Then, 0.75 mM of 3,4-DHPAA, 125 mM Borate buffer with a pH of 7.5, and 1.25 mM sodium periodate were added and incubated at 37 °C for 10 min. The mixture was maintained in an ice bath. The reaction was read using a microplate reader with the fluorescence mode at 375 and 465 nm. The positive control was ascorbic acid.

Antihyaluronidase Assay

The antihyaluronidase assay was performed according to Chaiyana et al. (2020) with some modifications [17] in Eppendorf tubes. The samples were mixed with 2 mg/mL of hyaluronidase enzyme and incubated for 10 min at 37 °C. Then, 0.03% hyaluronic acid was added before incubating at 37 °C for 45 min. After that, an acidic albumin solution with a pH of 3.4 was mixed and incubated at RT for 10 min. Then, 200 μL of the mixtures were pipetted in a 96-well plate, and the absorbance was read at 600 nm. Tannic acid was used as a positive control.

Antielastase Assay

According to the method of Theansungnoen et al. with modifications [18], the assay was performed in 0.2 M Tris–HCl buffer (pH of 8.0). Then, 0.04 U/mL of porcine pancreatic elastase was prepared in sterile water. After that, 1 mg/mL of N-Succinyl-Ala-Ala-Ala-*p*-nitroanilide (SANA) was used as a substrate. The sample solutions were mixed with elastase enzyme and Tris–HCl buffer before measuring the reaction spectrophotometrically at 410 nm using a microplate reader (BMGlabtech, Ortenberg, Germany) with the kinetic mode. EGCG represents a positive control.

2.5.3. Antityrosinase Assay

The inhibition of the tyrosinase activity was measured using L-tyrosine as a substrate, and the positive control was ascorbic acid. Tyrosinase enzyme, mixing with the samples, and 20 mM phosphate buffer with a pH of 6.8 were incubated at RT for 10 min. Then, the substrate was added and incubated for another 20 min at RT. The absorbance was read at 492 nm using a microplate reader. The % tyrosinase inhibition was calculated using the equation as mentioned above.

2.6. *Proteomics Analysis of the MRE*

2.6.1. Cell Preparation and Protein Extraction

To investigate protein expression of the MRE, 1×10^5 cells/mL of mouse fibroblast cells (L929) was seeded in a T-25 flask with RPMI 1640 medium containing 10% FBS and 1% penicillin/streptomycin. The cells were treated with the MRE at the concentration of 31.25 μg/mL for 24 h. After harvesting by directly scraping in sterile cold PBS, the cells were collected and centrifuged at 14,000 rpm for 20 min before washing with sterile cold PBS three times. The pellet was snap-frozen rapidly in liquid nitrogen and stored at −80 °C until the proteomic analysis.

2.6.2. LC-MS/MS Analysis

The harvested cells were lysed using lysis buffer (0.5% SDS, 5 mM DTT in 0.1× PBS, pH of 7.4) with 1 mm PMSF for the LC-MS/MS before being subjected to tandem–mass spectroscopy using a nano-LC system coupled with a high-resolution 6600 TripleTOF™ (AB-Sciex, Concord, ON, Canada). The conditions were as follows: mobile phase A consisted of 0.1% formic acid in water and mobile phase B comprised 95% acetonitrile with 0.1% formic acid. The LC parameters were composed of a 135 min long process for a single injection, and the analytical temperature was set at 55 °C. The MS scanned a mass ranging from 400 to 1600 m/z using the data-dependent acquisition mode. The top 20 most abundant peptide ions with charge states ranging from 2 to 5 (positive mode) for fragmentation were selected, and the dynamic-exclusion duration was set at 15 s.

2.6.3. Protein Identification

The raw data of the MS-spectra file were extracted, and protein sequences were annotated using the Paragon™ Algorithm by ProteinPilot™ Software. RStudio ver. 2023.03.0

Software was applied to analyze the MS data to identify proteins. Those proteins with a $\log_{10}$ (p value) greater than 1.3 were considered significant. The proteins with $\log_2$ fold changes less than −0.5 and higher than 0.5 were classified as downregulated and upregulated, respectively. Downregulation refers to a decrease in the production of a protein, while upregulation describes an increase in the production of a protein. In proteomics, downregulation and upregulation can be used to study the effects of stimuli on protein expression. STRING analysis was used to further investigate the interaction of these identified upregulated proteins with a high interaction value (0.7); the false discovery rate was set at less than 0.05.

2.7. Preparation of MRE-Loaded Nanoparticles (MNPs) by the Electrospraying Technique

Spraying solution was prepared by dissolving 2% w/v of Eudragit® L100 in ethanol and 0.8% w/v of the MRE in dimethylacetamide at ratio of 9:1. The polymer was stirred using a magnetic stirrer at RT until clear. To obtain an appropriate condition, the parameters, including the applying voltage (kV), flow rate (mL/min), and the distance between the tip and collector (cm), were varied. A foil-covered aluminum plate was used as a collector. The electrospraying process was performed by connecting the positive electrode of the voltage power supply to a metal needle tip, while the ground electrode was connected to the collector plate. The polymer solution was pumped through a 10 mL syringe with a 1.20 mm diameter 18G needle. The electrospraying was carried out under 30 to 40% RH. The particle size, polydispersity index (PDI), and zeta potential were studied in deionized water using a Malvern zetasizer (Malvern Instrument Ltd., Malvern, UK). Particle morphology was investigated using a JSM-IT300 scanning electron microscope (JEOL Ltd., Boston, MA, USA). The %EE was adapted with modifications [19] by dissolving the MNPs and MRE in a suitable solvent. The suspensions were centrifuged at 10,000 rpm for 30 min. Supernatants were filtered through a 0.22 micron filter, and the %EE was evaluated by HPLC following the formula below:

$$\%\ \text{Entrapment efficiency (EE\%)} = \frac{\text{Amount of drug in nanocarriers}}{\text{Initial amount of drug}} \times 100$$

2.8. Irritation Study

The irritation study of the MRE and MNPs were conducted using the hen's egg chorioallantoic membrane (HET-CAM) test according to Yeerong et al. (2021) [20]. The HET-CAM was prepared from 7 to 9 days using fertilized eggs. The air sac was cut with a rotating dentist saw blade; then, the eggshell was slowly peeled off. After adding normal saline solution, the eggs were incubated for 15 min. The inner-eggshell membrane was then removed with forceps, and 30 μL of the tested solution was applied. The reaction was observed for 5 min. Eggs treated with 1% w/v sodium lauryl sulfate (SLS) were set up as the positive control, and eggs treated with the 0.9% NaCl solution served as the negative control. The time to the onset of vascular hemorrhage, vascular lysis, and vascular coagulation, as well as the irritation score (IS), were observed using the formula below:

$$\text{IS} = \frac{(301 - \text{t(h)})}{300} \times 5 + \frac{(301 - \text{t(l)})}{300} \times 7 + \frac{(301 - \text{t(c)})}{300} \times 9$$

where t (h) = the time to the onset of vascular hemorrhage within 300 s after applying the test substance (s); t (l) = the time to the onset of vascular lysis within 300 s after applying the test substance (s); t (c) = the time to the onset of vascular coagulation within 300 s after applying the test substance (s). The irritation score (IS) was calculated and expressed as mean ± SD, and the test was performed in replicate. A score from 0.0 to 0.9 indicated no irritation, 1.0 to 4.9 indicated weak irritation, 5.0 to 8.9 indicated moderate irritation, and 9.0 to 21.0 indicated severe irritation.

2.9. Preparation of the Formulation Containing MNPs

Table 1 demonstrates the ingredients of the emulsion gel. Carbopol 940 was dispersed in DI water and stirred to form a gel. Then, glycerin was added to the gel base, and light-cream maker and jojoba oil were mixed. Cyclopentasiloxane and dimethicone cross-polymer (DC9045) was dissolved in cyclopentasiloxane and dimethicone, and then introduced in the gel base. Disodium EDTA dissolving in DI water and DMDM hydantoin were added, and the emulsion gel was adjusted to a pH from 5.5 to 6 with triethanolamine.

Table 1. MNPs containing emulsion-gel formulation.

Phase	Ingredient	Function	% *w/w*
A	DI water	Solvent	79.13
	Glycerin	Humectant	1.5
	Carbopol 940	Gelling agent	0.25
	MRE-loaded nanoparticles	Active agent	0.05
B	Sodium Polyacrylate (and) Polyisobutene (and) Water (Light-cream maker)	Polymeric emulsifier	1
	Jojoba oil	Emollient	0.5
C	Cyclopentasiloxane	Emollient	3
	Dimethicone	Emollient	1
	Cyclopentasiloxane and dimethicone cross polymer (DC 9045)	Slip modifier	3
D	Disodium EDTA in DI water	Chelator	0.02 10
E	DMDM hydantoin	Preservative	0.5
F	Triethanolamine	pH adjuster	0.1

The stability test of the formulations was evaluated by heating–cooling cycling. The emulsion gel base and emulsion gel containing MNPs were stored at 45 °C for 24 h and 4 °C for 24 h. Heating–cooling cycles were performed for six cycles.

2.10. Skin-Retention Study

Franz cell diffusion was performed to determine the permeability of the MRE solution and the emulsion gel containing the MNPs. The MNP-containing emulsion gel and the MRE solution were applied to the Strat-M® membrane (diameter 25 mm, thickness 300 µm). The membranes were mounted on Franz cells (Logan Instruments Corp, Somerset, NJ, USA), and the diffusion area was 1.76 cm^2. The receiver compartment was filled with PBS with a pH of 7.4 with a constant temperature of 32.0 °C. Then, 1 g of the MNP-loaded emulsion gel was placed in the donor compartment. Samples were taken from the receiver compartment at intervals of 30 min, and 1, 3, 6, 9, and 12 h, to analyze the mangiferin amount. After each interval, the receptor part was replaced with the same amount of fresh PBS. The membranes were cut into pieces and filled with solvent, then sonicated for 15 min. The samples were filtered through a 0.22 µm membrane filter for the HPLC analysis of the quantity of mangiferin.

2.11. Statistical Analysis

The results have been described as mean ± SD of three replicates. All statistical analyses were calculated using SPSS Software, Version 17.0, using a one-way ANOVA or the t-test at the 5% level ($p < 0.05$).

3. Results and Discussion

3.1. MRE Yield and Characterization

The result of the MRE yield depends on the ratio of solvents used. Extraction of mango "Talapnak" leaf powder with 50, 60, and 70% isopropanol by reflux extraction yielded 0.618 ± 0.054%, 0.881 ± 0.593%, and 1.073 ± 0.093%, respectively. The appearance of the extract was a fine pale-yellow powder.

The use of 70% isopropanol provided the greatest yield compared with the other ratios. Therefore, the extract from 70% isopropanol was chosen for further investigation. The purification process was performed to confirm the presence of mangiferin in the MRE. The pMRE yielded 0.042% with a 77.4% purity. The result of the characterization of the MRE and pMRE using a UV–Vis spectrophotometer showed absorption peaks at 241, 257.5, 315, and 365 nm, as shown in Figure 2a,b, which was related to absorption peaks presented in Vietnamese pharmacopeia in Figure 2c [21].

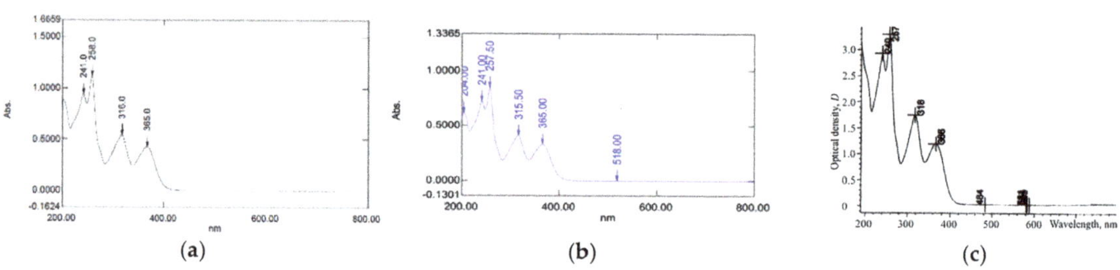

Figure 2. UV spectrum of the mangiferin–rich extract (MRE), (**a**) purified MRE (pMRE), (**b**) and the mangiferin spectrum presented in the Vietnamese pharmacopeia (**c**).

FT-IR was used to identify the chemical structure of the pMRE. Figure 3 demonstrates functional groups of mangiferin at the peaks 3370.08 cm^{-1} (O–H str), 2938.97 cm^{-1} (C–H str), 1649.67 cm^{-1} (C=O str), 1622.44 cm^{-1} (C=C str), 1494.06 cm^{-1} (CH–CH str), 1254.49 cm^{-1} (C–O str), and 1095.56 cm^{-1} (O–H deformation), which was consistent to relevant studies [16,22]. This result confirmed the presence of the mangiferin molecule in the MRE.

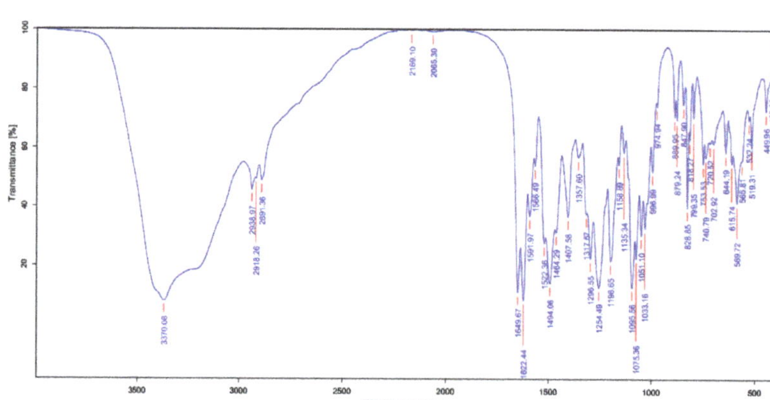

Figure 3. FTIR spectrum of the pMRE.

The DSC was investigated for the thermal transition of a material. Figure 4 depicts the DSC thermogram, revealing an endothermic peak of the MRE at 259.8 °C compared with the reference standard mangiferin at 272 °C. It could have been caused by the MRE containing other compounds, which lower those of the endothermic peak of the reference mangiferin.

Basically, pure substance melts at a sharp and highly defined temperature. On the other hand, contaminated substances typically display a large melting point range, and lower than that of a pure substance, due to melting-point depression. Because the impurities disrupt the crystal structure of the pure substance, resulting in the weaker of intermolecular forces, the melting point of the MRE displayed as lower than the reference mangiferin.

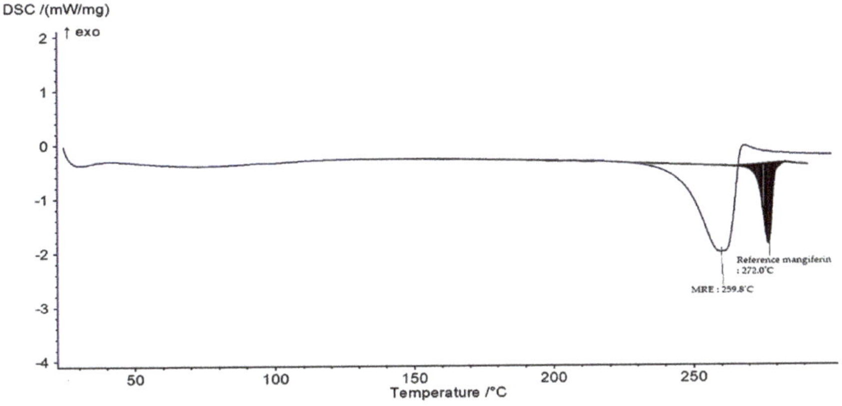

Figure 4. DSC thermogram of the MRE and reference mangiferin.

3.2. Standardization of the MRE

The mangiferin content in the samples was standardized using HPLC, as shown in Figure 5. Three of the ten MRE samples extracted with 70% IPA were chosen randomly to investigate the mangiferin content. The result demonstrated no significant difference among the samples ($p > 0.05$), as shown in Table 2. Therefore, each sample contained mangiferin over 50%, which was largely present in the extracts.

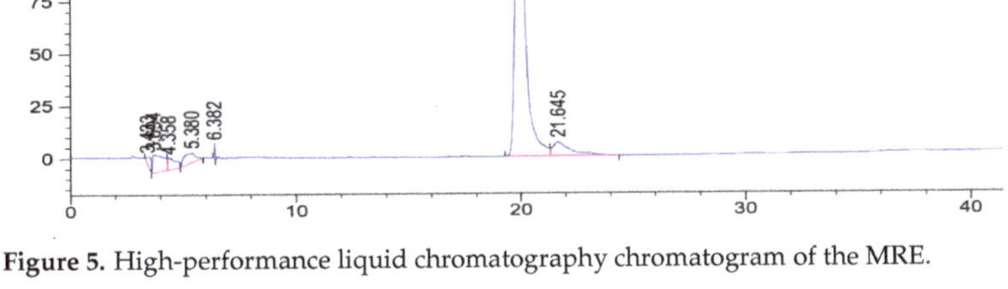

Figure 5. High-performance liquid chromatography chromatogram of the MRE.

Table 2. Mangiferin content in the random samples.

Sample	Mangiferin Content (mg/g)
1	528.62 ± 41.50 [a]
2	581.08 ± 27.89 [a]
3	509.23 ± 28.74 [a]

Data were expressed as mean ± SD ($n = 3$). [a] indicate a significant difference.

3.3. *Bioactivities Assessment of the MRE*

3.3.1. Antioxidant Activity

The DPPH radical scavenging assay and lipid peroxidation inhibition assay were conducted to measure the antioxidant activity of the MRE. The IC_{50} on the DPPH radical scavenging activity of the MRE was 4.967 ± 0.588 µg/mL, which differed insignificantly from the standard mangiferin (4.780 ± 1.147 µg/mL) and ascorbic acid (5.676 ± 0.398 µg/mL). According to Masibo and He (2008), mangiferin was more effective than ascorbic acid, which is related to the result [8]. On the other hand, the MRE at a concentration of 4.5 mg/mL exhibited a 45.673 ± 0.944% inhibition of lipid peroxidation. Nevertheless, Trolox and ascorbic acid at a concentration of 4.5 mg/mL could inhibit 98.005 ± 0.001% and 99.282 ± 0.406%, respectively. As a result, the MRE scavenged the DPPH radical better than it inhibited lipid peroxidation according to the four hydroxyl groups of mangiferin, which can donate hydrogen atoms to free radicals [23,24]. Furthermore, mangiferin is a phenolic compound that comprises a hydrophilic antioxidant. For this reason, the DPPH assay is a better choice for testing the antioxidant activity of mangiferin than the lipid peroxidation assay.

3.3.2. Anticollagenase Activity

Collagen is a structural protein containing amino acids in the dermis, accounting for 70 to 80% of the dry weight of the skin, and helping to strengthen, hydrate, and provide elasticity to the skin. However, as we age, the imbalance between collagen production and collagen breakdown leads to a loss of collagen. The IC_{50s} of the MRE, standard mangiferin, and ascorbic acid on collagenase inhibition were 53.63 ± 0.002 µg/mL, 30.06 ± 0.001 µg/mL, and 107.84 ± 0.001 µg/mL, respectively, as depicted in Figure 6a. The IC_{50} of the MRE significantly differed from that of the standard mangiferin and ascorbic acid ($p < 0.05$). Additionally, the MRE exhibited excellent collagenase inhibition activity. The result from the study of Ochocka et al. (2017) revealed that standard mangiferin could inhibit collagenase in a dose-dependent manner [25], which was comparable and consistent to our study.

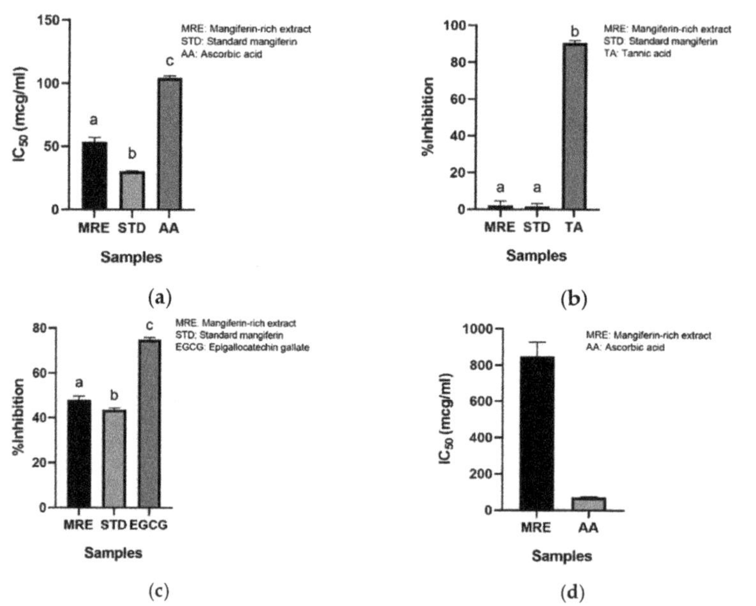

Figure 6. Anticollagenase activity of the MRE; (**a**) %Hyaluronidase inhibition of the MRE; (**b**) %Elastase inhibition of the MRE; (**c**) Antityrosinase activity of the MRE (**d**). All data are represented as mean ± SD ($n = 3$) and different alphabets indicate a significant difference.

3.3.3. Antihyaluronidase Activity

Hyaluronic acid is a glycosaminoglycan found in the skin that retains moisture content. It helps to hydrate the skin. Hyaluronic acid can be hydrolyzed by hyaluronidase. The MRE at a concentration of 40.0 μg/mL inhibited hyaluronidase by 2.25 ± 1.84%, while the standard mangiferin at 40.0 μg/mL could inhibit it by 1.72 ± 1.13%. Tannic acid, a positive control, exhibited 90.61 ± 1.04% of antihyaluronidase activity, as represented in Figure 6b. The % inhibition of the MRE and standard mangiferin on hyaluronidase inhibition significantly differed from tannic acid ($p < 0.05$). As a result, the MRE showed a very low inhibitory activity in the antihyaluronidase assay. Surprisingly, the previous study indicated a strong inhibition on hyaluronidase of the mangiferin extracts from honeybush [26]. This may be caused by using different plant sources and solvents in the isolation process, which led to the presence of substances antagonizing the antihyaluronidase effect of mangiferin.

3.3.4. Antielastase Activity

Elastase is an enzyme that degrades elastin, an extracellular matrix that provides elasticity to tissue. Figure 6c illustrates that EGCG, a positive control, showed excellent activity at 0.7 mg/mL, with 74.97 ± 0.82% elastase inhibition. Regarding the antielastase of the MRE and standard mangiferin, the percentage of elastase inhibition was 47.93 ± 1.34% and 43.57 ± 0.61%, respectively. The EGCG showed a significant difference compared to the MRE and standard mangiferin ($p < 0.05$). Our result was comparable to the prior study [26].

3.3.5. Antityrosinase Activity

The tyrosinase enzyme is involved in the melanogenesis pathway. This enzyme catalyzes tyrosine to melanin and other pigments. The IC_{50} of the MRE with the L-tyrosine substrate was 877.49 ± 0.063 μg/mL. Ascorbic acid showed an IC_{50} of 69.8 ± 0.018 μg/mL (Figure 6d). A significant difference was found between the IC_{50} of the MRE and that of ascorbic acid ($p < 0.05$). The result demonstrated that the MRE inhibited the tyrosinase enzyme, but not as effectively as ascorbic acid. The previous study demonstrated that the standard mangiferin had a stronger tyrosinase inhibition activity than the extract (IC_{50} = 195.50 ± 1.40 μg/mL), but not as much as kojic acid (IC_{50} = 24.65 ± 0.61 μg/mL). The inhibitory action of mangiferin may involve the carbonyl and phenolic groups in the structure, as it is a copper chelator. Because tyrosinase is a metalloenzyme, the copper chelator can inhibit tyrosinase by mimicking the substrate at the active sites [27].

3.4. Proteomics Analysis of the MRE

The proteomics of the MRE were analyzed using R-studio. Altogether, 779 proteins were found. Mostly, protein expression changes were downregulated (91.2%; 711 of 779), while 54 proteins were not significant (6.9%). The cutoff on the axis was chosen arbitrarily to identify the upregulated proteins. Among the 779 proteins, almost 14 proteins were observed to be upregulated (0.8%; 14 of 779), as shown in Figure 7. The description and function of these proteins are demonstrated in Table 3.

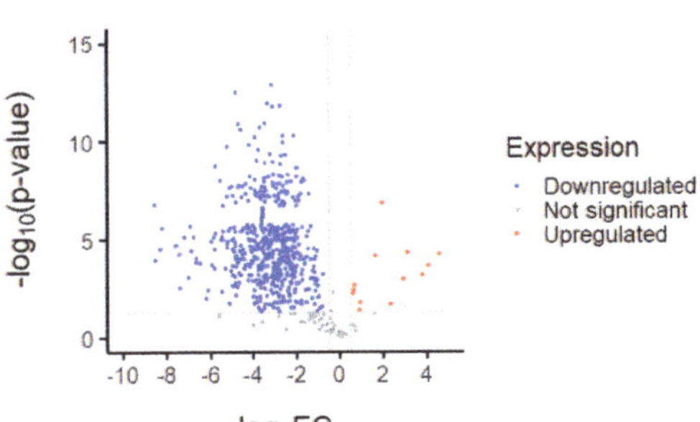

Figure 7. Volcano plot of the MRE analyzed using R-studio.

Table 3. Top 11 potential functions of the upregulated proteins identified in the MRE compared with the control.

GO Term	Functions	False Discovery Rate
GO:0006936	Actin-filament organization	8.33×10^{-5}
GO:0005200	Positive regulation of cytoskeleton organization	0.0103
GO:0051015	Rho protein signal transduction	0.0448
GO:0005198	Regulation of actin-filament polymerization	0.0448
GO:0044877	Cytoplasmic translation	0.0478
GO:0006936	Muscle contraction	0.0478
GO:1902905	Positive regulation of supramolecule fiber organization	0.0478
GO:0006412	Translation	0.0478
GO:0032956	Regulation of actin-cytoskeleton organization	0.0478
GO:0051493	Regulation of cytoskeleton organization	0.0478
GO:0050790	Regulation of catalytic activity	0.0478

These proteins are involved in actin-filament organization, cytoskeleton organization, Rho protein signal transduction, the regulation of actin-filament polymerization, cytoplasmic translation, muscle contraction, the positive regulation of supramolecule fiber organization, the translation and regulation of actin-cytoskeleton organization, the regulation of cytoskeleton organization, and the regulation of catalytic activity. It has been suggested that the actin cytoskeleton is essential for an arrangement of the collagenous extracellular matrix in embryonic tendons [28]. Actin disassembly reduced the TGF-β pathway, regulating collagen production in dermal fibroblasts [29]. Excess collagen production results in fibrosis, where mangiferin that has been shown to reduce BLM-induced pulmonary fibrosis in mice by blocking the TLR4/p65 and TGF-β1/Smad2/3 pathways, perhaps preventing pulmonary fibrosis in mice [30,31]. Another finding demonstrated that mangiferin decreased RAC1/WAVE2, belonging to the Rho proteins, assuming an important role in cytoskeletal rearrangement and promoting malignant cell migration and invasion, resulting in the inhibition of breast tumor progression [32]. Regulating the actin-filament polymerization in the nucleus of mangiferin causes the regulation of gene transcription and repairs damaged DNA. According to prior research, mangiferin reduced DNA damage and improved DNA repair in human peripheral blood lymphocytes exposed to 1–4 Gy γ-radiation [33]. The result of the MRE on collagenase and elastase activity inhibition is associated with regulating catalytic or metalloenzyme activity. One study examined that mangiferin reversibly inhibited elastase and collagenase in a noncompetitive manner. Mangiferin interacts with the free form of enzymes at a location other than the active site, and the enzyme–substrate complex [25]. The catalytic-activity regulation of the MRE is related to the antiaging activity. In addition, mangiferin is a potent antioxidant and probably reduces the activity of the enzyme by controlling free radicals induced by UV radiation.

3.5. Evaluation of MNPs

The 2% Eudragit® L100 was dissolved in ethanol. In all, 0.8% of the MRE in dimethylacetamide was added to the polymer solution. The appropriate condition involved applying voltage of 12 kV with 30 cm of distance between the needle tip and the collector at flow rate of 0.7 mL/min. The mixture was electrosprayed to create nanoparticles with a mean size of 247.8 nanometers and a polydispersity index (PDI) of 0.271. This indicates that the nanoparticles were homogenous in size. The zeta potential of the particles was −31.7 mV. Then, the % entrapment efficiency of the MNPs was 84.9%, and the morphology of nanoparticles was evaluated using SEM images, as shown in Figure 8.

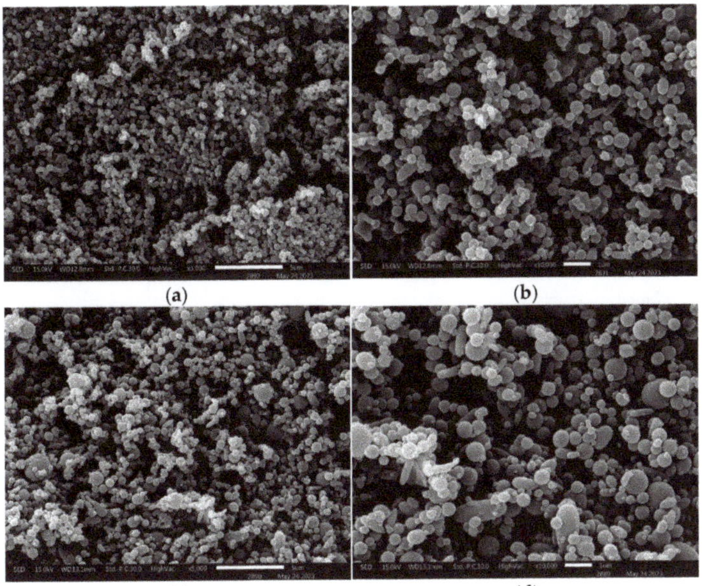

Figure 8. SEM images of the base polymer nanoparticles at 5000× (**a**) and 10,000× (**b**), and the MRE-loaded nanoparticles at 5000× (**c**) and 10,000× (**d**).

3.6. Irritation Study

To evaluate the irritation potential of the MRE and MNPs, the hen's egg chorioallantoic membrane (HET-CAM) test was used. The result showed that the 1% *w*/*v* SLS, a positive control, caused severe irritation, as displayed in Table 4. Vascular hemorrhage, vascular lysis, and vascular coagulation were onset within 5 min after application. The irritation score of 1% *w*/*v* SLS was 14.8 ± 2.2, describing severe irritation. Moreover, the 0.9% *w*/*v* NaCl, 0.5 mg/mL of MNP, and 0.5 mg/mL of the MRE were observed to be nontoxic, with an IS 0.0 ± 0, indicating no irritation. Overall, the MRE and MNPs are safe to use as an active ingredient in cosmetic formulations.

Table 4. Hen's egg chorioallantoic membrane before and at 5 min.

Tested Substance	Before	At 5 min
Positive control (1% *w*/*v* SLS)	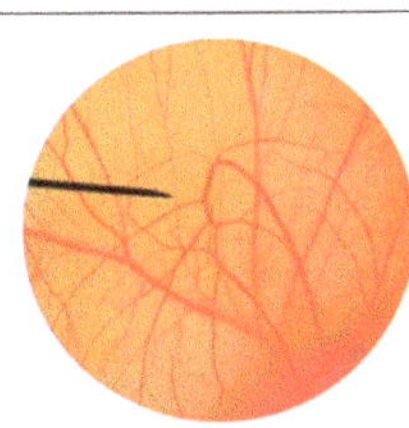	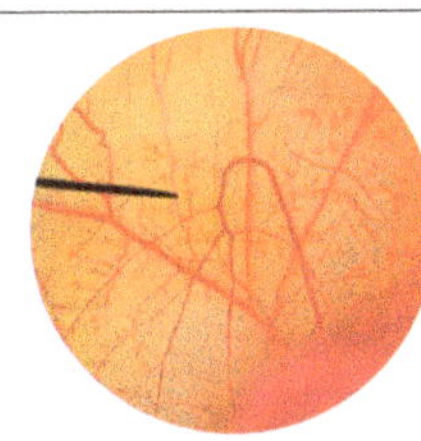
	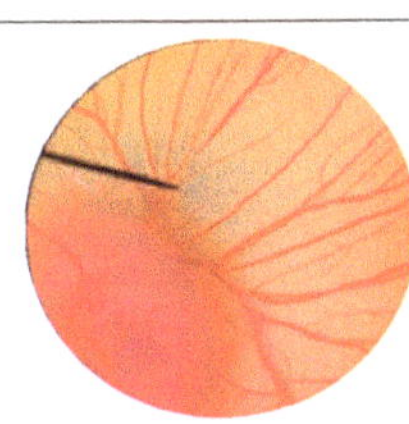	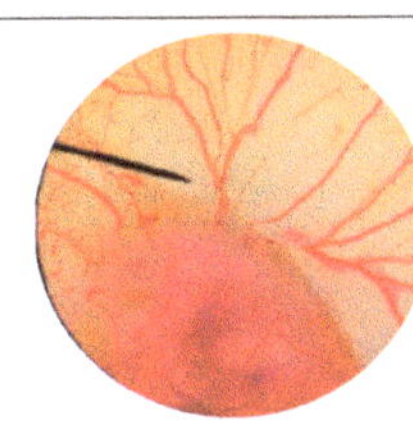

Table 4. *Cont.*

Tested Substance	Before	At 5 min
Negative control/Vehicle control (0.9% *w*/*v* NaCl)	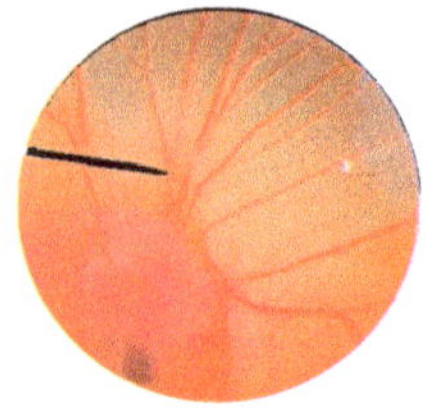	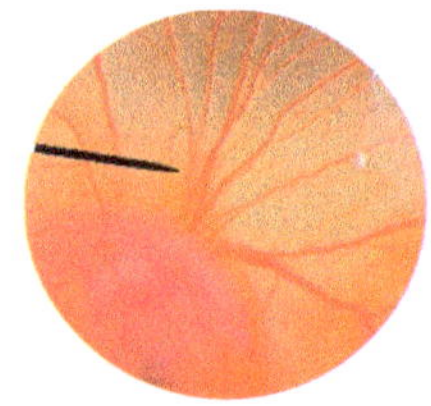
	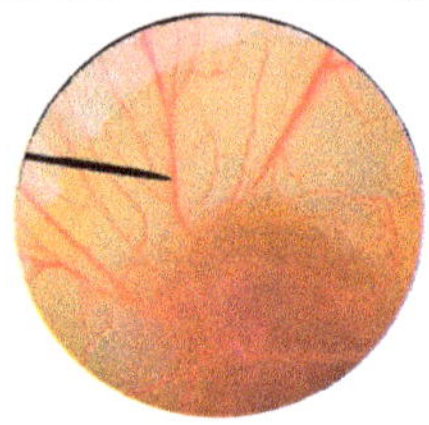	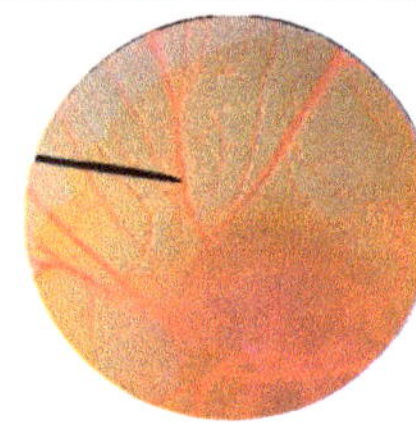
MNPs (0.5 mg/mL)	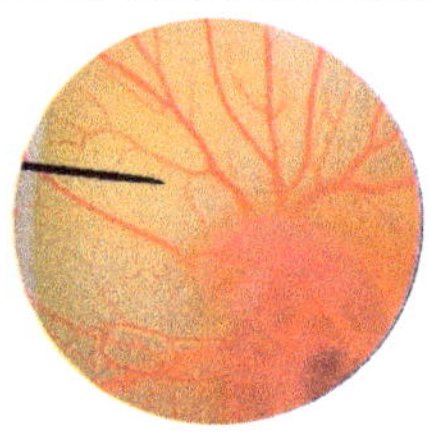	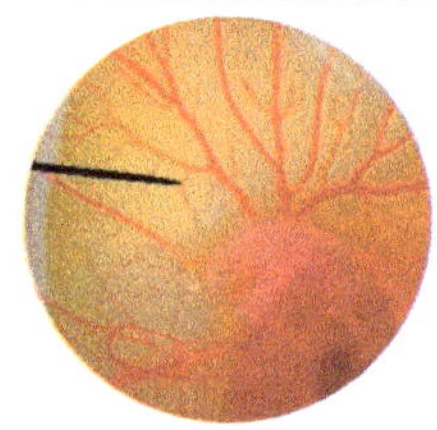
	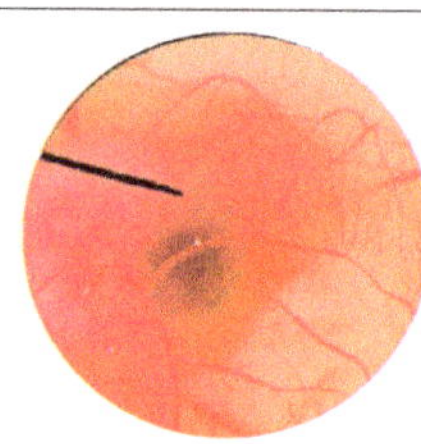	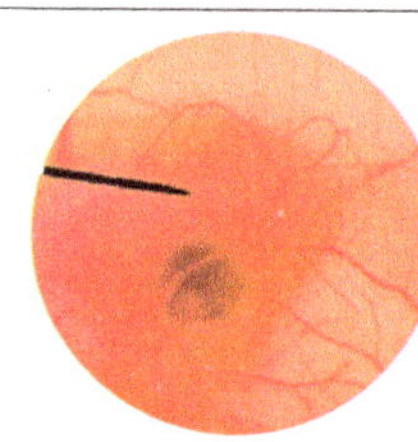
MRE (0.5 mg/mL)	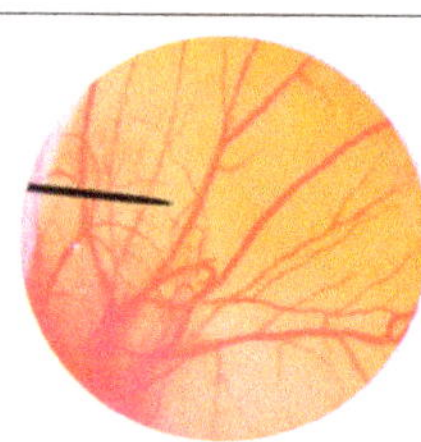	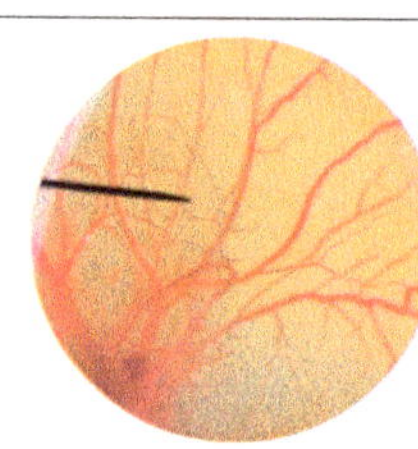
	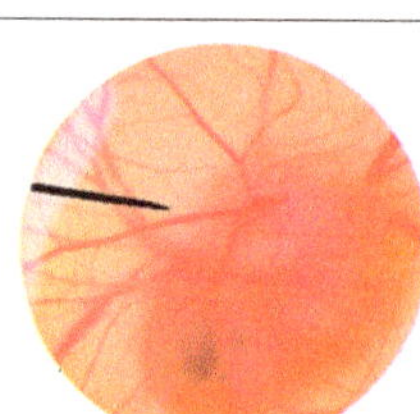	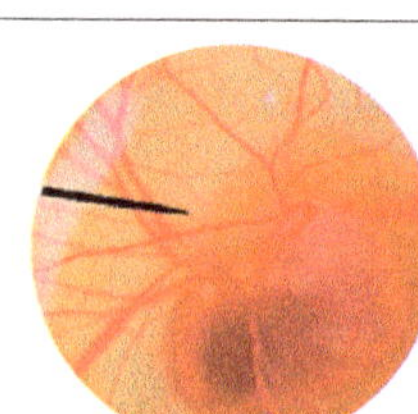

3.7. Preparation and Evaluation of the Emulsion Gel Containing MRE-Loaded Nanoparticles

The physical appearance of the emulsion gel was evaluated under the categories of color, pH, viscosity, and stability. Table 5 describes that the appearance of the formulation before and after heating–cooling cycles did not change. The viscosity of the base and MNP-loaded emulsion gel increased, probably caused by the evaporation of water in the formulations during the stability testing, but did not differ significantly. The % DPPH scavenging of the 1 g of MNPs containing the emulsion gel at day 0 was 85.35 ± 2.34%, while after heating–cooling, the MNP emulsion gel showed 93.81 ± 3.74%. The evaporation of water in the formulation after the heating–cooling test caused it to thicken, enhancing the mangiferin concentration. Moreover, Eudragit® L100 is a delayed-release polymer; therefore, the % DPPH scavenging of the emulsion gel after heating–cooling stability presented as higher than at day 0. The result indicates that the emulsion gel containing MRE-loaded nanoparticles is stable.

Table 5. Stability of emulsion gel in heating–cooling cycles.

Formulation	Condition	Appearance	pH	Viscosity (Pa. s)	% DPPH Inhibition
Emulsion gel base	Day 0	White, opaque, no precipitation, homogenous	6.04	1.639 ± 0.05	ND
	H-C 6 cycles	White, opaque, no precipitation, homogenous	6.06	1.826 ± 0.09	
Emulsion gel containing MNPs	Day 0	Pale greenish, opaque, no precipitation, homogenous	6.07	1.968 ± 0.09	85.35 ± 2.34%
	H-C 6 cycles	Pale greenish, opaque, no precipitation, homogenous	6.07	2.020 ± 0.07	93.81 ± 3.74%

ND = not done.

3.8. Skin-Retention Study

A test was conducted to determine whether the emulsion gel containing MNPs could penetrate the Strat-M® membrane, an artificial membrane that mimics human skin. The test results showed that no mangiferin content was found in the receptor fluid of the MRE solution and that of the emulsion gel containing MNPs at 30 min, and 1, 3, 6, 9, and 12 h, which indicated that the MRE and MNPs did not cross the Strat-M® membrane and eventually appeared in the bloodstream. It suggests that the MNP emulsion gel is unlikely to have any systemic side-effects. It also has been shown that mangiferin accumulated more in the MNP-emulsion-gel membranes than in the MRE solution membranes, as presented in Figure 9. It indicates that the MNP emulsion gel helps mangiferin to penetrate the skin.

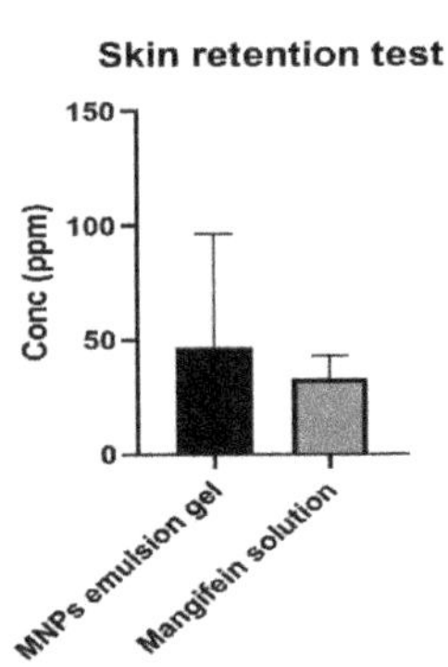

Figure 9. Mangiferin accumulation in the membranes of the skin-retention study. Data are represented as mean ± SD (n = 3).

4. Conclusions

The MRE from the mango leaves of the "Talapnak" variety, which are otherwise considered as an agricultural waste material after harvesting, proved to have beneficial antioxidant and anticollagenase actions. From the proteomic analysis, it was demonstrated that the MRE is mostly involved in actin-filament organization. The MRE also possesses catalytic activity, which is consistent with its anticollagenase and antielastase activities. The emulsion gel containing electrosprayed nanoparticles loaded with the MRE was found to retain in the skin and be stable after six heating–cooling cycles. Electrosprayed nanoparticles containing the MRE from mango leaves may thus be a promising source of cost-effective cosmeceutical active ingredients to slow down skin-aging processes.

Author Contributions: Formal analysis, P.S. (Padchanee Sangthong) and P.P.; Investigation, V.S. and C.S.; Writing—original draft, V.S.; Writing—review & editing, P.S. (Pawitrabhorn Samutrtai), P.L., C.S. and B.S. All authors have read and agreed to the published version of the manuscript.

Funding: The authors received financial support from the master's degree Program in Cosmetic Science, Faculty of Pharmacy, Chiang Mai University, Chiang Mai, Thailand, grant number 2564-068, under the CMU Presidential Scholarship, and received funding from the National Research Council of Thailand (NRCT).

Data Availability Statement: Data are contained within the article.

Acknowledgments: The authors are grateful for the financial support and funding. Also, we appreciate Evonik (Thailand) Ltd. for granting the use of Eudragit® L100. The authors acknowledge the Faculty of Pharmacy, Chiang Mai University, Chiang Mai, Thailand, for allowing the use of all its facilities.

Conflicts of Interest: The authors declare no conflict of interest.

Abbreviations

The following abbreviations are used in this manuscript:

MRE	Mangiferin-rich extract
pMRE	Purified mangiferin-rich extract
FT-IR	Fourier transform infrared spectrometer
DSC	Differential scanning calorimeter
EGCG	Epigallocatechin gallate
STD	Standard mangiferin
AA	Ascorbic acid
TA	Tannic acid
MNPs	MRE-loaded nanoparticles
HPLC	High-performance liquid chromatography
SEM	Scanning electron microscopy
PDI	Polydispersity index
IPA	Isopropyl alcohol
HET-CAM	Hen's egg chorioallantoic membrane
IS	Irritation score

References

1. Trade Policy and Strategy Office. *Regional Economic Report of April 2021*; Trade Policy and Strategy Office: Nonthaburi, Thailand, 2021; p. 2565.
2. Ediriweera, M.K.; Tennekoon, K.H.; Samarakoon, S.R. A Review on Ethnopharmacological Applications, Pharmacological Activities, and Bioactive Compounds of *Mangifera indica* (Mango). *Evid. Based Complement. Altern. Med.* **2017**, *2017*, 6949835. [CrossRef]
3. Shah, K.A.; Patel, M.B.; Patel, R.J.; Parmar, P.K. *Mangifera indica* (mango). *Pharmacogn. Rev.* **2010**, *4*, 42–48. [CrossRef] [PubMed]
4. Kumar, M.; Saurabh, V.; Tomar, M.; Hasan, M.; Changan, S.; Sasi, M.; Maheshwari, C.; Prajapati, U.; Singh, S.; Prajapat, R.K.; et al. Mango (*Mangifera indica* L.) Leaves: Nutritional Composition, Phytochemical Profile, and Health-Promoting Bioactivities. *Antioxidants* **2021**, *10*, 299. [CrossRef]

5. Ajila, C.M.; Bhat, S.G.; Prasada Rao, U.J.S. Valuable components of raw and ripe peels from two Indian mango varieties. *Food Chem.* **2007**, *102*, 1006–1011. [CrossRef]
6. Tacias-Pascacio, V.G.; Castañeda-Valbuena, D.; Fernandez-Lafuente, R.; Berenguer-Murcia, Á.; Meza-Gordillo, R.; Gutiérrez, L.-F.; Pacheco, N.; Cuevas-Bernardino, J.C.; Ayora-Talavera, T. Phenolic compounds in mango fruit: A review. *J. Food Meas. Charact.* **2022**, *16*, 619–636. [CrossRef]
7. Telang, M.; Dhulap, S.; Mandhare, A.; Hirwani, R. Therapeutic and cosmetic applications of mangiferin: A patent review. *Expert Opin. Ther. Pat.* **2013**, *23*, 1561–1580. [CrossRef] [PubMed]
8. Masibo, M.; He, Q. Major Mango Polyphenols and Their Potential Significance to Human Health. *Compr. Rev. Food Sci. Food Saf.* **2008**, *7*, 309–319. [CrossRef] [PubMed]
9. Imran, M.; Arshad, M.S.; Butt, M.S.; Kwon, J.H.; Arshad, M.U.; Sultan, M.T. Mangiferin: A natural miracle bioactive compound against lifestyle related disorders. *Lipids Health Dis.* **2017**, *16*, 84. [CrossRef] [PubMed]
10. Dar, A.; Faizi, S.; Naqvi, S.; Roome, T.; Zikr-ur-Rehman, S.; Ali, M.; Firdous, S.; Moin, S.T. Analgesic and antioxidant activity of mangiferin and its derivatives: The structure activity relationship. *Biol. Pharm. Bull.* **2005**, *28*, 596–600. [CrossRef]
11. Jagetia, G.C.; Venkatesha, V.A. Effect of mangiferin on radiation-induced micronucleus formation in cultured human peripheral blood lymphocytes. *Environ. Mol. Mutagen.* **2005**, *46*, 12–21. [CrossRef]
12. Kim, H.-S.; Song, J.H.; Youn, U.J.; Hyun, J.W.; Jeong, W.S.; Lee, M.Y.; Choi, H.J.; Lee, H.K.; Chae, S. Inhibition of UVB-induced wrinkle formation and MMP-9 expression by mangiferin isolated from *Anemarrhena asphodeloides*. *Eur. J. Pharmacol.* **2012**, *689*, 38–44. [CrossRef]
13. Wang, J.; Jansen, J.A.; Yang, F. Electrospraying: Possibilities and Challenges of Engineering Carriers for Biomedical Applications—A Mini Review. *Front. Chem.* **2019**, *7*, 258. [CrossRef] [PubMed]
14. Song, J.; Zhang, Z. Brinzolamide loaded core-shell nanoparticles for enhanced coronial penetration in the treatment of glaucoma. *J. Appl. Biomater. Funct. Mater.* **2020**, *18*, 2280800020942712. [CrossRef] [PubMed]
15. Al-Amrani, S.; Al-Jabri, Z.; Al-Zaabi, A.; Alshekaili, J.; Al-Khabori, M. Proteomics: Concepts and applications in human medicine. *World J. Biol. Chem.* **2021**, *12*, 57–69. [CrossRef] [PubMed]
16. Vo, T.H.T.; Nguyen, T.D.; Nguyen, Q.H.; Ushakova, N.A. Extraction of Mangiferin from the Leaves of the Mango Tree *Mangifera indica* and Evaluation of Its Biological Activity in Terms of Blockade of α-Glucosidase. *Pharm. Chem. J.* **2017**, *51*, 806–810. [CrossRef]
17. Chaiyana, W.; Sirithunyalug, J.; Somwongin, S.; Punyoyai, C.; Laothaweerungsawat, N.; Marsup, P.; Neimkhum, W.; Yawootti, A. Enhancement of the Antioxidant, Anti-Tyrosinase, and Anti-Hyaluronidase Activity of *Morus alba* L. Leaf Extract by Pulsed Electric Field Extraction. *Molecules* **2020**, *25*, 2212. [CrossRef]
18. Theansungnoen, T.; Nitthikan, N.; Wilai, M.; Chaiwut, P.; Kiattisin, K.; Intharuksa, A. Phytochemical Analysis and Antioxidant, Antimicrobial, and Antiaging Activities of Ethanolic Seed Extracts of Four *Mucuna* Species. *Cosmetics* **2022**, *9*, 14. [CrossRef]
19. Nguyen, D.N.; Clasen, C.; Van den Mooter, G. Encapsulating darunavir nanocrystals within Eudragit L100 using coaxial electrospraying. *Eur. J. Pharm. Biopharm.* **2017**, *113*, 50–59. [CrossRef]
20. Yeerong, K.; Sriyab, S.; Somwongin, S.; Punyoyai, C.; Chantawannakul, P.; Anuchapreeda, S.; Prommaban, A.; Chaiyana, W. Skin irritation and potential antioxidant, anti-collagenase, and anti-elastase activities of edible insect extracts. *Sci. Rep.* **2021**, *11*, 22954. [CrossRef]
21. Nguyen, T.A. *Vietnamese Pharmacopoeia*; Vietnamese Pharmacopoeia Commission: Hanoi, Vietnam, 2005.
22. Khurana, R.K.; Kaur, R.; Kaur, M.; Kaur, R.; Kaur, J.; Kaur, H.; Singh, B. Exploring and validating physicochemical properties of mangiferin through GastroPlus® software. *Future Sci. OA* **2017**, *3*, Fso167. [CrossRef]
23. Lizárraga-Velázquez, C.E.; Hernández, C.; González-Aguilar, G.A.; Heredia, J.B. Effect of hydrophilic and lipophilic antioxidants from mango peel (*Mangifera indica* L. cv. Ataulfo) on lipid peroxidation in fish oil. *CyTA J. Food* **2018**, *16*, 1095–1101. [CrossRef]
24. Stohs, S.; Swaroop, A.; Moriyama, H.; Bagchi, M.; Ahmad, T.; Bagchi, D. A Review on Antioxidant, Anti-Inflammatory and Gastroprotective Abilities of Mango (*Magnifera indica*) Leaf Extract and Mangiferin. *J. Nutr. Health Sci.* **2018**, *5*, 303.
25. Ochocka, R.; Hering, A.; Stefanowicz–Hajduk, J.; Cal, K.; Barańska, H. The effect of mangiferin on skin: Penetration, permeation and inhibition of ECM enzymes. *PLoS ONE* **2017**, *12*, e0181542. [CrossRef]
26. Hering, A.; Stefanowicz-Hajduk, J.; Gucwa, M.; Wielgomas, B.; Ochocka, J.R. Photoprotection and Antiaging Activity of Extracts from Honeybush (*Cyclopia* sp.)—In Vitro Wound Healing and Inhibition of the Skin Extracellular Matrix Enzymes: Tyrosinase, Collagenase, Elastase and Hyaluronidase. *Pharmaceutics* **2023**, *15*, 1542. [CrossRef]
27. Efendi, M.R.; Bakhtiar, A.; Rusdi, M.S.; Putra, D.P. Simplified isolation method of mangiferin from *Mangifera indica* L. leaves and evaluation of tyrosinase inhibitory activity. *Asia-Pac. J. Sci. Technol.* **2023**, *28*, APST-28-06-12.
28. Canty, E.G.; Starborg, T.; Lu, Y.; Humphries, S.M.; Holmes, D.F.; Meadows, R.S.; Huffman, A.; O'Toole, E.T.; Kadler, K.E. Actin filaments are required for fibripositor-mediated collagen fibril alignment in tendon. *J. Biol. Chem.* **2006**, *281*, 38592–38598. [CrossRef] [PubMed]
29. Qin, Z.; Fisher, G.J.; Voorhees, J.J.; Quan, T. Actin cytoskeleton assembly regulates collagen production via TGF-β type II receptor in human skin fibroblasts. *J. Cell. Mol. Med.* **2018**, *22*, 4085–4096. [CrossRef]
30. Jia, L.; Sun, P.; Gao, H.; Shen, J.; Gao, Y.; Meng, C.; Fu, S.; Yao, H.; Zhang, G. Mangiferin attenuates bleomycin-induced pulmonary fibrosis in mice through inhibiting TLR4/p65 and TGF-β1/Smad2/3 pathway. *J. Pharm. Pharmacol.* **2019**, *71*, 1017–1028. [CrossRef]

31. Zhu, X.; Cheng, Y.-Q.; Du, L.; Li, Y.; Zhang, F.; Guo, H.; Liu, Y.W.; Yin, X.X. Mangiferin Attenuates Renal Fibrosis through Down-Regulation of Osteopontin in Diabetic Rats. *Phytother. Res.* **2015**, *29*, 295–302. [CrossRef]
32. Deng, Q.; Tian, Y.X.; Liang, J. Mangiferin inhibits cell migration and invasion through Rac1/WAVE2 signalling in breast cancer. *Cytotechnology* **2018**, *70*, 593–601. [CrossRef] [PubMed]
33. Jagetia, G.C.; Venkatesha, V.A. Mangiferin protects human peripheral blood lymphocytes against γ-radiation–induced DNA strand breaks: A fluorescence analysis of DNA unwinding assay. *Nutr. Res.* **2006**, *26*, 303–311. [CrossRef]

polymers

MDPI

Article

Effect of Graphene Oxide-Modified CaAl-Layered Double Hydroxides on the Carbon Dioxide Permeation Properties of Fluoroelastomers

Chuanbo Cong *, Daigang Peng, Qingkun Liu, Mingyang Yuan, Xiaoyu Meng and Qiong Zhou

New Energy and Material College, China University of Petroleum, Beijing 102249, China; pengdg_123@163.com (D.P.); liuqingkun1992@163.com (Q.L.); yuanmy.apply@gmail.com (M.Y.); xymeng800418@sohu.com (X.M.); zhouqiong_cn@163.com (Q.Z.)

* Correspondence: congcb@cup.edu.com

Abstract: This work aimed to investigate the CO_2 gas barrier and mechanical properties of fluorine rubber nanocomposites filled with Ca/Al layered hydroxide (graphene oxide [GO]/LDH-Ca_2Al) modified by GO. GO/LDH-Ca_2Al nanocomposite fillers were prepared by depositing Ca/Al layered hydroxide (LDH-Ca_2Al) into the surface of alkalized GO (Al-GO). The prepared GO/LDH-Ca_2Al nanocomposite fillers and complexes were characterized by Fourier infrared spectroscopy (FTIR), X-ray diffraction (XRD), scanning electron microscopy (SEM), and transmission electron microscopy (TEM) for structural and micromorphological characterization. The results showed that GO/LDH-Ca_2Al was successfully prepared with strong interactions between Al-GO and LDH, and the compatibility of GO/LDH-Ca_2Al nanocomposite fillers with the polymer was significantly improved compared with that of LDH-Ca_2Al. Consequently, both the fracture strength (σ_b) and strain (ε_b) of GO/LDH-Ca_2Al nanocomplexes remarkably increased, and they exhibited excellent mechanical properties. Differential scanning calorimetry and thermogravimetric analysis were used to characterize the thermal stability of GO/LDH-Ca_2Al nanocomposite fillers, and GO/LDH-Ca_2Al nanocomposite fillers have better thermal stability than LDH-Ca_2Al. The reaction products (S-LDH-Ca_2Al and S-GO-Ca_2Al) of LDH-Ca_2Al and GO/LDH-Ca_2Al with CO_2 were characterized using XRD and TGA, respectively, and the results show that LDH-Ca_2Al reacts readily and chemically with CO_2, resulting in a lower diffusion coefficient of CO_2 in the LDH-Ca_2Al nanocomplexes than that of the GO/LDH-Ca_2Al nanocomplexes and leading to the destruction of the laminar structure of LDH-Ca_2Al, while GO/LDH-Ca_2Al has better CO_2 resistance stability. GO/LDH-Ca_2Al nanocomplexes exhibited a reduced content of hydroxyl groups with pro-CO_2 nature exposed on the surface of LDH-Ca_2Al, improving the interfacial interaction between the nanofillers and the rubber matrix and enhancing the dispersion of GO/LDH-Ca_2Al in the polymers. Moreover, CO_2 in the soluble GO/LDH-Ca_2Al nanocomposites was significantly reduced, while the diffusion properties demonstrated weak temperature dependence on solubility. The mechanism of the CO_2 gas barrier of polymers filled with GO/LDH-Ca_2Al was proposed on the basis of the Arrhenius equation.

Keywords: LDH; graphene oxide; permeability; carbon dioxide

1. Introduction

Climate change is one of the most serious problems facing mankind, and anthropogenic air pollutant emissions and climate change have a direct relationship in which carbon dioxide (CO_2) is the main air pollutant [1]. In the field of oilfield development, CO_2 is an attractive repellant for enhanced oil recovery, but some problems, such as CO_2 leakage during CO_2 transportation, are encountered; hence, a sealing material with excellent barrier performance needs to be developed [2].

Rubber is a highly elastic and amorphous polymer with good toughness, elasticity, and elongation, and they are also known as elastomers; for this class of material, a very small external force can produce a large deformation, which can be restored after removing the external force [3,4]. Rubber materials are widely used in tire liners, chemical protection products, medical packaging, automotive tanks and natural gas storage tanks, and natural long-distance pipeline valve sealing materials, and their gas barrier properties are very important [5]. The gas barrier properties of rubber materials are very important. Given the high free volume fraction between rubber molecular chains, gas molecules can easily diffuse from one end of the rubber material to the other, and most diene-based rubbers such as natural rubber (NR), ethylene–propylene rubber (EPDM), and styrene–butadiene rubber (SBR) have high permeability to gases [6].

The gas barrier properties can be effectively improved by filling a certain amount of nanoparticles into the rubber to form a filling network, thus creating distorted paths and inhibiting gas molecules from penetrating the rubber matrix. More importantly, the strong interface between the nanoparticles and the rubber molecules is significant in limiting chain migration and further reducing the free volume between the nanoparticles and the rubber molecules. Among a range of nanoparticles, layered particles such as layered double hydroxides (LDH) [7–9] and graphene (GE) or graphene oxide (GO) [10,11] have higher aspect ratios than spherical and fibrous particles, making them more conducive to enhancing the gas barrier properties of rubber composites [12]. In particular, LDH is considered to be a promising CO_2 adsorbent because of its controllable layer spacing and pro-CO_2 properties [13]. The chemical formula of LDH can be represented by the general expression $M^{2+}_{1-x}M^{3+}_x(OH)_2(A^{n-})_{x/n}{\cdot}yH_2O$ [14,15]. LDH consists of a hydromagnesite-like layer in which a small portion of octahedrally coordinated divalent metal cations is replaced by trivalent metal cations, resulting in a positively charged host layer [16]. Exchangeable inorganic or organic anions are accommodated in the interlayer channels to compensate for the positive charge. In addition, the hydroxyl groups of the host layer are connected to anions or water molecules via hydrogen bonding. The X value is equal to the molar ratio of $M^{3+}/(M^{2+}+M^{3+})$, and it ranges from 0.17 to 0.33 [16]. Given the controllable M^{2+}/M^{3+} molar ratio, the tunability of the metal cations, and the exchangeable charge compensating anions [17], the layer spacing of the LDH can be optimized, which can aid in the design of the lamellar packing that facilitates CO_2 intercalation, further forming a more complex pathway for CO_2 diffusion.

However, the thermal stability of LDH is poor, and it starts to undergo layer structure destruction at approximately 250 °C [18]. Most importantly, LDH is a polar inorganic material with limited compatibility with organic macromolecules such as natural rubber (NR), fluorine rubber (FKM), and nitrile rubber (NBR), leading to agglomeration in the polymer and the formation of weak phase interfaces [8,9]. Therefore, the dispersion and thermal stability of LDH in polymers can be improved by forming a compound with LDH along with new components. GO has high electronic conductivity, good mechanical strength, excellent thermal stability, large specific area, and abundant oxygen-containing groups on its surface, such as epoxide, hydroxyl, and carboxyl groups, which can form strong interactions with polymer molecules via hydrogen or ionic bonding [13–15]. It also has good descriptive properties with polymers. Moreover, GO is an almost monolayer layered structure with a large surface area [19]. GO is an ideal carrier for LDH because of its negatively charged layer surface and oxygen-containing groups, which can form hydrogen bonding and strong electrostatic interactions with the positively charged LDH on the surface of the host layer [20–22]. Yang et al. prepared FeNi-LDH/GO hybrid nanosheets by alternately stacking GO layers and FeNi double hydroxide ion layers [20].

To meet the application requirements for rubber of the tire, aerospace, and military fields, nanofillers such as graphene oxide, carbon nanotubes, carbon black, and montmorillonite have been widely used to improve the mechanical properties of rubber. The effects of various modifying additives and fillers on the physicochemical and mechanical properties of polymer composites are determined by many factors as follows: (1) filler–polymer interfacial interaction [23–25], (2) content of fillers [26], and (3) cross-linking density of polymers [27]. Among these, filler–polymer interfacial interaction exerts an extremely significant effect on the mechanical properties of polymer composites.

In this study, GO/LDH-Ca_2Al nanolamellar fillers with better thermal stability and better compatibility with polymers were prepared via a simple synthesis method (Scheme 1). The layer spacing of this nanofiller increased by approximately 0.04 nm compared with that of pristine LDH-Ca_2Al, thus increasing the chances of interlayer insertion of CO_2 gas molecules. At the same time, the resistance to CO_2 stability was significantly improved, and a stable layer structure can be maintained in a CO_2 environment. The GO/LDH-Ca_2Al nanofiller reduced the content of exposed hydroxyl groups of a pro-CO_2 nature on the surface of LDH-Ca_2Al, increased the interfacial interactions between LDH-Ca_2Al and the rubber matrix, and improved the dispersion of LDH-Ca_2Al in the polymer. Consequently, the CO_2 solubility of the nanocomposites was significantly reduced.

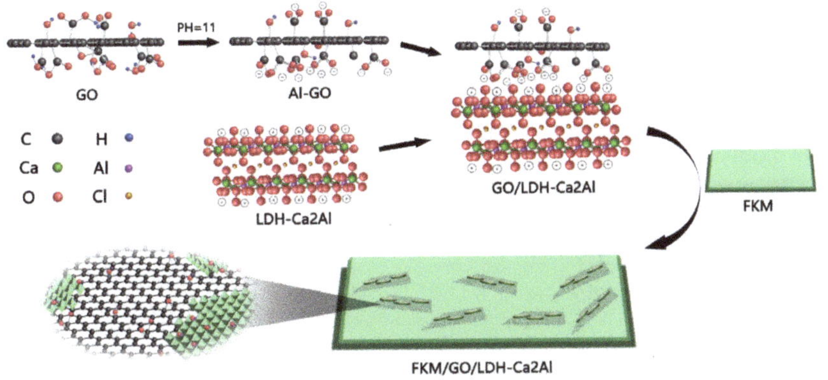

Scheme 1. Preparation flow of GO/LDH-Ca_2Al nanocomposite fillers.

2. Experimental

2.1. Materials

FKM 246, a terpolymer of vinylidene fluoride (VF2); hexafluoropropylene (HFP); and tetrafluoroethylene (TFE), with 68.5% fluorine content, density of 1.86 g/cm^3, Mooney viscosity, ML 1 + 10 at 121 °C = 25, and solubility parameter, δ = 19.7 MPa$^{1/2}$ were supplied by Shanghai Huayi 3F New Materials Co., Ltd. (Shanghai, China). 2,5-Dimethyl-2,5-di(tert-butylperoxy)hexane (Macklin), triallyl isocyanurate (TAIC, $\geq$98%, Aladdin), and calcium chloride dihydrate ($CaCl_2 \cdot 2H_2O$, $\geq$98%) were purchased from Beijing Yongchang Haoran Bio-technology Co., Ltd. (Beijing, China). Aluminum chloride hexahydrate ($AlCl_3 \cdot 6H_2O$, $\geq$98%) was obtained from Shanghai Taitan Technology Co., Ltd. (Shanghai, China). Graphene oxide (GO, >99%), sodium hydroxide (NaOH, $\geq$98%), and deionized water were purchased from Shanghai Boer Chemical Reagent Co. (Shanghai, China). All reagents were of analytical grade, and the purity of the CO_2 test gas used for the experiments was 99.9% (Beijing Chengweixin Industrial Gas Sales Center, Beijing, China).

2.2. Preparation of Layered LDH-Ca_2Al and GO/LDH-Ca_2Al

LDH-Ca_2Al was prepared via co-precipitation [28–30]. Approximately 15 g $CaCl_2 \cdot 2H_2O$ and 7.606 g $AlCl_3 \cdot 6H_2O$ were dissolved in 300 mL of deionized water to obtain a colorless and transparent Ca_2Al solution. Approximately 7.36 g NaOH was dissolved in 30 mL of deionized water to obtain a colorless and transparent NaOH solution, and the NaOH solution was added to the vigorously stirred (800 rpm) Ca_2Al solution at room temperature. The pH of the suspension was maintained at approximately 11, and stirring was maintained for 1 h. The sample was then filtered, washed with deionized water until the pH was approximately 10, and finally vacuum-dried at 60 °C to obtain the white powder LDH-Ca_2Al.

Approximately 0.5 g GO was added to 1050 mL of deionized water to obtain the GO suspension [31]. Then, an appropriate amount of NaOH was added to adjust the pH of GO suspension at approximately 11 (Al-GO), and 8.15 g LDH-Ca_2Al was added to the solution. The sample was sonicated for 10 min and then placed in a vacuum oven at 60 °C for 24 h to obtain GO/LDH-Ca_2Al suspension. Finally, it was filtered, washed with deionized water, and freeze-dried.

2.3. CO_2 Resistance Stability Testing of LDH-Ca_2Al and GO/LDH-Ca_2Al

S-LDH-Ca_2Al and S-GO/LDH-Ca_2Al, which were intercalated and reacted by CO_2, were obtained by simultaneously placing 1.5 g of LDH-Ca_2Al and GO/LDH-Ca_2Al in an autoclave of CO_2 gas as the ambient medium at 80 °C and 4.3 MPa, and the sample was taken out after 1 day.

2.4. Preparation of FKM/GO/LDH-Ca_2Al and FKM/LDH-Ca_2Al Composites

The raw rubber was sheared on the rolls for 2 min at room temperature by using an open double-roller mill (friction ratio: 1:1.4). Then, different proportions of GO/LDH-Ca_2Al were added for 3–4 min. Finally, 5 phr LUPEROX 101XL-50 and 5 phr TAIC were added to the sample, and the mixture was sheared and agitated for 3–4 min to obtain the unvulcanized FKM/GO/Ca_2Al composite. A rotorless vulcanometer was used to determine the optimal vulcanization time (tc90) of the unvulcanized FKM/GO/LDH-Ca_2Al composites at 160 °C, and was vulcanized to the optimal vulcanization time (tc90) by using a plate vulcanizer at 160 °C and 10 MPa. FKMs unfilled and filled with the same amount of LDH-Ca_2Al were prepared using the same method as the blank control group. Table 1 provides the formulation of FKM/GO/LDH-Ca_2Al composites. Different portions of GO/LDH-Ca_2Al-filled fluoroelastomers were denoted as FKM/GO/LDH-Ca_2Al-x, where x refers to the number of portions of nanofillers in 100 phr of fluoroelastomers.

Table 1. FKM nanocomposites formulation.

Samples	FKM 246	LUPEROX 101XL-50	TAIC	LDH-Ca_2Al [a]	GO/LDH-Ca_2Al [a]
FKM	100	5	5		
FKM/LDH-Ca_2Al-5	100	5	5	5	5
FKM/LDH-Ca_2Al-8	100	5	5	8	
FKM/LDH-Ca_2Al-10	100	5	5	10	
FKM/GO/LDH-Ca_2Al-5	100	5	5		5
FKM/GO/LDH-Ca_2Al-8	100	5	5		8
FKM/GO/LDH-Ca_2Al-10	100	5	5		10

[a] The densities of the materials FKM 246, LDH-Ca_2Al, and GO/LDH-Ca_2Al are 1.86, 1.65, and 2.01 g/cm^3, respectively.

2.5. Characterization

Fourier transform infrared spectroscopy (FT-IR) measurements were carried out on a TENSOR II spectrometer (Bruker, MA, USA) at the wave number range of 4000–400 cm^{-1} with a resolution of 4 cm^{-1} and a 64-scan signal via the KBr pellet technique. An X-ray diffractometer (D8 Focus, Bruker) was used to characterize LDH-Ca_2Al and GO/LDH-Ca_2Al before and after CO_2 immersion and GO, and the measurements were carried out at a wavelength of λ = 0.154056 nm with a step size of 0.01° under Cu/Kα radiation. LDH-Ca_2Al and GO/LDH-Ca_2Al before and after CO_2 immersion, FKM/LDH-Ca_2Al, and FKM/GO/LDH-Ca_2Al were subjected to morphological observation by using a SU8010 cold field emission scanning electron microscope (SU8010, Hitachi, Japan). The CO_2 gas permeability of FKM/LDH-Ca_2Al and FKM/GO/LDH-Ca_2Al composites were measured via differential pressure by using a homemade gas permeability tester according to ISO 2782-1:2022 [32]. The samples to be tested were cut into dumbbell-shaped specimens. The tensile test was carried out by using a universal testing machine at room temperature with a tensile rate of 500 mm/min, following the GB/T 528–2009 standard [33] and three parallel specimens were used for each group of materials. According to the ISO 7619-1:2010 standard [34], the Shore A hardness of the composites was tested using an LX A-type hardness tester. The samples were subjected to DMA testing by using a DMA (NETZSCH 242 C) in tensile mode at the frequency of 10 Hz. Each sample was scanned in the range of −50 °C to 60 °C at a heating rate of 3 °C/min^{-1}. The samples were subjected to thermogravimetric analysis (TGA) by using a Shimadzu DTG-60 thermogravimetric analyzer, and the weight of the test samples ranged from 8 mg to 10 mg. The samples were heated from 25 °C to 800 °C at a heating rate of 10 °C/min under a nitrogen atmosphere purge of 100 mL/min. A differential scanning calorimeter (DSC, NETZSCH-5) was used to characterize the LDH-Ca_2Al and GO/LDH-Ca_2Al thermal transition. The weight of the test samples ranged from 6 mg to 10 mg and was heated from 25 °C to 250 °C at a heating rate of 10 °C/min under a nitrogen atmosphere of 100 mL/min.

The crosslink density of FKM/LDH-Ca_2Al and FKM/GO/LDH-Ca_2Al was measured using the equilibrium dissolution method [27,35] according to the standard ASTM D6814-02 [36]. Approximately 0.2 g of the sample was placed in a sealed container containing 25 mL of acetone and held at 25 °C for 3 days, during which the acetone solvent was changed every 24 h. The mass of the swelling equilibrium was measured as m_S and dried in a blower oven at 70 °C for 16 h. Then, the weight of the dried rubber m_d was measured, and the density ρ_d was calculated. The crosslink density v_e was calculated as follows:

$$V_r = (m_d/\rho_d)/[m_d/\rho_d + (m_S - m_d)/\rho_S] \tag{1}$$

$$v_e = \frac{-\left[ln(1-V_r) + V_r + \chi_1 V_r^{\,2}\right]}{\left[V_1\left(V_r^{\,1/3} - V_r\right)/2\right]} \tag{2}$$

where V_r and ρ_S are the volume fraction of rubber and the density of the solvent (the density of acetone is 0.79 g/cm^3), respectively, V_1 is the molar volume of acetone solvent (73.53 cm^3/mol), and χ_1 is the interaction parameter of the polymer with the solvent, which is 0.358 in this case. According to the Bristow–Watson equation, the following expression can be obtained [37]:

$$\chi_1 = \beta_1 + \left(\frac{V_1}{RT}\right)(\delta_s - \delta_p)^2 \tag{3}$$

where β_1 is the lattice constant (typically 0.34), R is the gas constant, T is the absolute temperature, and δ_p and δ_s are the solubility parameters of FKM and solvent acetone, respectively.

3. Results and Discussion

3.1. Fourier Transform Infrared (FTIR) Spectroscopy

The IR spectra of GO, Al-GO, LDH-Ca_2Al, GO/LDH-Ca_2Al, S-LDH-Ca_2Al, and S-GO/LDH-Ca_2Al are shown in Figure 1. The broad and strong absorption peaks of the infrared spectra of GO near 3381 cm^{-1} represent the O–H stretching vibrations in carboxyl and hydroxyl groups, and the peaks at 1223 and 1733 cm^{-1} represent the out-of-plane bending vibrations of O–H in COOH and its C=O stretching vibrations [38,39]. Alkoxy (C–OH) stretching vibrations and hydroxyl deformation vibrations occurred at 1052 and 1390 cm^{-1}, respectively [40], and while that at 1620 cm^{-1} can be attributed to the conjugate bond of aromatic C=C-C=C [39]. The peaks of the C=O stretching vibration of COOH and O–H out-of-plane bending vibration of COOH on Al-GO disappeared, and the symmetric and antisymmetric peaks of COO^- appeared near 1636 and 1458 cm^{-1} [41]. The O–H stretching vibration peaks of Al–OH and Ca–OH of LDH-Ca_2Al, GO/LDH-Ca_2Al, S-LDH-Ca_2Al, and S-GO/LDH-Ca_2Al appeared at 3634 and 3474 cm^{-1}, respectively, the bending vibration of the interlayer H_2O molecule was observed at approximately 1621 cm^{-1}, and the absorption peaks at 787, 535, and 434 cm^{-1}, which are below 800 cm^{-1}, represent M-O vibrational peaks (M is Ca or Al) [18,41,42]. In the IR spectra of GO/LDH-Ca_2Al, Al-GO peaks appeared at 3741 and 1637 cm^{-1}, but these peaks appear to be blue-shifted, possibly because of the positively charged nature on the LDH-Ca_2Al layer and the presence of the electron-absorbing induced effect. The peaks at 1733 and 1223 cm^{-1}, which represent COOH, disappeared from the infrared spectra of GO/LDH-Ca_2Al and Al-GO, while symmetric and antisymmetric telescopic vibrational peaks near 1458 and 1636 cm^{-1} for COO^- appeared [41]. The symmetric and antisymmetric telescopic vibrational peaks of COO^- in the GO/LDH-Ca_2Al spectra also shifted to the high-frequency region, indicating the presence of a strong interaction between GO and LDH-Ca_2Al.

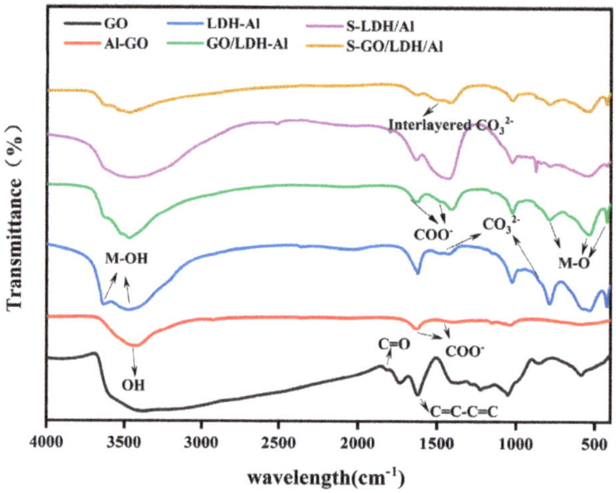

Figure 1. IR spectra of GO, Al-GO, LDH-Ca_2Al, GO/LDH-Ca_2Al, S-LDH-Ca_2Al, and S-GO/LDH-Ca_2Al.

The $CO_3{}^{2-}$ peaks at 1428 and 875 cm^{-1} represent small amounts of $CaCO_3$ impurities in GO/LDH-Ca_2Al with LDH-Ca_2Al [43]. Both LDH-Ca_2Al and GO/LDH-Ca_2Al showed weak interlayer $CO_3{}^{2-}$ absorption peaks at 2974 and 1486 cm^{-1}, but the interlayer $CO_3{}^{2-}$ absorption peaks of LDH-Ca_2Al of S-LDH-Ca_2Al and S-GO/LDH-Ca_2Al were significantly enhanced, suggesting that CO_2 can be intercalated in the interlayer of LDH-Ca_2Al mainly in the form of $CO_3{}^{2-}$ [18].

3.2. XRD Analysis of LDH-Ca_2Al and GO/LDH-Ca_2Al

The XRD results of GO, LDH-Ca_2Al, and GO/LDH-Ca_2Al are shown in Figure 2, where the appearance of $CaCO_3$ impurity (PDF #97-001-6710 and PDF #97-018-1959) peaks proves the accuracy of the FITR results (Figure 1). The X-ray diffraction pattern of GO/LDH-Ca_2Al (PDF#04-010-4677) does not show GO diffraction peaks, suggesting that the periodic stacking arrangement was not formed by the composite of GO and LDH-Ca_2Al [13]. The GO/LDH-Ca_2Al shows the typical Bragg reflection of LDH-Ca_2Al as the pristine LDH-Ca_2Al, but the diffraction peaks of GO/LDH-Ca_2Al at the (003) crystal plane shifted to a lower angle; according to Bragg equation ($2d\sin(2\theta) = n\lambda$) the calculated LDH-$Ca_2Al$ and GO/LDH-Ca_2Al basal spacings were approximately 2.587 and 2.623 nm, respectively, and the thickness of the LDH-like hydromagnesite layer was 0.48 nm [41]. The LDH-Ca_2Al and GO/LDH-Ca_2Al interlayer spacings were 2.107 and 2.143 nm, respectively. Consequently, Al-GO formed a complex with LDH-Ca_2Al, resulting in an increase in interlayer spacing CO_2 by approximately 0.036 nm, as shown in Figure 2b. This phenomenon can be attributed to the strong interactions between the Al-GO and the LDH-Ca_2Al layers, which weakened the electrostatic interactions of the anionic in the interlayer channel with the cationic on the layer, consistent with the FTIR results. Given the increase in GO/LDH-Ca_2Al layer spacing, the specific surface area also increased, which provided more active sites for CO_2 molecular adsorption and increased the possibility of CO_2 gas adsorption and intercalation.

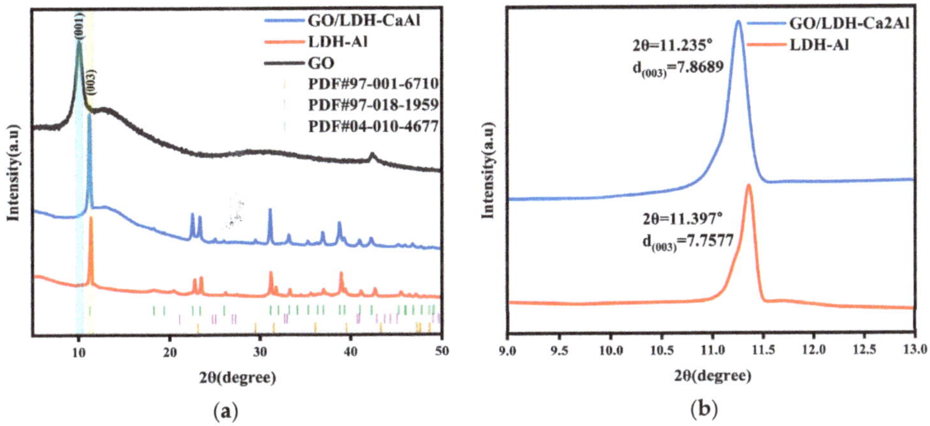

Figure 2. (**a**) XRD of GO, LDH-Ca_2Al, and GO/LDH-Ca_2Al and (**b**) localized enlargement near (003) of LDH-Ca_2Al and GO/LDH-Ca_2Al.

3.3. Structural Characteristics of LDH-Ca_2Al and GO/LDH-Ca_2Al

The SEM of images of LDH-Ca_2Al and GO/LDH-Ca_2Al are shown in Figure 3. LDH-Ca_2Al and GO/LDH-Ca_2Al have obvious lamellar structures, and the diameter of LDH-Ca_2Al flakes is in the range of 0.3–0.5 μm with a thickness of approximately 0.03 μm. The XRD of LDH-Ca_2Al showed a layer spacing of approximately 2.107 nm (Figure 2), indicating that LDH-Ca_2Al is composed of multilayered LDH-Ca_2Al lamellae stacked along the (003) direction. In GO/LDH-Ca_2Al, LDH-Ca_2Al lamellae are much smaller than Al-GO (S1), which is consistent with the TEM results of GO/LDH-Ca_2Al in Figure 4e. A closer look reveals that a large amount of LDH-Ca_2Al is loaded onto the surface of Al-GO because of the existence of a strong interaction force between LDH-Ca_2Al and Al-GO, which is consistent with the FITR results of GO/LDH-Ca_2Al (Figure 1).

Figure 3. SEM images of (**a**) LDH-Ca_2Al and (**b**) GO/LDH-Ca_2Al.

The TEM of GO, LDH-Ca_2Al, and GO/LDH-Ca_2Al is shown in Figures S1 and 4, and they show a lamellar structure, which is consistent with the SEM results. The HR-TEM of LDH-Ca_2Al in Figure 4d shows the appearance of obvious (110) and (104) lattice stripes, in which the spacing of the stripes are 0.286 and 0.378 nm, consistent with the results of (110) and (104) diffraction peaks in the XRD diffraction pattern. The SAED plot along the crystal axis (16, −7, 1) is shown in the lower right corner of Figure 4d, consistent with PDF#04-010-4677. A large amount of two-dimensional LDH-Ca_2Al grows on GO/LDH-Ca_2Al, and the HR-TEM of GO/LDH-Ca_2Al clearly demonstrates that the two-dimensional

GO in GO/LDH-Ca_2Al grows LDH-Ca_2Al (Figure 4e); a (110) lattice face of LDH-Ca_2Al was observed in the GO/LDH-Ca_2Al lattice streak, and the SAED image in this region shows LDH-Ca_2Al diffraction spots along the (17,16,2) directions [44], indicating that LDH-Ca_2Al is deposited on the GO surface and stacked along the (003) direction on the GO surface.

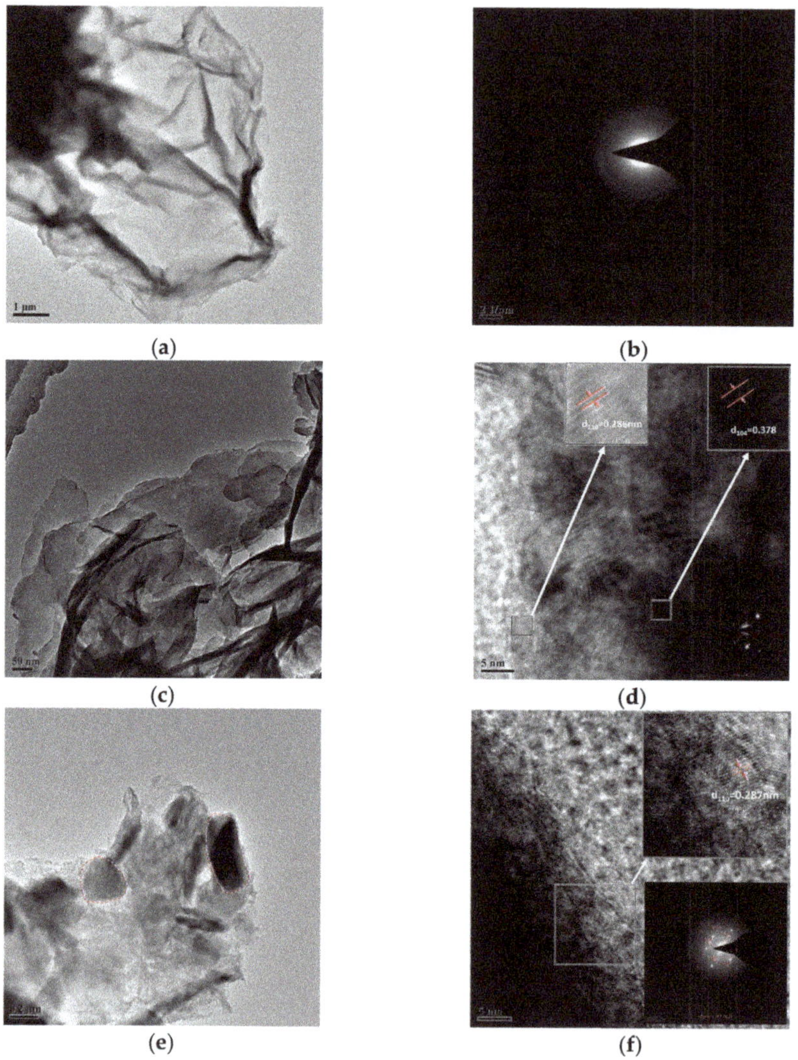

Figure 4. TEM of (**a**) GO, (**c**) LDH-Ca_2Al, and (**e**) GO/LDH-Ca_2Al (red circles—LDH-Ca_2Al), (**b**) SAED diagrams of GO, and (**d**) lattice fringes of LDH-Ca_2Al and (**f**) GO/LDH-Ca_2Al with SAED plots inset in the lower left corner.

3.4. Thermal Stability of GO, LDH-Ca_2Al, and GO/LDH-Ca_2Al

The DSC and TGA curves of GO, LDH-Ca_2Al, and GO/LDH-Ca_2Al are shown in Figure 5a–c, respectively. Figure 5a shows a physisorbed water removal near 125.4, 136.4, and 117.5 °C for GO, LDH-Ca_2Al, and GO/LDH-Ca_2Al, respectively. Meanwhile, GO shows the oxidation of oxygen-containing functional groups near 198 °C, indicating an exothermic reaction [45]. The removal of interlayer H_2O molecules from LDH-Ca_2Al and

GO/LDH-Ca_2Al occurred near 192.6 and 200.3 °C, respectively, suggesting that GO/LDH-Ca_2Al has better thermal stability than LDH-Ca_2Al. Meanwhile, GO/LDH-Ca_2Al did not show the exothermic reaction of GO near 198 °C, indicating that the thermal stability of GO in GO/LDH-Ca_2Al was significantly improved due to the presence of strong interaction between Al-GO and LDH-Ca_2Al.

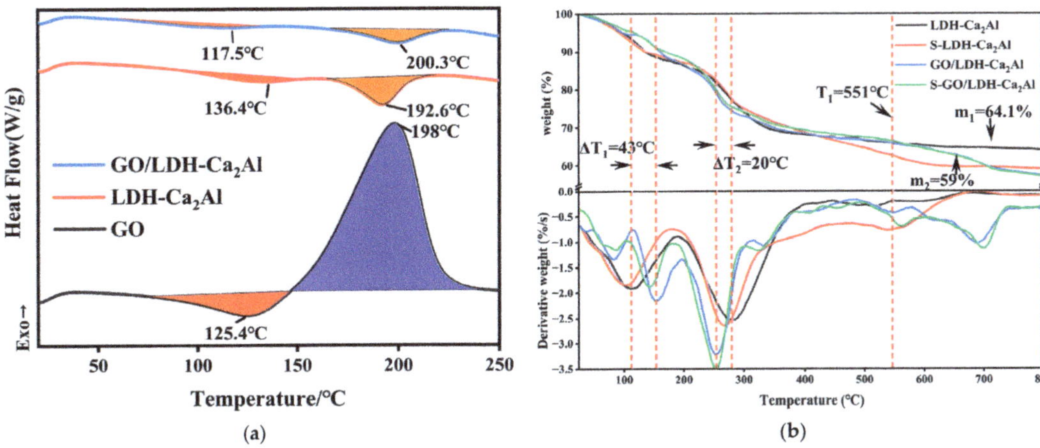

Figure 5. (**a**) DSC curves of GO, LDH-Ca_2Al, and GO/LDH-Ca_2Al and (**b**) TGA-DTG curves of LDH-Ca_2Al, S-LDH-Ca_2Al, GO/LDH-Ca_2Al, and S-GO/LDH-Ca_2Al.

The thermal stability of GO, LDH-Ca_2Al, and GO/LDH-Ca_2Al is also depicted in Figures S2 and 5b. Similar to the DSC results (Figure 5a), GO underwent three key weight loss steps at <125 °C corresponding to the removal of physically adsorbed water, 130–350 °C corresponding to the removal of oxygen functional groups, and 360–800 °C corresponding to the carbon framework of the oxidative pyrolysis [39]. Based on Figure 5b, three distinct weight loss steps can be observed for LDH-Ca_2Al at 30–100 °C corresponding to the loss of physisorbed water, 100–170 °C corresponding to the process of interlayer water removal, 215–400 °C corresponding to the dehydroxylation of the layers and the removal of interlayer chloride ions, and 440–550 °C corresponding to the further elimination of the hydroxyl groups on the layers [18,41,46]. Compared with LDH-Ca_2Al, the thermal stability of GO/LDH-Ca_2Al was significantly improved between 30 and 200 °C. Although the temperature at which approximately 1.2% of physisorbed water was lost was approximately 7 °C lower than that of LDH-Ca_2Al, the temperature at which the interlayer water was removed increased by nearly 43 °C (ΔT_1). Interestingly, the temperature of releasing interlayer chloride ions decreased by about 20 °C (ΔT_2), which was caused by the enlarged interlayer spacing of LDH-Ca_2Al. At the same time, the thermal stability of the GO layer on GO/LDH-Ca_2Al was improved, in which the carbon framework decomposition temperature increased by approximately 29 °C, respectively. The residual mass of S-LDH-Ca_2Al with a value of approximately 59% at 800 °C (m_2), which is approximately 5% higher than that of the LDH-Ca_2Al, can be attributed to the decomposition of the carbonates that occurred after 450 °C based on the XRD patterns of S-LDH-Ca_2Al; this finding revealed the presence of carbonates (Figure 6a) [18].

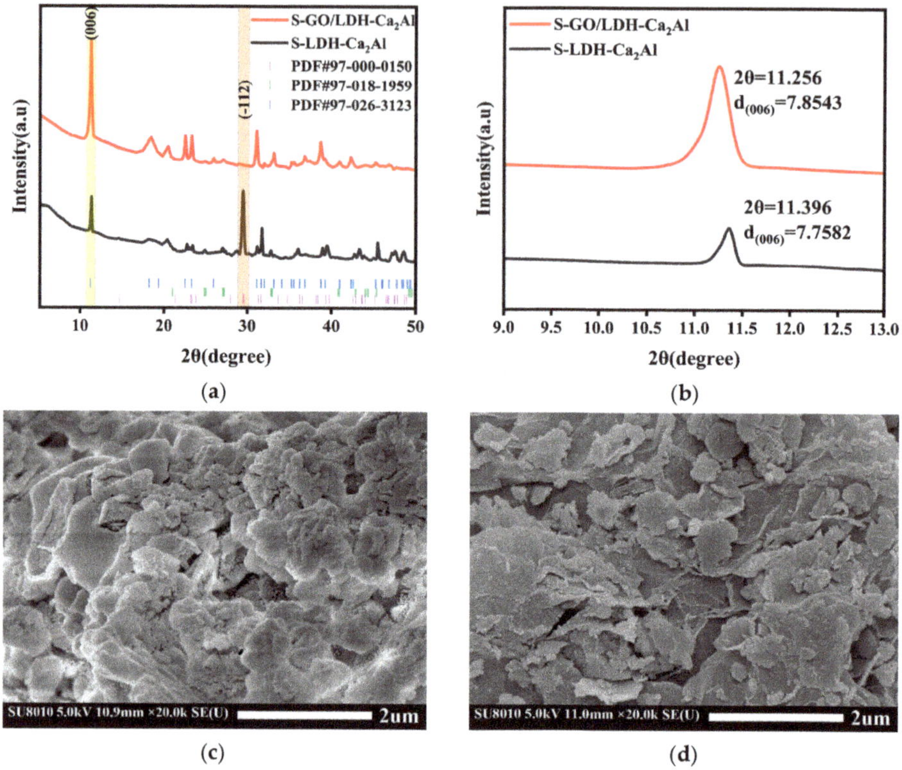

Figure 6. (**a**) XRD of S-LDH-Ca_2Al, S-GO/LDH-Ca_2Al, and (**b**) localized enlargement near (003) of LDH-Ca_2Al and GO/LDH-Ca_2Al. (**c**) SEM of S-LDH-Ca_2Al, and (**d**) S-GO/LDH-Ca_2Al.

3.5. Carbon Dioxide Resistance Stability of LDH-Ca_2Al and GO/LDH-Ca_2Al

The XRD of S-LDH-Ca_2Al and S-GO/LDH-Ca_2Al are shown in Figure 6a,b. LDH-Ca_2Al and GO/LDH-Ca_2Al were placed in a pure CO_2 environment at a pressure of 4 MPa and a temperature of 80 °C. After 1 day, the XRD diffraction peaks of the reaction products S-LDH-Ca_2Al and S-GO/LDH-Ca_2Al were very similar to those of Ca_4Al_2 $(OH)_{12}$ $(OH_{0\cdot4}$ $(CO_3)_{0.8}$ $(H_2O)_4$ compounds (PDF#97-026-3123). According to the Bragg equation ($2d\sin(2\theta) = n\lambda$), the calculated layer spacings of LDH-Ca_2Al, GO/LDH-Ca_2Al, S-LDH-Ca_2Al, and S-GO/LDH-Ca_2Al were 2.106, 2.143, 2.106, and 2.138 nm, respectively (Figures 2b and 6b) [41], and the kinetic diameter of CO_2 was approximately 0.33 nm [47]. The IR spectra of S-LDH-Ca_2Al and S-GO/LDH-Ca_2Al also showed significant interlayer $CO_3{}^{2-}$ absorption peaks (PDF #97-018-1959 and PDF #97-000-0150, Figure 1), suggesting that CO_2 was intercalated into the interlayer of LDH and appeared in the interlayer in the form of $CO_3{}^{2-}$. The intensity of the $CaCO_3$ diffraction peaks increased in the S-LDH-Ca_2Al, and the micro-morphology of the S-LDH-Ca_2Al showed a large number of massive structures (Figure 6c), indicating that CO_2 reacted with LDH-Ca_2Al, which led to an increase in $CO_3{}^{2-}$ content in S-LDH-Ca_2Al. By contrast, the intensity of the $CaCO_3$ diffraction peaks in the XRD pattern of S-GO/LDH-Ca_2Al decreased, no significant change was observed in the microscopic morphology (Figure 6d), and GO/LDH-Ca_2Al hardly reacted with CO_2, indicating that GO/LDH-Ca_2Al has a very good CO_2-resistant chemical stability, which is consistent with the TGA results (Figure 5). Table 2 also provides information on the change of $CaCO_3$ content in LDH-Ca_2Al, GO/LDH-Ca_2Al, S-LDH-Ca_2Al, and S-GO/LDH-Ca_2Al.

Table 2. Contents of $CaCO_3$ in LDH-Ca_2Al, GO/LDH-Ca_2Al, S-LDH-Ca_2Al, and S-GO/LDH-Ca_2Al [a].

Sample	Content ($CaCO_3$) (%)
LDH-Ca_2Al	0.23
GO/LDH-Ca_2Al	5.97
S-LDH-Ca_2Al	59.94
S-GO/LDH-Ca_2Al	2.37

[a] XRD data of LDH-Ca_2Al, GO/LDH-Ca_2Al, S-LDH-Ca_2Al, and S-GO/LDH-Ca_2Al were obtained via fine-tuning by using FullProf 2023.2 software.

3.6. Interfacial Interactions of FKM Matrix with LDH-Ca_2Al and GO/LDH-Ca_2Al

The interaction between the nanofillers and the polymer matrix influences the dispersion state of the nanofillers in the polymer, thus remarkably affecting the gas barrier properties and mechanical properties of the nanocomplexes [48–50]. GO/LDH-Ca_2Al nanofillers had an excellent dispersion state in the FKM/GO/LDH-Ca_2Al nanocomposites, and the energy storage modulus of the FKM/GO/LDH-Ca_2Al complexes substantially increased because of the existence of a good phase interface between the GO/LDH-Ca_2Al nanofillers and the rubber matrix compared with the LDH-Ca_2Al nanofillers (Figure 7a,b). This interfacial interaction will be quantitatively measured next using two-phase modeling [51]. The loss tangent relationship between the filler and the unfilled polymer matrix satisfies the following equation [27,51,52]:

$$tan\ \delta = \frac{tan\ \delta_m}{1 + 1.5B\varphi} \tag{4}$$

where $tan\ \delta_m$ and $tan\ \delta$ represent the tangent values of the filler and the unfilled polymer matrix, φ is the volume fraction of the filler, and B is a phenomenological interaction parameter, which can represent the strength of the interface interaction between the filler and the matrix; the larger the value of B, the stronger the interface interaction between the filler and the polymer matrix [27]. The $tan\ \delta_m$, $tan\ \delta$, T_g, and B values of FKM, FKM/LDH-Ca_2Al-8, and FKM/GO/LDH-Ca_2Al-8 are shown in Table 3. The loss tangent $tan\ \delta$ and T_g of FKM/GO/LDH-Ca_2Al shifted toward high temperature, and the B value of FKM/GO/LDH-Ca_2Al-8 is approximately 0.528, which is slightly higher than that of FKM/LDH-Ca_2Al-8 by 0.041. This finding indicates that the interfacial interactions between GO/LDH-Ca_2Al and the FKM matrix were enhanced compared with that of LDH-Ca_2Al, thus hindering the movement of the molecular chain segments. this finding explains that the crosslink density of FKM/GO/LDH-Ca_2Al is always higher than that of FKM/LDH-Ca_2Al for the fluoroelastomer composites filled with the same content of fillers, as shown in Figure 8. The SEM images of FKM/LDH-Ca_2Al-8 and FKM/GO/LDH-Ca_2Al-8 are shown in Figure 7a,b. The results indicate a good phase interface between GO/LDH-Ca_2Al and the rubber matrix, which is caused by the strong interfacial interaction between GO/LDH-Ca_2Al and the fluoroelastomer matrix than that of LDH-Ca_2Al.

Table 3. Phenomenological interaction parameter (B), $tan\ \delta_m$, $tan\ \delta$, T_g, and φ of FKM, FKM/LDH-Ca_2Al-8, and FKM/GO/LDH-Ca_2Al-8.

Sample	$tan\delta_m$	$tan\delta$	φ (%)	B	Tg (°C)
FKM	1.480	-	-	-	3.3
FKM/LDH-Ca_2Al-8	-	1.396	8.254	0.487	4.9
FKM/GO/LDH-Ca_2Al-8	-	1.404	6.88	0.528	5.6

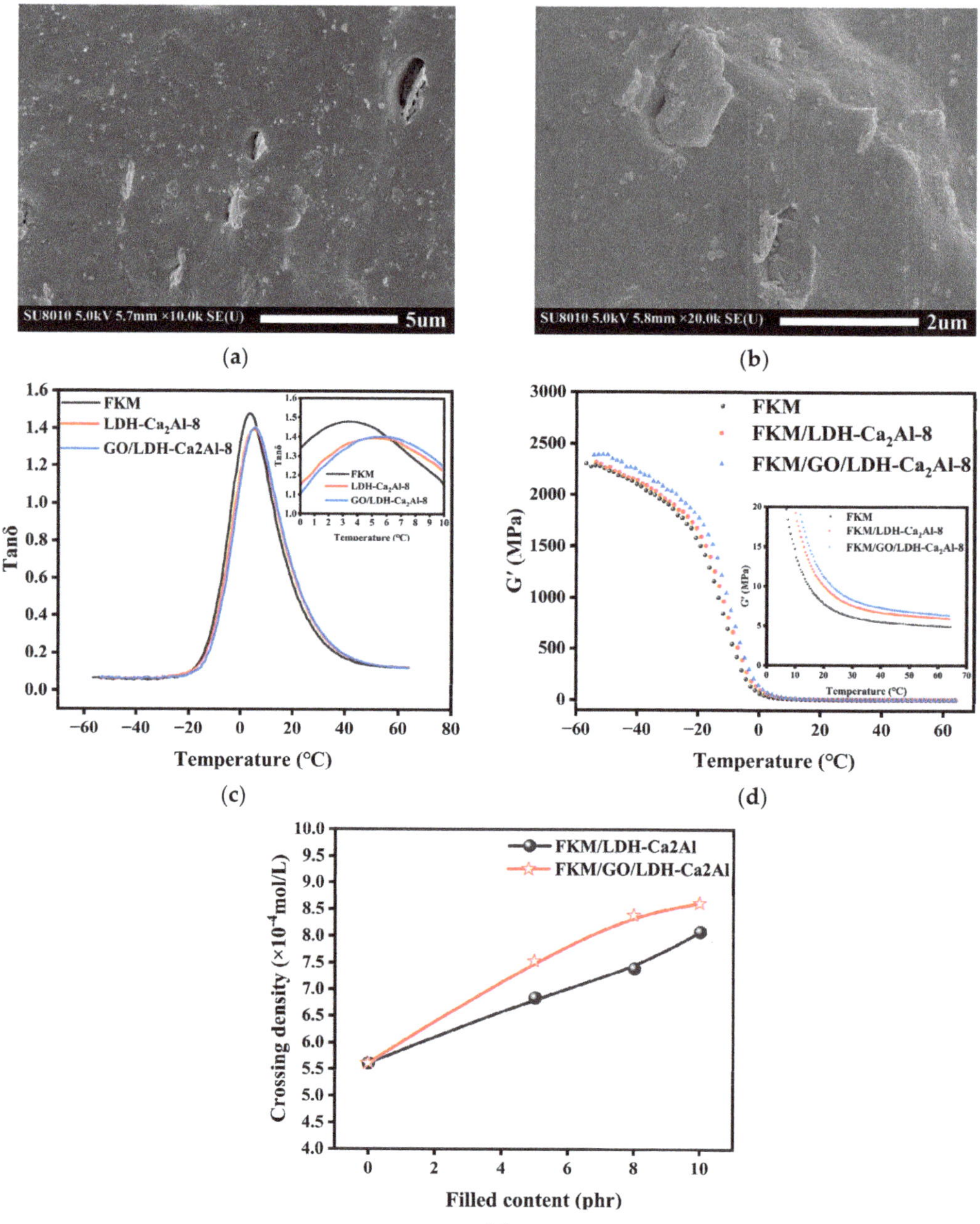

Figure 7. SEM images of (**a**) FKM/LDH-Ca_2Al-8 and (**b**) FKM/GO/LDH-Ca_2Al-8, (**c**,**d**) DMA of FKM, LDH-Ca_2Al-8, and GO/LDH-Ca_2Al-8, and (**e**) Crosslink density of FKM/LDH-Ca_2Al and FKM/GO/LDH-Ca_2Al with different filler parts.

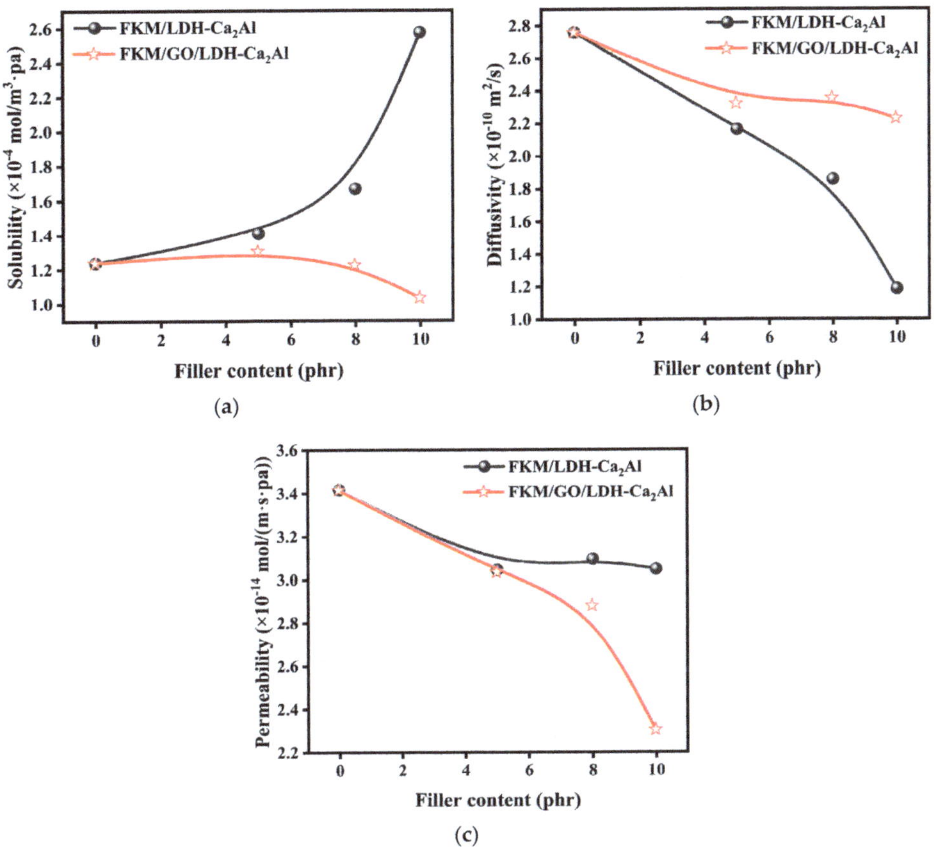

Figure 8. (**a**) Solubility (*S*), (**b**) diffusion coefficient (*D*), and (**c**) permeability coefficient (*Q*) of FKM, FKM/GO/LDH-Ca_2Al, and FKM/LDH-Ca_2Al as a function of filler content.

3.7. Gas Barrier Properties of FKM/LDH-Ca_2Al and FKM/GO/LDH-Ca_2Al Composites

3.7.1. Effect of Filler Content on the Gas Barrier Properties of FKM/LDH-Ca_2Al and FKM/GO/LDH-Ca_2Al Composites

The gas barrier properties of rubber composites, which were measured in terms of the gas permeability coefficient ($Q = DS$), are related to the shape, content, and degree of dispersion of the filler [5,12,53], and lamellar fillers with large aspect ratios have good gas barrier properties [54]. LDH-Ca_2Al and GO/LDH-Ca_2Al are lamellar fillers with large aspect ratios (Figure 3a,b). The aspect ratios of LDH-Ca_2Al are in the range of 10–16, and the *Q* and *D* of the rubber filled with 8 phr LDH-Ca_2Al fillers were reduced by nearly 9% and 33%, respectively. The relationships of permeability coefficient (*Q*), diffusion coefficient (*D*), and solubility (*S*) with the contents of LDH-Ca_2Al and GO/LDH-Ca_2Al are illustrated in Figure 8a–c. The results show that the permeability coefficients and diffusion coefficients of the rubber composites decrease significantly with the increase in filler content, which is consistent with the change in the crosslinking density of the rubber composites (Figure 7e). This phenomenon can be explained using the free volume theory [55,56]. The nanofillers can be approximated as physical cross-linking points and the cross-linking density increases with the increase in nanofiller content. The increase in cross-linking density shortens the distance between the rubber chains, reduces the mobility of the rubber chains, and leads to a decrease in the free volume fraction of the permeation path of the gas molecules [6]. The free volume fraction of the gas molecules' permeation path decreased.

The variation of solubility (S) with filler content is given in Figure 8a. The solubility of FKM/LDH-Ca_2Al increased significantly with the increase in LDH-Ca_2Al content, and the solubility of the rubber composite increased by 108% when filled with 10 phr of LDH-Ca_2Al. By contrast, the solubility of FKM/GO/LDH-Ca_2Al gradually decreased with the increase in GO/LDH-Ca_2Al content, and the solubility of the rubber composite decreased by about 19% when filled with 10 phr of GO/LDH-Ca_2Al. This finding was obtained because LDH-Ca_2Al is a nanofiller with high CO_2 absorption [48]. Moreover, the loading of LDH-Ca_2Al onto the GO surface reduces the content of exposed polar hydroxyl groups on the LDH-Ca_2Al layer, thus forming strong hydrogen bonding interactions with O in the CO_2 molecule and increasing the content of CO_2 adsorbed on the surface of the filler. In addition, compared with the LDH-Ca_2Al, the GO/LDH-Ca_2Al in the composites has good compatibility (Figure 7a,b), which consistently supports the interfacial interactions between GO/LDH-Ca_2Al and the rubber matrix (Table 3). Therefore, more constrained polymer regions are present near the interface between the rubber matrix and the filler [5,57], thus reducing the adsorption capacity of CO_2 molecules at the interface.

Figure 8b shows the schematic relationship between filler content and diffusion coefficient, in which the diffusion coefficient of the rubber composites decreased significantly with the increase in the filler content of LDH-Ca_2Al and GO/LDH-Ca_2Al, which is consistent with many previous findings [5,58–61]. This finding was obtained because the filler increases the gas molecule diffusion path and hinders the gas molecule movement [6], while LDH-Ca_2Al and GO/LDH-Ca_2Al are multilayer fillers (Figures 2–4) with a layer spacing of more than 2.1 nm, which is significantly higher than the kinetic diameter of the CO_2 molecule (~0.33 nm). Hence, the CO_2 molecule can be interpolated into the layer spacing of the filler, thus significantly improving the gas molecule diffusion path. Accordingly, the diffusion coefficient of the rubber composite material was significantly reduced. Moreover, when the filler content of the filler reached 10 phr, the diffusion coefficient of the rubber composite material was reduced by nearly 57%. However, the diffusion coefficient of FKM/LDH-Ca_2Al is lower than that of FKM/GO/LDH-Ca_2Al at the same filler content, in which even the diffusion coefficient of FKM/LDH-Ca_2Al-10 is approximately 47% lower than that of FKM/GO/LDH-Ca_2Al-10. This finding was obtained because CO_2 entered the spacing of the LDH-Ca_2Al layers and reacted with the LDH-Ca_2Al to generate inorganic compounds containing $CO_3{}^{2-}$ (Table 2), such as $CaCO_3$. Consequently, the diffusion activation energy of CO_2 gas remarkably decreased during diffusion in FKM/LDH-Ca_2Al (Table 4). For example, the diffusion activation energy of FKM/LDH-Ca_2Al-8 increased by approximately 27.6 J/mol. However, GO/LDH-Ca_2Al has better-stabilizing properties during CO_2 diffusion (Figures 2, 3b and 6), and the $CaCO_3$ content of S-GO/LDH-Ca_2Al did not change significantly compared with that of GO/LDH-Ca_2Al (Table 2). Hence, although FKM/GO/LDH-Ca_2Al lost some of the ability to reduce the diffusion coefficient brought by LDH-Ca_2Al, it improved the CO_2 resistance stability of FKM/GO/LDH-Ca_2Al.

Table 4. Activation energy of D and S of FKM, FKM/LDH-Ca_2Al-8, and FKM/GO/LDH-Ca_2Al-8.

Sample	Activation Energy of D(ΔE) [a]				Activation Energy of S(ΔH) [a]			
	D_0 [b]	ΔE_{D0}	ΔE_{Df}	ΔE_{Dc}	S_0 [c]	ΔH_{S0}	ΔH_{sf}	ΔH_{sc}
FKM	2.6×10^{-2}	−52.8	-	-	1.8×10^{-11}	44.5	-	-
FKM/LDH-Ca_2Al-8	1.0×10^{-6}	−25.2	22.4	5.2	2.2×10^{-6}	44.5	−32.2	-
FKM/GO/LDH-Ca_2Al-8	7.7×10^{-6}	−30.4	22.4	-	4.7×10^{-7}	44.5	−32.2	3.8

[a] the unit of activation energy is KJ/mol. [b] D_0 and [c] S_0 are the dimensionless constants.

3.7.2. Effect of Temperature on the Gas Barrier Properties of FKM/LDH-Ca_2Al and FKM/GO/LDH-Ca_2Al Composites

The barrier mechanism of GO/LDH-Ca_2Al and LDH-Ca_2Al nanofillers in FKM/GO/LDH-Ca_2Al with FKM/LDH-Ca_2Al composites against CO_2 was investigated by conducting a temperature-dependence study on the gas barrier properties of FKM/LDH-Ca_2Al and FKM/GO/LDH-Ca_2Al composites. The permeability coefficients (Q), diffusion coefficients (D), and solubility (S) of the rubber nanocomposites with temperature are given in Figure 9. The diffusion coefficients of FKM, FKM/GO/LDH-Ca_2Al-8, and FKM/LDH-Ca_2Al-8 remarkably increased with temperature, but the solubility remarkably decreased with the increase in temperature. Based on this phenomenon, the Arrhenius equation can be used to explain the relationship between the ln(S) and ln(D) of rubber nanocomposites and temperature (T), which can be expressed by Equations (5) and (6) as follows:

$$Ln(S) = Ln(S_0) + \left(\Delta H_{S0} + \Delta H_{sf} + \Delta H_{sc}\right)/RT \quad (5)$$

$$Ln(D) = Ln(D_0) + \left(\Delta E_{D0} + \Delta E_{Df} + \Delta E_{Dc}\right)/RT \quad (6)$$

where S_0 and D_0 are pre-finger factors, which are related to the nature of the nanocomposites; ΔH_{S0} and ΔE_{D0} represent the heat of dissolution and diffusion activation energy of the pure polymer; ΔH_{sf} and ΔE_{Df} is the packing effect caused by nanofillers, which are related to the interaction of nanofillers with polymers and diffusion gases; and ΔE_{Df} is related to factors such as the aspect ratio, size, and shape of the nanofiller. In this experiment, ΔH_{sc} refers to the change in the heat of dissolution caused by the compound effect of GO and LDH-Ca_2Al in GO/LDH-Ca_2Al. ΔE_{Dc} is the activation energy of the reaction between the nanofiller and the diffusion gas. Table 4 summarizes the relevant parameters of Equations (5) and (6) for the FKM, FKM/GO/LDH-Ca_2Al-8, and FKM/LDH-Ca_2Al-8 samples. The diffusion activation energy (ΔE) of the polymers is less than -25 KJ/mol, and the solubility activation energy (ΔH) is higher than 12.3 KJ/mol, which is consistent with the diffusion coefficient (D) and the solubility (S) with temperature. The ΔH_{sf} from the packing effect of LDH-Ca_2Al is -32.2 KJ/mol, resulting in a significant decrease in the heat of dissolution of CO_2 in the FKM/LDH-Ca_2Al nanocomposites to approximately 12.3 KJ/mol, which explains the strong interaction between CO_2 and LDH-Ca_2Al. This phenomenon led to an increase in the solubility of CO_2 in the FKM/LDH-Ca_2Al nanocomposites. However, the change in the heat of dissolution caused by the compound effect of GO with LDH-Ca_2Al (ΔH_{sc}) is 3.8 KJ/mol, which explains the existence of a strong interaction of GO/LDH-Ca_2Al with the polymer matrix that reduced the CO_2 adsorption effect on the nanofiller surface. The diffusion activation energy of the nanocomposites remarkably increased, which rendered the activation energy of the reaction of CO_2 with GO/LDH-Ca_2Al negligible, in which the value increased by approximately 42.5%. Hence, the diffusion coefficients of the FKM/LDH-Ca_2Al and FKM/GO/LDH-Ca_2Al nanocomposites were insensitive to the temperature variation. In addition to the packing effect caused by the filler (ΔE_{Df}), a chemical activation energy was observed between CO_2 and LDH-Ca_2Al ($\Delta E_{Dc} \approx 5.2$ KJ/mol) in the ΔE of the FKM/LDH-Ca_2Al nanocomposites. Therefore, it has a significantly lower diffusion coefficient than that of FKM/GO/LDH-Ca_2Al. The permeability coefficient of the composites is a combination of the diffusion coefficient and solubility ($Q = DS$) [5,48,62,63]. As shown in Figure 9c, the solubility of the nanocomposites substantially contributes to the permeability coefficient at 60–80 °C, resulting in the low permeability coefficient of FKM/GO/LDH-Ca_2Al. By contrast, the diffusion coefficient of the nanocomposites substantially contributes to the permeability coefficient at 80–100 °C. Therefore, the permeability coefficient of the FKM/LDH-Ca_2Al nanocomposites is lower than that of FKM/GO/LDH-Ca_2Al.

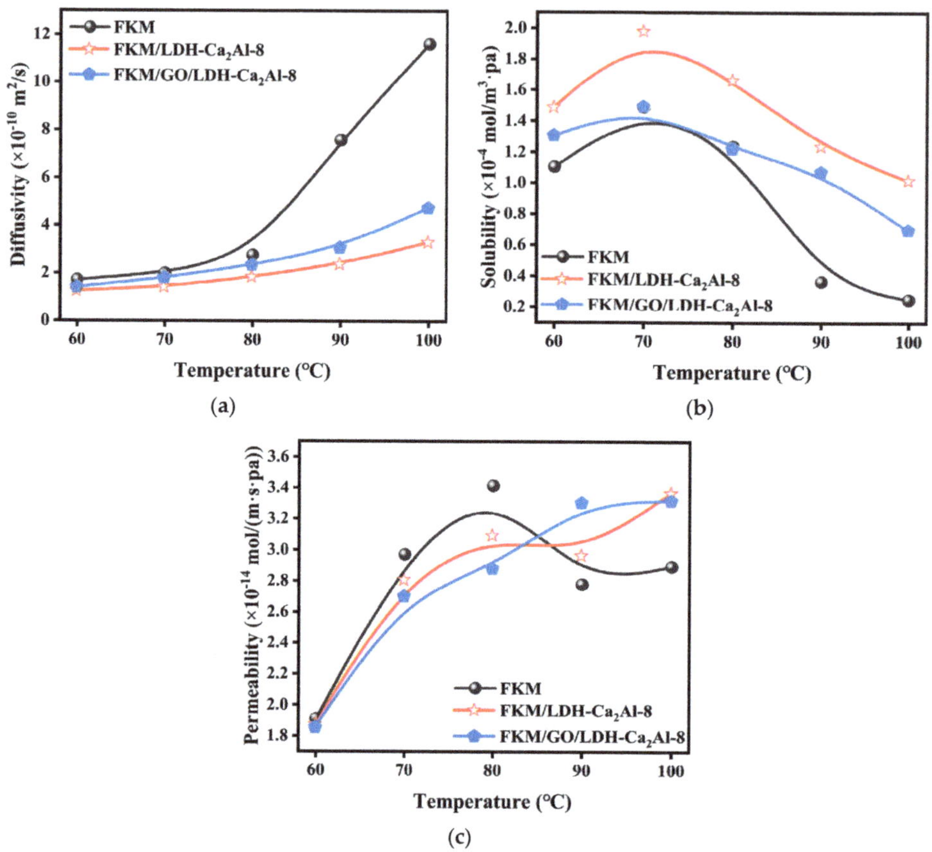

Figure 9. (**a**) Diffusion coefficients (D), (**b**) solubility (S), and (**c**) permeability coefficients (Q) as a function of temperature for FKM, FKM/GO/LDH-Ca_2Al-8, and FKM/LDH-Ca_2Al-8.

3.8. Mechanical Properties of FKM/LDH-Ca_2Al and FKM/GO/LDH-Ca_2Al Nanocomposites

The effect of LDH-Ca_2Al and GO/LDH-Ca_2Al on the mechanical properties of fluoroelastomers was quantified by determining the key mechanical parameters of the nanocomposites with different filler contents (5, 8, and 10 phr), including 100% modulus (*E100*), Shore A hardness, σ_b, and ε_b (Figure 10 and Table 5). The mechanical properties of the nanocomposites were significantly improved compared with those of FKM, and interestingly, the ε_b of the nanocomposites gradually increased with the content of nanofillers, in which the largest ε_b (ε_b > 287%, σ_b > 11.1 MPa) value was obtained for the 10 phr filled FKM/GO/LDH-Ca_2Al nanocomposites. This finding can be attributed to the orientation arrangement of the nanofillers with two-dimensional layered structures in the complexes when subjected to force, leading to the stable transfer of stress inside the material [9]. In addition, FKM/GO/LDH-Ca_2Al nanocomposites exhibit superior mechanical properties to FKM/LDH-Ca_2Al nanocomposites, in which σ_b, and ε_b increased by approximately 25% and 12% (Table 5) because of the strong interfacial interaction between GO/LDH-Ca_2Al and the polymer matrix, which promotes the efficient transfer of stress from the polymer matrix to GO/LDH-Ca_2Al, consistent with the results described by the phenomenological interaction parameters (Table 3) [19].

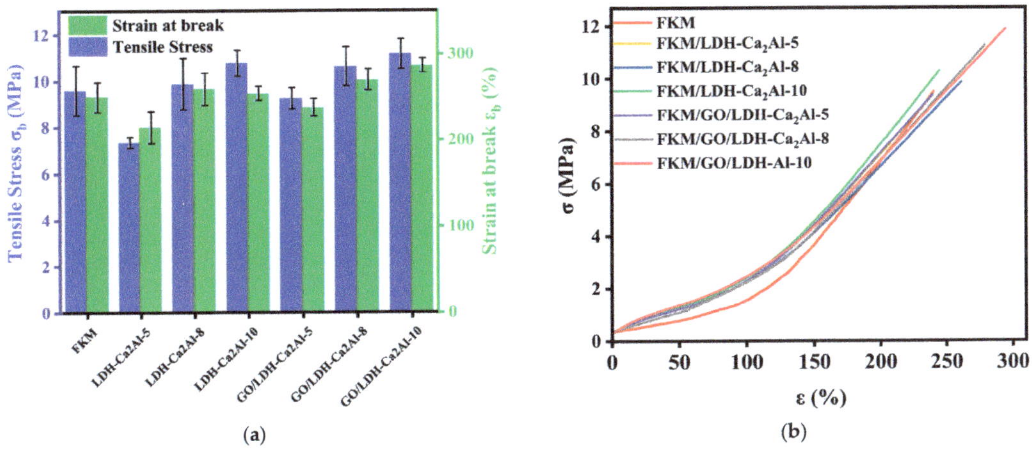

Figure 10. (**a**) Fracture strength (σ_b) and strain (ε_b). (**b**) stress–strain curves of FKM, FKM/LDH-Ca_2Al, and FKM/GO/LDH-Ca_2Al nanocomposites.

Table 5. Mechanical properties of FKM, FKM/LDH-Ca_2Al, and FKM/GO/LDH-Ca_2Al complexes.

Sample	100% Modulus, *E100*(MPa)	Tensile Strength, σ_b (MPa)	Strain at Break, ε_b (%)	Shore A Hardness
FKM	1.5 ± 0.1	9.6 ± 1.1	251 ± 17	50 ± 1
FKM/LDH-Ca_2Al-5	2.1 ± 0.2	7.3 ± 0.2	215 ± 18	55 ± 1
FKM/LDH-Ca_2Al-8	2.3 ± 0.08	9.9 ± 1.1	260 ± 19	56 ± 1
FKM/LDH-Ca_2Al-10	2.4 ± 0.03	10.8 ± 0.6	254 ± 8	58 ± 1
FKM/GO/LDH-Ca_2Al-5	2.3 ± 0.01	9.2 ± 0.5	238 ± 10	55 ± 1
FKM/GO/LDH-Ca_2Al-8	2.2 ± 0.03	10.6 ± 0.8	270 ± 12	56 ± 1
FKM/GO/LDH-Ca_2Al-10	2.4 ± 0.08	11.1 ± 0.6	287 ± 8	57 ± 1

4. Conclusions

In this paper, the CO_2 gas barrier and mechanical properties of nanocomposites filled with Ca/Al layered hydroxide modified by GO (GO/LDH-Ca_2Al) were investigated. The GO/LDH-Ca_2Al nanocomposite filler was prepared by depositing LDH-Ca_2Al on the surface of alkalized graphene oxide (Al-GO). This nanocomposite filler has good thermal stability compared with the pristine LDH-Ca_2Al, and the temperature of the stripped interlayer water increased by nearly 43 °C. At the same time, its CO_2-resistant stability was significantly improved, and almost no carbonate generation occurred at 80 °C and 4.3 MPa pure carbon dioxide environment, resulting in low CO_2 reactivity. The barrier mechanism of modified and unmodified LDH-Ca_2Al-filled nanocomposites to carbon dioxide was also proposed, and the diffusion activation energy and solubility activation energy of the nanocomposites consisted of three components: (1) ΔH_{S0} and ΔE_{D0}, which were determined by the nature of the matrix; (2) ΔH_{sf} and ΔE_{Df} due to packing effect; and (3) the compound effect ΔH_{sc} and reaction activation energy ΔE_{Dc}. In the nanocomposites, the multilayer stacked LDH-Ca_2Al acts as a CO_2 gas barrier to form a distorted diffusion path, and this effect induces an activation energy of ΔE_{D0}. Considering the pro-CO_2 multilayer structure and strong chemical reactivity of LDH-Ca_2Al, it provides the opportunity for CO_2 intercalation and reaction, resulting in the reaction activation energy of ΔE_{Dc}. These two factors led to an increase in the diffusion activation energy of FKM/LDH-Ca_2Al nanocomposites by approximately 42.5%. The solubility activation energy of FKM/GO/LDH-Ca_2Al nanocomposites increased by approximately 3.8 KJ/mol compared with that of FKM/LDH-Ca_2Al nanocomposites, which was caused by the compound effect of GO and LDH-Ca_2Al. This phenomenon explains the existence of strong interfacial interactions between GO/LDH-Ca_2Al and the polymer matrix, thus reducing the CO_2 adsorption effect on the nanofiller surface. The permeability coefficients of the nanocomposites to CO_2 gradually decreased, and the mechanical properties (Shore A, σ_b, and ε_b) were significantly improved with the increase in nanofiller content. The best CO_2 permeation resistance and mechanical properties were obtained for the FKM/GO/LDH-Ca_2Al nanocomposites due to the strong interfacial interactions between GO/LDH-Ca_2Al and FKM.

Supplementary Materials: The following supporting information can be downloaded at https://www.mdpi.com/article/10.3390/polym15204151/s1. Figure S1: SEM images of GO; Figure S2: TGA-DTG curves of GO.

Author Contributions: C.C.: Conceptualization, Formal analysis, Investigation, Resources, Writing—Review and Editing. D.P.: Methodology, Visualization, Investigation, Validation, Writing—Original Draft. Q.L.: Visualization, Investigation. M.Y.: Visualization, Investigation. X.M.: Visualization, supervision. Q.Z.: Formal analysis, Writing—Review and Editing, Supervision. All authors have read and agreed to the published version of the manuscript.

Funding: This study was supported financially by the China National Key R&D Program "Key Technology Research on Non-metallic Non-bonded Flexible Mixing Pipe for Deep Sea Mining" (2022YFC2803700): Topic 2 "Research and Development of Key Materials for Non-metallic Flexible Mixing Pipe for Deep Sea Mining" (2022YFC2803702).

Institutional Review Board Statement: Not applicable.

Data Availability Statement: The data presented in this study are available on request from the corresponding author.

Conflicts of Interest: The authors declare no conflict of interest.

References

1. Moreira, D.; Pires, J.C.M. Atmospheric CO_2 capture by algae: Negative carbon dioxide emission path. *Bioresour. Technol.* **2016**, *215*, 371–379. [CrossRef] [PubMed]
2. Dai, Z.; Viswanathan, H.; Middleton, R.; Pan, F.; Ampomah, W.; Yang, C.; Jia, W.; Xiao, T.; Lee, S.Y.; McPherson, B.; et al. CO_2 Accounting and Risk Analysis for CO_2 Sequestration at Enhanced Oil Recovery Sites. *Environ. Sci. Technol.* **2016**, *50*, 7546–7554. [CrossRef]
3. Sethulekshmi, A.S.; Saritha, A.; Joseph, K. A comprehensive review on the recent advancements in natural rubber nanocomposites. *Int. J. Biol. Macromol.* **2022**, *194*, 819–842. [CrossRef]
4. Gent, A.N. Rubber Elasticity: Basic Concepts and Behavior. In *The Science and Technology of Rubber*; Academic Press: New York, NY, USA, 2013; pp. 1–26.
5. Guo, B.; Tang, Z.; Zhang, L. Transport performance in novel elastomer nanocomposites: Mechanism, design and control. *Prog. Polym. Sci.* **2016**, *61*, 29–66. [CrossRef]
6. Zhang, C.; An, X.; Tang, Z.; Fang, S.; Guo, B.; Zhang, L.; Liu, F.; Liu, J.; Chen, Z. Creation of Tortuosity in Unfilled Rubber via Heterogeneous Cross-Linking toward Improved Barrier Property. *Macromolecules* **2021**, *54*, 11522–11532. [CrossRef]
7. Pradhan, B.; Srivastava, S.K.; Bhowmick, A.K.; Saxena, A. Effect of bilayered stearate ion-modified Mg—Al layered double hydroxide on the thermal and mechanical properties of silicone rubber nanocomposites. *Polym. Int.* **2012**, *61*, 458–465. [CrossRef]
8. Laskowska, A.; Zaborski, M.; Boiteux, G.; Gain, O.; Marzec, A.; Maniukiewicz, W. Ionic elastomers based on carboxylated nitrile rubber (XNBR) and magnesium aluminum layered double hydroxide (hydrotalcite). *Express Polym. Lett.* **2014**, *8*, 374–386. [CrossRef]
9. Wang, X.; Zhang, Y.; Li, K.; Kang, Q.; He, Q. Study on the effect of layered double hydroxide on the mechanical and thermal properties of nitrile butadiene rubber. *J. Appl. Polym. Sci.* **2023**, *140*, e54257. [CrossRef]
10. Scherillo, G.; Lavorgna, M.; Buonocore, G.G.; Zhan, Y.H.; Xia, H.S.; Mensitieri, G.; Ambrosio, L. Tailoring assembly of reduced graphene oxide nanosheets to control gas barrier properties of natural rubber nanocomposites. *ACS Appl. Mater. Interfaces* **2014**, *6*, 2230–2234. [CrossRef]
11. Raef, M.; Razzaghi-Kashani, M. The role of interface in gas barrier properties of styrene butadiene rubber-reduced graphene oxide composites. *Polymer* **2019**, *182*, 121816. [CrossRef]
12. Wolf, C.; Angellier-Coussy, H.; Gontard, N.; Doghieri, F.; Guillard, V. How the shape of fillers affects the barrier properties of polymer/non-porous particles nanocomposites: A review. *J. Membr. Sci.* **2018**, *556*, 393–418. [CrossRef]

13. Mohd Sidek, H.B.; Jo, Y.K.; Kim, I.Y.; Hwang, S.-J. Stabilization of Layered Double Oxide in Hybrid Matrix of Graphene and Layered Metal Oxide Nanosheets: An Effective Way to Explore Efficient CO_2 Adsorbent. *J. Phys. Chem. C* **2016**, *120*, 23421–23429. [CrossRef]
14. Goh, K.H.; Lim, T.T.; Dong, Z. Application of layered double hydroxides for removal of oxyanions: A review. *Water Res.* **2008**, *42*, 1343–1368. [CrossRef]
15. Zhang, R.; Ai, Y.; Lu, Z. Application of Multifunctional Layered Double Hydroxides for Removing Environmental Pollutants: Recent Experimental and Theoretical Progress. *J. Environ. Chem. Eng.* **2020**, *8*, 103908. [CrossRef]
16. Sideris, P.J.; Nielsen, U.G.; Gan, Z.; Grey, C.P. Mg/Al ordering in layered double hydroxides revealed by multinuclear NMR spectroscopy. *Science* **2008**, *321*, 113–117. [CrossRef] [PubMed]
17. Lu, X.; Xue, H.; Gong, H.; Bai, M.; Tang, D.; Ma, R.; Sasaki, T. 2D Layered Double Hydroxide Nanosheets and Their Derivatives toward Efficient Oxygen Evolution Reaction. *Nanomicro. Lett.* **2020**, *12*, 86. [CrossRef] [PubMed]
18. Qu, J.; Zhong, L.; Li, Z.; Chen, M.; Zhang, Q.; Liu, X. Effect of anion addition on the syntheses of Ca–Al layered double hydroxide via a two-step mechanochemical process. *Appl. Clay Sci.* **2016**, *124–125*, 267–270. [CrossRef]
19. Zheng, L.; Jerrams, S.; Xu, Z.; Zhang, L.; Liu, L.; Wen, S. Enhanced gas barrier properties of graphene oxide/rubber composites with strong interfaces constructed by graphene oxide and sulfur. *Chem. Eng. J.* **2020**, *383*, 123100. [CrossRef]
20. Long, X.; Li, J.; Xiao, S.; Yan, K.; Wang, Z.; Chen, H.; Yang, S. A strongly coupled graphene and FeNi double hydroxide hybrid as an excellent electrocatalyst for the oxygen evolution reaction. *Angew. Chem. Int. Ed. Engl.* **2014**, *53*, 7584–7588. [CrossRef]
21. Jia, Y.; Zhang, L.; Gao, G.; Chen, H.; Wang, B.; Zhou, J.; Soo, M.T.; Hong, M.; Yan, X.; Qian, G.; et al. A Heterostructure Coupling of Exfoliated Ni-Fe Hydroxide Nanosheet and Defective Graphene as a Bifunctional Electrocatalyst for Overall Water Splitting. *Adv. Mater.* **2017**, *29*, 1700017. [CrossRef]
22. Chen, S.; Duan, J.; Jaroniec, M.; Qiao, S.Z. Three-dimensional N-doped graphene hydrogel/NiCo double hydroxide electrocatalysts for highly efficient oxygen evolution. *Angew. Chem. Int. Ed. Engl.* **2013**, *52*, 13567–13570. [CrossRef]
23. Mostovoy, A.S.; Yakovlev, A.V.; Lopukhova, M.I. Directional control of physico-chemical and mechanical properties of epoxide composites by the addition of graphite-graphene structures. *Polym.-Plast. Technol. Mater.* **2019**, *59*, 874–883. [CrossRef]
24. Wang, L.; Fu, W.; Peng, W. Enhanced strength and toughness of polyurethane rubber by introducing hydrogen bond sacrificial units at rubber-graphene interfaces. *Polym. Compos.* **2019**, *41*, 1242–1254. [CrossRef]
25. Li, J.; Zhang, P.; Chen, L.; Li, G.; Chen, H.; Jia, C.; Wu, Y.; Chen, M.; Zhao, X.; Song, P. Strong, tough and healable elastomer nanocomposites enabled by a hydrogen-bonded supramolecular network. *Compos. Commun.* **2020**, *22*, 100530. [CrossRef]
26. Zhou, S.; Liu, L.; Hu, S.; Wen, S. Properties of LDHs/ENR/SBR Gas Barrier Composites. *China Rubber Ind.* **2019**, *66*, 284. [CrossRef]
27. Rooj, S.; Das, A.; Heinrich, G. Tube-like natural halloysite/fluoroelastomer nanocomposites with simultaneous enhanced mechanical, dynamic mechanical and thermal properties. *Eur. Polym. J.* **2011**, *47*, 1746–1755. [CrossRef]
28. Silva, A.F.d.; Duarte, J.L.d.S.; Meili, L. Different routes for MgFe/LDH synthesis and application to remove pollutants of emerging concern. *Sep. Purif. Technol.* **2021**, *264*, 118353. [CrossRef]
29. Zhang, J.; Lin, Z.; Lan, Y.; Ren, G.; Chen, D.; Huang, F.; Hong, M. A multistep oriented attachment kinetics: Coarsening of ZnS nanoparticle in concentrated NaOH. *J. Am. Chem. Soc.* **2006**, *128*, 12981–12987. [CrossRef] [PubMed]
30. He, F.; Yang, Z.; Zhao, F.; Repo, E.; Yang, W.; Liao, Q.; Si, M.; Zou, B.; Lin, Z. Unveiling the dual roles of the intercalation of $[MoS_4]^{2-}$ clusters in boosting heavy metal capture by Ca–Al layered double hydroxide. *Environ. Sci. Nano* **2023**, *10*, 190–202. [CrossRef]
31. Li, D.; Muller, M.B.; Gilje, S.; Kaner, R.B.; Wallace, G.G. Processable aqueous dispersions of graphene nanosheets. *Nat. Nanotechnol.* **2008**, *3*, 101–105. [CrossRef]
32. *ISO 2782-1:2022*; Rubber, Vulcanized or Thermoplastic—Determination of Permeability to Gases—Part 1: Differential-Pressure Methods. The International Organization for Standardization: Geneva, Switzerland, 2022.
33. *GB/T 528-2009*; Rubber, Vulcanized or Thermoplastic—Determination of Tensile Stress-Strain Properties. National Standards of the People's Republic of China: Beijing, China, 2009.
34. *ISO 7619-1:2010*; Rubber, Vulcanized or Thermoplastic—Determination of Indentation Hardness—Part 1: Durometer Method (Shore Hardness). The International Organization for Standardization: Geneva, Switzerland, 2010.
35. Yang, X.; Li, Q.; Li, Z.; Xu, X.; Liu, H.; Shang, S.; Song, Z. Preparation and Characterization of Room-Temperature-Vulcanized Silicone Rubber Using Acrylpimaric Acid-Modified Aminopropyltriethoxysilane as a Cross-Linking Agent. *ACS Sustain. Chem. Eng.* **2019**, *7*, 4964–4974. [CrossRef]
36. *ASTM D6814-02*; Standard Test Method for Determination of Percent Devulcanization of Crumb Rubber Based on Crosslink Density. ASTM: West Conshohocken, PA, USA, 2018. [CrossRef]
37. Sathi, S.G.; Jang, J.Y.; Jeong, K.U.; Nah, C. Thermally stable bromobutyl rubber with a high crosslinking density based on a 4,4′-bismaleimidodiphenylmethane curing agent. *J. Appl. Polym. Sci.* **2016**, *133*, 44092. [CrossRef]
38. Sarath, P.S.; Thomas, S.; Haponiuk, J.T.; George, S.C. Fabrication, characterization and properties of silane functionalized graphene oxide/silicone rubber nanocomposites. *J. Appl. Polym. Sci.* **2022**, *139*, e52299. [CrossRef]
39. Yap, P.L.; Kabiri, S.; Tran, D.N.H.; Losic, D. Multifunctional Binding Chemistry on Modified Graphene Composite for Selective and Highly Efficient Adsorption of Mercury. *ACS Appl. Mater. Interfaces* **2019**, *11*, 6350–6362. [CrossRef]

40. Costa, M.C.F.; Marangoni, V.S.; Ng, P.R.; Nguyen, H.T.L.; Carvalho, A.; Castro Neto, A.H. Accelerated Synthesis of Graphene Oxide from Graphene. *Nanomaterials* **2021**, *11*, 551. [CrossRef]
41. Jadam, M.L.; Syed Mohamad, S.A.; Zaki, H.M.; Jubri, Z.; Sarijo, S.H. Antibacterial activity and physicochemical characterization of calcium-aluminium-ciprofloxacin-layered double hydroxide. *J. Drug Deliv. Sci. Technol.* **2021**, *62*, 102314. [CrossRef]
42. Pérez-Barrado, E.; Pujol, M.C.; Aguiló, M.; Cesteros, Y.; Díaz, F.; Pallarès, J.; Marsal, L.F.; Salagre, P. Fast aging treatment for the synthesis of hydrocalumites using microwaves. *Appl. Clay Sci.* **2013**, *80–81*, 313–319. [CrossRef]
43. Santos, V.H.J.M.d.; Pontin, D.; Ponzi, G.G.D.; Stepanha, A.S.d.G.e.; Martel, R.B.; Schütz, M.K.; Einloft, S.M.O.; Dalla Vecchia, F. Application of Fourier Transform infrared spectroscopy (FTIR) coupled with multivariate regression for calcium carbonate ($CaCO_3$) quantification in cement. *Constr. Build. Mater.* **2021**, *313*, 125413. [CrossRef]
44. Song, F.; Hu, X. Exfoliation of layered double hydroxides for enhanced oxygen evolution catalysis. *Nat. Commun.* **2014**, *5*, 4477. [CrossRef] [PubMed]
45. Mehetre, S.S.; Maktedar, S.S.; Singh, M. Understanding the mechanism of surface modification through enhanced thermal and electrochemical stabilities of N-doped graphene oxide. *Appl. Surf. Sci.* **2016**, *366*, 514–522. [CrossRef]
46. Zheng, L.; Xia, S.; Lu, X.; Hou, Z. Transesterification of glycerol with dimethyl carbonate over calcined Ca-Al hydrocalumite. *Chin. J. Catal.* **2015**, *36*, 1759–1765. [CrossRef]
47. Golemme, G.; Santaniello, A. Perfluoropolymer/Molecular Sieve Mixed-Matrix Membranes. *Membranes* **2019**, *9*, 19. [CrossRef] [PubMed]
48. Matteucci, S.; Kusuma, V.A.; Swinnea, S.; Freeman, B.D. Gas permeability, solubility and diffusivity in 1,2-polybutadiene containing brookite nanoparticles. *Polymer* **2008**, *49*, 757–773. [CrossRef]
49. Matteucci, S.; Kusuma, V.A.; Sanders, D.; Swinnea, S.; Freeman, B.D. Gas transport in TiO_2 nanoparticle-filled poly(1-trimethylsilyl-1-propyne). *J. Membr. Sci.* **2008**, *307*, 196–217. [CrossRef]
50. Merkel, T.C.; Freeman, B.D.; Spontak, R.J.; He, Z.; Pinnau, I.; Meakin, P.; Hill, A.J. Ultrapermeable, Reverse-Selective Nanocomposite Membranes. *Science* **2002**, *296*, 519–522. [CrossRef]
51. Song, K. Interphase characterization in rubber nanocomposites. In *Progress in Rubber Nanocomposites*; Elsevier: Amsterdam, The Netherlands, 2017; pp. 115–152.
52. He, S.-J.; Wang, Y.-Q.; Feng, Y.-P.; Liu, Q.-S.; Zhang, L.-Q. The preparation of an elastomer/silicate layer nanocompound with an exfoliated structure and a strong ionic interfacial interaction by utilizing an elastomer latex containing pyridine groups. *Nanotechnology* **2010**, *21*, 115601. [CrossRef]
53. Chen, H.; Li, Y.; Wang, S.; Li, Y.; Zhou, Y. Highly ordered structured montmorillonite/brominated butyl rubber nanocomposites: Dramatic enhancement of the gas barrier properties by an external magnetic field. *J. Membr. Sci.* **2018**, *546*, 22–30. [CrossRef]
54. Gatos, K.G.; Karger-Kocsis, J. Effect of the aspect ratio of silicate platelets on the mechanical and barrier properties of hydrogenated acrylonitrile butadiene rubber (HNBR)/layered silicate nanocomposites. *Eur. Polym. J.* **2007**, *43*, 1097–1104. [CrossRef]
55. Choudalakis, G.; Gotsis, A.D. Free volume and mass transport in polymer nanocomposites. *Curr. Opin. Colloid Interface Sci.* **2012**, *17*, 132–140. [CrossRef]
56. Srithawatpong, R.; Peng, Z.L.; Olson, B.G.; Jamieson, A.M.; Simha, R.; McGervey, J.D.; Maier, T.R.; Halasa, A.F.; Ishida, H. Positron annihilation lifetime studies of changes in free volume on cross-linking cis-polyisoprene, high-vinyl polybutadiene, and their miscible blends. *J. Polym. Sci. Part B Polym. Phys.* **1999**, *37*, 2754–2770. [CrossRef]
57. Beall, G. New conceptual model for interpreting nanocomposite behavior. In *Polymer–Clay Nanocomposites*; Wiley: Hoboken, NJ, USA, 2000; pp. 267–279.
58. Choudalakis, G.; Gotsis, A.D. Permeability of polymer/clay nanocomposites: A review. *Eur. Polym. J.* **2009**, *45*, 967–984. [CrossRef]
59. Jung, J.K.; Baek, U.B.; Lee, S.H.; Choi, M.C.; Bae, J.W. Hydrogen gas permeation in peroxide-crosslinked ethylene propylene diene monomer polymer composites with carbon black and silica fillers. *J. Polym. Sci.* **2023**, *61*, 460–471. [CrossRef]
60. Jung, J.K.; Lee, C.H.; Baek, U.B.; Choi, M.C.; Bae, J.W. Filler Influence on H(2) Permeation Properties in Sulfur-CrossLinked Ethylene Propylene Diene Monomer Polymers Blended with Different Concentrations of Carbon Black and Silica Fillers. *Polymers* **2022**, *14*, 592. [CrossRef] [PubMed]
61. Smith, A.P.; Crossley, S.D.; Gruber, T.C. Computational Investigation of the Effects of Spherical Filler Morphology and Loading on Diffusion Tortuosity and Rubber Permeability. *Rubber Chem. Technol.* **2013**, *86*, 175–189. [CrossRef]
62. Cui, Y.; Kundalwal, S.I.; Kumar, S. Gas barrier performance of graphene/polymer nanocomposites. *Carbon* **2016**, *98*, 313–333. [CrossRef]
63. Yoo, B.M.; Shin, H.J.; Yoon, H.W.; Park, H.B. Graphene and graphene oxide and their uses in barrier polymers. *J. Appl. Polym. Sci.* **2014**, *131*, 39628. [CrossRef]

polymers

MDPI

Review

Cellulose-Based Intelligent Responsive Materials: A Review

Sisi Chang [1,†], Zhangzhao Weng [2,*,†], Chunmei Zhang [3], Shaohua Jiang [1,*] and Gaigai Duan [1,*]

1 Jiangsu Co-Innovation Center of Efficient Processing and Utilization of Forest Resources, International Innovation Center for Forest Chemicals and Materials, College of Materials Science and Engineering, Nanjing Forestry University, Nanjing 210037, China; sisi_chang_lab@126.com
2 Strait Institute of Flexible Electronics (SIFE, Future Technologies), Fujian Normal University, Fuzhou 350117, China
3 Institute of Materials Science and Devices, School of Materials Science and Engineering, Suzhou University of Science and Technology, Suzhou 215009, China; cmzhang@usts.edu.cn
* Correspondence: ifezzw@fjnu.edu.cn (Z.W.); shaohua.jiang@njfu.edu.cn (S.J.); duangaigai@njfu.edu.cn (G.D.)
† These authors contributed equally to this work.

Abstract: Due to the rapid development of intelligent technology and the pursuit of green environmental protection, responsive materials with single response and actuation can no longer meet the requirements of modern technology for intelligence, diversification, and environmental friendliness. Therefore, intelligent responsive materials have received much attention. In recent years, with the development of new materials and technologies, cellulose materials have become increasingly used as responsive materials due to their advantages of sustainability and renewability. This review summarizes the relevant research on cellulose-based intelligent responsive materials in recent years. According to the stimuli responses, they are divided into temperature-, light-, electrical-, magnetic-, and humidity-responsive types. The response mechanism, application status, and development trend of cellulose-based intelligent responsive materials are summarized. Finally, the future perspectives on the preparation and applications of cellulose-based intelligent responsive materials are presented for future research directions.

Keywords: cellulose; actuator; response; temperature; biomimetic robots

1. Introduction

At present, the rapid development of materials science and engineering, civil and structural construction engineering, and the manufacturing industry has greatly promoted the transformation of electronic technology from traditional rigid rectangular buildings to flexible, electronic, and various flexible building forms. Intelligent driving materials can convert external energy from external stimulus [1] into mechanical energy and bring shape changes [2]. It is widely used in soft robots [3], artificial muscles [4], actuators [5], and other fields. An actuator is a device that can convert universal input energy into mechanical motion. This broad definition has led to a wide variety of actuators with different classifications. According to the definition proposed by Constantinos et al. [6] in 1999, traditional actuators include electric, hydraulic, and pneumatic types. These actuators can utilize electromagnetic energy to provide rotary or linear motion (known since Faraday's experiments in 1821 [7]). In the initial phase, soft actuators are mainly based on block and plate structures exhibiting in-plane deformation in the presence of variations in volume, alignment, and permutation. Subsequently, a great deal of research has focused on soft actuators with dual wafer or gradient structures, which involve reversible deformation between two-dimensional (2D) shapes and three-dimensional (3D) shapes. However, in this case, a limited number of deformation modes can be used, such as bending and twisting [8].

In nature, many biological phenomena utilize the principle of actuators, such as sunflowers growing toward the sunlight; the leaves of plants converting light energy into organic matter through photosynthesis; and pine cones falling from trees, contracting when exposed to water. In nature, many organisms use one-dimensionally shaped structures to adapt or change their living environment. Typically, animals, especially mammals, execute multiple bodily movements by controlling the contraction or relaxation of muscles attached to bones or other organs; sunflower stems exhibit phototropic bending toward sunlight to harvest more energy. These natural phenomena have inspired the development of novel soft actuators. Inspired by intelligent systems in the natural environment, people are committed to converting external stimulus such as electricity, light, heat, humidity, or magnetism into energy. Electric response actuators have advantages in energy conversion efficiency, the amount of stored energy, and controllability [9]. Optical responsive actuators are an intangible and remote control method with advantages in energy, wavelength selectivity, and ecological friendliness [10]. In this type of actuation, light is absorbed and converted into thermal or electrical energy, achieving photothermal and optoelectronic actuation [11]. Thermal and humidity responsive actuation typically relies on the expansion or contraction of materials caused by changes in temperature or humidity [12–14]. In addition, by assembling multi-layer materials with different responsibilities, their performance can be further improved or new properties can be added. For the magnetic-responsive actuation, magnetic fillers are added to soft materials to generate shape-changing forces following changes in external magnetic fields [15]. As single actuators, thermal-, optical-, electrical-, and magnetic-responsive actuators can only convert one type of energy into another. However, recently, single actuators have been unable to meet human needs, and more scientists have begun to study multi-responsive actuators [16], which has become a development trend.

As a crucial component of any machine, the primary purpose of an actuator [17] is to transform an input energy or signal into another form of energy or signal, which is subsequently released. Typically constructed from multiple layers of thin films, an actuator utilizes the asymmetric deformations from each layer as the driving force for its operation, making it highly responsive to changes from external stimulus [18]. With their vast potential for use in artificial muscles [19], neuroprosthetics bionics [20], precise control [21], and smart homes [22], it is crucial to explore alternative materials to petroleum-based actuators since they are non-degradable and non-renewable, which is contrary to the principles of sustainable development.

As the most abundant natural polymer in the world, cellulose is a sustainable and eco-friendly material that is increasingly being utilized as an actuator material. Cellulose is a macromolecular polysaccharide composed of glucose, which is the main component of plant cell walls [23,24]. It is a sustainable and renewable raw material with characteristics such as reusability, non-toxicity, environmental friendliness, biocompatibility, and biodegradability. It is one of the most abundant and commonly used biopolymers on Earth. Cellulose macromolecules are composed of a repetitive β-D-glucose unit, which is covalently linked by glycosidic bonds between hydroxyl groups [25], forming a linear macromolecule. Cellulose has strong hydrogen bond and high crystallinity within and between molecules, thus having excellent mechanical properties. The polymer chains of cellulose in natural cellulose fibers form a layered network structure through intermolecular and intramolecular bonds that are created by the abundant hydroxyl groups in the cellulose macromolecules. This particular structure also grants cellulose fibers exceptional moisture absorption and swelling properties, leading their responses to water molecules [26].

So far, cellulose has been discovered as an intelligent material that can be used as sensors [27], actuators [28], smart devices [29], and so on. It can be used as Electroactive Paper (EAPap) [30], which has advantages such as light weight, flexibility, dryness, biodegradability, easy chemical modification, and cheapness [31]. Cellulose has become a green and renewable chemical resource, making great contributions to human development [32]. Figure 1 summarized the response styles and different applications of cellulose-based actuators.

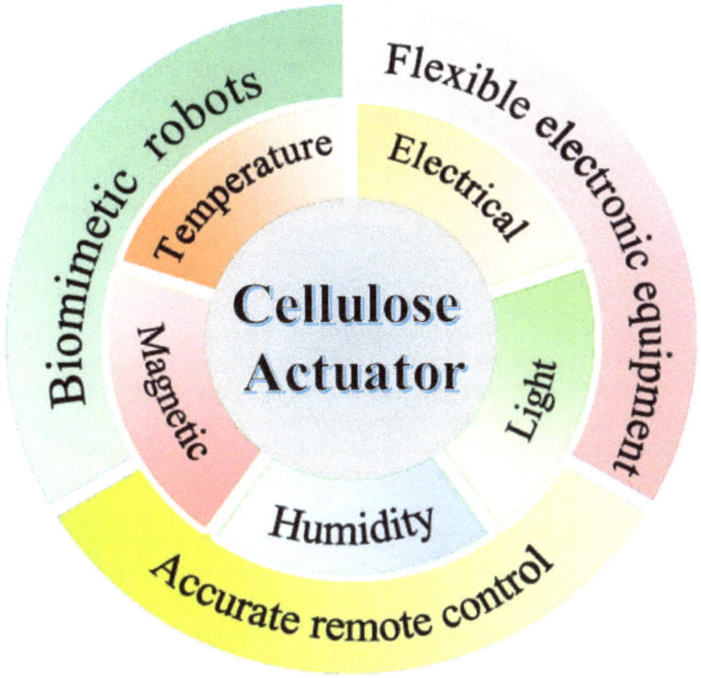

Figure 1. Response styles and applications of cellulose-based actuators.

2. Preparation of Cellulose-Based Responsive Materials

The structure of cellulose-based actuator polymers can be roughly divided into isotropic [33] and anisotropic structures [34]. Current methods to achieve complex deformation can be divided into two main categories. One is the non-homogeneous excitation of isotropic cellulose-based actuators or hydrogels, such as localized electric fields or localized light. The main principle is that the local stimulation for the complete cellulose-based actuator will only present locally uniform changes; however, multiple different degrees of stimulation will make the overall actuator uneven deformation. Through a reasonable structural design or program design, the complex deformation of the actuator can be realized.

The second is the anisotropic cellulose-based actuator, which can be used as the actuator layer therein by introducing the orientation structures or orientation patterns of the cellulose fibers. When the actuator is uniformly stimulated, the anisotropic structure causes the cellulose-based actuator to be subjected to different variations in each direction, generating forces of different strengths, which in turn cause the actuator to undergo complex deformation. Here we focus on the preparation method of anisotropic actuators that has been extensively studied. The anisotropic structures of cellulose-based actuators are mainly categorized into single-layer (also called gradient), bilayer, and multilayer structures.

2.1. *Single-Layer Structure*

An anisotropic monolayer actuator is an ideal material, as it eliminates the necessity for interface bonding or complexities in the production process commonly associated with multi-layer structures. Both the gradient distribution of some fillers and the gradient distribution of polymer chains can produce shape changes in cellulose-based actuators with gradient structures. Therefore, the ways to prepare cellulose-based actuators with a gradient structure can be broadly classified into two categories. (1) The construction of gradient structures is realized by generating an asymmetric distribution of polymer chains in vivo. For example, Roop sung et al. [35] employed electrophoresis to manufacture a single gradient hydrogel actuator utilizing cellulose nanocrystals (CNCs). As shown in Figure 2a, this hydrogel actuator showed temperature-responsive bending behavior, whereby the bending angle progressively increased over time, resulting in the formation of a semicircular shape, as the temperature rose from 25 °C to 50 °C within four minutes. As shown in Figure 2c, Schäfer et al. [36] developed a single-layer actuator consisting of polymer-modified cellulose paper, which shows a bending response when exposed to moisture, as shown in Figure 2d. The researchers showed that this design not only minimizes the likelihood of delamination, but also simplifies the overall design and construction processes in contrast to multilayer structures. (2) The second category is embedding stimuli-responsive nanoparticles in the polymerization process and utilizing their migration in an applied electric or magnetic field to form a gradient arrangement. For example, Wang et al. [37] immersed Fe_3O_4 nanoparticles into nitrocellulose fibers of chromatography paper. As shown in Figure 2b, they observed that the end of the fiber showed augmented deflection and bending angle as the magnetic field strength increased, indicating the reversibility of this effect. Additionally, the fiber exhibited the ability to pick up paper masses twice its own weight under magnetic stimulation.

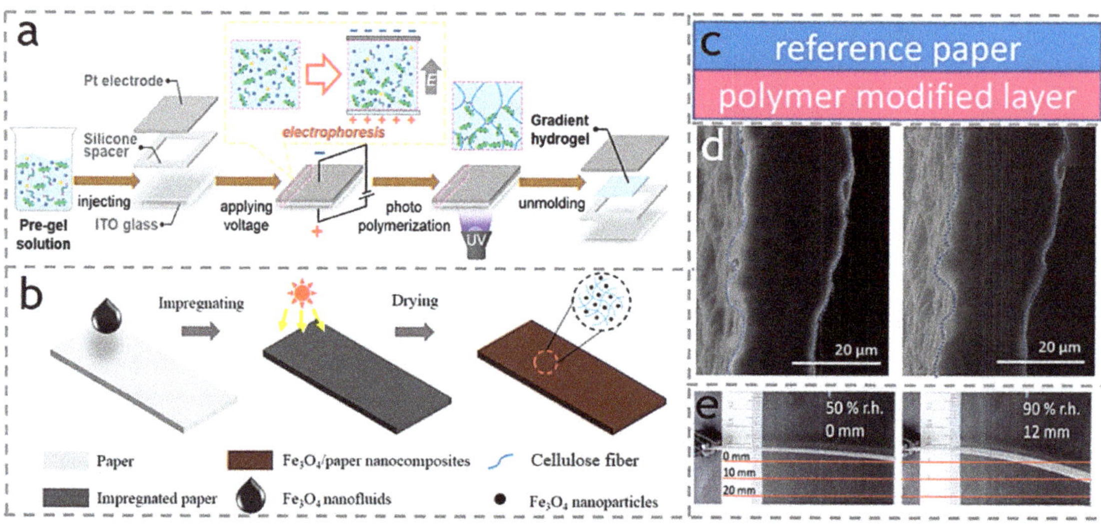

Figure 2. (**a**) A single gradient hydrogel actuator utilizing cellulose nanocrystals (CNCs). Reproduced with permission [35]. Copyright 2023. The Royal Society of Chemistry. (**b**) Fabrication process of Fe_3O_4/paper nanocomposite via a low-cost blending method. Reproduced with permission [37]. Copyright 2018. ELSEVIER. (**c**) Single-layer actuator consisting of polymer-modified paper. (**d**) Cross-sectional ESEM images of single-layer films at different RH. (**e**) polymer modified "layer" pointing down, in the 50% RH and 90% RH. Reproduced with permission [36]. Copyright 2023. Biomimetics.

Fibrous cellulose nanofibers (CNFs) possess a high aspect ratio, which facilitates the intercalation of water molecules within the cellulose network and enhances actuation performance. Lopes et al. [38] prepared CNF monolayer membrane actuators with the carboxymethylation of CNF. As shown in Figure 2d,e, upon observation of the film under 100% relative humidity, a substantial volume expansion was observed through microscopic morphology analysis.

2.2. Double-Layer/Multi-Layer Structure

At the present stage, the preparation of bilayer actuators mainly consists of two methods: layer-by-layer deposition [39] and reversible switch assembly [40]. In the preparation process of layer-by-layer deposition, the polymer of the second layer will penetrate into the first layer in a small amount, forming an interpenetrating network at the interface and thus making the two layers closely connected. For example, Liu et al. [41] electrospun PNIPAM and TPU into a bilayer-structured actuator, where the interfaces can be strengthened by the hot-pressed melting TPU, and the thermal responsiveness of the actuator can be modulated by adjusting the thickness of the two fiber membranes. For the reversible switch assembly method, generally the actuator layers are connected by host–guest interactions or hydrogen bonding, which makes the layers tightly connected together to form a bilayer structure actuator. For example, Ma et al. [42] constructed a bilayer-structured hydrogel actuator by a supramolecular building block assembly method, in which they prepared a responsive body hydrogel with a β-cyclodextrin structure and carboxyl groups and a non-responsive guest hydrogel containing ferrocene groups. After that they utilized such reversible host–guest interactions between β-cyclodextrin and ferrocene to assemble a hydrogel actuator that could bend with the changes of pH value and ionic strength.

In addition to the two methods mentioned above, many researchers have bonded hydrogels and polymers together by simple methods of preparing polymers, such as in in situ polymerization, which is one of the effective methods for synthesizing nanocomposites and appears to be particularly useful in the synthesis of composites based on monomers. As shown in Figure 3a, Wu et al. [43] prepared a paper/PNIPAM composite hydrogel actuator. Due to the in-situ polymerization of NIPAM monomer integrating the thin paper and hydrogel layers together, and the strong bonding of the bilayer structure, the hydrogel composite actuator can achieve a variety of controlled deformations, showing remarkable photothermal responsiveness. As shown in Figure 3b, Chen et al. [44] also used the in situ polymerization method by applying a N-isopropylacrylamide (NIPAM) monomer on the surface of a bamboo sheet to form a continuous layer, followed by UV polymerization to convert it into a PNIPAM hydrogel, resulting in an anisotropic bamboo/PNIPAM hydrogel complex (ABH) actuator. With the difference of adopting a morphogenetic approach based on a natural bio-template of bamboo, they demonstrated how its biomimetic structure can integrate the features of high-strength, fast response, and programmable composite deformation into a single device.

In general, the ability to respond to external stimuli of the single-layer cellulose-based actuator is limited. Hence, researchers are increasingly incorporating additional functional polymeric materials and anisotropic structural materials to enhance its performance [45]. For instance, the inclusion of conductive polymers can impart outstanding electroactive characteristics to the actuator. For example, as shown in Figure 3c, Wu et al. [46] successfully fabricated a three-layer cellulose paper actuator with a capacitor-type design. This actuator consists of two layers of doped poly (3-styrene sulfonate) (PEDOT/PSS) poly (4,4-vinylidene dithiophene) thin films, with another polyelectrolyte layer sandwiched in between. The actuator demonstrates a stable electromechanical deformation that is symmetric in both directions. This cellulose paper-based actuator, with a thickness of 48 μm, achieves a maximum displacement of 0.05 mm when operated at a voltage of 5.8 V and a frequency of 1.5 Hz. As shown in Figure 3d, Naohiro et al. [47] utilized a gel electrode composed of 2,2,6,6-tetramethylpiperidine-1-oxyl (TEMPO)-oxidized cellulose nanofibrils (TOCN), PEDOT:PSS, and an ionic liquid (IL) for the preparation of a novel actuator. This gel electrode actuator demonstrates exceptional electrical conductivity as well as ion conductivity properties.

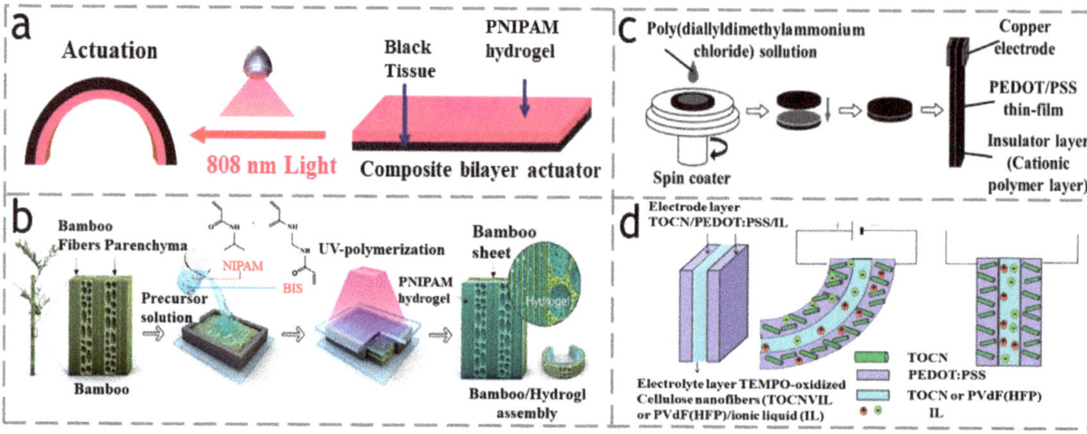

Figure 3. (**a**) A paper/PNIPAM composite hydrogel actuator. Reproduced with permission [43]. Copyright 2022. Polymers. (**b**) A bamboo/PNIPAM hydrogel composite actuator. Reproduced with permission [44]. Copyright 2021, AMER CHEMICAL SOC. (**c**) A single-layer actuator consisting of polymer-modified paper. Reproduced with permission [46]. Copyright 2020. Japan Oil Chemists' Society. (**d**) TOCN/PEDOT: PSS/IL electrode actuator. Reproduced with permission [47]. Copyright 2021. ELSEVIER.

3. Response Mechanism of Actuators

3.1. Temperature Response Mechanism

Temperature response refers to the transformation of a material's state through changes in the external environmental temperature [48,49], from the interaction between polymer and water to the interaction between polymer and polymer. The heat source is usually heating, electrothermal, and photothermal [50,51].

3.2. Light Response Mechanism

Photoactuation is a wireless- and remote-control method that has advantages in energy, wavelength selectivity, and ecological friendliness. In this type of actuation, light is absorbed and converted into thermal, chemical, or electrical energy, achieving photothermal, photochemical, and optoelectronic actuation.

The photo-thermal-responsive actuator, with its unique dual photo and thermal responses, can achieve process control and is remotely driven by a special light source, thus realizing the comprehensive regulation of the actuator. It is worth noting that for the photothermal response, the required light generally refers to the near-infrared light or lasers, which possess high energy and non-radiation penetration. Thus, the light-controlled regulation of the actuator is of great significance in areas that are inconvenient for people to access.

3.3. Electrical Response Mechanism

Actuation by electricity is working based on the electric energy, which can be converts into torque, then drives the actuator. Because of the fast conductive and clean energy of electricity, the electrical response has the advantages of being quick, precise, and clean. Many previous studies have found that cellulose has piezoelectric properties [52]. The high crystallinity and abundance of polar hydroxyl groups make cellulose contain a large number of dipoles with strong electron-donating ability, and thus cellulose can be used to fabricate electrically responsive actuators.

3.4. Magnetic Response Mechanism

Magnetic actuation is mainly from the addition of magnetic fillers to soft materials, which follow changes in the external magnetic field to produce changes in shape or torque. As is well known, the surface of cellulose contains abundant hydroxyl groups and the porosity of cellulose. Magnetic particles can be easily added with cellulose to produce composites. The commonly used ferromagnetic particles include Fe_3O_4 [53], Fe_2O_3 [54], $MnFe_2O_4$ [55], $CoFe_2O_4$ [56], and so on. When the external magnetic field changes, each magnetic particle will have a similar trend and undergo a relative adjustment. Due to this adjustment, the cellulose composites can elongate or contract, leading to the corresponding actuation.

3.5. Humidity Response Mechanism

It should be noted that cellulose contains a large number of hydrogen bonds, and the interaction of hydrogen bonds provides the polymer with excellent hygroscopicity. The absorption and desorption of water molecules by cellulose can significantly change its volume. At the same time, the abundant hydroxyl and carboxyl groups on the molecular chain of cellulose result in cellulose itself being a hygroscopic material. Humidity-responsive actuators have attracted great interest due to the presence of water in nature and the concept of green environmental protection. The design principle of humidity actuation [57] is that materials with different hydrophilicity have different responses to the humidity, and their materials will undergo varying degrees of deformation due to swelling. When external moisture or humidity increases, the materials will undergo a corresponding actuation. This material system has great application potential in multifunctional or intelligent actuators.

4. Cellulose-Based Temperature-Responsive Actuators

Temperature-responsive actuators can undergo a volumetric phase transition through the conversion of its own hydrophilicity.

Thermal-responsive materials are qualified candidate materials that can be directly used as actuators by thermal stimulus [58]. Therefore, this kind of hydrogel generally has a lower critical solution temperature (LCST) or a upper critical temperature (UCST) [59], and the temperature-responsive hydrogel will expand or shrink when the temperature is higher or lower than the critical temperature. For example, poly(N-isopropylacrylamide) (PNIPAM), as a typical thermal-responsive polymer [60–62], has been widely studied by researchers. The LCST of PNIPAM is approximately 32 °C [63], and the hydrophilic/hydrophobic state of PNIPAM will change nearby. Since the thermal-responsive PNIPAM micro gel was reported in 1986, PNIPAM hydrogel has been frequently used as the raw material of actuators.

In general, the pure PNIPAM hydrogel is isotropic, and only has a simple volume contraction or expansion under the condition of temperature change. Many researchers focus on the cellulose-based composites with PNIPAM to achieve a complex deformation. For example, as shown in Figure 4a,b, Liu et al. [64] have successfully prepared a bilayer hydrogel actuator with a vertically oriented anisotropic structure, which exhibits temperature-responsive behavior. The top layer of the hydrogel is constructed using a double network of poly(N-isopropylacrylamide)/cellulose nanofiber (PNIPAM/CNF), while the bottom layer consists of a double network of polyacrylamide-co-acrylic acid/cellulose nanofiber (PAM-AA/CNF). This bilayer hydrogel actuator demonstrates remarkable driving capability and mechanical performance, as it is capable of bending at a high speed of 70.10 (°/s) and can lift a weight that is 21 times its own weight, as shown in Figure 4c.

In another work, Wei et al. [65] prepared a complex hydrogel based on dextran and PNIPAM. When the temperature is higher than 32 °C, the PNIPAM molecular chain will shrink, thereby extruding the curcumin loaded in the hydrogel (Figure 5a). As shown in Figure 5b, Wang et al. [66] prepared a multifunctional composite nanocarrier poly(N-isopropylacrylamide)-modified graphene oxide (PNIPAM-GO), and through a simple physical adsorption process, load λ- Cyhalothrin (LC) onto PNIPAM GO nanocomposite carriers. However, it is a pesticide formulation that can adjust the pesticide release based on changes in the external temperature and has a certain intelligent effect on crop growth. As shown in Figure 5c, Chen et al. [67] integrated conductive hydrogel (carboxymethyl cellulose) and thermal-responsive poly(N-isopropylacrylamide) (PNIPAM) hydrogel to form a double-layer hydrogel, which has the ability to respond to environmental temperature changes. More interesting is that because the hydrogel has different degrees of expansion and contraction under different thermal stimuli, it can produce two-way bending. The dotted lines in Figure 5c indicated the transmittance of the hydrogel reversibly changes when the temperature changes.

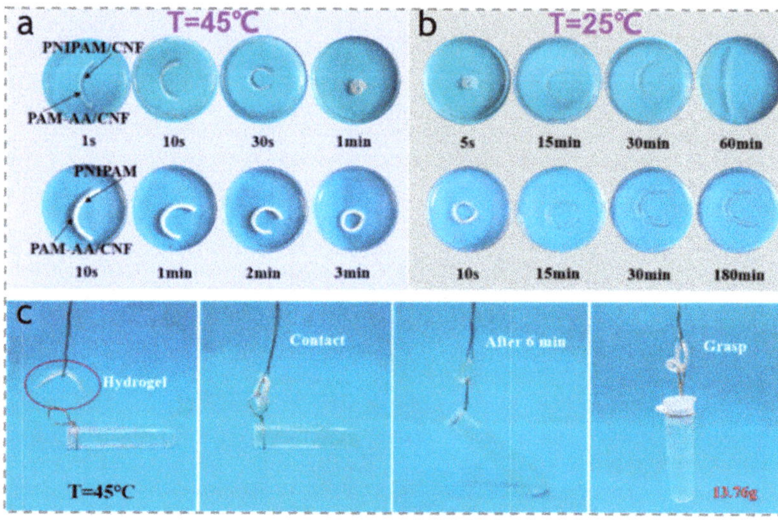

Figure 4. (**a**) Pictures showing the bending of PAM-AA/CNF + PNIPAM/CNF hydrogel strip and PAM-AA/CNF + PNIPAM hydrogel strip at 45 °C in water bath and (**b**) their unbending at 25 °C. (**c**) Demonstrations of bilayer hydrogel as temperature-controlled manipulators at 45 °C. Reproduced with permission [64]. Copyright 2022. ELSEVIER. The red circle indicates the bilayer hydrogel sample.

In addition, Joule heating is also a method of using electrical energy for heating, which involves using thermal conductive materials as an electric heating layer, such as carbon nanotubes [68,69], MXene [70,71], and so on. Conductive materials can be used as electric heating layers, and then compounded with natural or synthetic cellulose to form a double-layer actuator. For example, Wei et al. [72] reported a CNF (cellulose nanofiber)—MXene ($Ti_3C_2T_x$) nanosheet-TA (tannic acid) composite, in which MXene nanosheets have excellent conductivity, resulting in good Joule thermal performance of composite thin-film actuators.

Shape-memory polymer (SMP) can perform repeatable shape transformation under various stimulus; thus, it has great application prospects in the field of actuators and has been widely studied. As shown in Figure 6a, Bai et al. [73] prepared an electrospun cellulose acetate (CA)/carbon nanotube nanofiber composite (CAC) with both shape memory performance and self-powered sensing performance for the first time. The material was stretched to 90% strain at 50 °C (higher than its T_g), cooled to room temperature to fix the temporary shape, and then reheated to 90 °C to restore its original shape. As shown in Figure 6b, it is clearly observed that CAC has the ability to fully restore its original shape, while also having a shape fixation and recovery rate of over 95%. They also demonstrated that the shape memory properties of the material are not related to the manufacturing method and carbon nanotube content, but only to the CA itself.

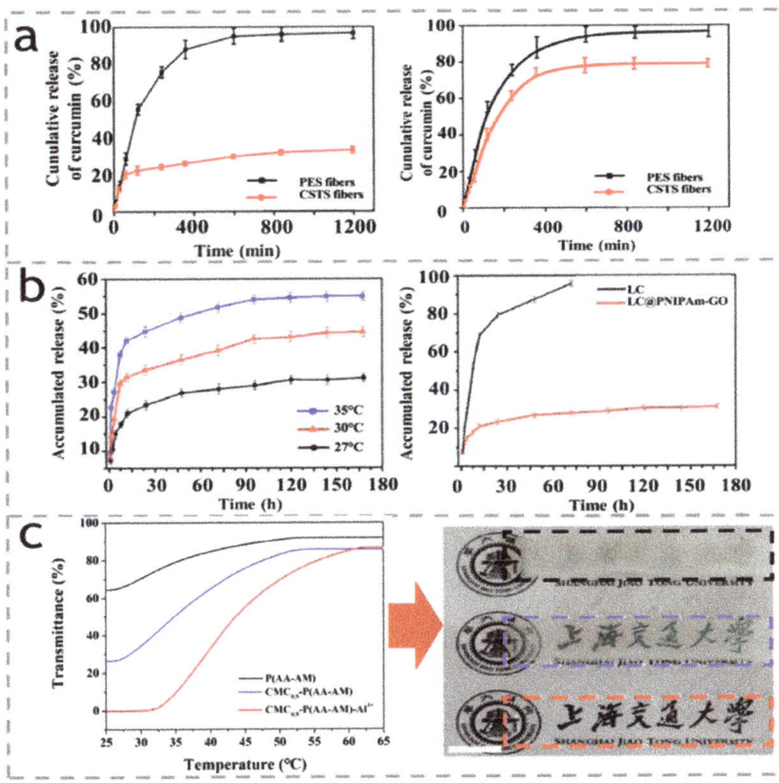

Figure 5. (**a**) Release profiles of curcumin from PES and CSTS ultrafine fibers at 20 °C and 60 °C; reproduced with permission [65]. Copyright 2019. ELSEVIER. (**b**) Effect of temperature on the release behavior of that, and the release behavior of free LC and LC@PNIPAm-GO at RT. Reproduced with permission [66]. Copyright 2021. ELSEVIER. (**c**) Temperature-transmittance curves of the P(AA-AA), $CMC_{0.5}$-P(AA-AM), and $CMC_{0.5}$-P(AA-AM)-Al^{3+} hydrogels (**left**) and the corresponding photographs of each hydrogel in a 25 °C water bath; the scale bar is 1 cm. Reproduced with permission [67]. Copyright 2020. American Chemical Society. The Chinese words indicated the name of Shanghai Jiaotong University.

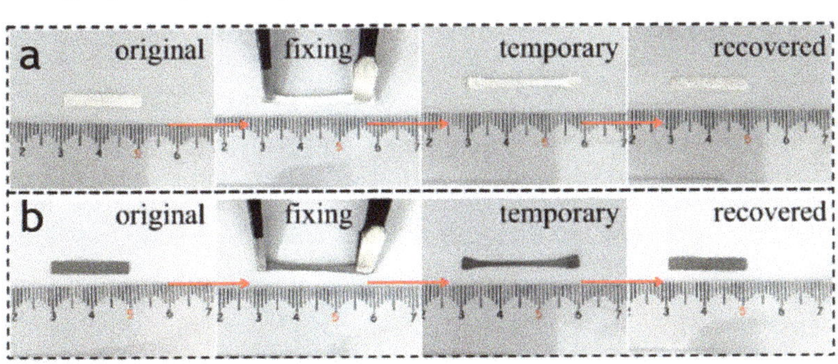

Figure 6. (**a**) Photothermal conversion performance of CA nanofoborous composites. (**b**) NIR light-induced shape memory behaviors of CAC. Reproduced with permission [73]. Copyright 2023. ELSEVIER.

5. Cellulose-Based Light-Responsive Actuators

Light-responsive actuators using light as a stimulus source can not only be process-controlled, but can also be remotely driven by a special light source, thus realizing the comprehensive regulation of the actuator. It is worth noting that, in some inconvenient entering areas, the intelligent light-controlled regulation of the actuator is very meaningful.

Many plants around the world have anisotropic cellulose structures, which play an important role in determining the different forms and colors of plants. Color variations in plants are mainly attributed to pigments and dye molecules. In addition to the structural color of the plant itself, pigments in plants can undergo color changes through selective absorption of light by their internal structures as well as scattering and refraction of light [74]. Cellulose nanocrystals (CNC) have emerged as prime candidates for manufacturing materials with unique optical characteristics [75], owing to their high aspect ratio and intermolecular interactions. Sun et al. [76] combined cellulose nanocrystals (CNCs) with different surface charge density and particle size with phenol glutaraldehyde resin and graphite oxide in a certain proportion to prepare composite materials of different colors, as shown in Figure 7a. After immersion in water for 25 s and in dimethyl sulfoxide (DMSO) for 20 min, the color of the films changed from green to red, but even after immersion for 3 months, other solutions with smaller polarity would not change color, indicating that it has a good degree of stability. This is a straightforward color change that only takes place when the substance is immersed in a solution. Inspired by the color-changing mechanism observed in neon lights, Li et al. [77] created a bionic smart hydrogel using cellulose nanocrystals. This hydrogel demonstrates a dual response to both pressure and temperature. As shown in Figure 7b, as vertical pressure is applied, the hydrogel's color changes from red to orange, yellow, green, and blue. Moreover, by maintaining a constant external pressure and varying the temperature, the hydrogel is capable of undergoing a comparable rainbow-like color transformation. Even more remarkable, Orelma et al. [78] developed optical cellulose fibers by coating regenerated cellulose long filaments with acetic acid fibers. These fibers possess mechanical properties that are comparable to textile fibers. However, when light is observed in the range of 500–1400 nm, the cellulose fiber exhibits an attenuation constant of 3.1300 dB/cm at 6 nm. When the fiber is immersed in water, a substantial reduction in light intensity is achieved, indicating the potential use of this optical cellulose fiber as water sensors.

In addition to the single temperature conversion mentioned above, photothermal conversion is another actuation mechanism. The photothermal response actuator can convert light energy into heat from carbon nanotubes [79], carbon nanorods [80], graphite oxide [81], and other materials. Due to the high photothermal conversion and absorbance of near-infrared (NIR) energy, the near-infrared laser is usually used as the light source. For example, Zhao et al. [82] prepared a new type of double-layered nano-composite hydrogel actuator containing nano fibrillar cellulose (NFC), poly (N-isopropylacrylamide)-clay, and GO. GO absorbed near-infrared laser in the hydrogel and converted it into energy leading to a temperature rise, thus making the volume of the composite hydrogel actuator shrink and bend. Li et al. [83] self-assembled cellulose nanocrystals (CNCs), thin films, and polyurethane (PU) substrates, as well as dispersed in situ synthesized silver nanoparticles (AgNPs) into the PU matrix. Due to the high photothermal conversion efficiency and thermal conductivity of AgNPs, the elastomer exhibits excellent photothermal conversion efficiency and thermal conductivity. As shown in Figure 8, Chen et al. [84,85] made wood/PNIPAM and bamboo/PNIPAM composite hydrogel actuators, respectively, by combining wood and bamboo with hydrogel through simple and rapid in situ polymerization. The natural anisotropy of wood and bamboo [86–89] shows different shapes of deformation under near-infrared light and can be remotely controlled by light response actuation.

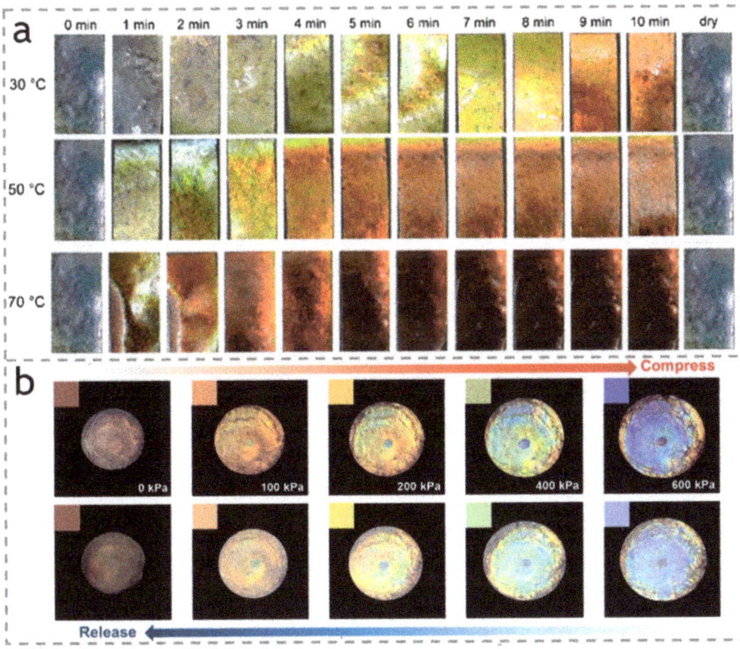

Figure 7. (**a**) Images of photonic film immersed in DMSO at various temperatures. Reproduced with permission [76]. Copyright 2023. Wiley-VCH Verlag. (**b**) Digital photos of the sample during compressing and releasing process. Reproduced with permission [77]. Copyright 2019. SPRINGER.

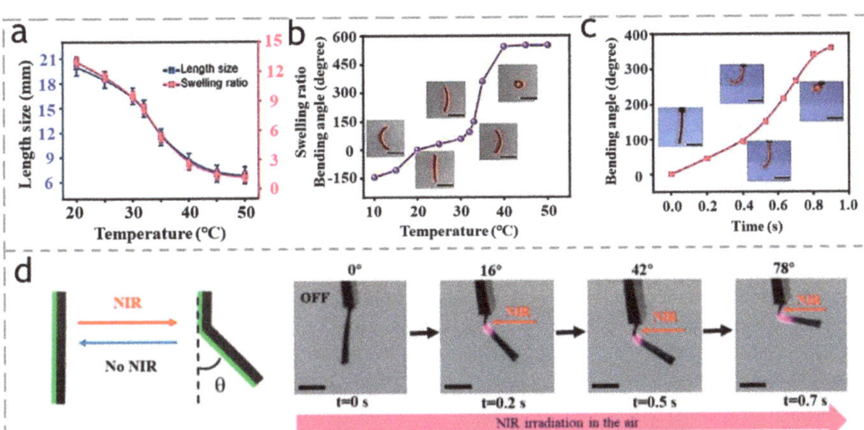

Figure 8. (**a**) The equilibrium swelling rate and length size of PNIPAM hydrogel at different temperatures. (**b**) The bending equilibrium angle of the wood-PNIPAM composite hydrogel actuator at different temperatures. (**c**) Curve of bending angle over time of the wood-PNIPAM composite hydrogel actuator in 40 °C warm water. Reproduced with permission [84]. Copyright 2022. ELSEVIER. (**d**) The illustration of folding angle for bamboo/PNIPAM composite hydrogel actuator and the optical photos of bamboo/PNIPAM composite hydrogel actuator locally irradiated by NIR light. Reproduced with permission [85]. Copyright 2022. ELSEVIER.

6. Cellulose-Based Electrically Responsive Actuators

Intelligent electrically responsive actuators have the advantage of simple operation and excellent controllability. These soft actuators can either convert electrical energy directly into mechanical energy or drive thermally responsive materials through Joule heating. Previously, it was reported that cellulose-based electroactive paper (EApap) [30] is used as an electric-responsive actuator. This type of actuator is made of a piece of cellulose paper, and its surface is composed of two electrodes. In fact, these two electrodes are basically used to determine the performance of cellulose paper [90]. The cellulose of the paper is arranged through the regeneration process, and usually the solid material undergoes Coulomb interaction at higher voltages, leading to bending and responses. Muhammad et al. [91] prepared a new type of composite gel from functionalized cellulose carboxylate nanocrystals (CCNs) and polyvinyl chloride (PVC). Under the excitation of the specified DC voltage (1000 V), the deformation of the actuator was about 36.7%, as shown in Figure 9a.

In addition to the direct conversion mentioned above, cellulose can also be combined with some conductive materials to form a composite film, achieving more efficient electroactive properties. For example, MXene [92,93] and graphene nanosheet (GN) [94] have excellent conductivity and chemical stability, reducing the Joule loss during conversion into electrical energy. Sen et al. [95] demonstrated that loading graphene nanosheets onto a mixture of microcrystalline cellulose powder (MCC), 1-butyl-3-methylimidazolium chloride (BMCl), and N,N-dimethylacetamide (DMA) results in better electroactive performance of the cellulose-based composite actuator, as shown in Figure 9b. Tang et al. [96] prepared a three-layer composite membrane as flexible actuator (called BMB). They sandwiched pure MXene membrane between polypropylene (BOPP) and bacterial cellulose (BC) after wet expansion and biaxial stretching. The BMB actuator could quickly lift a weight of 200 mg under a voltage of 4 V, as shown in Figure 9c, and gradually returned to its original state after the voltage was turned off. They also used tape to stick a pair of plastic butterfly wings (approximately 435 mg) to the free end of the RMB actuator. When the input voltage is turned off, the butterfly wings will sag due to their own weight. However, when the voltage of 4 V is turned on, the actuator quickly lifts the butterfly wings, and the work conducted when lifting is higher than most actuators, as shown in Figure 9d,e.

Conductive polymers (CPs) can also be used as another component of actuator systems [97]. Due to their low price, affordability, high conductivity doping state, and large storage capacity, they are very suitable for making electrochemical capacitors. Among them, poly (3,4-ethylenedioxythiophene) (PEDOT) is considered as a very useful CP due to its conductivity, stability, and processability [98]. In order to investigate the actuation mechanism of electric-responsive actuators, Chalil et al. [99] prepared a poly (3,4-ethylenedioxythiophene) polystyrene sulfonate (PEDOT:PSS)/bacterial cellulose bilayer actuator. The electric response was performed using HCL, KCl, and LiCl electrolytes. They found that the actuator was related to the hydrodynamic size of the cations presented in the electrolyte. The experimental results also indicate that the PEDOT: PSS/bacterial cellulose bilayer actuator has a cation-dominated driving mechanism. Driven by the external voltage, the internal ions will move in the active layer, resulting in interaction, cross-linking between groups, etc., finally leading to the change of material volume and deformation actuation. Therefore, many researchers prepare ionic conductive hydrogels or actuators through ionic liquids. Naohiro et al. [100] prepared a novel actuator with a cellulose nanofiber/poly (3,4-ethylenedioxythiophene): poly (4-styrene sulfonate)/ionic liquid (CNF/PEDOT: PSS/IL), which replaces carbon nanotubes (CNTs) with a CNF skeleton. The maximum strain and maximum generated stress exhibited simultaneously exceeded the strain and maximum stress by PEDOT: PSS/IL actuators without cellulose nanofibers (CNFs), indicating that cellulose-based materials exhibit better strain performance in actuator devices.

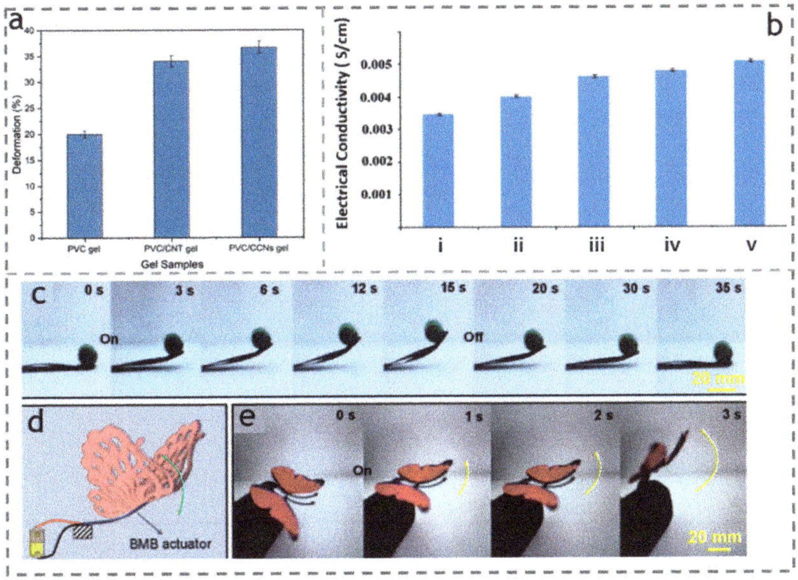

Figure 9. (**a**) Deformation of the pure PVC gel, PVC/CNT, and PVC/CCNs composite gel. Reproduced with permission [91]. Copyright 2023. AMER CHEMICAL SOC. (**b**) Effect of Gr loading on electrical conductivity of samples (i) MCC, (ii) MCC–BMCl, (iii) MCC–BMCl–0.1Gr, (iv) MCC–BMCl–0.25Gr, (v) MCC–BMCl–0.5Gr film. Reproduced with permission [95]. Copyright 2015. ELSEVIER. (**c**) Optical images showing the BMB actuator lifting up a 200 mg weight under an applied voltage of 4 V. (**d**) Schematic illustration showing the structure of the moveable butterfly. (**e**) Optical images of the butterfly during upward movement (driving voltage: 4 V). Reproduced with permission [96]. Copyright 2022. American Chemical Society.

In another report, Wang et al. [101] prepared a low-pressure ion actuator based on sulfonated cellulose nanocrystal (SCN), microfibrillated cellulose (MFC), graphene nanoplatelets (GN), and ionic liquid (IL, [EMIM][BF4]). This actuator exhibited a significant peak displacement of 0.1 mm at 1 V and 6.6 Hz. Furthermore, their team developed a high-fidelity bioelectronic muscle actuator based on a carboxymethyl cellulose bacterial cellulose (CBC) membrane using a simple zinc oxide (ZnO) particle leaching (PL) method. The membrane showed a remarkable improvement of 70.63%, 22.50%, and 18.2% in ionic liquid absorption capacity, ion exchange capacity, and ionic conductivity of CBC, respectively. As shown in Figure 10a, the bending deformation of the CZ-PL membrane-based actuator was 5.8 times larger compared to the pure CBC-based actuator [102].

Lai et al. [103] proposed pure cellulose nanocrystalline films as biomass actuators in ionic solutions. They expanded the CNC film in 0.1 M NaCl and cut it into rectangular cantilevers, which were installed in a simple dual motor test cell. A 10 V field was applied to NaCl electrolyte solutions of different intensities. The CNC cantilevers showed reversible driving with an alternating electric field potential for up to 20 min without mechanical failure; then, they compounded CNC with polyacrylamide. As shown in Figure 10b, this nanocomposite hydrogel showed reversible actuation at 1.4 °/s. Under 10 V, there was no mechanical failure during 60 min of experimental operation. At the same time, they also made a comparison. The sample without CNC did not show obvious actuation, which further confirmed that CNC had an electric-responsive actuation behavior. As shown in Figure 10c, Correia et al. [104] prepared cellulose nanocrystals with different surface charges (neutral, positive, and negative) and incorporated an increasing amount of IL 2-hydroxyethyl trimethylammonium dihydrogen phosphate ([Ch] [DHP]) (10 and 25 wt%) into the CNC master matrix. Regardless of the surface charge of the CNC, an increase in conductivity was observed when IL was incorporated into the CNC matrix. When a voltage of 100 V was applied at a frequency of 9 mHz, a maximum bending displacement of 8 mm was obtained.

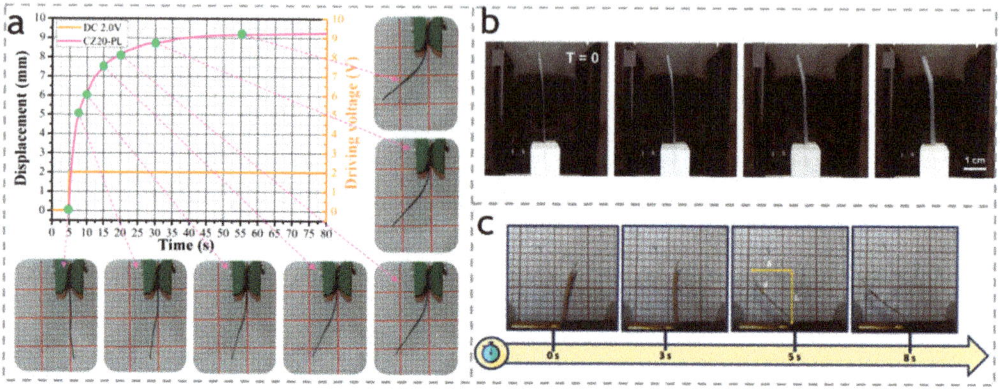

Figure 10. (**a**) Electromechanically deformed shapes of the CZ-PL muscular actuator under a step excitation of 2.0 V. Reproduced with permission [102]. Copyright 2017. ELSEVIER. (**b**) Swelling and disintegration of a pure CNC hydrogel cantilever during actuation testing in 3 mm NaCl electrolyte at T = 0 and after 1 min and 2 min at 10 V, and subsequently 1 min after switching voltage to −10 V. Reproduced with permission [103]. Copyright 2021. AMER CHEMICAL SOC. (**c**) Bending response as a function of time for NaCMC/[Ch][DHP] composite at a 100 mHz frequency and 8 V. Reproduced with permission [104]. Copyright 2020. MDPI.

7. Cellulose-Based Magnetic-Responsive Actuators

Intelligent magnetically responsive soft actuators for bio-inspired soft actuators offer unique advantages over other soft actuators. These advantages include: (1) non-contact remote control that can penetrate a wide range of materials; (2) precise manipulation by controlling the direction and strength of the magnetic field; (3) enhanced controllability by independently varying the magnetic field and gradient; (4) the ability to generate a magnetic alternating current (AC) for use in applications such as magnetothermal therapy; and (5) scalability and controllability of the actuator, which ranges from the nanoscale to the macroscopic level. Their disadvantage is that they require complex and bulky external devices to precisely control the magnetic field and its gradient.

As shown in Figure 11a, Wang et al. [37] made a cellulose paper/Fe_3O_4 nanocomposite actuator through a low-cost blending. The actuator is driven by a magnet with a magnetic strength of about 230 mT, and the maximum deformation and strain achieved by the actuator are 30 mm and 100%, respectively.

Cellulose can be modified with magnetic materials, making it have excellent performance in the magnetic field. As shown in Figure 11b, Chen et al. [105] developed a nanocomposite, which is composed of rod-like cellulose microcrystals with magnetite nanoparticles attached to its surface. This design takes advantage of the optical transparency and birefringence of cellulose microcrystals and their anisotropic shape. When their surfaces are modified by magnetite nanoparticles, their anisotropic shape enables them to effectively align. Only a small amount of magnetite nanoparticles is needed to conduct such a real-time and reversible orientation control of cellulose microcrystals, which thus ensuring the high transparency of the system, as shown in Figure 11c.

Tomás et al. [106] produced a hybrid of cellulose nanocrystals decorated with magnetic nanoparticles. This actuator can synergistically promote hASC tension through mechanical sensing mechanisms and may regulate its paracrine signal transduction that promotes healing, thereby jointly promoting the improvement of the regenerative potential of engineered tendon grafts. Kim et al. [107] synthesized a dense surface layer of cellulose nanofibers (made from Acetobacter xylinum), enclosing magnetic nanoparticles (MNPs) in a solid matrix to form an actuator string with response to external magnetic fields. The nanofiber actuator string is convertible in order to adapt to various shapes of tubular structures on the cross-section, reducing friction and stress on organ tissue walls due to its softness and plastic deformation. As shown in Figure 11d, such nanofiber skin strings can be bent at acute angles through magnetic actuation and can be used as endoscopic guides to reach targets deep in the kidney model.

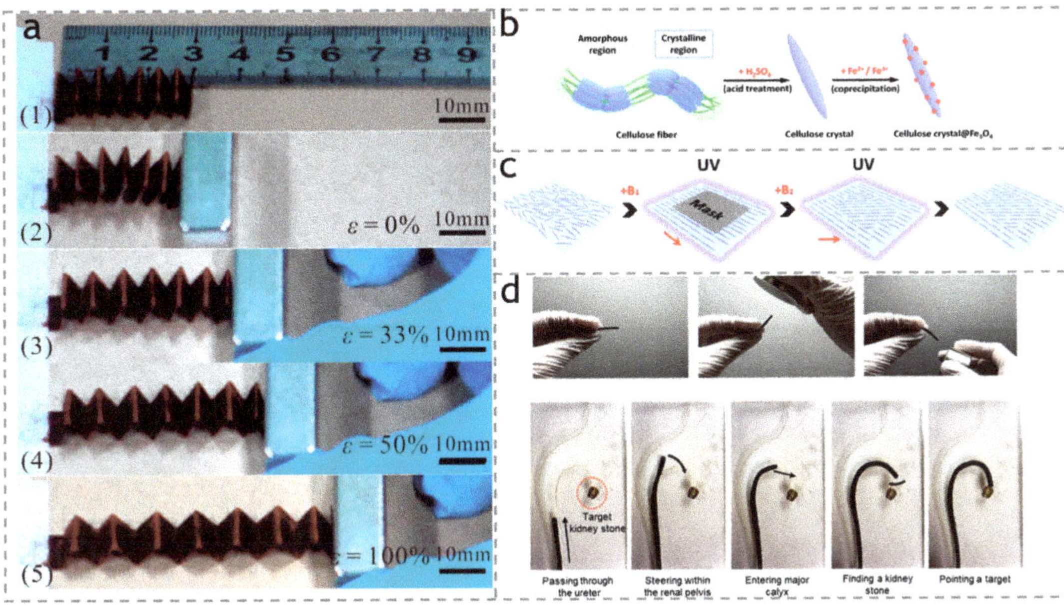

Figure 11. (**a**) Performance of accordion-shaped Fe_3O_4/paper actuator subjected to tunable magnetic field. (1–5) show actuator deformation magnetically actuated at specific normal strains. Reproduced with permission [37]. Copyright 2018. ELSEVIER. (**b**) Digital images of cellulose microcrystals containing Fe_3O_4 nanoparticles. (**c**) Scheme showing the lithography process for patterning a polymer film that contains magnetic cellulose microcrystals with different orientations in different regions. Reproduced with permission [105]. Copyright 2020. ELSEVIER. (**d**) Steering of the nanocellulose actuator string. Reproduced with permission [107]. Copyright 2021. American Chemical Society.

8. Cellulose-Based Humidity-Responsive Actuators

Intelligent humidity-responsive actuators can be used in the stimulus of liquid (moisture) or vapor (humidity). Moisture gradients and changes in humidity are common phenomena in nature. Various types of plants (e.g., mimosa and pinecones) respond to water stimuli. These water-responsive plants (also known as tidal grasses) attribute water-responsiveness and shape-altering properties to their bilayer structure, with each layer responding differently to water stimuli, resulting in anisotropic deformation of the bilayer structure.

Ge et al. [108] reported a photonic film based on cellulose nanocrystals (CNCS), in which poly (ethylene glycol) dimethacrylate (PEGDMA) converges into the chiral nematic structure of CNC through UV-triggered free radical polymerization in a N, N-dimethylformamide solvent system. When the vertically fixed-size photonic film is exposed to moisture on its left side for 4 s, the vertical bending angle of the composite film increases to 135 °, and the bending rate is 4.6 $°s^{-1}$, as shown in Figure 12a. Then, the moisture was removed and the film returned to its original vertical position for about 10 s. They further studied the reversibility of the film's actuation and found that it exhibited excellent stability and reversibility after alternating exposure to moisture on both sides of the film for 20 cycles. Matthew et al. [109] used the humidity-driven volume change in carboxymethyl cellulose (CMC) to manufacture the humidity response actuator on the glass fiber substrate. As shown in Figure 12b, CMC-coated fiber showed a linear dependence on humidity in the range of 5–40% relative humidity, and the response time was 1 min. Cellulose nanofibers (CNFs) are obtained through the mechanical fibrillation of fibers. Due to their high aspect ratio, they facilitate the formation of intercalation layers of water molecules in the fiber network. Therefore, using CNFs as a substrate can significantly enhance their actuation performance. As shown in Figure 12c, Lisa et al. [38] prepared thin films based on cellulose nanofibers (CNFs), which undergo varying degrees of distortion and expansion when in contact with water or organic solvents. Kuang et al. [110] designed a substrate micropattern actuator based on selectively arranged cellulose nanofibers (CNFs). Due to the presence of more intermolecular hydrogen bonds (H bonds) between CNFs, reversible fracture and recombination can occur under the erosion of water molecules, resulting in good controllability, fast response time (less than 1 s), and curl displacement actuation during dehydration, as shown in Figure 12d.

Now, with the in-depth research of transparent materials, and due to their unique transparency and invisibility, they have been widely used in stealth robots, biomimetic skin, and other fields. Transparent actuators have also been extensively studied. Li et al. [111] reported a transparent cellulose-based actuator driven by humidity and infrared (IR) light, which is made of two layers of regenerated cellulose film (RCF) and hydrophobic polytetrafluoroethylene (PTFE) film, as shown in Figure 12e. The principle of its actuation is that RCF is hydrophilic (due to the presence of abundant hydroxyl groups), while PTFE film is hydrophobic. This difference leads to the asymmetry of the bilayer membrane structure, thereby obtaining internal stress that responds to the changes in environmental humidity. When the relative humidity decreases to 29%, the RCF/PTFE actuator bends towards the side of the RCF film, with a maximum bending curvature of 2.5 cm. Then, the rate of curvature increase slows down. When the relative humidity increases to 50%, the actuator returns to its original flat state. On the contrary, when the relative humidity increased to 54%, the RCF/PTFE actuator showed significant bending towards the PTFE film side, as shown in Figure 12f.

Combining high-performance materials with traditional environmentally friendly cellulose-based materials is a universal strategy for developing intelligent, green materials and flexible equipment. Transition metal carbides including carbonitrides and nitrides (MXene) are two-dimensional layered materials with excellent photothermal conversion capabilities and rich surface-chemical properties. As shown in Figure 13a, Cao et al. [112] reported a two-dimensional MXenes/cellulose nanofiber (CNF)/biomacromolecule soft actuator using a bio-inspired mesoscale assembly strategy that combines nano-, micro-, and macro-length scales. When humidity increased, the actuator exhibited a direct rapid actuation rate of 34.2 $^{\circ}s^{-1}$. Under infrared light (IR) heating, the actuator's hygroscopicity and actuation performance were significantly improved. Inspired by the structure of the pearl layer of brick mortar, Wei et al. [113] prepared high-performance carboxymethyl cellulose-based actuators. MXene nanosheets and Al^{3+} were introduced into the CMC matrix through H and ion bonds, forming a multi-level pearl layer structure. The good wet strength in a high humidity environment is 2.97 MPa, with impressive flexibility, as shown in Figure 13b. More interestingly, due to the sensitive hygroscopicity of CMC molecules and MXene nanosheets, as shown in Figure 13c, the composite film can also exhibit a large bending angle (>180°), rapid response/recovery (2.3/3.0 s), and excellent cycling stability over 1500 cycles in response to humidity gradients, and simultaneously with reversible deformation. Yang et al. [114] prepared pearl-layered and layered composite films composed of MXene modified with polydopamine (PDA), and bacterial cellulose nanofibers, as shown in Figure 13d. The actuator has high conductivity (2848 S cm^{-1}), excellent tensile strength (406 MPa), and toughness (15.3 MJ m^{-3}). In addition, the actuator is highly sensitive to moisture (30% humidity), has a 'fast response (1.6 s), large deformation (176 °), and high driving power output (6.5 N m^{-2}), as shown in Figure 13e,f. Li et al. [115] prepared MXene/cellulose/polystyrene sulfonic acid (PSSA) composite membranes (MCPM), as shown in Figure 13g, which are highly sensitive to moisture and roll up immediately upon contact with the skin. When the relative humidity increases from 20% to 97%, the maximum bending angle of the thick MCPM actuator has increased from 28° to 130°, providing high sensitivity for humidity sensing, as shown in Figure 13h.

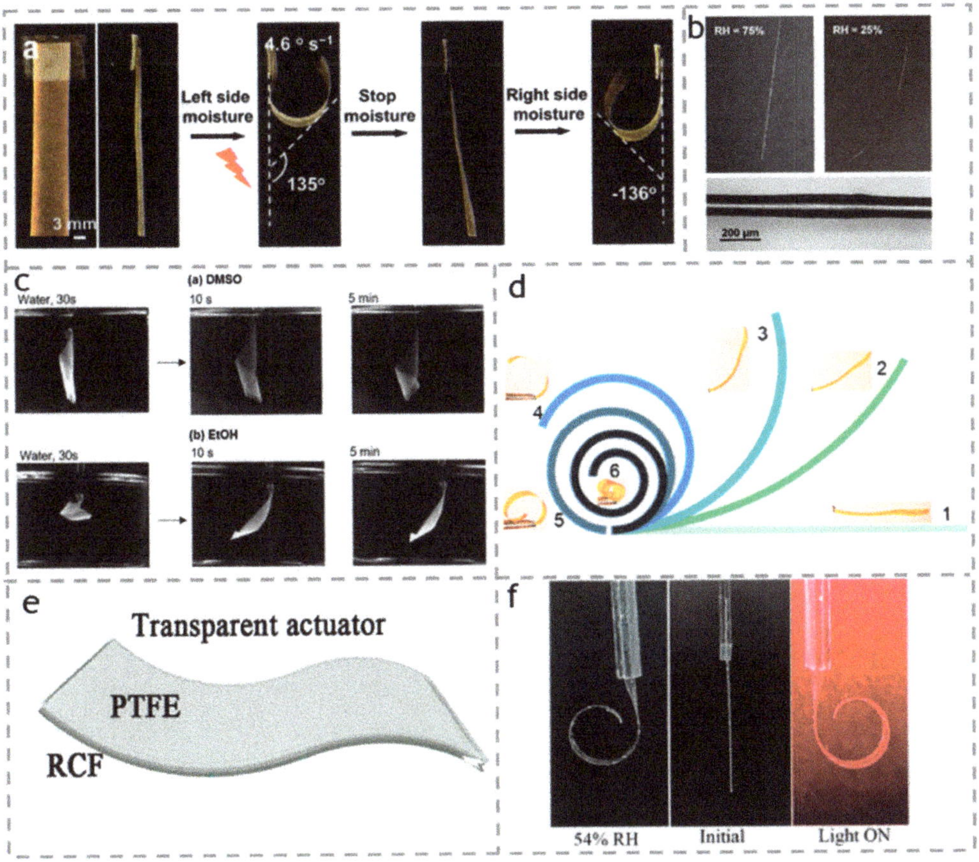

Figure 12. (**a**) A red rectangular CNC-PP film could rapidly carry out bidirectional moisture actuation at a speed of 4.6° s^{-1}. Reproduced with permission [108]. Copyright 2022. WILEY-V C H VERLAG GMBH. (**b**) Response of the CMC-fiber to humidity. Reproduced with permission [109]. Copyright 2018. ELSEVIER. (**c**) Photographs showing the shape recovery test of CNF/CMCNF9 bilayer film in DMSO and EtOH. Reproduced with permission [38]. Copyright 2030. SPRINGER. (**d**) Curling displacement of the CNF-soft actuator during the dehydration process; insets are the corresponding snapshots at different states. The numbers 1–5 indicated the different bending states of the samples. Reproduced with permission [110]. Copyright 2019. ELSEVIER. (**e**) Schematic of the RCF/PTFE composite film. (**f**) Ptical photos of the RCF/PTFE actuator at a low RH of 29% (**left panel**), room RH of 50% (**middle panel**), and high RH of 54% (**right panel**). Reproduced with permission [111]. Copyright 2021. ELSEVIER.

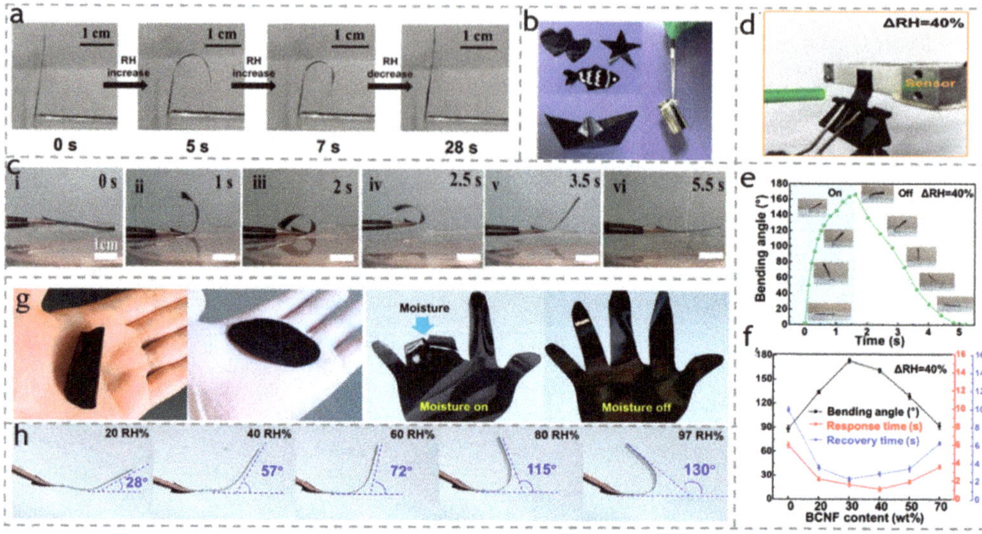

Figure 13. (**a**) Optical images of reversible hygroscopic bending actuation of the G-MXCP actuator under a relative humidity (RH) change from 40% to 90% at room temperature. Reproduced with permission [112]. Copyright 2020. AMER CHEMICAL SOC. (**b**) Digital pictures of CMC/MXene/Al^{3+} composite film, showing the excellent flexibility and mechanical strength. (**c**) The reversible deformation of CMC/MXene/Al^{3+} composite film upon exposure to humidity gradients. The sequential numbers i–vi indicated the sample with the different bending states. Reproduced with permission [113]. Copyright 2022. ELSEVIER. (**d**) Photographs of hygroscopic actuation force test of PDMM/BCNF film. (**e**) Bending angle–time curve of the PDMM/BCNF film (30 wt% BCNF) during a cyclic motion. (**f**) Bending angle and response/recovery time of PDMM/BCNF film with different BCNF content. Reproduced with permission [114]. Copyright 2021. ELSEVIER. (**g**) Optical images of an MCPM on a human hand with or without wearing a dry glove. (**h**) The maximum bending angles of an MCPM/PET actuator under various RH. Reproduced with permission [115]. Copyright 2021. American Chemical Society.

9. Application of Cellulose-Based Actuators

From the previous chapter, it can be observed that the driving forces of the cellulose-based actuators are diverse, and can even achieve multi stimulus responses. The combination of these characteristics makes the actuator have a wide range of application potential in the field of intelligence. In this chapter, the applications of four types of cellulose actuators are presented including biomimetic robots, flexible electronic devices, self-powered intelligent switches, and precise remote control.

9.1. Biomimetic Robots

Currently, the usage scenarios of robots are gradually tending towards precision, narrowness, and complexity. The demand for this task scenario will inevitably accelerate the transformation of biomimetic robots towards miniaturization. Through highly integrated components such as actuators and sensors, their overall miniaturization can be achieved. At the same time, the profiling of biomimetic robots has also become the main development trend.

Deforming the actuator and lifting heavy objects is currently the most widely studied and applied method, demonstrating excellent mechanical robustness. As shown in Figure 14a, Mu et al. [116] prepared multifunctional soft electromagnetic actuators (copper wire coils and central circular magnets) that can contract, expand, jump, and move. As shown in Figure 14b, Wei et al. [117] prepared CNF/GO/CNT composite membrane actuators (2 cm × 3 cm × 12 μm) containing 50% carbon nanotubes. Used as a crawling robot, it can gradually crawl forward on rough substrates in response to periodic water vapor changes. Interestingly, crawling robots can move heavy objects forward at an average speed of 2.6 mm/s through alternating humidity switches, which is likely to be used for crawling robots or humidity control switches. As shown in Figure 14c, the actuator has a good contact with the cylinder and repeatedly lifts a weight of 2 g under different currents and strains, perfectly demonstrating its operating ability under irregular geometric shapes. It can not only be used as an artificial muscle to make 3D printed polylactic acid (PLA) arms move normally, but can also lift a weight of 2 g weight in its expanded state. According to its two states (expansion and contraction), it can jump in the air at a height several times its own height. Finally, they use it as an attachment to the leg, where the actuator can crawl along the structured surface, crawling like a human leg. Kuang et al. [110] designed a micro-patterned substrate actuator based on selectively arranged cellulose nanofibers (CNFs). They made it into a 7 cm soft mechanical arm. After heating, the actuator strip quickly curled. With the length contraction, the attached object was lifted up. When the humidity in the environment was increased, the actuator strip quickly unfolded, leading to the release of suspended weights.

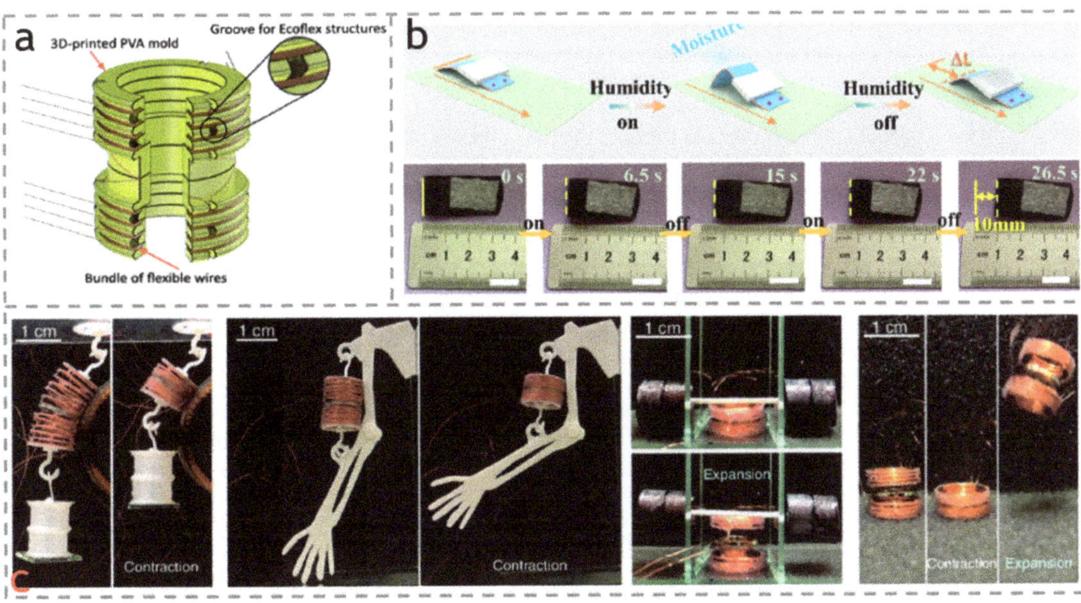

Figure 14. (**a**) Schematic of the soft actuator fabrication. Reproduced with permission [116]. Copyright 2023. WILEY. (**b**) Load-bearing smart robot crawling forward on a rough substrate under periodic humidity on and off. Reproduced with permission [117]. Copyright 2021. American Chemical Society. (**c**) Relaxed and excited states of a 12-ring actuator when lifting a 2 g weight while conforming to a cylindrically shaped object. Relaxed and excited states of a 12-ring actuator when used as an artificial muscle. Relaxed and excited states of an 8-ring actuator when lifting a 1.6 g weight by expansion. Hopping mode of actuation for an 8-ring actuator by contraction and subsequent expansion. Reproduced with permission [116]. Copyright 2023. WILEY.

Many researchers have been inspired by nature to create petal-shaped actuators that simulate the opening and closing of petals. Ge et al. [108] prepared flower-shaped actuators based on cellulose nanocrystals. As shown in Figure 15a, under the stimulation of water, the petals not only produce bending responses but also undergo corresponding color changes. As shown in Figure 15b, Yang et al. [118] used a GO-BC/PET/PEDOT: PSS actuator as the stem of a sunflower. After 15 s of near-infrared light exposure, the temperature rapidly increased, causing the biomimetic sunflower to bend towards the light. More interestingly, Wang et al. [119] simulated the opening and closing process of narcissus petals. As shown in Figure 15c, they simulated the opening and closing process of the petals under a sinusoidal voltage of 0 V and 1.8 Hz, and changed from a closed state to open state in 2 s, based on the fast electric response ionic actuator of the functional cellulose carboxylate nanofiber (CCNF) doped ionic liquid (IL), and Graphene nanosheet (GN). When the input voltage stops, the bionic flower can recover its state of nature in a short time. Cai et al. [120] innovatively crafted a composite material, consisting of MXene (Ti_3C_2Tx) and cellulose, referred to as MXCC. Alongside a polycarbonate membrane, the composite replicates multiple essential characteristics of natural leaves, ranging from microstructural components to their photosynthetic abilities, encompassing energy harvesting and conversion. In addition, Chen et al. [44] prepared bamboo/PNIPAM composite hydrogel actuators. As shown in Figure 15d, they used cutouts of different thicknesses and sizes of actuators as the petals of a lotus flower. As the temperature decreases, the lotus flower continuously blooms from the outside to the inside.

The biomimetic actuators discussed above have an inherent drawback of being opaque, limiting their use in discreet tasks. This has led to a growing interest among researchers to develop transparent actuators that offer high concealment. Such advancements in technology would enable the design of biomimetic animals, windows, and robots that can execute covert operations without being easily detected by humans. Rühlicke et al. [121] devised a cutting-edge cellulose blend with distinctive capabilities in specific environments, leading to the development of a clear, thin film actuator with the ability to self-heal and undergo shape deformation. On another note, as shown in Figure 15e, Li et al. [122] successfully created a cellulose composite thin film that is both visible and transparent, while also displaying a remarkable capacity for absorbing ultraviolet light. The actuator designed by Li et al. was modeled after the arm of an excavator and proved to be highly efficient, as it was able to lift an object 17 times its weight swiftly when exposed to infrared light. Upon turning off the light source, the object fell rapidly.

9.2. Flexible Electronic Equipment

Flexible electronic devices with high mechanical performance have been widely used in a large number of wearable/implantable devices, greatly enriching the collection of health data and disease detections. However, due to the monopoly of its core technology and the expensive raw materials, a main direction to search for simple, cost-effective and biodegradable technique is highly desirable.

Skin conductivity is a key feature in the design of electronic watches. By monitoring the pulse and other physiological indicators of the skin, these smart watches can provide accurate health data, helping users better to understand their physical conditions. As shown in Figure 16a, Lian et al. [123] have developed a cellulose-based ion-conductive hydrogel actuator (ICHs) using a simple one-pot method. This actuator is capable of accurately detecting the different pressure signals when humans write different numbers. Surprisingly, apart from monitoring complex finger movements, the actuator can also monitor complex muscle movements. In an experiment, participants were asked to say "apple" while drinking water, and the actuator detected different resistance signals by sensing the weak acoustic vibrations and water vibrations. This demonstrates the reliability of cellulose hydrogel actuators in detecting both sound signals and actions. Similarly, in order to improve the high sensitivity of the flexible pressure sensor, as shown in Figure 16b, Chen et al. [124] designed a cellulose ion-conductive hydrogel (ICH) using only cellulose as the sole material. They simulated the scenario of patient infusion and placed the sensor at different positions to monitor the patient's health status by detecting physiological signals such as pulse, respiration, and infusion.

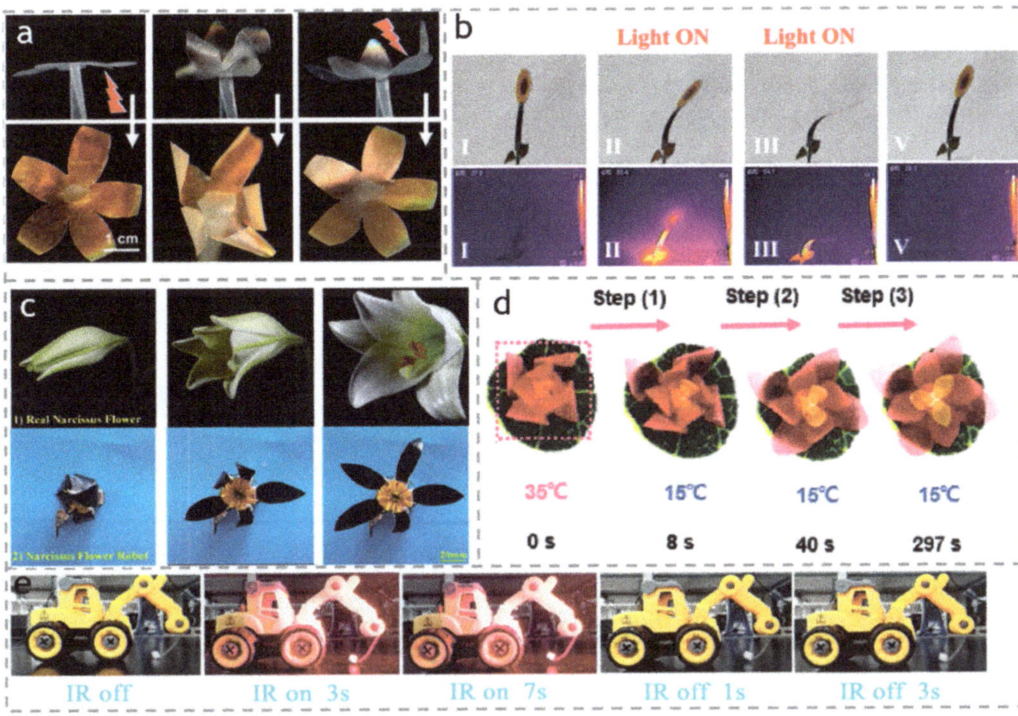

Figure 15. (**a**) Closing and opening on an artificial "flower" by moisture stimulation. Reproduced with permission [108]. Copyright 2022. WILEY-VCHVERLAG GMBH. (**b**) Optical photos of infrared thermal images. The sub-images I–III, V indicated the different states of the samples under the light on and off. Reproduced with permission [118]. Copyright 2022. Royal Society of Chemistry. (**c**) Petal opening of a bionic flower. Petal opening of a bionic flower. Reproduced with permission [119]. Copyright 2023. ELSEVIER. (**d**) The blooming of the lotus flower of ABH actuator. The red dots indicated the flower sample. The arrows indicated the blooming process. Reproduced with permission [44]. Copyright 2021, AMER CHEMICAL SOC. (**e**) Series of optical photographs demonstrating the arm of an excavator-shape actuator responding to IR light. Reproduced with permission [122]. Copyright 2023. ELSEVIER.

The above actuators did not consider the water resistance of the sensors. Yun et al. [125] developed a superhydrophobic and highly sensitive cellulose paper-based actuator (PB) that can quickly detect and perceive the bending of a finger. Impressively, the continuous dripping of water does not affect the sensing performance, which is advantageous for applications in wet or rainy conditions, as shown in Figure 16c.

Electrospinning is a simple and efficient fiber preparation method [126,127], and has various applications in different fields [128–130]. Electrospinning technology can integrate actuation and sensing functions into a functional material. It is understood that electrospun fibrous membrane has the characteristics of a large surface area, multi-level porous structure, strong flexibility, and easy shape control [131–134]. Therefore, Bai et al. [73] used electrospun cellulose acetate (CA)/carbon nanotube nanofiber composite membrane to prepare a friction generator with a polyvinylidene fluoride nanofiber membrane (PVDF). This membrane can detect the human movement of fingers, wrists, elbows, and other joints. In addition, when it is installed on the sole of the shoe, people can distinguish the walking state (slow walking and running) by the change of voltage and frequency.

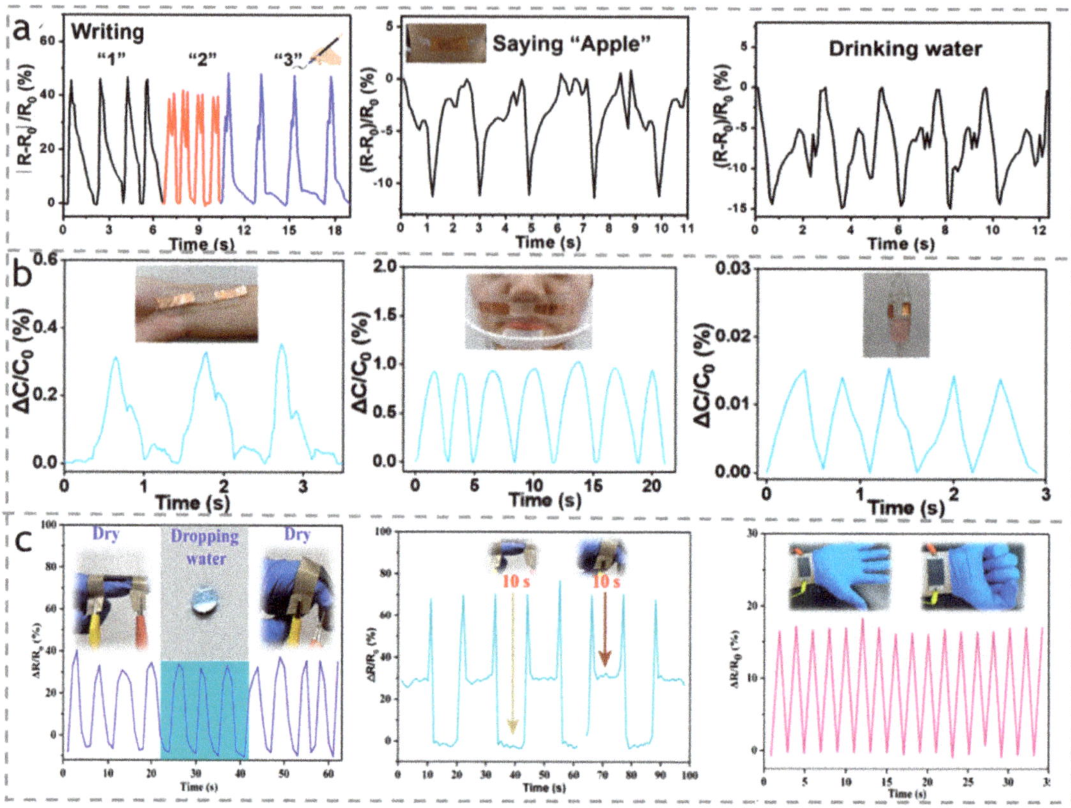

Figure 16. (**a**) The relative resistance changes by writing a different number, saying "apple", and drinking water. Reproduced with permission [123]. Copyright 2023. ELSEVIER. (**b**) The corresponding pulse signals, breathing in relaxation, and the drip pot's little vibration when liquid drips into it. Reproduced with permission [124]. Copyright 2023. ROYAL SOC CHEMISTRY. (**c**) The resistance output changes in the state of finger movement. Insets: the PB strain sensor attached to the finger was repeatedly bent under normal conditions and in the presence of water droplets. The resistance output changes of periodically bending the index finger at a fixed angle, with fist opening and closing. Reproduced with permission [125]. Copyright 2023. ELSEVIER.

9.3. Accurate Remote Control

With the rapid development of modern society's technology, mechanical equipment plays an important role in the development of society's technology. However, most construction machinery requires manual operation by skilled workers. Many workplaces are mostly high-risk or toxic environments, and the remote control of mechanical equipment is particularly important.

The bamboo/PNIPAM composite hydrogel actuator prepared by Chen et al. [84,85] is designed as a manipulator with three claws for fixed-point control, as shown in Figure 17a. It can lift objects that are dozens of times heavier than the actuator under the control of near-infrared light. They place the manipulator in water, irradiate the right paw, left paw, and rear paw with infrared light in sequence. The paws shrink in sequence, and start to grasp the heavy objects. After turning off the light source, the paws return to its original shape; as a result, the heavy object falls down, which perfectly demonstrates the precise remote control. Similarly, they also made a wood/PNIPAM composite hydrogel actuator [84,85]. Through a similar operation, it was also proved that the actuator can easily grasp and place objects according to external changes. Wei et al. [117] prepared CNF/GO/CNT composite membrane actuators containing 50% carbon nanotubes. When the actuator was connected to a plastic straw, the resultant simple gripper can perform tasks such as grasping, lifting, transporting, and releasing according to the changes in humidity, as shown in Figure 17b. The plastic foam can be even be bent and grasped by applying water vapor through the plastic straw. Then, the holder can lift the foam (about 0.12 g, 9 times of the actuator) and carry 82 mm horizontally under the stimulation of continuous water vapor. Finally, when the water vapor is removed, the gripper can quickly release foam. Such an actuator is expected to be applied to mechanical grippers due to its large deformation, fast response time, and good cycle stability.

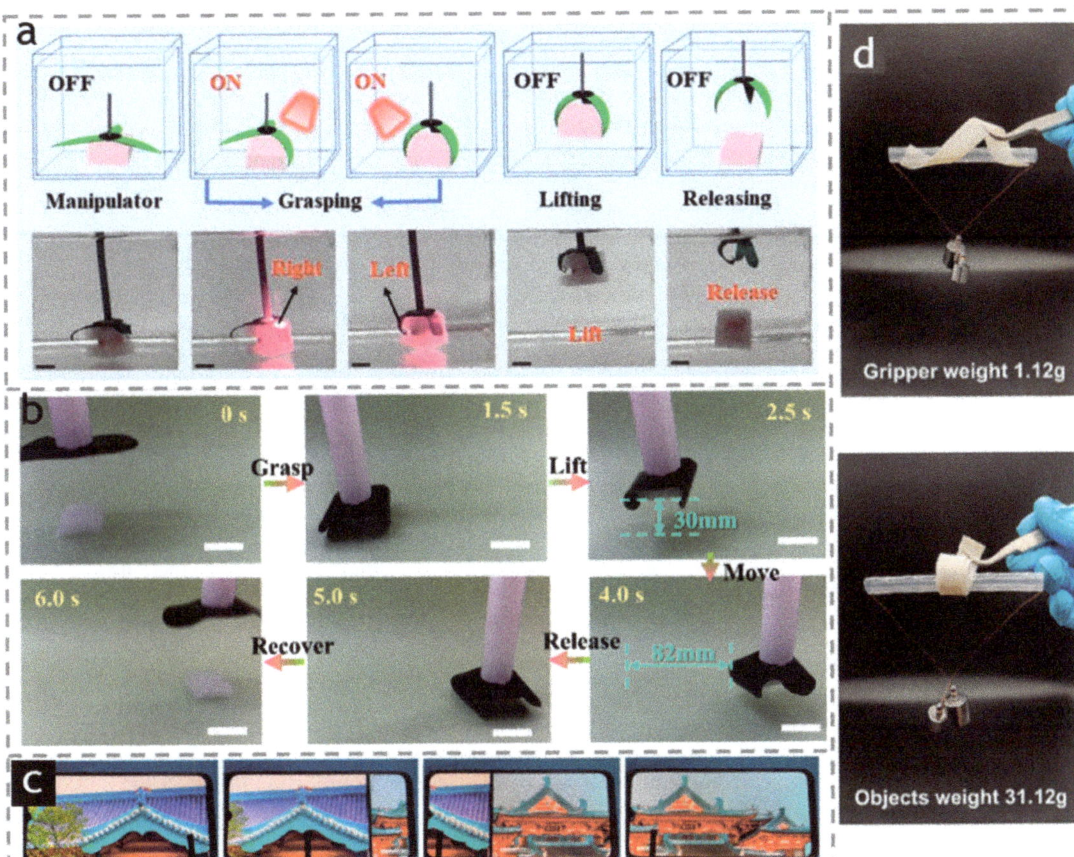

Figure 17. (**a**) Demonstration of precise control performance of bamboo/PNIPAM composite hydrogel actuator-manipulator grasping object. Reproduced with permission [84,85]. Copyright 2022. ELSEVIER. (**b**) Schematic illustration and photographs of the mechanical gripper for transferring a plastic foam. Reproduced with permission [84,85]. Copyright 2022. ELSEVIER. (**c**) A bionic finger sliding photos on the electronic screen. Reproduced with permission [135]. Copyright 2023. SPRINGER. (**d**) Soft grippers withstanding 30 times their weight. Reproduced with permission [136]. Copyright 2023. ELSEVIER.

In addition to accurately grasping heavy objects, people have even studied using actuators as human fingers to control smartphones. Wang et al. [135] used CCNF-IL-GN actuators to simulate human fingers sliding on smartphones and switching photos. When a

4 V DC voltage is applied, the bionic fingers move on the track to touch the screen, as shown in Figure 17c, causing them to slide through albums. In addition, the bionic fingers can also accurately control the opening or closing status of flashlights. This high-performance bionic actuator can be used in the field of artificial muscle.

Shape memory polymers (SMPs) have found extensive application in the field of 3D printing. Furthermore, with advancements in intelligent deformations, 4D printing has begun to mature. Gu et al. [136] performed an experiment to showcase soft grippers that were inspired by plants and achieved through 4D printing with different filling angles. They utilized a straightforward fabrication approach to create thermo-responsive shape memory polymer composites by reinforcing poly ε-caprolactone (PCL)/polybutylene adipate-co-terephthalate (PBAT) with cellulose nanofibers (CNF). By subjecting the samples to thermal stimulation at a temperature of 3 °C, they observed a rapid deformation and wrapping around objects within 60 s. These spiral and bending deformations enabled the grippers to hold objects weighing up to 30 times their own weight, as shown in Figure 17d.

10. Conclusions and Future Prospects

With the aggravation of the energy crisis and environmental degradation, the development and utilization of renewable biomass resources has become the main content of research worldwide [137–141]. The development and utilization of cellulose, the most abundant biomass resource in nature, in order to build new materials, conforms to the future development trend. This article reviews the latest progress and achievements of cellulose-based actuators, discussing them from three aspects: actuation mechanism, driving force, and applications. Due to the significant advantages of cellulose itself, temperature-sensitive materials, conductive materials, and magnetic materials are often introduced to endow cellulose-based materials with different functions, to promote the speed and stability of their response to external stimuli. This has wide applications in fields such as biomimetic robots, flexible electronic wearable devices, and remote control.

There has been significant progress in cellulose-based actuators, but there are still some challenges. Firstly, cellulose fiber, as a textile fiber, has long permeated people's daily lives. However, fibers produced by the adhesive method still dominate the market, requiring the use of various toxic and harmful substances in the production process, causing great environmental pollution. The further development and improvement of production processes of cellulose fibers is highly required.

Secondly, the lifespan of actuators that are produced through cellulose degradation is relatively short. Moreover, biomimetic materials are frequently derived from materials that are sensitive to environmental variations. While this characteristic may prove useful, it also limits their applicability in certain settings. Hydrogels are a notable example, as they necessitate an aqueous environment and cannot be utilized in dry applications. Presently produced actuators do not possess the ability to adapt to intricate environments. Consequently, it is recommended to conduct further research into their multifunctional capabilities, such as luminescence, anti-freezing properties, signal sensing, and adsorption.

Finally, although functional materials based on cellulose have shown promising practical applications, most of them are still in the stage of basic research and are limited in size, which hinders their practical applications on a large scale. Further research and development is required to enhance the experimental equipment, broaden the scope of applications, and promote practical utilization, thereby creating substantial economic and social benefits.

Despite the aforementioned limitations, cellulose-based actuators present several advantages such as low production cost, utilization of eco-friendly raw materials, and versatility in various fields such as bionic robots, wearable electronic devices, and remote-controlled gripping. Consequently, with the swift advancement of technology, the research on cellulose-based actuators is steadily progressing towards maturity, and their incorporation into flexible electronic devices will undoubtedly take center stage in the forthcoming years.

Author Contributions: Conceptualization, Z.W., G.D. and S.J.; methodology, S.C. and C.Z.; formal analysis, S.C. and C.Z.; investigation, S.C. and C.Z.; data curation, S.C. and C.Z.; writing—original draft preparation, S.C. and C.Z.; writing—review and editing, Z.W., G.D. and S.J.; supervision, Z.W., G.D. and S.J. All authors have read and agreed to the published version of the manuscript.

Funding: This research received no external funding.

Institutional Review Board Statement: Not applicable.

Informed Consent Statement: Not applicable.

Data Availability Statement: The data presented in this study are available on request from the corresponding author. The data are not publicly available due to the demand from our further research.

Conflicts of Interest: The authors declare no conflict of interest.

References

1. Wu, Y.; Li, Y.; Tao, Y.; Sun, L.; Yu, C. Recent advances in the material design for intelligent wearable devices. *Mater. Chem. Front.* **2023**, *7*, 3278–3297. [CrossRef]
2. Wang, J.; Li, S.; Gao, D.; Xiong, J.; Lee, P.S. Reconfigurable and programmable origami dielectric elastomer actuators with 3D shape morphing and emissive architectures. *NPG Asia Mater.* **2019**, *11*, 71. [CrossRef]
3. Wang, F.; Wang, L.; Wang, Y.; Wang, D. Highly bendable ionic electroactive polymer actuator based on carboxylated bacterial cellulose by doping with MWCNT. *Appl. Phys. A* **2022**, *128*, 911. [CrossRef]
4. Nevstrueva, D.; Murashko, K.; Vunder, V.; Aabloo, A.; Pihlajamäki, A.; Mänttäri, M.; Pyrhönen, J.; Koiranen, T.; Torop, J. Natural cellulose ionogels for soft artificial muscles. *Colloids Surf. B Biointerfaces* **2018**, *161*, 244–251. [CrossRef] [PubMed]
5. Chen, W.; Sun, B.; Biehl, P.; Zhang, K. Cellulose-Based Soft Actuators. *Macromol. Mater. Eng.* **2022**, *307*, 2200072. [CrossRef]
6. Constantinos, M.; Pfeiffer, C.; Mosley, M. Conventional actuators, shape memory alloys and electrorheological fluids. *Autom. Miniat. Robot. Sens. Non-Destr. Test. Eval.* **1999**, *1*, 26.
7. Al-Khalili, J. The birth of the electric machines: A commentary on Faraday (1832) 'Experimental researches in electricity'. *Philos. Trans. R. Soc. A Math. Phys. Eng. Sci.* **2015**, *373*, 20140208. [CrossRef]
8. Yu, Y.; Wang, J.; Han, X.; Yang, S.; An, G.; Lu, C. Fiber-Shaped Soft Actuators: Fabrication, Actuation Mechanism and Application. *Adv. Fiber Mater.* **2023**, *5*, 868–895. [CrossRef]
9. Kim, H.; Lee, J.A.; Ambulo, C.P.; Lee, H.B.; Kim, S.H.; Naik, V.V.; Haines, C.S.; Aliev, A.E.; Ovalle-Robles, R.; Baughman, R.H.; et al. Intelligently Actuating Liquid Crystal Elastomer-Carbon Nanotube Composites. *Adv. Funct. Mater.* **2019**, *29*, 1905063. [CrossRef]
10. Hwang, I.; Kim, H.J.; Mun, S.; Yun, S.; Kang, T.J. A Light-Driven Vibrotactile Actuator with a Polymer Bimorph Film for Localized Haptic Rendering. *ACS Appl. Mater. Interfaces* **2021**, *13*, 6597–6605. [CrossRef]
11. Xiong, R.; Sauvage, F.; Fraire, J.C.; Huang, C.; De Smedt, S.C.; Braeckmans, K. Photothermal Nanomaterial-Mediated Photoporation. *Acc. Chem. Res.* **2023**, *56*, 631–643. [CrossRef] [PubMed]
12. Chen, L.; Weng, M.; Huang, F.; Zhang, W. Light- and humidity-driven actuators with programmable complex shape-deformations. *Sens. Actuators B Chem.* **2019**, *282*, 384–390. [CrossRef]
13. Bhatti, M.R.A.; Bilotti, E.; Zhang, H.; Varghese, S.; Verpaalen, R.C.P.; Schenning, A.; Bastiaansen, C.W.M.; Peijs, T. Ultra-High Actuation Stress Polymer Actuators as Light-Driven Artificial Muscles. *ACS Appl. Mater. Interfaces* **2020**, *12*, 33210–33218. [CrossRef] [PubMed]
14. Han, D.D.; Zhang, Y.L.; Ma, J.N.; Liu, Y.Q.; Han, B.; Sun, H.B. Light-Mediated Manufacture and Manipulation of Actuators. *Adv. Mater.* **2016**, *28*, 8328–8343. [CrossRef] [PubMed]
15. Quashie, D., Jr.; Benhal, P.; Chen, Z.; Wang, Z.; Mu, X.; Song, X.; Jiang, T.; Zhong, Y.; Cheang, U.K.; Ali, J. Magnetic bio-hybrid micro actuators. *Nanoscale* **2022**, *14*, 4364–4379. [CrossRef] [PubMed]
16. Xu, H.; Xu, X.; Xu, J.; Dai, S.; Dong, X.; Han, F.; Yuan, N.; Ding, J. An ultra-large deformation bidirectional actuator based on a carbon nanotube/PDMS composite and a chitosan film. *J. Mater. Chem. B* **2019**, *7*, 7558–7565. [CrossRef]
17. Miriyev, A.; Stack, K.; Lipson, H. Soft material for soft actuators. *Nat Commun* **2017**, *8*, 596. [CrossRef]
18. Yang, H.; Leow, W.R.; Wang, T.; Wang, J.; Yu, J.; He, K.; Qi, D.; Wan, C.; Chen, X. 3D Printed Photoresponsive Devices Based on Shape Memory Composites. *Adv. Mater.* **2017**, *29*, 1701627. [CrossRef]
19. Zhao, P.; Xu, B.; Zhang, Y.; Li, B.; Chen, H. Study on the Twisted and Coiled Polymer Actuator with Strain Self-Sensing Ability. *ACS Appl. Mater. Interfaces* **2020**, *12*, 15716–15725. [CrossRef]
20. Liu, W.; Cheng, Y.; Liu, N.; Yue, Y.; Lei, D.; Su, T.; Zhu, M.; Zhang, Z.; Zeng, W.; Guo, H.; et al. Bionic MXene actuator with multiresponsive modes. *Chem. Eng. J.* **2021**, *417*, 129288. [CrossRef]
21. Melling, D.; Martinez, J.G.; Jager, E.W.H. Conjugated Polymer Actuators and Devices: Progress and Opportunities. *Adv. Mater.* **2019**, *31*, 1808210. [CrossRef]

22. Wehner, M.; Truby, R.L.; Fitzgerald, D.J.; Mosadegh, B.; Whitesides, G.M.; Lewis, J.A.; Wood, R.J. An integrated design and fabrication strategy for entirely soft, autonomous robots. *Nature* **2016**, *536*, 451–455. [CrossRef] [PubMed]
23. Basu, S.; Omadjela, O.; Gaddes, D.; Tadigadapa, S.; Zimmer, J.; Catchmark, J.M. Cellulose Microfibril Formation by Surface-Tethered Cellulose Synthase Enzymes. *ACS Nano* **2016**, *10*, 1896–1907. [CrossRef] [PubMed]
24. Sinha, A.K.; Bhattacharya, S.; Narang, H.K. Abaca fibre reinforced polymer composites: A review. *J. Mater. Sci.* **2020**, *56*, 4569–4587. [CrossRef]
25. Turner, S.; Kumar, M. Cellulose synthase complex organization and cellulose microfibril structure. *Philos. Trans. R. Soc. A Math. Phys. Eng. Sci.* **2018**, *376*, 20170048. [CrossRef]
26. Che, X.; Wu, M.; Yu, G.; Liu, C.; Xu, H.; Li, B.; Li, C. Bio-inspired water resistant and fast multi-responsive Janus actuator assembled by cellulose nanopaper and graphene with lignin adhesion. *Chem. Eng. J.* **2022**, *433*, 133672. [CrossRef]
27. Zhang, D.; Jian, J.; Xie, Y.; Gao, S.; Ling, Z.; Lai, C.; Wang, J.; Wang, C.; Chu, F.; Dumont, M.-J. Mimicking skin cellulose hydrogels for sensor applications. *Chem. Eng. J.* **2022**, *427*, 130921. [CrossRef]
28. Yang, B.; Bi, W.; Zhong, C.a.; Huang, M.; Ni, Y.; He, L.; Wu, C. Moisture-triggered actuator and detector with high-performance: Interface engineering of graphene oxide/ethyl cellulose. *Sci. China Mater.* **2018**, *61*, 1291–1296. [CrossRef]
29. Lin, F.; Wang, Z.; Shen, Y.; Tang, L.; Zhang, P.; Wang, Y.; Chen, Y.; Huang, B.; Lu, B. Natural skin-inspired versatile cellulose biomimetic hydrogels. *J. Mater. Chem. A* **2019**, *7*, 26442–26455. [CrossRef]
30. Khan, A.; Khan, F.R.; Kim, H.S. Electro-Active Paper as a Flexible Mechanical Sensor, Actuator and Energy Harvesting Transducer: A Review. *Sensors* **2018**, *18*, 3474. [CrossRef]
31. Lee, S.-W.; Kim, J.-H.; Kim, J.; Kim, H.S. Characterization and sensor application of cellulose electro-active paper (EAPap). *Chin. Sci. Bull.* **2009**, *54*, 2703–2707. [CrossRef]
32. Yang, H.; Wu, F.; Zhu, G.; Li, H.; Jiang, S. Recent progress of modification and industrialization for nanocellulose towards green building materials. *J. For. Eng.* **2023**, *8*, 11–20. [CrossRef]
33. Liang, X.; Chen, G.; Lin, S.; Zhang, J.; Wang, L.; Zhang, P.; Lan, Y.; Liu, J. Bioinspired 2D Isotropically Fatigue-Resistant Hydrogels. *Adv. Mater.* **2022**, *34*, 2107106. [CrossRef] [PubMed]
34. Sano, K.; Ishida, Y.; Aida, T. Synthesis of Anisotropic Hydrogels and Their Applications. *Angew. Chem. Int. Ed.* **2018**, *57*, 2532–2543. [CrossRef]
35. Roopsung, N.; Sugawara, A.; Hsu, Y.I.; Asoh, T.A.; Uyama, H. Cellulose Nanocrystal-Based Gradient Hydrogel Actuators with Controllable Bending Properties. *Macromol. Rapid Commun.* **2023**, *16*, 2300205. [CrossRef]
36. Schafer, J.L.; Meckel, T.; Poppinga, S.; Biesalski, M. Chemical Gradients in Polymer-Modified Paper Sheets-Towards Single-Layer Biomimetic Soft Robots. *Biomimetics* **2023**, *8*, 43. [CrossRef]
37. Wang, X.; Han, B.; Yu, R.-P.; Li, F.-C.; Zhao, Z.-Y.; Zhang, Q.-C.; Lu, T.J. Magnetic-responsive Fe3O4 nanoparticle-impregnated cellulose paper actuators. *Extrem. Mech. Lett.* **2018**, *25*, 53–89. [CrossRef]
38. Lopes da Costa, L.; Moreau, C.; Lourdin, D.; Cathala, B.; Villares, A. Shape-recovery in organic solvents of water-responsive cellulose nanofiber actuators. *Cellulose* **2023**, *30*, 5811–5824. [CrossRef]
39. Maki, Y. Poly(N,N-dimethylacrylamide)-clay nanocomposite hydrogels with patterned mechanical properties. *Colloid Polym. Sci.* **2019**, *297*, 587–594. [CrossRef]
40. Zhang, E.; Wang, T.; Hong, W.; Sun, W.; Liu, X.; Tong, Z. Infrared-driving actuation based on bilayer graphene oxide-poly(N-isopropylacrylamide) nanocomposite hydrogels. *J. Mater. Chem. A* **2014**, *2*, 15633–15639. [CrossRef]
41. Liu, L.; Jiang, S.; Sun, Y.; Agarwal, S. Giving Direction to Motion and Surface with Ultra-Fast Speed Using Oriented Hydrogel Fibers. *Adv. Funct. Mater.* **2015**, *26*, 1021–1027. [CrossRef]
42. Ma, C.; Li, T.; Zhao, Q.; Yang, X.; Wu, J.; Luo, Y.; Xie, T. Supramolecular Lego Assembly Towards Three-Dimensional Multi-Responsive Hydrogels. *Adv. Mater.* **2014**, *26*, 5665–5669. [CrossRef] [PubMed]
43. Wu, Q.; Ma, C.; Chen, L.; Sun, Y.; Wei, X.; Ma, C.; Zhao, H.; Yang, X.; Ma, X.; Zhang, C.; et al. A Tissue Paper/Hydrogel Composite Light-Responsive Biomimetic Actuator Fabricated by In Situ Polymerization. *Polymers* **2022**, *14*, 5454. [CrossRef] [PubMed]
44. Chen, L.; Zhang, K.; Ahn, J.; Wang, F.; Sun, Y.; Lee, J.; Young Cheong, J.; Ma, C.; Zhao, H.; Duan, G.; et al. Morph-genetic bamboo-reinforced hydrogel complex for bio-mimetic actuator. *Chem. Eng. J.* **2023**, *463*, 142391. [CrossRef]
45. Huang, B.; Lin, F.; Tang, L.; Lu, Q.; Lu, B. Research advances of functional cellulose-based hydrogels and its applications. *J. For. Eng.* **2022**, *7*, 1–13. [CrossRef]
46. Wu, Y.; Minamikawa, H.; Nakazumi, T.; Hara, Y. Actuation Properties of Paper Actuators Fabricated Using PEDOT/PSS Electrode Films. *J. Oleo. Sci.* **2020**, *69*, 1331–1337. [CrossRef]
47. Terasawa, N. High-performance TEMPO-oxidised cellulose nanofibre/PEDOT:PSS/ionic liquid gel actuators. *Sens. Actuators B Chem.* **2021**, *343*, 130105. [CrossRef]
48. Chen, Y.; Kuang, P.; Shen, X.; Lv, X.; Wang, Y.; Yin, W.; Zou, T.; Wang, B.; Liu, Y.; Fan, Q. Lignin precipitation-driven fabrication of gradient porous hydrogel actuator with temperature response. *Smart Mater. Struct.* **2023**, *32*, 035037. [CrossRef]
49. Gao, S.; Zhou, A.; Cao, B.; Wang, J.; Li, F.; Tang, G.; Jiang, Z.; Yang, A.; Xiong, R.; Lei, J.; et al. A tunable temperature-responsive and tough platform for controlled drug delivery. *New J. Chem.* **2021**, *45*, 13056–13063. [CrossRef]
50. Ghosh, R.; Misra, A. Carbon Nanotube-Based Hierarchical Paper Structure for Ultra-high Electrothermal Actuation in a Wide Humidity Range. *ACS Appl. Electron. Mater.* **2021**, *3*, 1260–1267. [CrossRef]

51. Wei, Y.; Huo, H.; Huang, C.; Zhang, Q.; Hoogenboom, R.; Liu, F. Supramolecular control over self-assembly and double thermoresponsive behavior of an amphiphilic block copolymer. *Eur. Polym. J.* **2020**, *125*, 109537. [CrossRef]
52. Toron, B.; Szperlich, P.; Koziol, M. SbSI Composites Based on Epoxy Resin and Cellulose for Energy Harvesting and Sensors-The Influence of SBSI Nanowires Conglomeration on Piezoelectric Properties. *Materials* **2020**, *13*, 902. [CrossRef] [PubMed]
53. Liu, S.; Yu, B.; Wang, S.; Shen, Y.; Cong, H. Preparation, surface functionalization and application of Fe(3)O(4) magnetic nanoparticles. *Adv. Colloid Interface Sci.* **2020**, *281*, 102165. [CrossRef] [PubMed]
54. Vivekananthan, V.; Chandrasekhar, A.; Alluri, N.R.; Purusothaman, Y.; Khandelwal, G.; Pandey, R.; Kim, S.-J. Fe2O3 magnetic particles derived triboelectric-electromagnetic hybrid generator for zero-power consuming seismic detection. *Nano Energy* **2019**, *64*, 103926. [CrossRef]
55. Xu, W.; Zheng, X.; Shangguan, Z.; Qu, J.; Zhang, W. A low-cost magnetic catalyst (MnFe2O4) for ciprofloxacin degradation via periodate activation: The synergistic effect of Mn and Fe. *Chem. Eng. J.* **2023**, *464*, 142562. [CrossRef]
56. Erdem, D.; Bingham, N.S.; Heiligtag, F.J.; Pilet, N.; Warnicke, P.; Heyderman, L.J.; Niederberger, M. CoFe2O4and CoFe2O4-SiO2Nanoparticle Thin Films with Perpendicular Magnetic Anisotropy for Magnetic and Magneto-Optical Applications. *Adv. Funct. Mater.* **2016**, *26*, 1954–1963. [CrossRef]
57. Li, Z.; Wang, J.; Xu, Y.; Shen, M.; Duan, C.; Dai, L.; Ni, Y. Green and sustainable cellulose-derived humidity sensors: A review. *Carbohydr. Polym.* **2021**, *270*, 118385. [CrossRef]
58. Zhou, S.; Kong, X.; Strømme, M.; Xu, C. Efficient Solar Thermal Energy Conversion and Utilization by a Film of Conductive Metal–Organic Framework Layered on Nanocellulose. *ACS Mater. Lett.* **2022**, *4*, 1058–1064. [CrossRef]
59. Tian, Y.; Liu, Y.; Ju, B.; Ren, X.; Dai, M. Thermoresponsive 2-hydroxy-3-isopropoxypropyl hydroxyethyl cellulose with tunable LCST for drug delivery. *RSC Adv.* **2019**, *9*, 2268–2276. [CrossRef]
60. Mo, K.; He, M.; Cao, X.; Chang, C. Direct current electric field induced gradient hydrogel actuators with rapid thermo-responsive performance as soft manipulators. *J. Mater. Chem. C* **2020**, *8*, 2756–2763. [CrossRef]
61. Michalska-Walkowiak, J.; Forster, B.; Hauschild, S.; Forster, S. Bistability, Remanence, Read/Write-Memory, and Logic Gate Function via a Stimuli-Responsive Polymer. *Adv. Mater.* **2022**, *34*, 2108833. [CrossRef] [PubMed]
62. Jiang, S.; Helfricht, N.; Papastavrou, G.; Greiner, A.; Agarwal, S. Low-Density Self-Assembled Poly(N-Isopropyl Acrylamide) Sponges with Ultrahigh and Extremely Fast Water Uptake and Release. *Macromol. Rapid Commun.* **2018**, *39*, 1700838. [CrossRef] [PubMed]
63. Kajornprai, T.; Katesripongsa, P.; Nam, S.Y.; Hamid, Z.A.A.; Ruksakulpiwat, Y.; Suppakarn, N.; Trongsatitkul, T. Potential Applications of Thermoresponsive Poly(N-Isoproplacrylamide)-Grafted Nylon Membranes: Effect of Grafting Yield and Architecture on Gating Performance. *Polymers* **2023**, *15*, 497. [CrossRef] [PubMed]
64. Liu, W.; Geng, L.; Wu, J.; Huang, A.; Peng, X. Highly strong and sensitive bilayer hydrogel actuators enhanced by cross-oriented nanocellulose networks. *Compos. Sci. Technol.* **2022**, *225*, 109494. [CrossRef]
65. Wei, Z.; Liu, Z.; Wang, X.; Long, S.; Yang, J. Smart carrier from electrospun core-shell thermo-sensitive ultrafine fibers for controlled drug release. *Eur. Polym. J.* **2019**, *114*, 1–10. [CrossRef]
66. Wang, Y.; Song, S.; Chu, X.; Feng, W.; Li, J.; Huang, X.; Zhou, N.; Shen, J. A new temperature-responsive controlled-release pesticide formulation—*Poly*(N-isopropylacrylamide) modified graphene oxide as the nanocarrier for lambda-cyhalothrin delivery and their application in pesticide transportation. *Colloids Surf. A Physicochem. Eng. Asp.* **2021**, *612*, 125987. [CrossRef]
67. Zhen, C.; Jing, L.; Yujie, C.; Xu, Z.; Hezhou, L.; Hua, L. Multiple-Stimuli-Responsive and Cellulose Conductive Ionic Hydrogel for Smart Wearable Devices and Thermal Actuators. *ACS Appl. Mater. Interfaces* **2020**, *13*, 1353–1366. [CrossRef]
68. Zong, R.; Hu, X.; Shang, M.; Wu, C.; Shentu, B. Phase Morphology and Conductive Properties of PBT/POE-g-GMA/PP/CNT Nanocomposites with a Tri-Continuous Structure via Thermal Annealing. *Ind. Eng. Chem. Res.* **2023**, *62*, 8289–8296. [CrossRef]
69. Mao, F.; Fan, X.; Long, L.; Li, Y.; Chen, H.; Zhou, W. Constructing 3D hierarchical CNTs/VO2 composite microspheres with superior electromagnetic absorption performance. *Ceram. Int.* **2023**, *49*, 16924–16931. [CrossRef]
70. Ji, C.; Wang, Y.; Ye, Z.; Tan, L.; Mao, D.; Zhao, W.; Zeng, X.; Yan, C.; Sun, R.; Kang, D.J.; et al. Ice-Templated MXene/Ag-Epoxy Nanocomposites as High-Performance Thermal Management Materials. *ACS Appl. Mater. Interfaces* **2020**, *12*, 24298–24307. [CrossRef]
71. Mao, F.; Long, L.; Pi, W.; Li, Y.; Zhou, W. X-band electromagnetic absorption and mechanical properties of mullite/Ti3AlC2 composites. *Mater. Chem. Phys.* **2022**, *292*, 126819. [CrossRef]
72. Wei, J.; Jia, S.; Wei, J.; Ma, C.; Shao, Z. Tough and Multifunctional Composite Film Actuators Based on Cellulose Nanofibers toward Smart Wearables. *ACS Appl. Mater. Interfaces* **2021**, *13*, 38700–38711. [CrossRef] [PubMed]
73. Bai, Y.; Zhou, Z.; Zhu, Q.; Lu, S.; Li, Y.; Ionov, L. Electrospun cellulose acetate nanofibrous composites for multi-responsive shape memory actuators and self-powered pressure sensors. *Carbohydr. Polym.* **2023**, *313*, 120868. [CrossRef] [PubMed]
74. Almeida, A.P.C.; Canejo, J.P.; Almeida, P.L.; Godinho, M.H. Cholesteric-type cellulosic structures: From plants to applications. *Liq. Cryst.* **2019**, *46*, 1937–1949. [CrossRef]
75. Chang, T.; Wang, B.; Yuan, D.; Wang, Y.; Smalyukh, I.; Zhou, G.; Zhang, Z. Cellulose nanocrystal chiral photonic micro-flakes for multilevel anti-counterfeiting and identification. *Chem. Eng. J.* **2022**, *446*, 136630. [CrossRef]
76. Juanjuan, S.; Jiayin, L.; Weidong, Y.; Jialing, T.; Yunjie, Y.; Chaoxia, W. Chiral Nematic Solvent-Responsive Actuator Based on a Cellulose Nanocrystal Template. *ACS Appl. Polym. Mater.* **2022**, *12*, 8952–8961. [CrossRef]

77. Li, X.; Liu, J.; Zhang, X. Pressure/Temperature Dual-Responsive Cellulose Nanocrystal Hydrogels for On-Demand Schemochrome Patterning. *Adv. Funct. Mater.* **2023**, *33*, 2306208. [CrossRef]
78. Orelma, H.; Hokkanen, A.; Leppänen, I.; Kammiovirta, K.; Kapulainen, M.; Harlin, A. Optical cellulose fiber made from regenerated cellulose and cellulose acetate for water sensor applications. *Cellulose* **2019**, *27*, 1543–1553. [CrossRef]
79. Ren, Y.; Fang, L.; Liu, Y.; Shen, C.; Wu, L.; Zheng, X.; Fang, K. Warming performance of far infrared lights electrogenerated by carbon nanotubes composite fabrics. *Mater. Lett.* **2022**, *319*, 132220. [CrossRef]
80. Loeb, S.; Li, C.; Kim, J.H. Solar Photothermal Disinfection using Broadband-Light Absorbing Gold Nanoparticles and Carbon Black. *Env. Sci. Technol.* **2018**, *52*, 205–213. [CrossRef]
81. Simayee, M.; Esfandiar, A. Synergistic effect of reduced graphene oxide and carbon black as hybrid light absorber for efficient and antifouling texture-based solar steam generator. *Sol. Energy* **2022**, *238*, 226–237. [CrossRef]
82. Zhao, Q.; Liang, Y.; Ren, L.; Yu, Z.; Zhang, Z.; Qiu, F.; Ren, L. Design and fabrication of nanofibrillated cellulose-containing bilayer hydrogel actuators with temperature and near infrared laser responses. *J. Mater. Chem. B* **2018**, *6*, 1260–1271. [CrossRef] [PubMed]
83. Xinkai, L.; Jize, L.; Dongdong, L.; Shaoquan, H.; Kai, H.; Xinxing, Z. Bioinspired Multi-Stimuli Responsive Actuators with Synergistic Color- and Morphing-Change Abilities. *Adv. Sci.* **2021**, *8*, 2101295. [CrossRef]
84. Chen, L.; Wei, X.; Wang, F.; Jian, S.; Yang, W.; Ma, C.; Duan, G.; Jiang, S. In-situ polymerization for mechanical strong composite actuators based on anisotropic wood and thermoresponsive polymer. *Chin. Chem. Lett.* **2022**, *33*, 2635–2638. [CrossRef]
85. Chen, L.; Wei, X.; Sun, Y.; Xue, Y.; Wang, J.; Wu, Q.; Ma, C.; Yang, X.; Duan, G.; Wang, F.; et al. A bamboo/PNIPAM composite hydrogel assembly for both programmable and remotely-controlled light-responsive biomimetic actuations. *Chem. Eng. J.* **2022**, *446*, 137072. [CrossRef]
86. Wang, F.; Lee, J.; Chen, L.; Zhang, G.; He, S.; Han, J.; Ahn, J.; Cheong, J.Y.; Jiang, S.; Kim, I.-D. Inspired by Wood: Thick Electrodes for Supercapacitors. *ACS Nano* **2023**, *17*, 8866–8898. [CrossRef]
87. Zhang, Y.; Xiao, H.; Xiong, R.; Huang, C. Xylan-based ratiometric fluorescence carbon dots composite with delignified wood for highly efficient water purification and photothermal conversion. *Sep. Purif. Technol.* **2023**, *324*, 124513. [CrossRef]
88. Qu, Q.; Zhang, J.; Chen, X.; Ravanbakhsh, H.; Tang, G.; Xiong, R.; Manshian, B.B.; Soenen, S.J.; Sauvage, F.; Braeckmans, K.; et al. Triggered Release from Cellulose Microparticles Inspired by Wood Degradation by Fungi. *ACS Sustain. Chem. Eng.* **2021**, *9*, 387–397. [CrossRef]
89. Chen, L.; Sun, Y.; Wang, J.; Ma, C.; Peng, S.; Cao, X.; Yang, L.; Ma, C.; Duan, G.; Liu, Z. A wood-mimetic porous MXene/gelatin hydrogel for electric field/sunlight bi-enhanced uranium adsorption. *e-Polymers* **2022**, *22*, 468–477. [CrossRef]
90. Yun, G.-Y.; Kim, J.-H.; Kim, J. Dielectric and polarization behaviour of cellulose electro-active paper (EAPap). *J. Phys. D Appl. Phys.* **2009**, *42*, 082003. [CrossRef]
91. Irfan, M.; Ali, I.; Ali, A.; Ahmed, M.; Soomro, T.A.; Yang, W.; Rahman, S.; Faraj Mursal, S.N.; Jalalah, M.; Jazem Ghanim, A.A. Analysis of the Performance of a Gel Actuator Made of Plasticized Polyvinyl Chloride/Carboxylated Cellulose Nanocrystals. *ACS Omega* **2023**, *8*, 17976–17982. [CrossRef] [PubMed]
92. Zhu, H.; Dai, W.; Wang, L.; Yao, C.; Wang, C.; Gu, B.; Li, D.; He, J. Electroactive Oxidized Alginate/Gelatin/MXene (Ti(3)C(2)T(x)) Composite Hydrogel with Improved Biocompatibility and Self-Healing Property. *Polymers* **2022**, *14*, 3908. [CrossRef] [PubMed]
93. Deng, W.; Xu, Y.; Zhang, X.; Li, C.; Liu, Y.; Xiang, K.; Chen, H. (NH4)2Co2V10O28·16H2O/(NH4)2V10O25·8H2O heterostructure as cathode for high-performance aqueous Zn-ion batteries. *J. Alloys Compd.* **2022**, *903*, 163824. [CrossRef]
94. Wu, M.-S.; Lyu, L.-J.; Syu, J.-H. Copper and nickel hexacyanoferrate nanostructures with graphene-coated stainless steel sheets for electrochemical supercapacitors. *J. Power Sources* **2015**, *297*, 75–82. [CrossRef]
95. Sen, I.; Seki, Y.; Sarikanat, M.; Cetin, L.; Gurses, B.O.; Ozdemir, O.; Yilmaz, O.C.; Sever, K.; Akar, E.; Mermer, O. Electroactive behavior of graphene nanoplatelets loaded cellulose composite actuators. *Compos. Part B Eng.* **2015**, *69*, 369–377. [CrossRef]
96. Tang, Z.H.; Zhu, W.B.; Mao, Y.Q.; Zhu, Z.C.; Li, Y.Q.; Huang, P.; Fu, S.Y. Multiresponsive Ti(3)C(2)T(x) MXene-Based Actuators Enabled by Dual-Mechanism Synergism for Soft Robotics. *ACS Appl. Mater. Interfaces* **2022**, *14*, 21474–21485. [CrossRef]
97. Zhou, J.; Mulle, M.; Zhang, Y.; Xu, X.; Li, E.Q.; Han, F.; Thoroddsen, S.T.; Lubineau, G. High-ampacity conductive polymer microfibers as fast response wearable heaters and electromechanical actuators. *J. Mater. Chem. C* **2016**, *4*, 1238–1249. [CrossRef]
98. Kayser, L.V.; Lipomi, D.J. Stretchable Conductive Polymers and Composites Based on PEDOT and PEDOT:PSS. *Adv. Mater.* **2019**, *31*, e1806133. [CrossRef]
99. Najathulla, B.C.; Kumar, S.; Deshpande, A.S.; Khandelwal, M. PEDOT:PSS-bacterial cellulose bilayer actuators: From the movement of ions to deflection. *Polym. Adv. Technol.* **2023**, *34*, 2407–2413. [CrossRef]
100. Terasawa, N.; Asaka, K. Self-standing cellulose nanofiber/poly(3,4-ethylenedioxythiophene):poly(4-styrenesulfonate)/ionic liquid actuators with superior performance. *RSC Adv.* **2018**, *8*, 33149–33155. [CrossRef]
101. Wang, F.; Wang, L.; Wu, Z.; Wang, W. A low-voltage electro-ionic soft actuator based on graphene nanoplatelets-sulfonated cellulose nanowhisker combined with microfibrillated cellulose. *J. Mater. Sci.* **2023**, *58*, 466–477. [CrossRef]
102. Wang, F.; Jin, Z.; Zheng, S.; Li, H.; Cho, S.; Kim, H.J.; Kim, S.-J.; Choi, E.; Park, J.-O.; Park, S. High-fidelity bioelectronic muscular actuator based on porous carboxylate bacterial cellulose membrane. *Sens. Actuators B Chem.* **2017**, *250*, 402–411. [CrossRef]
103. Reid, L.; Hamad, W.Y. Electro-osmotic Actuators from Cellulose Nanocrystals and Nanocomposite Hydrogels. *ACS Appl. Polym. Mater.* **2021**, *4*, 598–606. [CrossRef]
104. Correia, D.M.; Lizundia, E.; Meira, R.M.; Rincón-Iglesias, M.; Lanceros-Méndez, S. Cellulose Nanocrystal and Water-Soluble Cellulose Derivative Based Electromechanical Bending Actuators. *Materials* **2020**, *13*, 2294. [CrossRef]

105. Chen, X.; Ye, Z.; Yang, F.; Feng, J.; Li, Z.; Huang, C.; Ke, Q.; Yin, Y. Magnetic cellulose microcrystals with tunable magneto-optical responses. *Appl. Mater. Today* **2020**, *20*, 100749. [CrossRef]
106. Tomás, A.R.; Gonçalves, A.I.; Paz, E.; Freitas, P.; Domingues, R.M.A.; Gomes, M.E. Magneto-mechanical actuation of magnetic responsive fibrous scaffolds boosts tenogenesis of human adipose stem cells. *Nanoscale* **2019**, *11*, 18255–18271. [CrossRef] [PubMed]
107. Kim, J.; Hyun, J. Soft Magnetostrictive Actuator String with Cellulose Nanofiber Skin. *ACS Appl. Mater. Interfaces* **2021**, *13*, 43904–43913. [CrossRef]
108. Ge, W.; Zhang, F.; Wang, D.; Wei, Q.; Li, Q.; Feng, Z.; Feng, S.; Xue, X.; Qing, G.; Liu, Y. Highly Tough, Stretchable, and Solvent-Resistant Cellulose Nanocrystal Photonic Films for Mechanochromism and Actuator Properties. *Small* **2022**, *18*, 2107105. [CrossRef]
109. Hartings, M.; Douglass, K.O.; Neice, C.; Ahmed, Z. Humidity Responsive Photonic Sensor based on a Carboxymethyl Cellulose Mechanical Actuator. *Sens. Actuators B Chem.* **2018**, *265*, 335–338. [CrossRef]
110. Kuang, Y.; Chen, C.; Cheng, J.; Pastel, G.; Li, T.; Song, J.; Jiang, F.; Li, Y.; Zhang, Y.; Jang, S.-H.; et al. Selectively aligned cellulose nanofibers towards high-performance soft actuators. *Extrem. Mech. Lett.* **2019**, *29*, 100463. [CrossRef]
111. Yinan, L.; Jun, W.; Huixin, L.; Liulian, H.; Lihui, C.; Yonghao, N.; Qinghong, Z. Highly transparent RCF/PTFE humidity and IR light dual-driven actuator with high force density, sensitivity and stability. *Appl. Surf. Sci.* **2021**, *572*, 151502. [CrossRef]
112. Cao, J.; Zhou, Z.; Song, Q.; Chen, K.; Su, G.; Zhou, T.; Zheng, Z.; Lu, C.; Zhang, X. Ultrarobust Ti(3)C(2)T(x) MXene-Based Soft Actuators via Bamboo-Inspired Mesoscale Assembly of Hybrid Nanostructures. *ACS Nano* **2020**, *14*, 7055–7065. [CrossRef] [PubMed]
113. Jie, W.; Shuai, J.; Chao, M.; Jie, G.; Chunxia, Y.; Libin, Z.; Ziqiang, S. Nacre-inspired composite film with mechanical robustness for highly efficient actuator powered by humidity gradients. *Chem. Eng. J.* **2022**, *451*, 138565. [CrossRef]
114. Yang, L.; Cui, J.; Zhang, L.; Xu, X.; Chen, X.; Sun, D. A Moisture-Driven Actuator Based on Polydopamine-Modified MXene/Bacterial Cellulose Nanofiber Composite Film. *Adv. Funct. Mater.* **2021**, *31*, 2101378. [CrossRef]
115. Li, P.; Su, N.; Wang, Z.; Qiu, J. A Ti3C2Tx MXene-Based Energy-Harvesting Soft Actuator with Self-Powered Humidity Sensing and Real-Time Motion Tracking Capability. *ACS Nano* **2021**, *15*, 16811–16818. [CrossRef]
116. Mohammadi, M.; Berggren, M.; Tybrandt, K. Versatile Ultrasoft Electromagnetic Actuators with Integrated Strain-Sensing Cellulose Nanofibril Foams. *Adv. Intell. Syst.* **2023**, *5*, 2200449. [CrossRef]
117. Wei, J.; Jia, S.; Guan, J.; Ma, C.; Shao, Z. Robust and Highly Sensitive Cellulose Nanofiber-Based Humidity Actuators. *ACS Appl. Mater. Interfaces* **2021**, *13*, 54417–54427. [CrossRef]
118. Yang, K.; Cai, W.; Lan, M.; Ye, Y.; Tang, Z.; Guo, Q.; Weng, M. Multi-responsive and programmable actuators made with nacre-inspired graphene oxide-bacterial cellulose film. *Soft Matter* **2022**, *18*, 9057–9068. [CrossRef]
119. Wang, F.; Huang, D.; Li, Q.; Wu, Y.; Yan, B.; Wu, Z.; Park, S. Highly electro-responsive ionic soft actuator based on graphene nanoplatelets-mediated functional carboxylated cellulose nanofibers. *Compos. Sci. Technol.* **2023**, *231*, 109854. [CrossRef]
120. Cai, G.; Ciou, J.H.; Liu, Y.; Jiang, Y.; Lee, P.S. Leaf-inspired multiresponsive MXene-based actuator for programmable smart devices. *Sci. Adv.* **2019**, *5*, 7956. [CrossRef]
121. Rühlicke, S.; Zhang, K. Synthesis of novel cellulose mixesters for transparent responsive films with switchable mechanical properties. *Mater. Today Commun.* **2020**, *24*, 101015. [CrossRef]
122. Li, Y.; Wang, J.; Guo, J.; Fu, C.; Huang, L.; Chen, L.; Ni, Y.; Zheng, Q. UV and IR dual light triggered cellulose-based invisible actuators with high sensitivity. *Int. J. Biol. Macromol.* **2023**, *238*, 124031. [CrossRef] [PubMed]
123. Shu, L.; Zhang, X.F.; Wu, Y.; Wang, Z.; Yao, J. Facile fabrication of strong and conductive cellulose hydrogels with wide temperature tolerance for flexible sensors. *Int. J. Biol. Macromol.* **2023**, *240*, 124438. [CrossRef]
124. Chen, M.; Wan, H.; Hu, Y.; Zhao, F.; An, X.; Lu, A. Rationally designed cellulose hydrogel for an ultrasensitive pressure sensor. *Mater. Horiz.* **2023**, *8*, 2618. [CrossRef] [PubMed]
125. Yun, T.; Du, J.; Ji, X.; Tao, Y.; Cheng, Y.; Lv, Y.; Lu, J.; Wang, H. Waterproof and ultrasensitive paper-based wearable strain/pressure sensor from carbon black/multilayer graphene/carboxymethyl cellulose composite. *Carbohydr. Polym.* **2023**, *313*, 120898. [CrossRef] [PubMed]
126. Cui, J.; Li, F.; Wang, Y.; Zhang, Q.; Ma, W.; Huang, C. Electrospun nanofiber membranes for wastewater treatment applications. *Sep. Purif. Technol.* **2020**, *250*, 117116. [CrossRef]
127. Deng, Y.; Lu, T.; Cui, J.; Keshari Samal, S.; Xiong, R.; Huang, C. Bio-based electrospun nanofiber as building blocks for a novel eco-friendly air filtration membrane: A review. *Sep. Purif. Technol.* **2021**, *277*, 119623. [CrossRef]
128. Deng, Y.; Lu, T.; Zhang, X.; Zeng, Z.; Tao, R.; Qu, Q.; Zhang, Y.; Zhu, M.; Xiong, R.; Huang, C. Multi-hierarchical nanofiber membrane with typical curved-ribbon structure fabricated by green electrospinning for efficient, breathable and sustainable air filtration. *J. Membr. Sci.* **2022**, *660*, 120857. [CrossRef]
129. Cui, J.; Lu, T.; Li, F.; Wang, Y.; Lei, J.; Ma, W.; Zou, Y.; Huang, C. Flexible and transparent composite nanofibre membrane that was fabricated via a "green" electrospinning method for efficient particulate matter 2.5 capture. *J. Colloid Interface Sci.* **2021**, *582*, 506–514. [CrossRef]
130. Lu, T.; Cao, W.; Liang, H.; Deng, Y.; Zhang, Y.; Zhu, M.; Ma, W.; Xiong, R.; Huang, C. Blow-Spun Nanofibrous Membrane for Simultaneous Treatment of Emulsified Oil/Water Mixtures, Dyes, and Bacteria. *Langmuir* **2022**, *38*, 15729–15739. [CrossRef]

131. Deng, Y.; Zhu, M.; Lu, T.; Fan, Q.; Ma, W.; Zhang, X.; Chen, L.; Min, H.; Xiong, R.; Huang, C. Hierarchical fiber with granular-convex structure for highly efficient PM2.5 capture. *Sep. Purif. Technol.* **2023**, *304*, 122235. [CrossRef]
132. Cui, J.; Wang, Y.; Lu, T.; Liu, K.; Huang, C. High performance, environmentally friendly and sustainable nanofiber membrane filter for removal of particulate matter 1.0. *J. Colloid Interface Sci.* **2021**, *597*, 48–55. [CrossRef] [PubMed]
133. Ma, W.; Ding, Y.; Li, Y.; Gao, S.; Jiang, Z.; Cui, J.; Huang, C.; Fu, G. Durable, self-healing superhydrophobic nanofibrous membrane with self-cleaning ability for highly-efficient oily wastewater purification. *J. Membr. Sci.* **2021**, *634*, 119402. [CrossRef]
134. Deng, Y.; Lu, T.; Cui, J.; Ma, W.; Qu, Q.; Zhang, X.; Zhang, Y.; Zhu, M.; Xiong, R.; Huang, C. Morphology engineering processed nanofibrous membranes with secondary structure for high-performance air filtration. *Sep. Purif. Technol.* **2022**, *294*, 121093. [CrossRef]
135. Wang, F.; Huang, D.; Wu, Y.; Wang, D. Ecofriendly low voltage high-performance ionic artificial muscles based on bacterial cellulose nanofibers reinforced with polyvinyl alcohol. *J. Mater. Sci. Mater. Electron.* **2023**, *34*, 123. [CrossRef]
136. Gu, T.; Bi, H.; Sun, H.; Tang, J.; Ren, Z.; Zhou, X.; Xu, M. Design and development of 4D-printed cellulose nanofibers reinforced shape memory polymer composites: Application for self-deforming plant bionic soft grippers. *Addit. Manuf.* **2023**, *70*, 103544. [CrossRef]
137. Wang, X.; Tang, S.; Wu, Z.; Fang, J.; Qin, X.; Wei, L. Research status of biomass-based composite films with high barrier properties. *J. For. Eng.* **2021**, *6*, 13–22. [CrossRef]
138. Mao, F.; Long, L.; Zeng, G.; Chen, H.; Li, Y.; Zhou, W. Achieving excellent electromagnetic wave absorption property by constructing VO2 coated biomass carbon heterostructures. *Diam. Relat. Mater.* **2022**, *130*, 109422. [CrossRef]
139. Xu, J.; Zhai, Q.; Long, F.; Jiang, X.; Han, S.; Jiang, J. Electrocatalytic oxidation and reduction of biomass-derived chemicals: A review. *J. For. Eng.* **2022**, *7*, 11–22. [CrossRef]
140. Deng, W.-N.; Li, Y.-H.; Xu, D.-F.; Zhou, W.; Xiang, K.-X.; Chen, H. Three-dimensional hierarchically porous nitrogen-doped carbon from water hyacinth as selenium host for high-performance lithium–selenium batteries. *Rare Met.* **2022**, *41*, 3432–3445. [CrossRef]
141. Guo, Z.; Han, X.; Zhang, C.; He, S.; Liu, K.; Hu, J.; Yang, W.; Jian, S.; Jiang, S.; Duan, G. Activation of biomass-derived porous carbon for supercapacitors: A review. *Chin. Chem. Lett.* **2023**, *in press*. [CrossRef]

materials

MDPI

Article

The Influence of Polyvinyl Alcohol Porogen Addition on the Nanostructural Characteristics of Hydroxyapatite

Indrika Novella [1], Bedah Rupaedah [2], Diana Rakhmawaty Eddy [1], Suryana [3], Ferli Septi Irwansyah [1,4] and Atiek Rostika Noviyanti [1,*]

1 Department of Chemistry, Faculty of Mathematics and Natural Sciences, Padjajaran University, Jl. Raya Bandung-Sumedang Km. 21, Bandung 45363, Indonesia; indrika21001@mail.unpad.ac.id (I.N.); diana.rahmawati@unpad.ac.id (D.R.E.); ferli@uinsgd.ac.id (F.S.I.)
2 Research Center for Applied Microbiology, National Research and Innovation Agency, Jl. Raya Jakarta-Bogor Km. 46, Bogor 16911, Indonesia; beda001@brin.go.id
3 Department of Biology, Faculty of Mathematics and Natural Sciences, Padjajaran University, Jl. Raya Bandung-Sumedang Km. 21, Bandung 45363, Indonesia; suryana.bio@unpad.ac.id
4 Department of Chemistry Education, UIN Sunan Gunung Djati, Bandung Jl. A.H. Nasution No. 105, Bandung 40614, Indonesia
* Correspondence: atiek.noviyanti@unpad.ac.id

Abstract: Hydroxyapatite (HA) is a porous material widely developed in various research fields because of its high biodegradability, biocompatibility, and low toxicity. In this research, HA was synthesized using a hydrothermal method with chicken eggshells as a calcium source and various concentrations of polyvinyl alcohol as a porogen (2.5%, 5.0%, and 7.5% by wt). The structure and morphology of HA were determined by X-ray diffraction (XRD) and scanning electron microscope (SEM), respectively. HA was obtained with varying concentrations of polyvinyl alcohol (PVA) porogen according to Inorganic Crystal Structure Database (ICSD) standard. Based on analysis using a refinement method, changes in unit cell parameters (cell volume and lattice strain) of HA synthesized using PVA porogen compared to the standard, the chi square (χ^2) and index of R values were relatively low, validating the acceptable of the data. In addition, HA [$Ca_{10}(PO_4)_6(OH)_2$] with hexagonal structure and the P63/m space group was successfully obtained. Morphological analysis of HA by SEM found that HA has a spherical shape, and the porosity of HA increases with increasing concentrations of polyvinyl alcohol. The highest porosity was obtained with an addition of 5.0 wt% of PVA porogen (HAP3), reaching 69.53%.

Keywords: hydroxyapatite; polyvinyl alcohol; porogen

1. Introduction

Hydroxyapatite (HA), with the chemical formula $Ca_{10}(PO_4)_6(OH)_2$, is a biomaterial that has garnered significant attention in recent years due to its desirable properties, such as high biodegradability and biocompatibility. Additionally, hydroxyapatite can be resorbed in a physiological environment. It is non-toxic and non-immunogenic [1,2]. The most commonly encountered crystal structure of HAs is hexagonal with the space group *P63/m* [3]. In addition, HA is found in another structure, namely monoclinic, with *P21/b* space group. The most common difference in hexagonal and monoclinic structures is the orientation of hydroxyl groups [4]. HA has numerous applications in various fields, including as a nanocarrier for drug delivery [4], environmental remediation [5], and enzyme immobilization [6].

Hydroxyapatite (HA) as a biomaterial for various applications requiring interaction with different biological tissues in living organisms has been discussed. Due to its biocompatibility and physicochemical similarity, HA has been examined as a material for various applications including bone implants in humans, dental adhesive materials, and bacterial carriers [7]. HA has also been used as a carrier for controlled release of compounds because it is the most stable calcium phosphate at ambient temperature and under physiological conditions. The synthesis procedure of nano HA significantly influences the shape, crystal structure, and phase integrity of hydroxyapatite particles, ultimately governing the mechanical characteristics of materials designed for biomedical purposes [8]. In drug delivery systems, HA has also been used for controlled and targeted drug release with enhanced adsorption across biological barriers [9]. The utilization of HA in these applications often requires specific pore designs, where the porosity needs to be optimized for specific applications. This has led to the development of various synthesis processes for modifying the porosity of hydroxyapatite biomaterials and it can be controlled and determined based on the fabrication methods. Various methods have been used for the synthesis of HA, such as hydrothermal [10], precipitation [11], solid state reaction [12], and sol-gel methods [13]. The hydrothermal method is one of the most widely used methods for synthesis material, including HA. Since this method also has good variable control of temperature, pressure, and composition/precursor, it can also be used for almost all forms of material, such as single crystals, nanoparticles, films, powders, fibers, polymers, and also ceramics [14]. In addition, the hydrothermal method is a simple method. Its ease of use also offers good control over material morphology and size [2]. Through the hydrothermal technique, hydroxyapatite (HA) nanopowder has been effectively synthesized, offering potential applications in both biomedical contexts and scientific investigations. This successful synthesis method opens avenues for utilizing the HA nanopowder in various biomedical applications and serves as a valuable resource for scientific research endeavors [15].

Increasing the porosity of hydroxyapatite will enhance its functionality by allowing its interactions with other substances to be more pronounced. Due to its similarity to natural bone mineral and its biocompatibility, the functionalization of hydroxyapatite can be tailored with suitable properties for various applications, such as drug delivery, tissue engineering, and bone regeneration. Processes to enhance the applications of porous hydroxyapatite include surface modification, bioactive molecules [15], loading and release of drugs [1], surface coatings [16], electrochemical functionalization [3], polymer coatings, metal ion doping [17], and photocatalytic functionalization [18]. The selection of the functionalization approach relies on the intended usage and the particular properties that are needed. The method chosen for functionalization is contingent upon the specific requirements and goals of the application. The functionalization technique to be employed is determined by the desired application and the distinct properties that are necessary for it. The decision regarding the method of functionalization is guided by the intended purpose and the specific properties that are deemed essential. The design and characterization of hydroxyapatite must be conducted carefully to ensure its effectiveness and safety for the intended purposes. The addition of a porogen is commonly carried out to enhance the porosity of a material, including hydroxyapatite.

A porogenic agent, or porogen, one of the components responsible for pore formation, played a crucial role in achieving controlled characteristics of HA pores [19]. Porogen can be in the form of crystals, particles, or fibers that readily evaporate or dissolve, such as paraffin, starch, sucrose, cellulose derivatives, water soluble polymers such as polyvinyl alcohol (PVA), polyvinyl pyrrolidone (PVP), and other polymers. Material porosity typically occurs due to the release of easily removable compounds, leaving pores behind [20]. PVA exhibits many versatile physicochemical properties; therefore, PVA is used in various fields of research. PVA is widely used as a porogenic agent for material pore modification because it is a biocompatible polymer, soluble in water, and it is frequently employed in synthesizing biomaterials due to its non-toxic nature. This polymer has a low melting point compared to the densification temperature of HA, allowing it to evaporate during the sintering process and create pores or cavities in HA nanoparticles [21,22].

Nowadays, the utilization of porogenic agents, especially polymers for designing the porosity of HA, has been widely explored. For instance, carbon has been used as a porogen, affecting the microstructure and enhancing the porosity of calcium phosphate granules [23].

Polyvinyl alcohol (PVA), polyvinyl pyrrolidone (PVP), and polyethylene oxide (PEO) have also been employed as a porogen in the fabrication process of HA scaffolds synthesized from golden snail shells as precursors. The porogen concentrations used are 0, 5, 10, and 15 wt%; however, the outcome is HA that exhibits higher agglomeration and high impurities due to an excessive porogen concentration. Therefore, in this study, lower porogen concentrations are utilized [21]. Paraffin has also been used as a porogenic agent in synthesizing porous HA scaffolds for bone tissue engineering applications [24].

The effect of porogen concentration on the properties of hydroxyapatite will significantly influence the porosity and structure of the resulting material. When porogens are added during the fabrication process of hydroxyapatite, they create voids or pores within the material which are subsequently removed. This porosity can have a notable impact on the characteristics and performance of the material, including porosity, surface area and osteoconductivity [25], mechanical properties and cell interactions [26], also drug delivery [27]. Control and optimization of porogen concentration can enable the customization of hydroxyapatite properties to meet desired outcomes for specific applications.

The focus of the present study was on the use of a hydrothermal method to investigate the influence of various concentrations (0, 2.5%, 5.0%, and 7.5% by wt%) of PVA as a porogen on the structural characteristics and porosity of HA synthesized from utilization of eggshell waste.

2. Materials and Methods

In this research, there are four main stages, including the isolation of calcium oxide from chicken eggshells as a calcium source, the synthesis of porous hydroxyapatite (HA) with the addition of polyvinyl alcohol (PVA) porogen, and the characterization of the structure and morphology of porous HA.

2.1. Materials

The chicken eggshells were collected from the Jatinangor region (utilization chicken eggs) as a source of calcium, along with diammonium hydrogen phosphate (Merck, Germany), polyvinyl alcohol, and distilled water.

2.2. Methods

2.2.1. Calcium Preparation

Calcium was isolated from chicken eggshells, where the eggshells were cleaned of any remaining debris and membranes and then dried. The dried eggshells were ground using a ball mill and sieved through a 100-mesh sieve to obtain a powder. The eggshell powder was then calcined for 5 h at 1000 °C, producing white calcium powder [28].

2.2.2. Hydroxyapatite Synthesis

Hydroxyapatite (HA) synthesis was based on previous research [21,28]. A total of 2.7925 g of calcium oxide and 3.9663 g of diammonium phosphate with a molar ratio of Ca/P = 1.67 were dissolved in 50 mL of distilled water. The mixture was then transferred to an autoclave and heated at 230 °C for 48 h. The resulting precipitate was filtered, washed with distilled water until reaching a pH of 7, and dried at 110 °C for 2 h. Next, the obtained HA powder was modified using polyvinyl alcohol (PVA) porogen, the concentration of PVA porogen was calculated based on Equation (1) [24].

$$\text{wt\% of polyvinyl alcohol (PVA)} = \frac{\text{PVA mass}}{\text{PVA mass} + \text{HA mass}} \quad (1)$$

Varying concentrations of polyvinyl alcohol (PVA) solution (0%, 2.5%, 5.0%, and 7.5% by *wt*) were prepared by dissolving PVA in 50 mL of distilled water and heating it at 125 °C for 2 h. The PVA solution was mixed with the HA powder and stirred until homogeneous. The mixture was subsequently dried at 100 °C for 2 h, and the resulting dried precipitate was sintered at 1000 °C for 6 h. The synthesized hydroxyapatite samples were labeled as HAP1 (without porogen addition), HAP2 (2.5 wt%), HAP3 (5.0 wt%), and HAP4 (7.5 wt%).

2.3. Characterization

2.3.1. Structure Characterization

The synthesized hydroxyapatite (HA) structure was characterized using X-ray diffraction (XRD) analysis conducted at the Advanced Characterization Laboratory 1, Physics-National Research and Innovation Agency. The XRD measurements were performed using a PANalytical AERIS instrument (Malvern, Worcestershire, United Kingdom) equipped with a copper (Cu) anode, voltage (40 kV), a PIXcel1D-Medipix3 detector, and a monochromator with a $K\alpha_1$ wavelength of 1.540598 Å with an angle of 2θ from 5° to 85°. The obtained data were processed using Origin 8.5.1 (OriginLab, Northampton, MA, USA) and Rietica v4.2 software.

2.3.2. Morphology Characterization

The morphology of the HA samples was analyzed using a scanning electron microscope (SEM) (JSM-6510LA) (JEOL, Akishima, Tokyo, Japan) with accelerated voltage (20 kV) at the Polymer Laboratory, National Research and Innovation Agency. Before imaging, the samples were coated with platinum (35 s; 30 mA) to enhance conductivity. SEM imaging was performed at magnifications ranging from 5000× to 15,000×. Furthermore, the captured SEM images were analyzed to determine the particle size distribution, average particle diameter, and material porosity using ImageJ 1.51 and Origin 8.5.1 (OriginLab, Northampton, USA) software [29].

2.3.3. Physicochemical Properties

Physicochemical characteristics of HA samples synthesized using a porogen were examined using a digital pH meter (MP220 pH Meter, Mettler Toledo, Greifensee, Switzerland) and electrical conductometer (MC 226 Conductivity Meter, Mettler Toledo, Greifensee, Switzerland) [30].

3. Results

3.1. Hydroxyapatite Synthesis

Calcium oxide was successfully obtained with chicken eggshell as a source. The weight of chicken eggshells obtained before and after calcination was 98.7548 g and 65.5125 g; at a temperature of 750 °C, $CaCO_3$ started to decompose into CaO and was completely decomposed at 1000 °C. Mass reduction was caused by the presence of water and organic compounds which evaporated in the calcination process [24]. The calcination reaction that occurred from the chicken eggshells which convert $CaCO_3$ into CaO was based on Equation (2):

$$CaCO_{(s)} \rightarrow CaO_{(s)} + CO_{2(g)} \quad (2)$$

Hydroxyapatite (HA) powder was successfully synthesized using a hydrothermal method at a temperature of 230 °C for 48 h. The formation of HA in the hydrothermal synthesis process was based on the following reaction:

$$10\ CaO_{(s)} + 6(NH_4)2HPO_{4(s)} + 4H_2O_{(l)} \rightarrow Ca_{10}(PO_4)_6(OH)_2(s) + 12NH_4OH_{9(aq)} \quad (3)$$

The formation of the HA phase occurs gradually during the solvothermal reaction. Porous HA was formed as a result of modification with a polyvinyl alcohol (PVA) porogen, where PVA infiltrated into HA has a lower melting point compared to the densification temperature of HA. Therefore, during the calcination process at 1000 °C, PVA will vaporize and leave pores in HA [31,32].

3.2. Characterization

3.2.1. Characterization of HA Structure Using XRD

Analysis of modified HA diffractograms with various concentrations of PVA has been conducted to determine the structure and phases of HA, followed by analysis using Origin 8.5.1 (OriginLab, Northampton, MA, USA) and Rietica v4.2 software. The diffraction patterns of the samples can be seen in Figure 1.

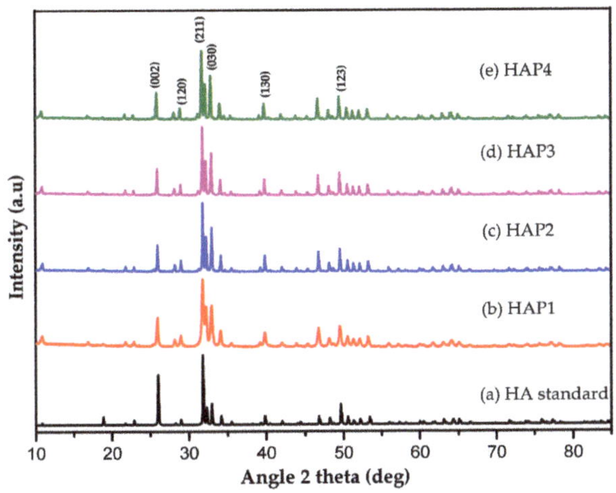

Figure 1. XRD pattern of: (**a**) HA standard (ICSD code 157481), (**b**) HAP1, (**c**) HAP2, (**d**) HAP3, and (**e**) HAP4.

The XRD pattern of HA synthesized with PVA as a porogen is shown in Figure 1 and corresponds to the ICSD code 157481 standard with *P63/m* space group, number of space groups 176 (release 2007) [33], indicating the successful synthesis of hydroxyapatite. Based on Figure 1, the XRD pattern shows the typical peak of HA appears at 2θ: 25.87°; 28.9°; 31.77°; 32.80°; 39.72°; and 49.54° with the *hkl* of (002); (120); (211); (030); (130); and (123), and the highest peak in HA appears at 31.77° [34]. The presence of these peaks revealed that the HA was successfully synthesized using this method. The diffractogram pattern exhibits the same diffraction pattern as the HA synthesized without PVA as a porogen, and the highest HA peak for all variations of porogen addition was the same as the highest HA peak without a porogen. This indicates that all PVA porogens were removed from the HA during the calcination process, but there was a peak shift caused by changes in lattice parameters and lattice strain [10,21,35]. Based on the XRD pattern of HA without the addition of PVA porogen, HA can be indexed to the hexagonal HA phase (*P63/m* space group) with lattice parameters $a = b = 9.412°$ and $c = 6.853$ Å (ICSD code 157481).

The synthesized compounds from the X-ray diffraction (XRD) pattern were analyzed using the *Le Bail* refinement method to study the phase formations and crystal structure in detail. The compound was analyzed using Rietica v4.2 software and the phase used for refinement in this study was the ICSD (Inorganic Crystal Structure Database) code 157481 standard of the structural data (release 2007), which has a *P63/m* space group with a space group number 176 with a hexagonal crystal structure and lattice parameters ($a = b = 9.412$ Å, and $c = 6.853$ Å). Refinement was carried out until a close fit between observed and calculated pattern was observed. Visual result of refined plots (Figure 2) showed the accomplishment of Le Bail refinement method and also showed a small difference between observed and calculated diffraction. It can be observed that the analysis result of the sample matches the standard, resulting in a red plot (calculated by the software) that aligns well with the black plot (sample), as evidenced by the blue plot (Bragg positions). The purity of the phase can be determined by the presence of blue lines following each peak.

Following the use of the Le Bail refinement method for the XRD analysis of the obtained samples, indicating that a single phase of the formed compound is $Ca_{10}(PO_4)_6(OH)_2$ with a hexagonal structure and the *P63/m* space group without revealing another supplemented phase in the samples [36].

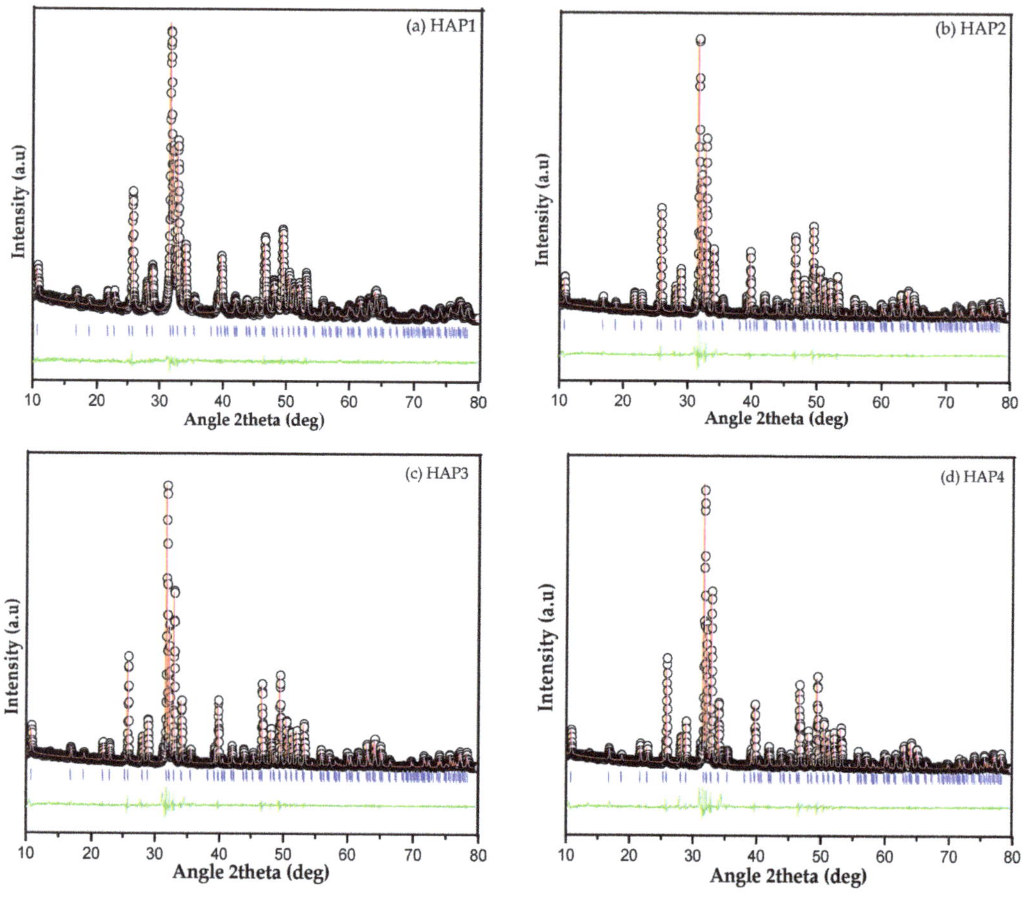

Figure 2. Le Bail refinement fit to SRD data for HA samples: (**a**) HAP1, (**b**) HAP2, (**c**) HAP3, and (**d**) HAP4. The black plot marks the experimentally observed pattern; the red plot marks the calculated diffraction pattern. The position of the Bragg peaks is marked with blue plot (vertical lines), and the difference between observed and calculated patterns is represented by the green plot at the bottom of the figure.

3.2.2. Characterization of HA Morphology Using SEM

Morphological observations of hydroxyapatite (HA) samples were conducted using a scanning electron microscope (SEM) (JEOL, Akishima, Tokyo, Japan) for all samples (HAP1, HAP2, HAP3, and HAP4) at a magnification of 5000× to 15,000×. The average particle size and particle size distribution were analyzed using image analysis software called ImageJ 1.51 and Origin 8.5.1 (OriginLab, Northampton, MA, USA). In SEM images, HA was found in spherical shape, which differs from the previously obtained granular form of HA [21]. The difference could be attributed to differences in the precursor CaO source employed in the HA synthesis process and the morphology of HA differed from that of the modified HA, indicating that the calcination process released PVA and created voids, as can be seen in Figure 3. The morphology observations of HA revealed that the addition of PVA (2.5%, 5.0%, and 7.5% by *wt*) resulted in the formation of pores within the hydroxyapatite. These pores appeared as black spots or areas in the SEM images and the pores increased with increasing the concentration of PVA porogen [32]. The addition of PVA porogen also resulted in a more uniform particle distribution as can be seen in bar charts. As shown by analysis of the SEM results using ImageJ 1.51 software, modified HA with the PVA porogen had an average diameter particle of 68.08, 69.32, 50.63, and 86.92 nm, respectively, for each sample (HAP1, HAP2, HAP3, and HAP4).

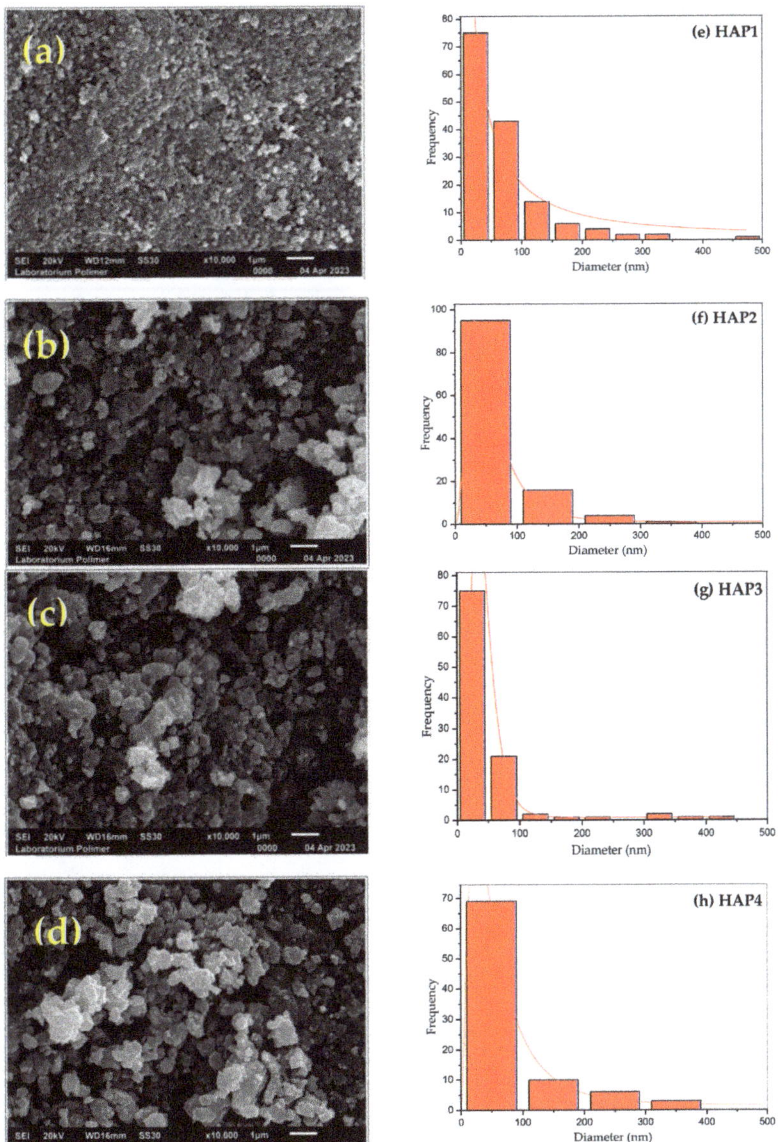

Figure 3. SEM image of HA samples: (**a**) HAP1, (**b**) HAP2, (**c**) HAP3, (**d**) HAP4 and particle size distribution of HA samples (**e**) HAP1, (**f**) HAP2, (**g**) HAP3, and (**h**) HAP4.

3.2.3. Physicochemical Characteristics

The physicochemical characteristics of the synthesized samples are presented in Table 1, illustrating a correlation between increased porogen concentration and elevated pH values. Furthermore, as the pH shifts toward alkalinity, a concurrent improvement in electrical conductivity is observed, and the concentration of ions in alkaline solutions affects the value of electrical conductivity [37].

Table 1. Physicochemical characteristics of HA.

Parameter	HAP1	HAP2	HAP3	HAP4
pH	8.95	9.89	9.85	9.79
EC (μs/cm)	55.4	80.50	81.45	64.15

The data represent the average of two replicates.

4. Discussion

The present result shows that polyvinyl alcohol (PVA) porogen use in hydroxyapatite (HA) synthesis influenced the characteristic of HA. The crystal size (D) and the lattice strain (ε) of the samples (HAP1, HAP2, HAP3, and HAP4) were changed by addition of PVA (0%, 2.5%, 5.0%, and 7.5% by wt), which caused a small shift peak in the XRD pattern of HA. The crystal size (D) was calculated based on the Scherrer Equation (4), and lattice strain (ε) of each sample was calculated based on Bragg's law (5), which can be seen in Table 2 [38,39].

$$D = \frac{K\lambda}{\beta \cos \theta} \tag{4}$$

$$\varepsilon = \frac{\beta}{4 \times \tan \theta} \tag{5}$$

where K is a constant (0.89), λ is the X-ray wavelength in nanometers (0.154 nm), β is the peak width of the diffraction peak at half maximum height, and the symbol θ can be degrees or radians where cos θ and tan θ refer to the θ values.

Table 2. Crystalline size and lattice strain of HA.

No.	Sample	Peak Position (°)	FWHM (Radian)	Crystalline Size (nm)	Lattice Strain
1.	HAP1	31.7729	0.2384	34.27	0.0369
2.	HAP2	31.8256	0.1631	50.10	0.0253
3.	HAP3	31.7988	0.1632	50.06	0.0252
4.	HAP4	31.7496	0.1640	49.79	0.0254

The average crystalline size of the HA samples was 46.06 nm. The crystalline size of the modified HA with addition of PVA porogen (0%, 2.5%, 5.0%, and 7.5% by *wt*) was increased, while a decrease of the lattice strain was observed with increasing PVA concentration and the shift peak of HAP2, HAP3, and HAP4 occurred from changes in the lattice strain of the HA samples.

The Le Bail refinement analysis of the XRD pattern revealed changes in the unit cell parameters (lattice parameter and cell volume) of the HA sample, as shown in Table 3.

Table 3 shows that the unit cell parameters (space group, lattice parameter, cell volume), *R* factors (*Rp*, *Rwp*, and *Rexp*), and the goodness of fit (Chi square, χ^2) of the HA sample analyzed using Le Bail refinement method, where the *GOF* (goodness of fit) which represented the χ^2 value was calculated according to Equation (6) [40].

$$GOF = Rwp/Rexp \tag{6}$$

Table 3. Unit cell parameters of the samples were analyzed by the Le Bail refinement method.

No.	Cell Parameter	HAP1	HAP2	HAP3	HAP4
1.	*Space group*	*P63/m*	*P63/m*	*P63/m*	*P63/m*
2.	$a = b$ (Å)	9.4316	9.4244	9.4222	9.4211
3.	c (Å)	6.8813	6.8832	6.8820	6.8814
4.	V (Å^3)	530.1233	529.4619	529.1135	528.9485
5.	*Rp* (%)	4.21	6.07	6.17	6.89
6.	*Rwp* (%)	5.53	8.37	8.83	10.20
7.	*Rexp* (%)	4.14	4.23	4.25	4.20
8.	χ^2	1.34	1.98	2.08	2.43

The alterations in lattice parameters of the samples, as demonstrated in Table 3, exhibit slight deviations when contrasted with the ICSD code 157481 standard values ($a = b = 9.412$ Å, and $c = 6.853$ Å) [33]. These deviations, however, remain relatively inconspicuous despite the introduction of varying concentrations of PVA porogen and the influence of the sintering procedure. For the analyzed samples, an increase in the network parameter "a" and parameter "c" with increasing polyvinyl alcohol (PVA) concentration was observed. This finding was consistent with previous reports [41,42]. The correctness of the Le Bail refinement of the data obtained from the samples is monitored by a reliability index parameter such as the index of the pattern (*Rp*), the weighted pattern (*Rwp*), *Rexp*, and the index of χ, where the index of χ (goodness of fit) is given by the ratio of Rwp and *Rexp* factors. The obtained χ^2 (Chi square) values were relatively low (< 4%) which was considered validating the acceptability of the data. In addition, the low values of the index of *R* factors (*Rp* and *Rwp*), below 20%, indicate a good agreement between the calculated data from the theoretical model and the observed pattern, thus validating the acceptance of the data [43].

The porosity of the HA material was analyzed and plotted in a 3D graph using Origin 8.5.1 software based on SEM images obtained and can be seen in Figure 4. This involves converting the image into a data format using the Origin software to generate a 2D matrix. The analysis of material volume was carried out in accordance with the application's instructions. The volume of space beneath the sample surface, or the material pore volume, is determined using mathematical equations, while the total material volume is calculated based on the projected area of the sample surface in the x–y plane. Porosity is computed from the total volume and the volume of space beneath the surface. Finally, the data are visualized through a 3D surface projection [29].

It was evident that the blue colored area represents the solid volume of a material. In contrast, the white colored area represents the space/voids within the material. The addition of PVA 2.5 wt% resulted in a uniform pore distribution represented in blue and yielded a porosity of 67.10% (Figure 4b). The hydroxyapatite porosity increased to 69.53% with the supplemental PVA 5.0 wt%, the allocation of pores was still uniform, and the allocation of solid particles decreased (Figure 4c). By the supplemental PVA 7.5 wt%, the porosity decreased to 63.53%. Additionally, based on the image, it can be observed that with an increase in the porogen concentration from 2.5% to 7.5% by *wt*, the pores' volume also increased compared to the volume of pores in HA without PVA porogen. This finding aligns with a previous report [32]. However, with the addition of 7.5 wt% PVA (HAP4), the porosity of the HA material decreased compared to concentrations of 2.5 wt% and 5.0 wt% (Figure 4b,c), although it still increased compared to HAP1 (Figure 4a).

The rise in porosity within HA material was linked to the introduction of the porogen. The incorporation of the porogen into the mixture caused a deceleration in the densification process, leading to an incomplete interlinkage among particles. This outcome is attributed to the presence of the porogen, which contributed to heightened porosity in the HA material due to the hindered densification process and incomplete particle interconnection. As a result, the porogen underwent evaporation upon the sample's calcination, resulting in the creation of voids or gaps within the material. Thus, the porogen dissipated as the sample underwent calcination, leaving behind empty spaces within the material. This led to the formation of empty spaces within the material, as the porogen evaporated during the calcination process [19,21]. In this study, the HA material synthesized with a lower PVA concentration demonstrated higher porosity percentage compared to the earlier research conducted with a higher porogen concentration, incorporating PVA, paraffin wax and natural porogen from honeycomb [24,32,44].

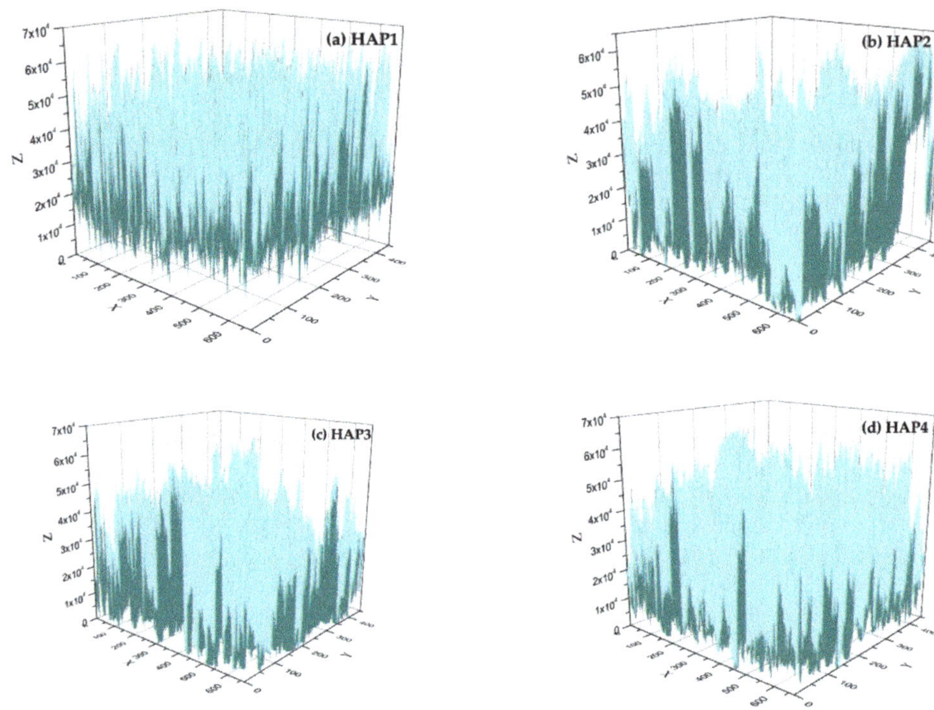

Figure 4. Three-dimensional graphics and porosity of the HA samples: (**a**) HAP1, (**b**) HAP2, (**c**) HAP3, and (**d**) HAP4.

Building upon the information provided, HAP samples modified with porogens yield materials with structure, morphologies, porosities, and physicochemical properties capable of creating a conducive environment for the growth of living organisms, facilitating the delivery of diverse biological molecules, and serving bone tissue engineering applications [24,26,45]. Nevertheless, achieving precise control and optimal alignment between porosity and physicochemical properties needs to be carried out for more specialized applications.

5. Conclusions

Hydroxyapatite (HA) has been successfully synthesized using a hydrothermal method using polyvinyl alcohol (PVA) as a porogenic agent. PVA is one of the porogens that influences the nanostructure of HA. Based on the characterization results of HA using X-ray diffraction (XRD), it was known that the resulting HA structure underwent changes in the unit cell parameters (lattice parameter and cell volumes), and based on refinement analysis, low chi square (χ^2) and the index of R factors (*Rp* and *Rwp*) values indicate a good agreement between theoretical model and experiment pattern and validating the acceptability of the data. Morphological analysis of the sample using scanning electron microscope (SEM) revealed a more uniform particle distribution for PVA-modified HA than unmodified HA and the addition of PVA porogen resulted in the formation of pores. Additionally, the porosity of HA increased with the modification of HA using PVA porogen. The highest porosity was achieved with a modification of 5.0 wt% PVA porogen, reaching 69.53%. Materials possessing such properties can be utilized in carriers for delivering a diverse range of biological molecules or even as construction material and bone tissue engineering.

Author Contributions: Conceptualization, I.N.; writing—original draft, data curation, visualization, I.N., B.R., D.R.E., S., F.S.I. and A.R.N.; methodology, B.R., D.R.E., S. and A.R.N.; resources, B.R. and A.R.N.; investigation, editing, funding acquisition, A.R.N. and F.S.I.; Investigation, review. All authors have read and agreed to the published version of the manuscript.

Funding: This research was funded by Academic Leader Grant (ALG) from Universitas Padjajaran under contract number 1549/UN6.3.1/PT.00.2023. This work is also supported by RP ORNM BRIN 2023 code number 76 under contract number 3/III.10/HK/2023.

Institutional Review Board Statement: Not applicable.

Informed Consent Statement: Not applicable.

Data Availability Statement: Not applicable.

Acknowledgments: The researchers would like to thank the administrative and technical support in Physical and Inorganic Chemistry Laboratory, Department of Chemistry, Faculty of Mathematics and Natural Sciences, Padjdjaran University, and the researchers and also technical support in the Research Center for Applied Microbiology, National Research and Innovation Agency, Serpong, Indonesia.

Conflicts of Interest: The authors state that they do not have any known conflict of interest that could have influenced the findings reported in this article.

References

1. Lara, W.O.; Ochoa, S.L.; Beltran, C.E.G. Hydroxyapatite nanoparticles in drug delivery: Physicochemistry and applications. *Pharmaceutics* **2021**, *13*, 1642. [CrossRef] [PubMed]
2. Pandharipande, S.L.; Sondawale, S.S. Review on synthesis of hydroxyapatite and its bio-composites. *Int. J. Sci. Eng. Technol.* **2016**, *5*, 3410–3416.
3. Awasthi, S.; Pandey, S.K.; Arunan, E.; Srivastava, C. A Review on hydroxyapatite coatings for the biomedical applications: Experimental and theoretical perspectives. *J. Mater. Chem. B.* **2021**, *9*, 228–249. [CrossRef] [PubMed]
4. Lin, K.; Chang, J. Structure and properties of hydroxyapatite for biomedical applications. In *Hydroxyapatite (Hap) for Biomedical Application*; Woodhead Publishing: Sawston, UK, 2023; pp. 3–19.
5. Lee, Y.J.; Jung, K.W.; Lee, S.Y.; Choi, J.W. A facile one-pot hydrothermal synthesis of hydroxyapatite/biochar nanocomposites: Adsorption behavior and mechanisms for the removal of copper(II) from aqueous media. *Chem. Eng. J.* **2019**, *369*, 529–541.
6. Farinas, C.S.; Coutinho, T.C.; Malafatti, J.O.D.; Paris, E.C.; Tardioli, P.W. Hydroxyapatite-$CoFe_2O_4$ magnetic nanoparticle composites for industrial enzyme immobilization, use, and recovery. *ACS. Appl. Nano. Mater.* **2020**, *3*, 12334–12345.
7. Abdulrahman, I.; Tijani, H.I.; Mohammed, B.A.; Saidu, H.; Yusuf, H.; Jibrin, M.N.; Mohammed, S. From garbage to biomaterials: An overview on egg shell based hydroxyapatite. *J. Mater.* **2014**, *2014*, 802467. [CrossRef]
8. Irwansyah, F.S.; Noviyanti, A.R.; Eddy, D.R.; Risdiana, R. Green Template-Mediated Synthesis of Biowaste Nano-Hydroxyapatite: A Systematic Literature Review. *Molecules* **2022**, *27*, 5586. [CrossRef]
9. Yeo, Y.; Lee, J.H. Controlled drug release from pharmaceutical nanocarriers. *Chem. Eng. Sci.* **2015**, *125*, 75–84.
10. Noviyanti, A.R.; Akbar, N.; Deawati, Y.; Ernawati, E.E.; Malik, Y.T.; Fauzia, R.P. A novel hydrothermal synthesis of nanohydroxyapatite from eggshell-calcium-oxide precursors. *Heliyon* **2020**, *6*, e03655. [CrossRef]
11. Cahyaningrum, S.E.; Herdayastuty, N.; Devina, B.; Supangat, D. Synthesis and characterization of hydroxyapatite powder by wet precipitation method. *IOP Conf. Ser. Mater. Sci. Eng.* **2018**, *299*, 012039. [CrossRef]
12. Wu, S.C.; Hsu, H.C.; Hsu, S.K.; Chang, Y.C.; Ho, W.F. Synthesis of hydroxyapatite from eggshell powders through ball milling and heat treatment. *J. Asian Ceram Soc.* **2016**, *4*, 85–90. [CrossRef]
13. Bakan, F.; Laçin, O.; Sarac, H. A novel low temperature sol-gel synthesis process for thermally stable nano crystalline hydroxyapatite. *Powder Technol.* **2013**, *233*, 295–302. [CrossRef]
14. Shandilya, M.; Rai, R.; Singh, J. Review: Hydrothermal technology for smart materials. *Adv. Appl. Ceram.* **2016**, *115*, 354–376. [CrossRef]
15. Fatah, S.K.; Kareem, M.M. Synthesis and characterisation of hydroxyapatite nanopowders by hydrothermal method. *Zanco J. Pure Appl Sci.* **2017**, *29*, s227–s233.
16. Rial, R.; Hassan, N.; Liu, Z.; Ruso, J.M. The design and green nanofabrication of noble hydrogel systems with encapsulation of doped bioactive hydroxyapatite toward sustained drug delivery. *J. Mol. Liq.* **2021**, *343*, 117598. [CrossRef]
17. Kottegoda, N.; Sandaruwan, C.; Priyadarshana, G.; Siriwardhana, A.; Rathnayake, U.A.; Berugoda Arachchige, D.M.; Kumarasinghe, A.R.; Dahanayake, D.; Karunaratne, V.; Amaratunga, G.A. Urea-Hydroxyapatite nanohybrids for slow release of nitrogen. *ACS Nano* **2017**, *11*, 1214–1221. [CrossRef] [PubMed]

18. Fatimah, I.; Fahrani, D.; Harmawantika, T.; Sahroni, I.; Kamari, A.; Sa'bana Rahmatillah, C.; Nurillahi, R. Functionalization of hydroxyapatite derived from cockle (Anadara granosa) shells into hydroxyapatite–nano TiO_2 for photocatalytic degradation of methyl violet. *Sustain. Environ. Res.* **2019**, *1*, 40. [CrossRef]
19. Turnbull, G.; Clarke, J.; Picard, F.; Riches, P.; Jia, L.; Han, F.; Li, B.; Shu, W. 3D bioactive composite scaffolds for bone tissue engineering. *Bioact. Mater.* **2018**, *3*, 278–314. [CrossRef]
20. Dorozhkin, S.V. Calcium orthophosphate-based bioceramics. *Materials* **2013**, *6*, 3840–3942. [CrossRef]
21. Mawuntu, V.J.; Yusuf, Y. Porous structure engineering of bioceramic hydroxyapatite-based scaffolds using PVA, PVP, and PEO as polymeric porogens. *J. Asian. Ceram. Soc.* **2019**, *7*, 161–169. [CrossRef]
22. Jalageri, M.B.; Kumar, G.C.M. Hydroxyapatite reinforced polyvinyl alcohol /polyvinyl pyrrolidone based hydrogel for cartilage replacement. *Gels* **2022**, *8*, 555. [CrossRef] [PubMed]
23. Effendi, M.D.; Gustiono, D. Effect of porogen agent on microstructure of CaP granules using the gelation of alginate droplet formation. *IOP Conf. Ser. Mater. Sci. Eng.* **2020**, *857*, 012026. [CrossRef]
24. Yusuf, Y.; Sari, M.; Aminatun; Suciati, T.; Sari, Y.W. Porous carbonated hydroxyapatite-based paraffin wax nanocomposite scaffold for bone tissue engineering: A physicochemical properties and cell viability assay analysis. *Coatings* **2021**, *11*, 1189.
25. Nie, L.; Han, Y.; Wei, Q.; Chang, P.; Hu, K.; Okoro, O.V.; Shavandi, A. Three-Dimensional printing of hydroxyapatite composites for biomedical application. *Crystals* **2021**, *11*, 353.
26. Lett, J.A.; Sundareswari, M.; Ravichandran, K. Porous hydroxyapatite scaffolds for orthopedic and dental applications-the role of binders. *Mater. Today Proc.* **2016**, *3*, 1672–1677. [CrossRef]
27. Munir, M.U.; Salman, S.; Javed, I.; Bukhari, S.N.A.; Ahmad, N.; Shad, N.A.; Aziz, F. Nano-hydroxyapatite as a delivery system: Overview and advancements. *Artif. Cells Nanomed. Biotechnol.* **2021**, *49*, 717–727. [CrossRef]
28. Yusuf, A.; Noviyanti, A.R.; Muhammad, N.M. The effect of temperature synthesis on the purity and crystallinity of hydroxyapatite. *Key Eng. Mater.* **2020**, *860*, 228–233. [CrossRef]
29. Abdullah, M.; Sciences, N. A Simple method for determining surface porosity based On. *Indones. J. Phys.* **2009**, *20*, 37–40. [CrossRef]
30. Sohaib, M.; Zahir, Z.A.; Khan, M.Y.; Ans, M.; Asghar, H.N.; Yasin, S.; Al-Barakah, F.N.I. Comparative evaluation of different carrier-based multi-strain bacterial formulations to mitigate the salt stress in wheat. *Saudi J. Biol. Sci.* **2020**, *27*, 777–787. [CrossRef]
31. Grzybek, P.; Jakubski, Ł.; Dudek, G. Neat chitosan porous materials: A review of preparation, structure characterization and application. *Int. J. Mol. Sci.* **2022**, *23*, 9932. [CrossRef]
32. Yusuf, Y.S.; Sari, M.; Hening, P.; Chotimah, A.I.D. Bioceramic hydroxyapatite-based scaffold with a porous structure using honeycomb as a natural polymeric porogen for bone tissue engineering. *Biomater. Res.* **2021**, *25*, 2.
33. Mostafa, N.Y.; Brown, P.W. Computer simulationof stoichiometric hydroxyapatite: Structure and substitutions. *J. Phys. Chem. Solids.* **2007**, *68*, 431–437. [CrossRef]
34. Noviyanti, A.R.; Rahayu, I.; Fauzia, R.P. The effect of Mg concentration to mechanical strength of hydroxyapatite derived from eggshell. *Arab. J. Chem.* **2021**, *14*, 103032. [CrossRef]
35. Mawuntu, V.J.; Yusuf, Y. Porous-structure engineering of hydroxyapatite-based scaffold synthesized from Pomacea canaliculata shell by using polyethylene oxide as polymeric porogen. *IOP Conf. Ser. Mater. Sci. Eng.* **2018**, *432*, 012045. [CrossRef]
36. Tahir, D.; Heryanto; Abdullah, B. Quantitative analysis of X-Ray diffraction spectra for determine structural properties and deformation energy of Al, Cu and Si. *J. Phys. Conf. Ser.* **2019**, *1317*, 012052.
37. Lin, W.; Yang, B.; Han, B.; Hu, X. A review of subsurface electrical conductivity anomalies in magnetotelluric imaging. *Sensors* **2023**, *23*, 1803. [CrossRef]
38. Farias, K.A.S.; Sousa, W.J.B.; Cardoso, M.J.B.; Lima, R.J.S.; Rodriguez, M.A.; Fook, M.V.L. Obtaining hydroxyapatite with different precursors for application as a biomaterial. *Ceramica* **2019**, *65*, 99–106. [CrossRef]
39. Monshi, A.; Foroughi, M.R.; Monshi, M.R. Modified scherrer equation to estimate more accurately nano-crystallite size using XRD. *World. J. Nano. Sci. Eng.* **2012**, *2*, 154–160. [CrossRef]
40. Idrisi, B.C.E.E.; Yacoubi, A.E.; Massit, A.; Moutaoikel, S.E.; Rezzouk, A. Rietveld refinement of the crystal structure of hydroxyapatite using X-ray Powder Diffraction. *Am. J. Mater. Sci. Eng.* **2017**, *5*, 1–5.
41. Singh, R.P.; Singh, M.; Verma, G.; Shukla, S.; Singh, S.; Singh, S. Structural analysis of silver doped hydroxyapatite nanopowders by rietveld refinement. *Trans Indian Inst. Met.* **2017**, *70*, 1973–1980. [CrossRef]
42. Predoi, D.; Iconaru, S.L.; Predoi, M.V.; Motelica-Heino, M.; Guegan, R.; Buton, N. Evaluation of antibacterial activity of zinc-doped hydroxyapatite colloids and dispersion stability using ultrasounds. *Nanomaterials* **2019**, *9*, 515. [CrossRef] [PubMed]
43. Kisi, E.H.; Howard, C.H. *Applications of Neutron Powder Diffraction*; Oxford Series on Neutron Scattering in Condensed Matter; Oxford University Press: Oxford, UK, 2008.
44. Sari, Y.W.; Sari, M. The effect of polyvinyl alcohol on the multicomponent biopolymer/hydroxyapatite/ carbonated hydroxyapatite composite. *Res. Sq.* **2023**, 1–21. [CrossRef]
45. Thunshirn, P.; Wenzel, W.W.; Pfeifer, C. Pore characteristics of hydrochars and their role as a vector for soil bacteria: A critical review of engineering options. *Crit. Rev. Environ. Sci. Technol.* **2022**, *52*, 4147–4171. [CrossRef]

nanomaterials

MDPI

Article

Hybrid Chitosan–TiO_2 Nanocomposite Impregnated in Type A-2186 Maxillofacial Silicone Subjected to Different Accelerated Aging Conditions: An Evaluation of Color Stability

Faten K. Al-Kadi [1,*], Jwan F. Abdulkareem [1] and Bruska A. Azhdar [2]

1 Department of Prosthodontics, College of Dentistry, University of Sulaimani, Sulaymaniyah 46001, Iraq; jwan.abdulkareem@univsul.edu.iq

2 Nanotechnology Research Laboratory, Department of Physics, College of Science, University of Sulaimani, Sulaymaniyah 46001, Iraq; bruska.azhdar@univsul.edu.iq

* Correspondence: faten.ali@univsul.edu.iq; Tel.: +964-(0)7703472240

Abstract: This study explores the impact of the incorporation of a chitosan–TiO_2 nanocomposite on the color stability of pigmented room-temperature vulcanization maxillofacial silicone under various accelerated aging conditions. Five hundred disk-shaped specimens were formed with type A-2186 silicone elastomer, and they were distributed into groups based on pigment types and nanoparticle treatments. The color difference (ΔE) was assessed using a colorimeter in the CIELAB color system before and after exposure to aging conditions, including UV-accelerated aging and outdoor weathering. ANOVA, Dennett's T3, and Tukey HSD tests revealed significant color alterations across all silicone types, with the most pronounced being in the red-colored 3% chitosan specimens and the least pronounced being in the 2% TiO_2 specimens that underwent UV-accelerated aging. Outdoor weathering consistently increased the ΔE values across all categories. This study suggests that while nanoparticles may offer some resistance against accelerated aging, they fall short in adequately defending against UV radiation during outdoor weathering.

Keywords: synthesized chitosan–TiO_2 nanocomposite; core–shell method; room-temperature vulcanization (RTV) of maxillofacial silicone; color stability; accelerated aging conditions

1. Introduction

The psychosocial consequences of visible aesthetic differences among individuals are determined through a combination of individual and social factors. These differences can render individuals temporarily or permanently vulnerable, depending on the nature of the impact. The utilization of surgery or prosthetics for rehabilitation serves as a crucial means of adaptation in order to address these challenges [1].

For nearly fifty years, elastomers have been utilized in the production of facial prostheses for individuals with missing facial features. In 1986, Factor II introduced A-2186, the first commercially available platinum-catalyzed silicone elastomer. A-2186 is a pourable silicone that is clear to translucent and consists of two parts (with a base-to-catalyst ratio of 10:1) [2]. The elastomeric element of A-2186 comprises a dimethylsiloxane polymer, a reinforcing silica, and a platinum catalyst. Concurrently, the curing agent comprises a dimethylsiloxane polymer, an inhibitor, and a siloxane crosslinker. Extensive investigations conducted to compare the physical traits of A-2186 with other maxillofacial silicon variants have reported a notable increase in both the tensile strength and tear resistance of A-2186. Notably, A-2186 exhibits a viscous state at the time of loading. The heightened mechanical durability exhibited by A-2186 is likely attributable to its superior filler loading and/or the high molecular weight of the dimethylsiloxane polymer within its composition [3].

Among anaplastologists, maxillofacial prosthodontists, and dental technicians involved in fabricating facial prostheses, RTV silicone elastomer products are commonly used. A-2186 silicone, specifically, has been widely employed for intrinsic or base color packing due to its superior tensile strength, tear strength, and softer, more realistic, and skin-like surface texture [2].

Despite its widespread use in maxillofacial prostheses, silicone has certain limitations. One significant drawback is its short shelf life, which is primarily due to the rapid degradation of its physical properties and color stability. The overall appearance and aesthetics of a prosthesis play a crucial role in determining the patient's wellbeing and social acceptance [4]. Clinical studies have indicated that the average lifespan of maxillofacial prostheses is approximately one and a half to two years, with many patients experiencing issues such as discoloration and decreased satisfaction within the first three years of use [5].

To achieve a lifelike appearance for facial prostheses, pigments are used as opacifiers and colorants for intrinsic and extrinsic coloring [2]. To create the primary skin shade, basic colors such as red, blue, and yellow are employed in maxillofacial prosthetic silicone elastomers. The amount of each color used depends on factors such as the darkness of the complexion, the complexity of the anatomy, and the desired skin tone. Red is used more for darker complexions, while blue is required more for lighter skin tones. Yellow is used to match moderate to substantial differences in skin color [6,7].

Previous studies have utilized dry earth pigments as opacifiers or intrinsic colors in their research. For instance, Beatty et al. [8] examined the color changes in dry-pigmented maxillofacial elastomer when exposed to ultraviolet light. They discovered that the red cosmetic dry earth pigment exhibited significant color changes after 400 h of exposure, while the cosmetic yellow ochre remained color-stable even after 1800 h. Similarly, Kiat-amnuay et al. [9] investigated the use of dry earth cosmetic pigments mixed with silicone A-2186 and opacifiers. Their findings aligned with those of Beatty et al. [8], demonstrating that the red pigment groups experienced the most pronounced color changes.

In recent years, significant research has been dedicated to developing a novel industrial process for incorporating nanoparticles into a polymeric matrix. This advancement has led to the emergence of a new class of polymeric materials that combine the strength of nano-oxides with the flexibility of an organic polymer matrix. Adding nanoparticles to a polymeric matrix enhances its properties due to the particles' higher surface energy and chemical reactivity. This enables them to interact with the silicone elastomer matrix, forming a three-dimensional network within the polymer chain. By utilizing nanoparticles at the nanoscale, it becomes possible to fine-tune specific attributes of individual particles while exerting control over a range of properties, such as biological, mechanical, electrical, magnetic, and optical characteristics. For example, researchers have discovered that nanosized rutile TiO_2 particles possess exceptional ultraviolet (UV) absorption and scattering capabilities, effectively safeguarding against UV light. These nanoparticles demonstrate diminutive dimensions, extensive surface areas, active functionality, and robust interfacial interactions with organic polymers. Consequently, they enhance the polymer's physical and optical properties while fortifying its resilience against aging triggered by environmental stressors [10,11].

Nanoparticles serve as effective barriers against UV radiation because their size is smaller than the wavelength of UV light. When exposed to UV radiation, the electrons in nanoparticles vibrate, leading to the absorption and dissipation of a portion of the light. Consequently, the smaller the nanoparticles, the more efficient they are at shielding against solar radiation. Maintaining an appropriate filler content is crucial when incorporating nanosized oxide particles into the silicone elastomer matrix. This is due to these particles' higher surface energy and chemical reactivity, which can lead to agglomeration if the content exceeds a certain level. The presence of agglomerated particles within the silicone elastomer matrix causes them to act as stress-concentrating centers when subjected to external forces, thus decreasing the elastomer's mechanical strength [12].

Considering the findings of Han et al. [12] regarding the color stability of maxillofacial prostheses and previous studies on their mechanical properties, it was determined that nano-TiO_2 at content levels of 2.0% and 2.5% proved to be the most effective opacifier when used with A-2186 silicone maxillofacial elastomer. In addition, with 2% nanosized

TiO_2, there is an increased tendency to agglomerate, as an increase in particle size and concentration results in suboptimal dispersion of nano-oxides within the matrix. An augmented propensity for agglomeration further diminishes UV-protective capabilities [10].

To overcome the limitations of nanoaggregation, researchers have investigated the use of TiO_2 support material composites. By combining TiO_2 nanoparticles (NPs) with chitosan coatings, nanocomposite films with enhanced properties, such as mechanical strength, swelling properties, and thermal stability, can be achieved. Chitosan is crucial in preventing agglomeration and facilitating the dispersion of metal oxide particles within a composite. This approach offers a potential solution for overcoming the limitations associated with nanoaggregation [13].

Chitosan, a natural polysaccharide derived from the chitin shells of crustaceans, is highly valued for its effectiveness in various biomedical applications. It is obtained through treating chitin with alkaline substances such as sodium hydroxide. Chitosan possesses several desirable qualities, such as biodegradability, biocompatibility, non-toxicity, and antibacterial and hydrating properties. One of its unique features is its cationic nature, which results from the presence of amino and hydroxyl groups in its structure. This cationic nature allows chitosan to interact with other compounds through electrostatic forces, through hydrogen bonds, or by incorporating the compounds into its polymeric matrix, thus enhancing its mechanical and biological properties. Furthermore, chitosan exhibits a strong affinity for negatively charged compounds, particularly under low-pH conditions. These exceptional properties, along with its biocompatibility, biodegradability, and lack of toxicity, have contributed to its widespread utilization in various fields [14–16].

Incorporating chitosan microparticles into a matrix can provide strengthening properties, as the microcracks between the matrix and chitosan can absorb energy. However, when the concentration of chitosan exceeds a certain threshold of 3.0–3.5%, the particles tend to come closer or aggregate, transforming the microcracks into macrodefects. Consequently, this results in a decrease in the material's tensile strength [17,18].

Core–shell nanoparticles, which are constructed with one material encased in another, offer significant advantages in biological applications over standard nanoparticles. These benefits include reduced cytotoxicity, enhanced bio- and cytocompatibility, and improved stability, both thermally and chemically. Additionally, these nanoparticles can be more readily conjugated with other bioactive substances. The effectiveness of core–shell nanoparticles in biomedical applications is largely influenced by their surface chemistry, which enhances their ability to bond with drugs and receptors. This has inspired the creation of innovative nanoparticles that interact harmoniously with biological systems, unlike their bulk material counterparts. Furthermore, the enhanced biocompatibility and cytocompatibility of core–shell nanoparticles contribute to their therapeutic potential. This opens up exciting opportunities for the development of novel drug carriers that demonstrate improved characteristics such as longer residence times, higher bioavailability, and the ability to lower the required dosage and frequency. All of these benefits, in combination with their increased specificity, make core–shell nanoparticles a promising area of focus for future advancements in the field [19].

The hybrid composite of CS-TiO_2 has been extensively utilized in numerous technical applications because of its potential advantages. These applications range from acting as antimicrobial agents to being used in packaging materials, and in the production of films or coatings aimed at preserving food. Furthermore, this hybrid composite has proven useful in promoting wound healing and skin regeneration, detecting glucose and alpha-fetoprotein, and even degrading water pollutants [15].

Over time, sweat and sebum can be absorbed into extraoral silicone when a prosthesis is in contact with human skin. However, the presence of UV radiation can have detrimental effects. While it increases crosslinking, it also breaks down bonds within the polymer matrix, resulting in a slower polymerization rate and silicone degradation. These processes ultimately accelerate color changes and material deterioration [20].

Silicone elastomers have been subjected to controlled experiments to replicate the effects of environmental and human conditions experienced by prostheses during their use. These experiments involve exposing the elastomers to simulated conditions, such as sebum solution or acidic perspiration, artificial daylight radiation for accelerated aging, outdoor natural weathering, and treatment with silicone cleaning solutions. By subjecting silicone elastomers to these conditions, researchers aim to closely mimic real-world scenarios and evaluate the performance and longevity of the materials used in prosthetic applications [21].

For the engineering of a color-stable prosthesis, thoroughly understanding the effects imposed by various environmental variables is of utmost importance. Although accelerated aging tests can offer indications regarding the outdoor performance of polymers and provide estimates of their service life, they can also impact the degradation mechanism and potentially yield inaccurate estimations of the actual lifespan of the polymers. Hence, it is essential to comprehensively comprehend the individual influences of environmental variables to effectively engineer a color-stable prosthesis with prolonged durability [22,23].

Although numerous research studies have focused on investigating the impact of the incorporation of different nanoparticles on the color stability of pigmented and non-pigmented maxillofacial silicone under various accelerated aging conditions [5,12,22,24–27], there is a notable absence of research exploring the effect of incorporating hybrid nanocomposites on the color stability of silicone. Specifically, not a single study has examined the influence of hybrid nanocomposites on the color stability of silicone materials.

This study aims to investigate the impact of incorporating chitosan-TiO_2 nanocomposite on the color stability of pigmented maxillofacial silicone that has undergone room-temperature vulcanization (RTV) when exposed to various accelerated aging conditions. The null hypothesis posits that the color stability of the pigmented maxillofacial A-2186 silicone elastomer would remain unaffected by impregnation with a chitosan-TiO_2 nanocomposite, even after undergoing different accelerated aging procedures.

2. Materials and Methods

2.1. Materials

One hundred groups ($n = 5$; 500 specimens in total) of disk-shaped specimens were prepared (twenty-two millimeters in diameter, two millimeters thick) [9,10,28,29] using A-2186 room-temperature-vulcanized (RTV) maxillofacial silicone elastomer (Factor II Inc., Lakeside, AZ, USA). The dimensions of the samples were designed using AutoCAD 2013 and processed using a CNC (Computer Numerical Control) machine (Jinan, Shandong, China). Transparent acrylic sheets were used to fabricate molds; these plastic molds were made with specified dimensions that were suitable for digital portable colorimeter devices. Each mold comprised a base, frame, and cover parts with exact measurements.

The specimens were divided into four main groups of equal size (25 specimens/group) according to the pigments that had been combined with the silicone (brilliant red, blue, and yellow dry intrinsic pigments at 0.2% concentration by weight) [12] (Technovent Ltd., Bridgend, UK), and the fourth group was without the addition of pigment (non-pigmented group). Each group was subdivided into five divisions representing the categories of silicone. Figure 1 illustrates the study design.

Five categories of silicone were made, and the silicone without incorporation of any nanoparticles was considered as the control; the second category contained titanium dioxide nanoparticles that were impregnated into the silicone at a concentration of 2% by weight [12,30] and were obtained from Sigma-Aldrich (CAS Number 718467)(Sigma-Aldrich Chemie GmbH Eschenstrasse 5 D-82024 TAUFKIRCHEN, Germany) with a primary particle size of 21 nm; the third category was impregnated with chitosan at a concentration of 3% by weight [17] and was obtained from Sigma-Aldrich (CAS Number 448869) with a 75–85% degree of acetylation and low molecular weight; the fourth category was reinforced with hybrid combination of nano-TiO_2 and chitosan (TC) at a concentration of 1% (0.5% TiO_2 + 0.5% chitosan) by weight; the fifth category contained silicone impregnated with a synthesized nanocomposite of chitosan-TiO_2 powder.

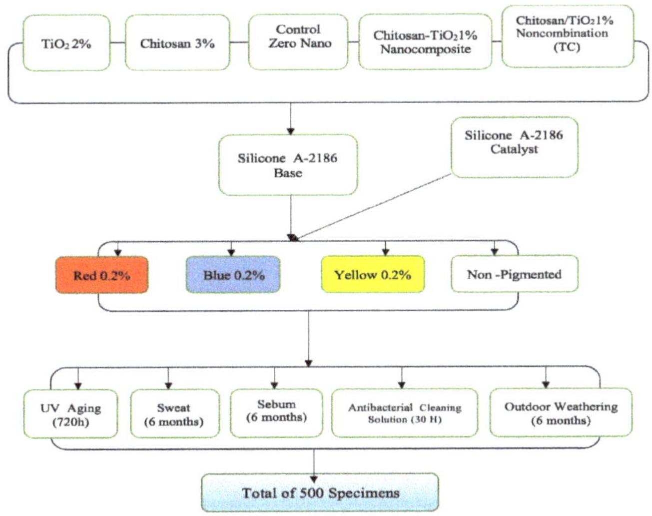

Figure 1. Study design.

Absolute ethanol 99.5% (EM- PARTA ACS; Merck-KGaA, 64271 Darmstadt, Germany) was used to assist in the dispersion of nanocomposite chitosan–TiO_2 powder in the silicone base [31].

2.2. Methods

2.2.1. Pilot Study

A pilot study was conducted to determine the most suitable concentration of the synthesized chitosan-TiO_2 nanocomposite for use by evaluating its color stability after subjecting it to different media for accelerated aging. At concentrations of 0.5%, 1%, and 1.5% by weight, the chitosan–TiO_2 nanocomposite was impregnated into red-pigmented silicone (at a concentration of 0.2% by weight of the brilliant red intrinsic pigment) and dispersed well with the assistance of absolute ethanol. The de-achromatization of weathering conditions for the red-pigmented silicone elastomers occurred at a higher rate than for those with yellow pigments and the un-pigmented samples [7,9,28,32–37].

Based on the results of the pilot study, the most suitable percentage of the chitosan-TiO_2 nanocomposite was 1%, which resulted in the least change in the red color of the silicone after being subjecting to accelerated aging conditions.

2.2.2. Preparation of the Nanocomposite

In a typical experiment with the core–shell method [38], 2 g of TiO_2 was dispersed in 200 mL of 1% (*v/v*) acetic acid (obtained from Merck, Rahway, NJ, USA (Cat. No. 100063)) and sonicated for thirty minutes at room temperature (Q700 Sonicator; Qsonica LLC, Newtown, CT, USA), where the TiO_2 was changed into Ti4 ions. In addition, 2 g of chitosan was dispersed in 200 mL of 1% (*v/v*) acetic acid and sonicated for thirty minutes at room temperature. Then, 200 mL TiO_2 + 200 mL chitosan was mixed and continuously stirred until a clear solution was obtained at room temperature. NaOH solution (1 *M*) was added dropwise until the solution reached a pH of 10. The solution was then separated, and the residue was filtrated with a Buchner funnel and washed with excess distilled water thrice until a pH of 7 was obtained from the chitosan-TiO_2 solution. After that, the mixture was dried in a vacuum oven at 60 °C overnight. Furthermore, the composite was dried using a magnetic stirrer (LabTech LMS-1003; Daihan Labtech Co., Ltd., Namyangju, Republic of Korea) for two hours at 80 °C. The dried composite was then crushed using a mortar to obtain a fine powder. An analysis of the morphological characteristics was conducted with a FESEM system, XRD, FTIR, and EDX.

2.2.3. Preparation of the Control Group Specimens

According to the manufacturer's instructions, the mixing ratio of the base to the catalyst for the RTV silicone type A-2186 was 10 to 1. The mixing of the control group began with the addition of the base to the container of an electronic balance that was sensitive to an accuracy of 0.0001 g (Nimbus Analytical, Adam Equipment, Oxford, CT, USA). The catalyst was added and mixed for five minutes with a vacuum mixer (AX-2000, Aixin Medical Equipment Co., Ltd., Tianjin, China) at a speed of 360 rpm and under a vacuum of −0.09 bar. For the intrinsically pigmented groups (brilliant red, blue, and yellow), pigment powder was added at a concentration of 0.2% [12] by weight to the container of the electronic balance, followed by the addition of the base; then, this was mixed without a vacuum for three minutes, followed by vacuum mixing for seven minutes. Next, the catalyst was added, mixed in a vacuum for the remaining five minutes, and placed in a vacuum chamber for fifteen minutes to eliminate air bubbles. Then, the mixed material was left for three minutes in the chamber without a vacuum to allow the settlement of the material. Afterward, the mixed silicone was poured into the plastic molds. G-clamps were used to tighten the mold so that the excess silicone material would flow outside its border. Then, the material was left to cure at room temperature for twenty-four hours.

2.2.4. Preparation of the Experimental Group Specimens

The experimental group specimens were fabricated by combining one type of nanoparticle (TiO_2, chitosan, and the hybrid chitosan–TiO_2 hybrid nanocombination (TC)) at 2%, 3%, and 1%, respectively, either alone or with one of the three dry intrinsic pigments (brilliant red, blue, and yellow) at a concentration of 0.2% [12] by weight; the base of the silicone to be mixed was as same as that in the control specimens, as stated in the manufacturer's instructions.

2.2.5. Preparation of the Experimental Group Specimens Reinforced with the Synthesized Chitosan–TiO_2 Nanocomposite

The synthesized chitosan–TiO_2 nanocomposite powder was combined with ethanol at weight percentages to fabricate the specimens. The volume of ethanol used depended on the amount of silicone base required to create the five specimens. The ethanol and chitosan–TiO_2 nanocomposite mixture was subjected to sonication using a Q700 Sonicator from Qsonica LLC. The sonication process lasted for thirty minutes at room temperature with continuous cooling. A titanium alloy probe with a standard half-diameter of 136 × 13 mm was utilized, and an amplitude power of four hundred watts was employed with a pulse-on time of five seconds and a pulse-off time of two seconds.

Next, the synthesized chitosan-TiO_2 nanocomposite and ethanol mixture was added to the RTV silicone type A-2186 base to prepare the specimens of the experimental group. The mixture was thoroughly mixed in a vacuum mixer for ten minutes. The vacuum mixer jar was placed on a magnetic hotplate stirrer (LabTech LMS-1003; Daihan Labtech Co., Ltd.) and connected to a vacuum rotary pump (EuroVac; Thompson CSF, La Défense, France) to evaporate the ethanol under a pressure of −0.075 MPa. The evaporation process took 120 min. To ensure the homogeneous dispersion of the synthesized chitosan–TiO_2 nanocomposite within the silicone, the concoction was mixed every three minutes for one minute during the two-hour period. Subsequently, the mixture was cooled to room temperature before the catalyst was added. The mixture was mixed for an additional five minutes in the vacuum mixer. Finally, the prepared mixture was loaded into molds and transferred to a vacuum chamber for two minutes to eliminate any air bubbles. G-clamps were used to tighten the mold so that the excess silicone material would flow outside its border. Then, the material was left to cure at room temperature for twenty-four hours. All of the specimens were then trimmed and cut-marked to classify the groups.

2.2.6. Color Stability Test

The prepared specimens were subjected to a color stability test that was carried out using a colorimeter (Portable Color Analyzer, Precise Digital Colorimeter/Color Meter Lab Tester with a four-millimeter measurement aperture—FRU WR10QC, China).

The evaluation of color differences was performed using the CIELAB (Commission Internationale de l'Eclairage) colorimetric system, which is shown in Figure 2. This system utilizes three parameters to define color: L*, a*, and b*. The L* axis corresponds to brightness, ranging from zero (black) to one hundred (pure white). The a* coordinate indicates the presence of red (positive values) or green (negative values), while the b* coordinate represents the amount of yellow (positive values) or blue (negative values).

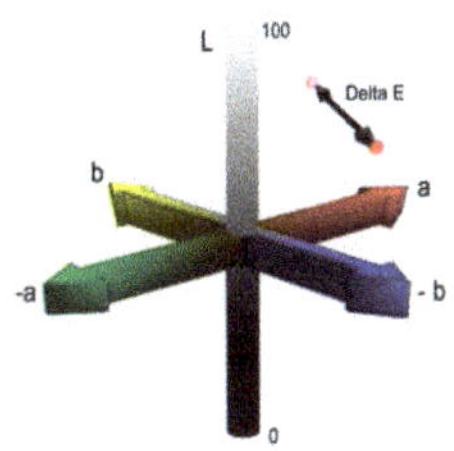

Figure 2. CIELAB color system [5].

The CIE colorimetric system allows for the calculation of the average ΔE (color variation) between two measurements using a specific formula:

$$\Delta E = [(\Delta L^*) + (\Delta a^*) + (\Delta b^*)]^{1/2} \quad (1)$$

This formula enables the determination of the mean color variation (ΔE) value by comparing the L*, a*, and b* values of two readings. Equation (1) was used.

Before weathering, all test specimens were washed, cleaned with distilled water, and gently dried with a clean paper towel. When readings were taken, a white card served as the background [39]. The L*, a*, and b* color coordinates for each sample prior to exposure to weathering were documented. Three measurements were taken and averaged, and the mean for each sample was determined using the CIELAB uniform color scale. L*, a*, and b* were the differences in the respective values before and after aging.

In the context of the LAB color space, the ΔE measure, which is determined as the Euclidean distance between two LAB vectors, is commonly employed to assess color perception. A lower value suggests that it is harder for an observer to differentiate between two colors, with 3.00 as the threshold value for human color discrimination.

Values of ΔE between zero and one represent unnoticeable color differences, whereas values between two and three represent color differences that are just noticeable. When values of ΔE exceed or are equal to 3.3, the color difference is visually noticeable and clinically unacceptable.

Distinguishing between perceptible and acceptable color changes is common in various studies. The variation in acceptability thresholds among these studies can be attributed to the variations in pigments and silicones used. In the context of this study, alterations in coloration that are below three units of magnitude are perceived to be visually detectable and are regarded as clinically tolerable [40].

2.2.7. Conditioning Modes

The specimens were randomly assigned to five groups, with each representing a different conditioning method. The color stability of each of the five categories of pigmented and non-pigmented silicone was evaluated according to the color difference before and after they were subjected to five different accelerated aging conditions. The first of these conditions was storage in a commercial antibacterial silicone-cleaning solution (B-200-12,

Daro Inc., Lakeside, AZ, USA) for thirty hours. The second condition was storage in a simulated sebum solution for six months; the sebum was prepared by dissolving 10% palmitic acid with 2% glyceryl tripalmitate in 88% linoleic acid (all *w/w*) [30,40–42]. The third condition was storage in simulated acidic perspiration (sweat) for six months; the sweat was prepared according to the specifications of the International Organization for Standardization (ISO 105-E04:96) [43] and comprised the following components (per liter of distilled water): 0.5 g of L-histidine monohydrochloride monohydrate, 5 g of sodium chloride, and 2.2 g of sodium dihydrogen orthophosphate dehydrate [30,40–42]. The fourth group of specimens was subjected to exposure to accelerated weathering with artificial ultraviolet (UV) radiation for 720 h. This was facilitated by an accelerated UV aging chamber equipped with two lightbulb fixtures positioned directly above the specimens. These fixtures provided ultraviolet light exposure equivalent to 720 $KJ/m^2/h$. In addition, the aging chamber was maintained at a consistent temperature of 60 °C and relative humidity of 80% [22,32]. The fifth group of specimens was subjected to outdoor weather conditioning for six months.

The environmental exposure of the samples to atmospheric conditions was conducted in accordance with the standards set forth in Designation G7-8.3.1 [44] of the American Society for Testing and Materials. The specimens that underwent natural outdoor weathering were meticulously suspended from stainless steel racks with ligature wire. The entire assembly was deliberately positioned on the roof of the College of Dentistry, University of Sulaimani, for six months [30,40,45,46] from February 2022 to August 2022. The monthly average high and low temperatures and climatic data were documented during the outdoor weathering, as shown in Table 1. The specimens were left uncovered and exposed to environmental conditions during the weathering exposure. They were suspended in such a way that there were no barriers or walls from the back, front, right, or left side, thus ensuring optimal exposure to environmental conditions. The specimens were checked daily to ensure that their position was not changed. Before the specimens were evaluated, they were cleaned for ten minutes in distilled water, wiped dry, and tested.

Table 1. Monthly average climatic data during the outdoor weathering experiment in 2022 in Sulaimani City in the region of Kurdistan in northern Iraq.

Date (2022)	Temperature °C			Average Humidity %	Pressure (mbar)
	Max	Min	Average		
February	15.2	5.4	10.3	64.0	917.4
March	15.3	6.4	10.8	59.6	915.9
April	26.2	14.6	20.4	43.5	914.5
May	28.2	17.3	22.8	41.8	913.2
June	37.4	24.6	31.0	28.3	910.1
July	40.7	26.8	33.7	24.2	906.4
August	42.1	28.0	35.0	23.9	908.5

The durations of conditioning were chosen to mimic the use of silicone prostheses over a period of twelve to eighteen months. Typically, patients use their prosthesis for eight to twelve hours a day; during this time, the prosthesis is anticipated to be exposed to one to three hours of daylight, regular environmental conditions, and constant sebum and sweat when worn on the defect site. Furthermore, patients typically dedicate around five minutes to cleansing their prostheses before bedtime. Hence, the duration of one month of service corresponds to approximately thirty to ninety hours [22] of daylight aging, along with ten to fifteen days of storage in sebum or acidic solutions (sweat), and one hundred and fifty minutes of storage in cleaning solutions [40].

2.3. Statistical Analysis

A comprehensive statistical analysis was carried out to concisely summarize the findings related to each variable that was studied. Descriptive statistics were calculated and presented for continuous variables, including the mean and standard deviation. A one-way analysis of variance (ANOVA) was utilized with a significance level of $p < 0.05$ to evaluate significant differences among the five accelerated aging conditions across the five categories of silicone specimens.

Since ANOVA operates on the assumption of equal variances across specimens, Levene's test of homogeneity of variance ($\alpha = 0.05$) was applied to all data, assuming equal variance. Consequently, when the equal-variance assumption was rejected, Dennett's T3 multiple comparison test was used to compare the groups ($p < 0.05$). On the other hand, when the assumption of equal variance was accepted ($p < 0.05$), Tukey's HSD multiple comparison tests were used to compare the groups.

A one-sample *t*-test was performed within each accelerated aging condition to examine the effects of the modes on color change, with a significance level of $p < 0.05$.

The finding that the variable in this analysis was normally distributed was obtained by applying the Shapiro–Wilk and Kolmogorov–Smirnov tests. Then, a *t*-test and one-way ANOVA were performed, and version 27.0 of the SPSS program for Windows (SPSS for Windows) was utilized.

3. Results

3.1. Morphological Characteristics

3.1.1. Scanning Electron Microscopy (SEM)

SEM was used to study the surface morphologies of the various coatings of nanoparticles; the SEM analysis revealed that the average particle size of the synthesized chitosan-TiO_2 nanocomposite was from 31.52 to 60.07 nm. Figure 3. Field-emission scanning electron microscopy (FESEM) was used to acquire information about the external morphology of the nanocomposites. In this research, FESEM confirmed the adequate homogenous distribution of nanosized and less-aggregated TiO_2 particles within the chitosan (CS) matrix. The irregular spherical morphologies were due to the interlocking interaction between TiO_2 and CS, which consequently improved the homogeneity of the two.

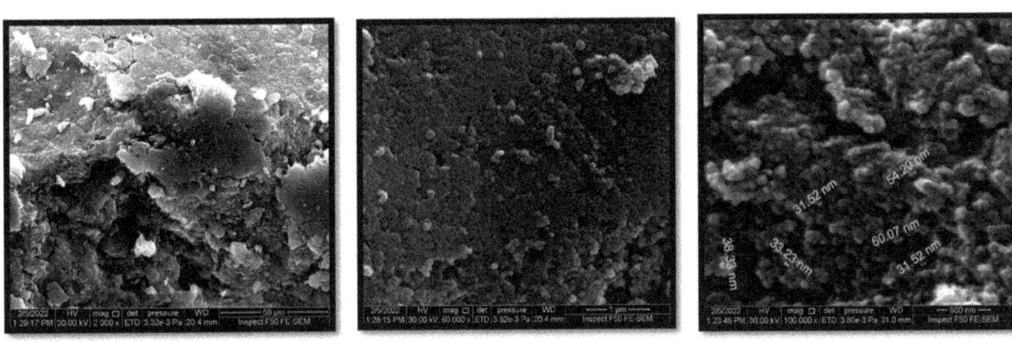

(**a**) 50 µm resolution (**b**) 1 µm resolution (**c**) 500 nm resolution

Figure 3. SEM of the synthesized chitosan-TiO_2 nanocomposite.

3.1.2. X-ray Diffraction (XRD)

The crystalline structure of the synthesized nanocomposites was analyzed via XRD. XRD patterns of the synthesized chitosan-TiO_2 nanocomposite, pure TiO_2, and chitosan are demonstrated in Figure 4. The XRD pattern of chitosan displayed small, broad peaks at (11.8°) and (20.0°), indicating its semi-crystalline structure. This suggests that the incorporation of TiO_2 into the chitosan matrix predominantly takes place in the semi-crystalline region of chitosan. In contrast, both chitosan and TiO_2 exhibited distinct diffraction peaks in the XRD pattern of the nanocomposite. The TiO_2 nanoparticles displayed the presence of anatase and rutile forms, as evidenced in Figure 4c. The coexistence of these mixed phases of TiO_2 is advantageous in minimizing the recombination of photogenerated electrons and holes, thereby enhancing the photocatalytic activity of titanium.

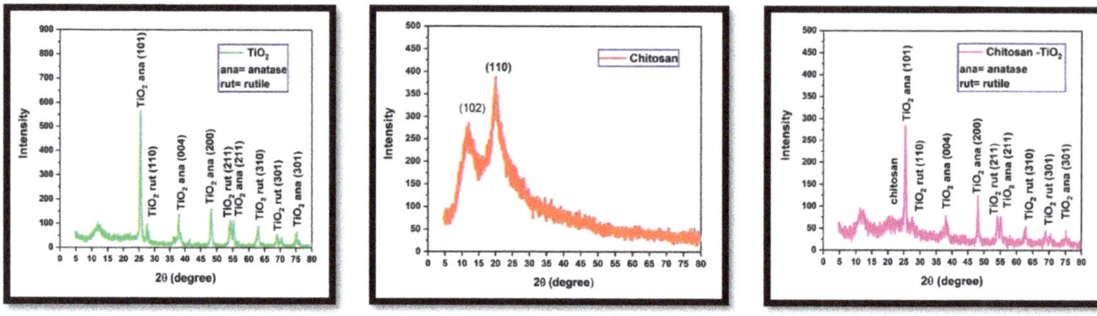

(a) (b) (c)

Figure 4. X-ray diffraction: (**a**) XRD of TiO_2; (**b**) XRD of chitosan (**c**) XRD of chitosan-TiO_2.

3.1.3. Fourier-Transform Infrared Spectroscopy (FTIR)

The FTIR analysis, Figure 5, revealed the presence of characteristic bands corresponding to chitosan, TiO_2, and the chitosan-TiO_2 nanocomposite. In the FTIR spectra of chitosan, Figure 5a, the peaks around (915, 1029), (1250), (1340, 2888), (1550), and (3460) cm^{-1} are related to (C-O-C), (C-O-H), (C-H), (C=O), and (OH−), respectively. In the FTIR spectra of TiO_2, Figure 5b, the peaks around (530, 654) are assigned to the bending vibration of (Ti-O-Ti) bonds, while that at 3500 cm^{-1} corresponds to the hydroxyl group (OH−).

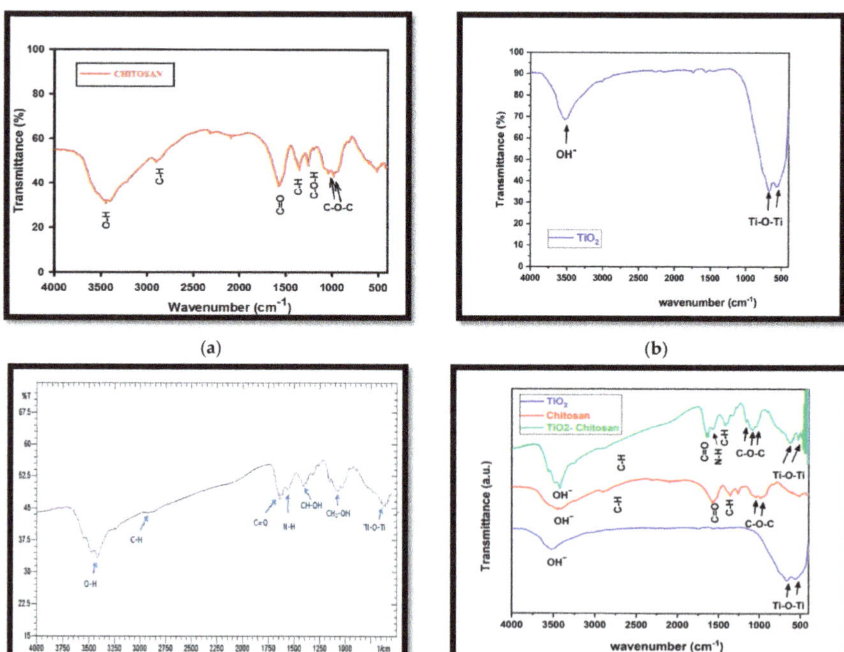

(a) (b)

(c) (d)

Figure 5. Fourier-transform infrared spectroscopy (FTIR); (**a**) FTIR of chitosan, (**b**) FTIR of TiO_2, (**c**) FTIR of synthesized chitosan-TiO_2 nanocomposite, (**d**) FTIR of chitosan, TiO_2, and chitosan-TiO_2.

Figure 5c,d demonstrate the FTIR spectra of the synthesized nanocomposite of chitosan-TiO_2. In the spectra, the peaks around (3441, 3437), (2855, 2926), and (1621, 1619) cm^{-1} are related to (O–H), (C–H), and (C=O) groups, respectively. The peaks around (1575, 1500), (1425, 1352), and (1112, 1040) cm^{-1} are related to (N–H), (CH–OH), and (CH2–OH) groups, respectively. The fingerprint band between 700 and 400 cm^{-1} shows stretching vibrations of (Ti-O-Ti), which indicates the immobilization of TiO_2 onto the chitosan matrix.

3.1.4. Energy-Dispersive X-ray (EDX)

Energy-dispersive X-ray (EDX) spectroscopy is frequently utilized to determine the elemental distribution within a material's composition. In this research, the presence of TiO_2 within the nanocomposite was examined using EDX. The existence and dispersion of TiO_2 throughout the composite were confirmed via the EDX spectrum, as shown in Figure 6. From the data presented in Table 2, it can be inferred that TiO_2 has been incorporated into the composite, and its particulates have been uniformly dispersed.

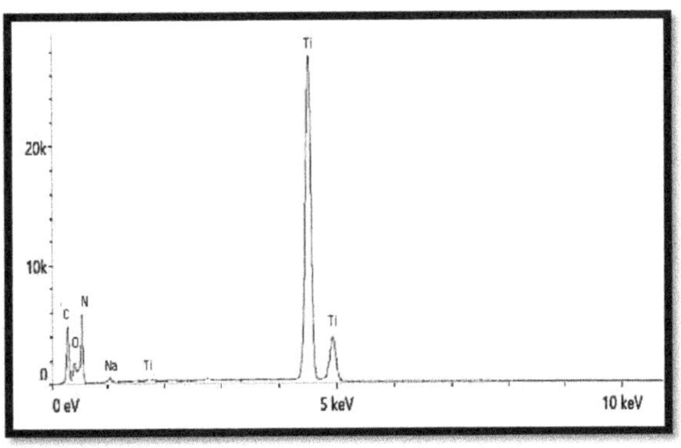

Figure 6. EDX spectra for nanocomposite chitosan-TiO_2.

Table 2. EDX results for nanocomposite chitosan-TiO_2.

Element	Weight %	Atomic %
C	12.9	22.0
N	3.7	5.4
O	42.9	54.8
Na	1.3	1.1
Si	0.4	0.3
Ti	38.8	16.5

3.2. *Pilot Study*

The mean, standard deviation, and *p*-value for ΔE in the pilot study are listed in Table 3; the red-pigmented experimental silicone specimens that were modified with 1% chitosan-TiO_2 and chitosan-TiO_2 1.5% nanocomposites exhibited significantly less of a color change ($p < 0.05$) than the chitosan-TiO_2 0.5% specimens did when exposed to outdoor weather for six months.

Moreover, as shown in the statistical analysis, the ΔE of the red color in the 1% chitosan–TiO_2 specimens was found to be significantly lower ($p < 0.05$) than that in the 0.5% chitosan–TiO_2 and 1.5% chitosan–TiO_2 specimens when subjected to sebum. These findings led to the selection of the 1% chitosan–TiO_2 nanocomposite as the percentage for use in the specimens of the subsequent experimental group.

Table 3. Mean values and standard deviations of ΔE for the brilliant-red-pigmented silicone categories under different conditions (pilot study).

Groups	Brilliant Red ΔE (Pilot Study)			
	1% Chitosan–TiO_2	1.5% Chitosan–TiO_2	0.5% Chitosan–TiO_2	*p*-Value
Sweat (6 months)	6.66 ± 3.60	5.20± 0.36	4.12± 0.87	0.305
Antibacterial cleaning solution (30 h)	3.31 ± 1.67	2.99 ± 2.02	1.28 ± 0.64	0.179
Outdoor weather (6 months)	16.62 ± 6.32 [b]	17.59 ± 1.68 [b]	24.64 ± 0.68	0.035
UV weather (1 month; 720 h)	27.92 ± 1.24	26.45 ± 1.11	25.61 ± 2.64	0.162
Sebum (6 months)	6.24 ± 1.66 [a,b]	10.09 ± 2.96	11.14 ± 0.92	0.01
p-value	0.000 *	0.000 *	0.000 *	

[a]: In comparison with 1.5% chitosan–TiO_2. [b]: In comparison with 0.5% chitosan–TiO_2. Different superscript letters in the same row indicate that the significant differences in ΔE ($p < 0.05$) were present only after applying one-way ANOVA, Dennett's T3 test, and Tukey's HSD multiple comparison tests. * There were overall significant differences among the conditions in the color changes in all silicone specimens.

3.3. Color Stability Results

The mean, standard deviation, and significant differences (*p*-value) for the brilliant red, blue, yellow, and non-pigmented ΔE among all silicone categories after being subjected to different conditions are demonstrated in Tables 4–7, respectively.

3.3.1. The Red Color

For the brilliant red pigments, within all silicone specimens, the UV weather (720 h) induced a prominent alteration in color (ΔE = 27.93, 31.38, 27.36, 41.19, and 40.84) ($p < 0.05$), with the greatest ΔE being observed in the 3% chitosan silicone specimens (ΔE = 41.19). However, smaller color changes were observed in all silicone specimens after immersion in antibacterial cleaning solution (30 h), with the smallest color changes being observed for 2% TiO_2 and 1% TC (ΔE = 1.03 and 1.35), respectively.

According to the statistical analysis, all conditioning modes provoked visually detectable color changes (ΔE > 3) in the 1% chitosan–TiO_2, 3% chitosan, and control (zero nanocomposite) silicone specimens. The color changes in all specimens were significantly smaller ($p < 0.05$) than those in the control (zero nanocomposite) specimens when they were exposed to sebum. Consistently, the ΔE values of the 1% TC and 2% TiO_2 categories were significantly lower ($p < 0.05$) than those of the 3% chitosan and control (zero nanocomposite) silicone specimens when exposed to the antibacterial cleaning solution.

Table 4. Mean values and standard deviations of the ΔE values of the brilliant red silicone categories under different conditions.

Groups	Brilliant Red ΔE					
	1% Chitosan–TiO_2	1% TC	2% TiO_2	3% Chitosan	Control (Zero Nanocomposite)	*p*-Value
Sweat (6 months)	6.66 ± 3.60	6.14 ± 0.38	5.45 ± 0.58	5.01 ± 2.02	5.15 ± 1.43	0.561
Antibacterial cleaning solution (30 h)	3.31 ± 1.67	1.35 ± 0.49 [c,d]	1.03 ± 0.98 [c,d]	4.62 ± 2.39	4.66 ± 1.11	0.000
Outdoor weather (6 months)	16.61 ± 6.32	21.19 ±1.70	19.74 ± 2.95	20.87 ± 1.03	18.91 ± 9.90	0.695
UV weather (1 month 720 h)	27.93 ± 1.24 [a,c]	31.38 ± 0.6 [b,c]	27.36 ± 0.71 [c]	41.19 ± 2.6	40.84 ± 7.15	0.000
Sebum (6 months)	6.24 ± 1.66 [d]	7.48 ± 1.66 [d]	6.25 ± 2.34 [d]	9.26 ± 1.85 [d]	14.44 ± 2.36	0.000
p-value	0.000 *	0.000 *	0.000 *	0.000 *	0.000 *	

[a]: In comparison with 1% TC. [b]: In comparison with 2% TiO_2. [c]: In comparison with 3% chitosan. [d]: In comparison with the control (zero nanocomposite). Different superscript letters in the same row indicate that significant differences in ΔE ($p < 0.05$) were present only after applying one-way ANOVA, Dennett's T3 test, and Tukey's HSD multiple comparison tests. * There were overall significant differences among conditions in the color changes in all silicone specimens.

3.3.2. The Blue Color

Significant color changes (ΔE = 9.08, 8.92, 10.65, 7.73, and 5.71) ($p < 0.05$) were most remarkable in the blue-pigmented specimens across all silicone categories when subjected to outdoor weather for six months. Specifically, the 2% TiO_2 specimens exhibited the greatest color changes (ΔE = 10.65). However, when these 2% TiO_2 specimens were exposed to sweat for six months and an antibacterial cleaning solution for thirty hours, they displayed significantly smaller color changes (ΔE = 0.96 and 0.74, respectively).

However, when exposed to all tested conditions except for the antibacterial cleaning solution, the 1% chitosan–TiO_2 and 3% chitosan specimens showed visually detectable color changes ($\Delta E > 3$).

Statistically, the color change, as indicated by the ΔE value, of the 1% chitosan–TiO_2 specimen was significantly greater ($p < 0.05$) than that in the control (zero nanocomposite) specimens when exposed to UV weathering for 720 h. Similarly, the color change in the 3% chitosan specimens was significantly greater ($p < 0.05$) than that in the control (zero nanocomposite) specimens when exposed to sweat, outdoor weathering, and UV weathering for 720 h.

On the other hand, 1% TC and 2% TiO_2 specimens displayed significantly smaller color changes ($p < 0.05$) than the 3% chitosan specimens when exposed to both sweat and sebum.

Table 5. Mean values and standard deviations of ΔE for the blue silicone categories under different conditions.

Groups	Blue ΔE					p-Value
	1% Chitosan-TiO_2	1% TC [a]	2% TiO_2 [b]	3% Chitosan	Control (Zero Nanocomposite)	
Sweat (6 months)	3.49 ± 2.57	1.18 ± 0.82 [c]	0.96 ± 0.26 [c]	3.46 ± 1.02 [d]	1.50 ± 0.55	0.001
Antibacterial cleaning solution (30 h)	1.21 ± 0.67	1.37 ± 0.93	0.74 ± 0.43 [c]	2.94 ± 1.39	1.32 ± 0.31	0.001
Outdoor weather (6 months)	9.08 ± 1.44 [d]	8.92 ± 0.59 [d]	10.65 ± 0.25 [d,c]	7.73 ± 0.66 [d]	5.71 ± 1.57	0.000
UV weather 1 month (720 h)	3.73 ± 0.89 [d]	2.67 ± 0.58	2.67 ± 0.37	3.63 ± 0.93 [d]	1.88 ± 0.86	0.005
Sebum (6 months)	4.21 ± 1.79	2.84 ± 1.34 [c]	3.09 ± 1.36 [c]	5.76 ± 2.19	4.85 ± 1.26	0.018
p-value	0.000 *	0.000 *	0.000 *	0.000 *	0.000 *	

[a]: In comparison with 1% TC. [b]: In comparison with 2% TiO_2. [c]: In comparison with 3% chitosan. [d]: In comparison with the control (zero nanocomposite). Different superscript letters in the same row indicate that significant differences in ΔE ($p < 0.05$) were present only after applying one-way ANOVA, Dennett's T3 test, and Tukey's HSD multiple comparison tests. * There were overall significant differences among conditions in the color changes in all silicone specimens.

3.3.3. The Yellow Color

Among all of the silicone specimens with yellow pigment, the most significant color change was observed when they were subjected to outdoor weather for six months. The highest ΔE value of 3.04 was recorded for the control (zero nanocomposite) specimens. Conversely, the 2% TiO_2 category specimens displayed the smallest color change after being subjected to sweat, with a ΔE value of 0.48.

In the statistical analysis, it was found that under most of the conditions, excluding that of sweat, there was not a significant color change in any of the specimens when they were compared with each other. However, when exposed to sweat, a significantly greater color change ($p < 0.05$) was seen in the 1% chitosan–TiO_2 and 1% TC specimens than in the 2% TiO_2 specimen. In contrast, with the same sweat exposure, the 2% TiO_2 specimen showed a significantly lower ΔE value ($p < 0.05$) than both the 3% chitosan and control (zero nanocomposite) specimens; the observed color changes were within the clinically accepted limits, with $\Delta E < 3$.

Table 6. Mean values and standard deviations of the ΔE values of the yellow silicone categories under different conditions.

Groups	Yellow ΔE					p-Value
	1% Chitosan–TiO_2	1% TC [a]	2% TiO_2	3% Chitosan	Control (Zero Nanocomposite)	
Sweat (6 months)	1.34 ± 0.48 [b]	1.69 ± 0.85 [b]	0.48 ± 0.23 [c,d]	1.5 ± 0.60	2.21 ± 1.24	0.006
Antibacterial cleaning solution (30 h)	1.84 ± 0.50	1.14 ± 0.35	0.86 ± 0.38	1.41 ± 0.74	1.45 ± 0.78	0.103
Outdoor weather (6 months)	2.61 ± 0.56	2.64 ± 0.63	2.72 ± 0.74	2.72 ± 0.40	3.04 ± 0.61	0.803
UV weather (1 month; 720 h)	0.95 ± 0.26	0.7 ± 0.26	1.02 ± 0.16	0.89 ± 0.32	1.13 ± 0.43	0.267
Sebum (6 months)	1.07 ± 0.36	1.49 ± 0.87	1.38 ± 0.50	1.32 ± 0.40	1.6 ± 1.29	0.788
p-value	0.000 *	0.002 *	0.000 *	0.000 *	0.035 *	

[a]: In comparison with 1% TC. [b]: In comparison with 2% TiO_2. [c]: In comparison with 3% chitosan. [d]: In comparison with the control (zero nanocomposite). Different superscript letters in the same row indicate that significant differences in ΔE ($p < 0.05$) were only present after applying one-way ANOVA, Dennett's T3 test, and Tukey's HSD multiple comparison tests. * There were overall significant differences among conditions in the color changes in all silicone specimens.

3.3.4. The Non-Pigmented Group

In the non-pigmented group, the most substantial color changes across all silicone categories occurred when the specimens were exposed to sebum, with the control (zero nanocomposite) specimens showing a ΔE value of 9.02. On the other hand, the UV weathering for 720 h resulted in significant color changes across all categories when they were compared with each other. Interestingly, the antibacterial cleansing solution induced the smallest color change across all silicone categories, with a low ΔE value of 0.76 being recorded for the 2% TiO_2 silicone variant.

From the statistical analysis, it emerged that the silicone specimens in the 3% chitosan category had a significantly greater color change ($p < 0.05$) than those in the control (zero nanocomposite) category when exposed to sweat for six months and UV weathering for 720 h. However, when exposed to outdoor weathering for six months, the specimens in the 1% chitosan–TiO_2 category exhibited a significantly higher ΔE value (7.77) ($p < 0.05$) than those of all other silicone categories.

Moreover, when exposed to sebum, the color change in the 1% chitosan–TiO_2 and 1% TC categories was significantly higher ($p < 0.05$) than that in the 2% TiO_2 silicone category. However, when exposed to the same conditions, the 2% TiO_2 category was found to have a lower ΔE value ($p < 0.05$) than the control (zero nanocomposite) silicone category. Additionally, with exposure to sweat, the ΔE values of all specimens were significantly lower ($p < 0.05$) than those of the 3% chitosan silicone category.

Table 7. Mean values and standard deviations of the ΔE values of the non-pigmented silicone categories under different conditions.

Groups	Non-Pigmented ΔE					p-Value
	1% Chitosan–TiO_2	1% TC	2% TiO_2	3% Chitosan	Control (Zero Nanocomposite)	
Sweat (6 months)	1.12 ± 0.41 [c]	1.70 ± 0.99 [c]	1.26 ± 0.59 [c]	7.63 ± 3.5 [d]	1.39 ± 0.46	0.000
Antibacterial cleaning solution (30 h)	1.24 ± 0.48	1.10 ± 0.37	0.76 ± 0.37	1.17 ± 0.32	0.95 ± 0.39	0.294
Outdoor weather (6 months)	7.77 ± 0.36 [a,b,c,d]	3.42 ± 0.45	4.11 ± 0.93	3.13 ± 1.61	4.67 ± 2.54	0.000
UV weather (1 month 720 h)	3.81 ± 0.35 [a,b,d]	2.00 ± 0.51 [c]	1.8 ± 0.34 [c]	5.04 ± 1.21 [d]	1.88 ± 0.99	0.000
Sebum (6 months)	4.81 ± 0.92 [b]	4.84 ± 0.92 [b]	2.55 ± 0.52 [d]	5.5 ± 4.28	9.02 ± 3.96	0.009
p-value	0.000 *	0.000 *	0.000 *	0.014 *	0.000 *	

[a]: In comparison with 1% TC. [b]: In comparison with 2% TiO_2. [c]: In comparison with 3% chitosan. [d]: In comparison with the control (zero nanocomposite). Different superscript letters in the same row indicate that significant differences in ΔE ($p < 0.05$) were only present after applying one-way ANOVA, Dunnett's T3 test, and Tukey's HSD multiple comparison tests. * There were overall significant differences among conditions in the color changes in all silicone specimens.

4. Discussion

In this study, all silicone categories, whether pigmented or non-pigmented, underwent different amounts of color change regardless of their aging conditions. The results of this study support the rejection of the null hypothesis that the color stability of the pigmented maxillofacial A-2186 silicone elastomer was affected by impregnation with the chitosan–TiO_2 nanocomposite after being subjected to different accelerated aging conditions.

The deterioration of color in facial prostheses can be attributed to environmental factors, such as solar radiation, temperature, and water. Solar radiation consists of ultraviolet (UV), visible, and infrared radiation, with UV radiation significantly impacting color stability. The depletion of the ozone layer since the 1970s has raised concerns about the effects of UV radiation on facial prostheses [12].

Additionally, routine cleaning and disinfection procedures can lead to color alterations in maxillofacial silicone prostheses due to the high permeability of silicone. While various cleansing agents, such as water, neutral soap, and chlorhexidine, are recommended, they should be used cautiously to prevent adverse effects on the material's physical properties [47].

In alignment with the study by Hatamleh et al. [40], the findings of this research also demonstrated that the antibacterial cleaning solution ensured the greatest color stability across all silicone categories, regardless of whether they were pigmented or not.

This study introduced the development of a novel three-phase composite by merging nanoparticles with polymeric silicone, a technique that has not been previously employed. This composite was successfully created by meticulously integrating two different particles—nanoparticles (TiO_2) and microparticles (chitosan)—in specific proportions. As a result, the overall qualities of the silicone polymers saw significant enhancements. This progressive development is a substantial stride toward endowing maxillofacial prostheses with ideal properties and realistic attributes.

Hybrid nanoparticles, such as chitosan–TiO_2 composites, have generated significant interest, as they merge the properties of organic and inorganic components, resulting in novel materials with enhanced and unique characteristics [15,48].

The core–shell mixing method was effectively employed in this study to prepare a TiO_2-supported chitosan nanocomposite. Comprehensive morphological characterization utilizing scanning electron microscopy (SEM) (Figure 3), X-ray diffraction (XRD) (Figure 4), and Fourier-transform infrared spectroscopy (FTIR) (Figure 5) confirmed the successful adsorption of the TiO_2 nanopowder into the chitosan matrix with excellent dispersion. These findings demonstrate the meticulous preparation and integration of the TiO_2 and chitosan components within the nanocomposite.

Achieving proper dispersion of chitosan–TiO_2 nanopowder in a polymer matrix, such as silicone, poses a significant challenge in nanocomposite production due to nanoparticle aggregation. However, achieving uniform dispersion is crucial for polymers and nanocomposites that require enhanced color stability. Ethanol, as a polar solvent that possesses hydroxyl (OH) groups, exhibits strong reactivity with ions, resulting in a prolonged dispersion effect. Its use as a stabilizer facilitates the sustained dispersion of a nanopowder within the ethanol solvent, thereby enabling a more uniform dispersion within the polymer matrix [48].

A novel method was employed in this study to achieve improved dispersion of synthesized chitosan–TiO_2 nanocomposite powder within a silicone matrix. The nanocomposite powder was dispersed in ethanol through sonication and subsequently mixed with the silicone base. The silicone mixture was subjected to heating and vacuuming to remove the ethanol. This approach intended to achieve a higher level of dispersion without the need for any additional third-party materials that could potentially compromise the properties of the silicone. According to Abdalqadir et al. [31], ethanol was found to have no detrimental effects on the structure and integrity of silicone elastomers.

Based on the current study's findings, the red-colored samples in all silicone categories showed drastic color alterations after undergoing 720 h of UV-accelerated aging. The most substantial change was observed in the 3% chitosan samples (41.19), and the smallest was observed in the 2% TiO_2 samples (27.36). Similarly, outdoor weathering over a period of six months led to a steep increase in ΔE values, which exceeded 21.19.

From a visual perspective, the pigments transitioned from a brilliant red to a lighter pink, and the pigmented elastomers indicated near-total color depletion. The findings are analogous to those of prior research conducted by Beatty et al. [8] and Kiat-Amnuay et al. [9].

Subsequently, the ΔE values for the brilliant-red pigment in all silicone categories exhibited discernible color shifts (ΔE > 3) in each instance of the accelerated aging processes, apart from the ΔE values of 2% TiO_2 and 1% TC, which demonstrated minimal color transformations after being submerged in an antibacterial solution for thirty hours, registering values of 1.03 and 1.35, respectively.

These observations align with the findings of earlier research by Beatty et al. [8] and Kiat-Amnuay et al. [33], who pointed out that using an intrinsic red pigment may contribute to color degradation. This could be due to the organic nature of such red pigments, which are more susceptible to the effects of irradiation. Organic colorants, depending on whether double and triple bonds are used for color provision, tend to be relatively reactive and less stable. Conversely, inorganic pigments usually offer greater color stability. However, they are often not favored due to their lower brightness compared with that of organic pigments, thus presenting a challenge in achieving a good color match.

Silicone is a polymer that is commonly available in the form of a moderately viscous liquid and is characterized by weak molecular interactions and difficulties in incorporating pigments. This property leads to chromatic alterations in the material, as smaller pigment particles tend to aggregate, while larger particles tend to separate from the polymer. Moreover, the polymerization process of silicone elastomers can be influenced by organic materials, such as makeup powder. Consequently, the presence of pigments may delay the polymerization process and exacerbate the effects of accelerated aging on the materials [11].

Among the various pigments, the cosmetic powder displayed the greatest degree of color alterations. This phenomenon could be attributed to factors such as the incorporated particles' magnitude or the impact of aging. Silicone, which is known for its lower cohesive energy, exhibits weaker molecular interactions. Consequently, diminutive particles have a propensity to cluster, while their larger counterparts often disengage from the polymer, providing no reinforcement for the material's structure. The particulate components of the cosmetic powder, which were possibly of a larger size, were more prone to separation from the polymer chain, which could have contributed to the increased color instability within these materials. Additionally, pigments derived from organic sources, such as makeup powder, undergo amplified degradation with aging, often dissolving upon interaction with ultraviolet light [27]. In addition, due to their larger size, dry earth pigments, which bear a structural similarity to cosmetic powders, tend to remain dispersed instead of becoming integrated within the polymer matrix. This characteristic potentially increases their vulnerability to UV degradation [44].

The susceptibility of polymeric biomaterials to deterioration under environmental conditions stems from their inadequate resistance to significant thermal changes and sunlight. Elements of weather, such as temperature, solar radiation, and moisture, can impact the properties of silicone elastomers by triggering chemical alterations. These changes, in turn, result in modifications to the functional features of these materials. The effective performance of silicone elastomers in response to extraoral factors can be ascertained by conducting tests that accurately simulate the conditions associated with outdoor exposure [45].

This study revealed that both outdoor weathering and UV-accelerated artificial weathering had a notable impact on the color of all types of silicone, irrespective of whether they were pigmented or not. Silicone, as a polymer, possesses aromatic rings and C=C bonds, which can be susceptible to the effects of UV light, resulting in color instability. When these functional groups in a polymer absorb UV light, they become energetically unstable. However, this excess energy can be mitigated through various means, such as by transferring the excitation to other molecules for stabilization. The excited groups can then return to their original state by releasing the excess energy in the form of light with a longer wavelength or heat. Failure to dissipate the excess energy can lead to photochemical degradation, causing detrimental effects such as color or brightness loss, reduced opacity, and material stiffness. Consequently, the presence of aromatic rings and C=C bonds in polymers allows UV-light-induced degradation, leading to adverse changes such as altered color and brightness, decreased opacity, crack formation, and increased rigidity [22,49].

As demonstrated by the findings of this study, the samples that underwent artificial aging showed more considerable color changes (ΔE) in comparison with those that were exposed to outdoor weathering. In outdoor weathering, the samples were subjected to natural conditions, yet this approach lacked the ability for precise control and was deemed subjective. In such weathering experiments, accurate regulation cannot be achieved for factors contributing to degradation, including geographic location, seasonal changes, specific weather conditions, time of day, and exposure duration [22].

The observed outcome could be attributed to the occurrence of post-polymerization crosslinking triggered by light irradiation, which leads to alterations in the structure of the polymer network. These modifications may involve changes in the polymer chain length, intermolecular bonding, and spatial arrangement of the polymer chains. Consequently, these changes affect light transmission through the maxillofacial material and contribute to the degradation of the polymer's color shade [22,49].

This study declares that integrating nanosized particles, whether they were utilized independently or as a composite, exhibited insufficient efficacy in safeguarding silicone against color degradation. The influences of all types of weathering on silicone color, particularly under outdoor weathering conditions, were notably significant.

Nano-TiO_2 is commonly used as an inorganic UV absorber due to its high thermal stability and photostability, unlike organic UV absorbers, which tend to migrate within the polymeric matrix and are less stable. When nanoparticles are exposed to UV light, their electrons vibrate, leading to a combination of scattering and absorption of UV radiation. The UV-shielding ability of nanoparticles is a result of this combined effect. Smaller particle sizes and lower nano-oxide concentrations enhance their dispersion within the elastomer matrix, thereby improving UV shielding [10,12,50].

According to Bangera et al. [51], aggregates of nano-oxide particles typically range between 30 and 150 nm, representing the most minor units in sunscreen formulations. These aggregates are formed through clustering primary particles, creating tightly bound structures that are larger than the individual building blocks. Increased particle size and concentration result in poor dispersion and a higher tendency for particle agglomeration, reducing the effectiveness of UV shielding. This is supported by studies conducted by Akash et al., Han et al., and Bishal et al. [10,12,50].

The decline in color stability observed in this study following exposure of the elastomer to outdoor conditions can be attributed to the incorporation of a nanosized composite comprising chitosan–TiO_2 particles with sizes ranging from 31 to 60 nm. The aggregation of these particles played a significant role in this reduced stability. These findings align with the research conducted by Bangera et al. [51], further supporting the notion that particle aggregation negatively impacts color stability in silicone elastomers subjected to outdoor conditions.

All silicone categories, whether pigmented or non-pigmented, demonstrated considerable color alterations upon exposure to outdoor weathering. Notably, the greatest color modification was detected in the blue-pigmented silicone with a ΔE value of 10.65—specifically, in the 2% TiO_2 variant. Similarly, significant color shifts were observed in the yellow-pigmented silicone, with the control variant registering a maximum ΔE value of 3.04. Furthermore, non-pigmented silicone was not immune to these changes; the chitosan–TiO_2 variant showed the most substantial color change, with a ΔE value of 7.77. These findings underline the broad impact of outdoor weathering on color stability across all silicone types.

The exposure of elastomers to outdoor conditions led to a slight whitening and yellowing phenomenon. This was caused by the photo-oxidative degradation of the polymer, which occurred due to the combined action of oxygen and sunlight. The process involves the formation of free radicals, reactions with oxygen, and the subsequent production of polymer oxy and peroxy radicals, resulting in chain scission. Additionally, crosslinking can occur through reactions between different free radicals or bonding between existing monomers and chains. When a polymer molecule absorbs ultraviolet light, it becomes molecularly unstable. Excess energy can be transferred between molecules, allowing the molecules to regain stability. However, the release of excess energy leads to photochemical degradation, contributing to the deterioration of the polymer molecules. This degradation process causes modifications in the polymer network structure, thus affecting the number of polymer chain units, their bonding, and their spatial arrangement. Consequently, this affects light transmission through the maxillofacial material and degrades the polymer's color shade. These findings are supported by research conducted by Malavazi et al. [52] and Hatamelh et al. [40], who emphasized the significance of photochemical degradation and crosslinking in the color changes observed in polymers exposed to outdoor environments.

In addition, facial prostheses are in direct contact with human skin for extended periods, and during this time, they can absorb perspiration and sebum. The absorbed secretions can potentially cause changes in the structure of the deteriorating elastomer, ultimately contributing to the overall deterioration of the prosthesis [5,40].

During six months of exposure to aging conditions with artificial sebum, the non-pigmented silicone specimens exhibited a notable increase in ΔE values across all categories of silicone. The control category, which lacked nanoparticles, experienced the most significant shift, reaching a ΔE value of 9.02. These findings suggest that the incorporation of nanoparticles, either alone or as composites, can potentially serve as a protective measure against color degradation in silicone materials when subjected to immersion in sebum.

Furthermore, the presence of fatty acids from sebaceous skin secretions in combination with environmental factors can cause silicone polymers to be partially decomposed by breaking down the bonds within the polymer chains. This degradation process may result in the continuous release of byproducts and alterations in the chromatic properties of the silicone. These observations are supported by the studies conducted by Hatamleh et al. and Al-Harbi et al. [40,45].

The color change (ΔE) in the yellow-pigmented silicone specimens with no nanocomposites (control) was significantly higher when exposed to sweat than that in the other silicone categories. Additionally, the color changes in the 1% chitosan–TiO_2 and 1% TC specimens were significantly greater than those in the 2% TiO_2 specimens. On the other hand, the ΔE value of the 2% TiO_2 specimens was significantly lower than those of the 3% chitosan and control specimens. However, all ΔE values of the yellow-pigmented silicone were within the clinically accepted limit of $\Delta E < 3$.

This finding suggests that immersion in acidic sweat may catalyze crosslinking reactions, leading to the formation of an additional polymer network in the silicone [41]. This contradicts the findings of Han et al. [12], who discovered that yellow silicone pigment mixed with nano-oxides significantly impacted the color stability of an A-2186 silicone elastomer.

The color changes observed in the silicone materials may have been triggered by various factors, as the study by Haug et al. [53] suggested. These factors could include impurities introduced during the manufacturing process, reaction byproducts, initiators, or other mechanisms. Understanding the underlying cause of these transformations could contribute to the development of more stable silicone formulations. By advancing the longevity of prosthetic devices, this research could extend their service life.

Additionally, Kiat-Amnauy et al. [33] highlighted that both the duration of exposure and the specific type of silicone elastomer employed significantly influenced the color stability. This implies that the length of time for which the silicone elastomer was exposed to particular conditions, along with its specific formulation, played a crucial role in determining its color stability.

Considering these factors is critical when creating a maxillofacial prosthesis. The environmental conditions of the Middle East—specifically, in the region of Kurdistan/northern Iraq—are characterized by intense heat, dryness, strong sunlight during the summer, and significant climatic shifts in winter. These conditions present challenges for the longevity and color stability of maxillofacial silicones. Extreme environmental conditions, such as elevated temperatures and ultraviolet radiation, can alter silicone elastomers' mechanical properties and color stability. Further research is needed to delve into the other impacts of incorporating chitosan–TiO_2 nanocomposites, such as studies of their antimicrobial capabilities and potential allergenicity through cytotoxicity tests with various types of maxillofacial silicone elastomers over a span of one year. Given the crucial role of mechanical properties in dental and maxillofacial materials, an all-encompassing examination of the investigated maxillofacial elastomers should incorporate both a color stability assessment and rigorous mechanical evaluations. A multidimensional approach of this nature would yield a more holistic view of the material's comprehensive performance and appropriateness for deployment in a demanding therapeutic context.

5. Conclusions

1. The core–shell mixing method was proven to be a suitable methodology for synthesizing chitosan–TiO_2 nanocomposites, resulting in a novel material with enhanced and unique characteristics.

2. The addition of nanosized particles, whether used alone or as a composite, effectively protected the silicone elastomer from accelerated aging conditions, with the exception of outdoor weathering. The significant color changes observed were primarily induced by UV radiation during outdoor exposure.

3. A remarkable alteration in color was predominantly observed in brilliant-red dry pigments. This substantial color change can primarily be attributed to post-polymerization crosslinking caused by the effects of UV radiation from the irradiating light. Conversely, the yellow and blue pigments exhibited color stability across the various conditions. Nevertheless, the degree of color degradation can vary based on several factors, including the specific silicone elastomer, pigment type, opacifier concentration, and specific aging method employed.

4. In comparison with artificial weathering, outdoor weathering induced more significant changes in the ΔE values across all categories of silicone, regardless of whether they were pigmented or non-pigmented.

5. The findings of this research hold significant potential as a useful guide for different uses within maxillofacial prosthetics and silicone-based medical device sectors. The developers and researchers focused on the creation of RTV silicone products that might benefit from integrating chitosan-TiO_2 nanocomposites to improve color stability under certain aging conditions, ensuring better aesthetic outcomes for patients over time. However, outdoor weathering remains a challenge for all silicone categories, necessitating further research for enhanced resilience.

Author Contributions: Conceptualization, F.K.A.-K. and J.F.A.; methodology, F.K.A.-K.; software, F.K.A.-K.; validation, F.K.A.-K. and J.F.A.; formal analysis, F.K.A.-K.; investigation, F.K.A.-K.; resources, F.K.A.-K.; data curation, F.K.A.-K.; writing—original draft preparation, F.K.A.-K.; writing—review and editing, F.K.A.-K.; visualization, F.K.A.-K.; supervision, J.F.A. and B.A.A.; project administration, F.K.A.-K., J.F.A. and B.A.A. All authors have read and agreed to the published version of the manuscript.

Funding: This research received no external funding.

Data Availability Statement: The data that support the finding of this study are available on request from the corresponding author [Al-Kadi, F.K.].

Conflicts of Interest: The authors declare no conflict of interest.

References

1. Rangel Goulart, D.; Sigua-Rodriguez, E.A.; Alvarez-Pinzón, N.; Rocha Fernandes, A.Ú.; Queiroz, E. Quality of life of patients with facial prosthesis. *Rev. Fac. Odontol.* **2017**, *29*, 131–147. [CrossRef]
2. Montgomery, P.C.; Kiat-Amnuay, S. Survey of currently used materials for fabrication of extraoral maxillofacial prostheses in North America, Europe, Asia, and Australia. *J. Prosthodont.* **2010**, *19*, 482–490. [CrossRef] [PubMed]
3. Lai, J.H.; Hodges, J.S. Effects of processing parameters on physical properties of the silicone maxillofacial prosthetic materials. *Dent. Mater.* **1999**, *15*, 450–455. [CrossRef] [PubMed]
4. Kotra, R.; Malhotra, P.; Yadav, B.; Bhardwaj, H.; Yadav, T. Color in Maxillofacial Prosthetics. *Indian J. Health Sci. Care* **2016**, *3*, 72–77. [CrossRef]
5. Eleni, P.N.; Krokida, M.K.; Polyzois, G.L.; Bisharat, G.I. Color stability of facial silicon prosthetic elastomer after artificial weathering. *Dent. Res. J.* **2008**, *5*, 71–79.
6. Keith, M. Industrial applications of colour science. *Phys. Technol.* **1988**, *19*, 51. [CrossRef]
7. Rashid, F.; Barman, A.; Farook, T.H.; Jamayet, N.B.; Yhaya, M.F.B.; Alam, M.K. Factors affecting color stability of maxillofacial prosthetic silicone elastomer: A systematic review and meta-analysis. *J. Elastomers Plast.* **2020**, *53*, 698–754. [CrossRef]
8. Beatty, M.W.; Mahanna, G.K.; Dick, K.; Jia, W. Color changes in dry-pigmented maxillofacial elastomer resulting from ultraviolet light exposure. *J. Prosthet. Dent.* **1995**, *74*, 493–498. [CrossRef]
9. Kiat-amnuay, S.; Lemon, J.C.; Powers, J.M. Effect of opacifiers on color stability of pigmented maxillofacial silicone A-2186 subjected to artificial aging. *J. Prosthodont.* **2002**, *11*, 109–116. [CrossRef]
10. Akash, R.N.; Guttal, S.S. Effect of Incorporation of Nano-Oxides on Color Stability of Maxillofacial Silicone Elastomer Subjected to Outdoor Weathering. *J. Prosthodont.* **2015**, *24*, 569–575. [CrossRef]
11. Han, Y.; Kiat-amnuay, S.; Powers, J.M.; Zhao, Y. Effect of nano-oxide concentration on the mechanical properties of a maxillofacial silicone elastomer. *J. Prosthet. Dent.* **2008**, *100*, 465–473. [CrossRef]
12. Han, Y.; Zhao, Y.; Xie, C.; Powers, J.M.; Kiat-amnuay, S. Color stability of pigmented maxillofacial silicone elastomer: Effects of nano-oxides as opacifiers. *J. Dent.* **2010**, *38* (Suppl. S2), e100–e105. [CrossRef]
13. Prameshwari, F.; Karlina, E.; Hasratiningsih, Z. Synthesis of Ca-Psz nanoparticles using sol-gel technique with chitosan as a dispersant for raw materials restoration and dental rehabilitation equipment. *Padjadjaran J. Dent.* **2013**, *25*, 185–189. [CrossRef]
14. Jaiswal, T.; Pisulkar, S.; Kambala, S. Evaluation of antifungal effect of maxillofacial silicone after incorporation of chitosan nanoparticles: Evidence in pharmaceutical therapeutics. *J. Datta Meghe Inst. Med. Sci. Univ.* **2020**, *15*, 63–67. [CrossRef]
15. Anaya-Esparza, L.M.; Ruvalcaba-Gómez, J.M.; Maytorena-Verdugo, C.I.; González-Silva, N.; Romero-Toledo, R.; Aguilera-Aguirre, S.; Pérez-Larios, A.; Montalvo-González, E. Chitosan-TiO_2: A Versatile Hybrid Composite. *Materials* **2020**, *13*, 811. [CrossRef] [PubMed]
16. Castellano, M.; Dodero, A.; Scarfi, S.; Mirata, S.; Pozzolini, M.; Tassara, E.; Sionkowska, A.; Adamiak, K.; Alloisio, M.; Vicini, S. Chitosan&Collagen Electrospun Nanofibers Loaded with Curcumin as Wound-Healing Patches. *Polymers* **2023**, *15*, 2931.
17. Sonnahalli, N.K.; Chowdhary, R. Effect of nanoparticles on color stability and mechanical and biological properties of maxillofacial silicone elastomer: A systematic review. *J. Indian Prosthodont. Soc.* **2020**, *20*, 244–254. [CrossRef]
18. Ibrahim, A.H.J.; Al-Judy, H.J. Mechanical properties of chitosan incorporated in maxillofacial silicone and its anti candidal activity in vitro. *J. Res. Med. Dent. Sci.* **2018**, *6*, 101–107.
19. Chatterjee, K.; Sarkar, S.; Jagajjanani Rao, K.; Paria, S. Core/shell nanoparticles in biomedical applications. *Adv. Colloid Interface Sci.* **2014**, *209*, 8–39. [CrossRef]
20. Gandhi, D.S.; Sethuraman, R. Comparative evaluation of tensile strength, tear strength, color stability and hardness of conventional and 1% trisnorbornenylisobutyl polyhedralsilsesquioxane modified room temperature vulcanizing maxillofacial silicone after a six month artificial aging period. *J. Indian Prosthodont. Soc.* **2022**, *22*, 328–337. [CrossRef]
21. Hatamleh, M.M.; Polyzois, G.L.; Nuseir, A.; Hatamleh, K.; Alnazzawi, A. Mechanical Properties and Simulated Aging of Silicone Maxillofacial Elastomers: Advancements in the Past 45 Years. *J. Prosthodont.* **2016**, *25*, 418–426. [CrossRef] [PubMed]
22. Nair, A.; Saratchandran, S. Comparative Evaluation of Color Stability of Maxillofacial Silicones Following Accelerated Aging Conditions. *Face* **2022**, *3*, 362–368. [CrossRef]
23. Hatamleh, M.M.; Polyzois, G.L.; Silikas, N.; Watts, D.C. Effect of extraoral aging conditions on mechanical properties of maxillofacial silicone elastomer. *J. Prosthodont.* **2011**, *20*, 439–446. [CrossRef] [PubMed]
24. dos Santos, D.M.; Goiato, M.C.; Moreno, A.; Pesqueira, A.A.; Haddad, M.F. Influence of pigments and opacifiers on color stability of an artificially aged facial silicone. *J. Prosthodont.* **2011**, *20*, 205–208. [CrossRef]
25. Sethi, T.; Kheur, M.; Coward, T.; Patel, N. Change in color of a maxillofacial prosthetic silicone elastomer, following investment in molds of different materials. *J. Indian Prosthodont. Soc.* **2015**, *15*, 153–157. [CrossRef]
26. Polyzois, G.L. Color stability of facial silicone prosthetic polymers after outdoor weathering. *J. Prosthet. Dent.* **1999**, *82*, 447–450. [CrossRef]
27. Mancuso, D.; Goiato, M.; Dekon, S.; Gennari-Filho, H. Visual evaluation of color stability after accelerated aging of pigmented and nonpigmented silicones to be used in facial prostheses. *Indian J. Dent. Res.* **2009**, *20*, 77–80. [CrossRef]
28. Kiat-amnuay, S.; Johnston, D.A.; Powers, J.M.; Jacob, R.F. Color stability of dry earth pigmented maxillofacial silicone A-2186 subjected to microwave energy exposure. *J. Prosthodont.* **2005**, *14*, 91–96. [CrossRef]

29. Chamaria, A.; Aras, M.A.; Chitre, V.; Rajagopal, P. Effect of Chemical Disinfectants on the Color Stability of Maxillofacial Silicones: An In Vitro Study. *J. Prosthodont.* **2019**, *28*, e869–e872. [CrossRef]
30. Nada, H.E.; Ahmad, M.A.; Moustafa, N.A. Evaluation of intrinsic color stability of Facial silicone elastomer reinforced with Different nanoparticles. *Alex. Dent. J.* **2016**, *41*, 50–54. [CrossRef]
31. Abdalqadir, M.; Faraj, S.; Azhdar, B. An evaluation of a technique to improve the mechanical properties of maxillofacial silicone elastomers with zinc oxide nanoparticles. *J. Prosthet. Dent.* **2022**, *128*, 531–538. [CrossRef] [PubMed]
32. Beatty, M.W.; Mahanna, G.K.; Jia, W. Ultraviolet radiation-induced color shifts occurring in oil-pigmented maxillofacial elastomers. *J. Prosthet. Dent.* **1999**, *82*, 441–446. [CrossRef] [PubMed]
33. Kiat-Amnuay, S.; Mekayarajjananonth, T.; Powers, J.M.; Chambers, M.S.; Lemon, J.C. Interactions of pigments and opacifiers on color stability of MDX4-4210/type A maxillofacial elastomers subjected to artificial aging. *J. Prosthet. Dent.* **2006**, *95*, 249–257. [CrossRef] [PubMed]
34. Kiat-amnuay, S.; Beerbower, M.; Powers, J.M.; Paravina, R.D. Influence of pigments and opacifiers on color stability of silicone maxillofacial elastomer. *J. Dent.* **2009**, *37* (Suppl. S1), e45–e50. [CrossRef]
35. Kantola, R.; Lassila, L.V.; Tolvanen, M.; Valittu, P.K. Color stability of thermochromic pigment in maxillofacial silicone. *J. Adv. Prosthodont.* **2013**, *5*, 75–83. [CrossRef]
36. Kheur, M.G.; Kakade, D.; Trevor, C.J.; Lakha, T.A.; Sethi, T. Effect of newly developed pigments and ultraviolet absorbers on the color change of pigmented silicone elastomer. *J. Indian Prosthodont. Soc.* **2017**, *17*, 395–400. [CrossRef]
37. Cevik, P.; Yildirim-Bicer, A.Z. Effect of different types of disinfection solution and aging on the hardness and colour stability of maxillofacial silicone elastomers. *Int. J. Artif. Organs* **2017**, *41*, 108–114. [CrossRef]
38. Haldorai, Y.; Shim, J.-J. Novel chitosan-TiO2 nanohybrid: Preparation, characterization, antibacterial, and photocatalytic properties. *Polym. Compos.* **2014**, *35*, 327–333. [CrossRef]
39. Farah, A.; Sherriff, M.; Coward, T. Color stability of nonpigmented and pigmented maxillofacial silicone elastomer exposed to 3 different environments. *J. Prosthet. Dent.* **2018**, *120*, 476–482. [CrossRef]
40. Hatamleh, M.M.; Watts, D.C. Effect of extraoral aging conditions on color stability of maxillofacial silicone elastomer. *J. Prosthodont.* **2010**, *19*, 536–543. [CrossRef]
41. Polyzois, G.L.; Tarantili, P.A.; Frangou, M.J.; Andreopoulos, A.G. Physical properties of a silicone prosthetic elastomer stored in simulated skin secretions. *J. Prosthet. Dent.* **2000**, *83*, 572–577. [CrossRef] [PubMed]
42. Stefaniak, A.B.; Harvey, C.J. Dissolution of materials in artificial skin surface film liquids. *Toxicol. Vitro* **2006**, *20*, 1265–1283. [CrossRef] [PubMed]
43. *ISO 105-E04:1996*; Textiles–Tests for Color Fastness. Part E04: Color Fastness to Perspiration. ISO: Geneva, Switzerland, 1996.
44. *ASTM G7-97*; Standard Practice for Atmospheric Environmental Exposure Testing of Nonmetallic Materials. ASTM International: West Conshohocken, PA, USA, 1997.
45. Al-Harbi, F.A.; Ayad, N.M.; Saber, M.A.; ArRejaie, A.S.; Morgano, S.M. Mechanical behavior and color change of facial prosthetic elastomers after outdoor weathering in a hot and humid climate. *J Prosthet Dent* **2015**, *113*, 146–151. [CrossRef]
46. Salloum, M.G.; Ganji, K.K.; Aldajani, A.M.; Sonune, S. Colour Stability of Two Commercially Available Maxillofacial Prosthetic Elastomers after Outdoor Weathering in Al Jouf Province. *Materials* **2023**, *16*, 4331. [CrossRef] [PubMed]
47. Chotprasert, N.; Shrestha, B.; Sipiyaruk, K. Effects of Disinfection Methods on the Color Stability of Precolored and Hand-Colored Maxillofacial Silicone: An In Vitro Study. *Int. J. Biomater.* **2022**, *2022*, 7744744. [CrossRef]
48. Hood, M.A.; Mari, M.; Munoz-Espi, R. Synthetic Strategies in the Preparation of Polymer/Inorganic Hybrid Nanoparticles. *Materials* **2014**, *7*, 4057–4087. [CrossRef]
49. dos Santos, D.M.; Goiato, M.C.; Sinhoreti, M.A.; Fernandes, A.U.; Ribeiro Pdo, P.; Dekon, S.F. Color stability of polymers for facial prosthesis. *J. Craniofac. Surg.* **2010**, *21*, 54–58. [CrossRef]
50. Bishal, A.K.; Wee, A.G.; Barao, V.A.R.; Yuan, J.C.; Landers, R.; Sukotjo, C.; Takoudis, C.G. Color stability of maxillofacial prosthetic silicone functionalized with oxide nanocoating. *J. Prosthet. Dent.* **2019**, *121*, 538–543. [CrossRef]
51. Bangera, B.S.; Guttal, S.S. Evaluation of varying concentrations of nano-oxides as ultraviolet protective agents when incorporated in maxillofacial silicones: An in vitro study. *J. Prosthet. Dent.* **2014**, *112*, 1567–1572. [CrossRef]
52. Marrega Malavazi, E.; Dos Santos, D.M.; de Moraes Melo Neto, C.L.; Pereira de Caxias, F.; Freitas da Silva, E.V.; Bannwart, L.C.; Pesqueira, A.A.; de Melo Moreno, A.L.; de Magalhaes Bertoz, A.P.; Goiato, M.C. Influence of Different Pigmentations and Accelerated Aging on the Hardness and Tear Strength of the A-2186 and MDX4-4210 Silicones. *Int. J. Dent.* **2020**, *2020*, 8492091. [CrossRef]
53. Haug, S.P.; Moore, B.K.; Andres, C.J. Color stability and colorant effect on maxillofacial elastomers. Part II: Weathering effect on physical properties. *J. Prosthet. Dent.* **1999**, *81*, 423–430. [CrossRef] [PubMed]

applied nano

MDPI

Article

Production and Properties of Quercetin-Loaded Liposomes and Their Influence on the Properties of Galactomannan-Based Films

Priscilla Barbosa Sales de Albuquerque [1,2,3,*], Marthyna Pessoa de Souza [4], Ana Isabel Bourbon [5], Miguel A. Cerqueira [5], Lorenzo Pastrana [5], Paula Jauregi [6,†,‡], José A. Teixeira [2] and Maria das Graças Carneiro-da-Cunha [1,7]

1 Laboratório de Imunopatologia Keizo Asami, Universidade Federal de Pernambuco, Campus Universitário, s/n, Cidade Universitária, Recife 50670-901, PE, Brazil
2 Center of Biological Engineering, Universidade do Minho, Campus de Gualtar, 4710-057 Braga, Portugal
3 Department of Medicine, Universidade de Pernambuco—UPE, R. Cap. Pedro Rodrigues, Garanhuns 55294-902, PE, Brazil
4 Centro de Saúde e Tecnologia Rural, Universidade Federal de Campina Grande, Av. Universitária s/n, Patos 58700-970, PB, Brazil
5 International Iberian Nanotechnology Laboratory, Av. Mestre José Veiga s/n, 4715-330 Braga, Portugal
6 Department of Food and Nutritional Sciences, University of Reading, Harry Nursten Building, Reading RG6 6DZ, UK
7 Departamento de Bioquímica, Universidade Federal de Pernambuco, Av. Prof. Moraes Rego, s/n, Cidade Universitária, Recife 50670-420, PE, Brazil
* Correspondence: priscilla.albuquerque@upe.br; Tel.: +55-87-37618210
† Current address: Ikerbasque, Basque Foundation for Science, Bilbao 48013, Spain.
‡ Current address: AZTI, Food Research, Basque Research and Technology Alliance (BRTA), Parque Tecnlógico de Bizkaia, Astondo Bidea, Edificio 609, 48160 Derio, Spain.

Abstract: The objective of this work was to prepare different concentrations of liposomes based on lecithin containing quercetin, and evaluate their effect on the properties of galactomannan films obtained from *Cassia grandis* seeds. Quercetin-loaded lecithin liposomes (QT-LL) were obtained by the ethanol injection method by incorporating quercetin in different concentrations in a previously prepared suspension of lecithin liposomes in water. Following characterization of QT-LLs by zeta potential and dynamic light scattering, QT-LL with 75 μg quercetin/mL suspension was incorporated at different concentrations in galactomannan films. The films obtained were characterized for color, solubility, moisture content (MC), water vapor permeability (WVP), scanning electron microscopy (SEM), X-ray diffraction (XRD), and Fourier-transform infrared (FTIR) spectroscopy. The size of lecithin liposomes with no quercetin was statistically than those containing quercetin above 50 μg/mL. All the QT-LLs presented a low polydispersity index, even considering their significant differences and similar values for zeta potential. The films displayed a rough surface and the galactomannan structure was confirmed by FTIR. Additionally, the amorphous nature of the polysaccharide was observed by XRD. The films were luminous, with a predominant yellow tendency and low opacity. The incorporation of QT-LL in galactomannan films did not lead to statistical differences for solubility and MC, while significant differences were observed for WVP. Galactomannan films were shown to be a promising structure for the incorporation of lecithin liposomes loaded with quercetin, pointing at promising applications for different applications.

Keywords: bioactive compounds; biopolymers; heath applications; encapsulation; polysaccharides

1. Introduction

Natural or synthetic polymers are molecules whose chains are long enough to produce different scaffolds, such as films, membranes, and coatings. The preparation of films from biodegradable polymers has attracted the interest of the scientific community in recent years, mainly due to the importance given to the replacement of synthetic polymers. Films developed from polysaccharides act as excellent barriers to oxygen due to the packing of molecules, forming a structural network ordered through hydrogen bonds with potential for many applications; however, their hygroscopic characteristics can reduce this potential and should be studied [1].

Galactomannans are polysaccharides mainly found as an energy reserve in leguminous seeds, being composed of a central core of (1→4)-linked D-mannopyranose units to which (1→6)-linked D-galactopyranose units are attached. They are an essential hydrophilic material with a variety of properties, such as low- or non-toxicity, biodegradability, film forming capacity, and ready availability. The galactomannan obtained from the seeds of *Cassia grandis* collected in the northeast of Brazil was extensively characterized [2] and used for the production of films containing different concentrations of bioactive compounds [3]. Subsequently, the same galactomannan was reported to show the ability to accelerate the stages of wound healing and improve the remodeling of the extracellular matrix of topical wounds produced on rats [4].

Quercetin, chemically named as 3,3′,4′,5′-7-pentahydroxy flavone, is one of the most abundant flavonoids in plants. Despite the wide spectrum of its pharmacological properties, the use of quercetin in the pharmaceutical field is limited due to its low aqueous solubility and instability in physiological medium, resulting in poor bioavailability and permeability, in addition to instability, thus hindering its absorption through the skin. One way to circumvent these problems is to entrap/adsorb the quercetin into nanocarriers because of their ability to encapsulate hydrophobic and hydrophilic moieties [5].

The first-developed closed bilayer systems are called liposomes, which were also the first nanomedicines to enter the market and be applied to human patients [6]. Regarding their structure, liposomes are colloidal systems formed by the self-assembly of amphiphilic lipid molecules in solution; their polar head and long-chain hydrocarbon tail provide a unique configuration that is capable of encapsulating both hydrophilic and lipophilic molecules [7]. Because of this, they have found applications, for example, as antitumor [8,9] and anti-ischemic stroke neuroprotective [10] agents, in the management of psoriasis [11], as vaccines to combat COVID-19 [6], and also as an approach to improve quercetin delivery to the skin [12].

The most common methods for preparing liposomes require organic solvents, for example, ethanol and diethyl ether, complex multistage processes, and high energy costs. Despite the utilization of methods such as thin-film hydration, reverse-phase evaporation, and freeze-drying in industry, the technique extensively used in scaled-up manufacturing was the ethanol injection method followed by extrusion [13]. In the ethanol injection method, phospholipids are dissolved in ethanol and rapidly injected into water to produce liposomes; this leads to the formation of encapsulated vesicles without an external energy source [14].

Lecithin, a mixture of phospholipids of vegetable origin comprised mainly of phosphatidylcholine and phosphatidyl-ethanolamine, is considered a safe and biocompatible excipient for humans. It is commonly used in many formulations [15–17]; for example, Lecinova®, Lipoid S45, and Lipoid S75 are soybean-derived types of lecithin widely used in the development of liposomes as they are easily available and less expensive than the pure phospholipids [18]. More specifically, Lipoid S45 was recently used for the production of liposomes, by ethanol injection or otherwise, and suggested for different applications: the development of an innovative antifibrotic pirfenidone-loaded lecithin core nanocapsule [19]; the production of melatonin-loaded lecithin-chitosan nanoparticles as wound-healing agents in diabetic rats [20]; and the encapsulation of eugenol by spray-drying in lecithin and maltodextrin matrices for food applications [21]. However, to the best of our knowledge, no work has described the ethanol injection method for the development of quercetin-loaded liposomes.

Considering the above, the technological development of a simple, quickly-prepared, and nanometric vesicle of lecithin emerges as a promising strategy to deliver quercetin. In addition, the galactomannan from *C. grandis* could be considered an efficient support to immobilize the liposome for a prolonged period of time; due to the inherent characteristics of the polysaccharide and the ability to promote the controlled delivery of quercetin from the film, interesting applications can be suggested for the system. During this work, quercetin was incorporated in lecithin liposomes and characterized by particle size and zeta potential measurements, in addition to encapsulation efficiency and loading. The best nanometric formulation was incorporated in galactomannan films and the film´s properties were evaluated. Moisture content, solubility, water vapor permeability, water contact angle, color, FTIR, XRD, and SEM were performed on the films, aiming at the development of a matrix valuable for medicine and biology, especially in the nanopharmacology field.

2. Materials and Methods

2.1. Materials

Soya lecithin, type Lipoid S45 (fat free soybean lecithin with 45% phosphatidylcholine, agglomerates), was kindly provided by Lipoid (Ludwigshafen, Germany), and ethanol (99.8%) and quercetin ≥95% were purchased from Sigma-Aldrich (Steinheim, Germany). The pods of *C. grandis* were collected in Pernambuco state, Brazil, in the city of Recife (8°02′50.0″ S 34°57′00.1″ W). Acetone (PA) and sodium chloride were obtained from Vetec Fine Chemicals Ltd. (Rio de Janeiro, Brazil). The solvents used were of analytical grade and all other reagents were purchased from Sigma-Aldrich (Steinheim, Germany).

2.2. Development of the Liposomes

Liposomes were produced by the ethanol injection method. An amount of 2 mL of an ethanolic solution of Lipoid S45 (25 mg/mL) was injected, drop by drop, during 2 min, into 23 mL of distilled water under magnetic stirring (600 rpm). The suspension was left in magnetic stirring for 8 min and then submitted to a vigorous shaking on a Vortex mixer (Thermo Fisher Scientific, Langenselbold, Germany) for 2 min at room temperature (20 ± 2 °C). The developed system was called LL (lecithin liposome). The incorporation of quercetin at different concentrations (0–150 μg/mL) was performed by prior dissolution in the ethanolic solution of Lipoid S45 and subsequent addition into distilled water as described above. Finally, quercetin-loaded lecithin liposomes (QT-LLs) were named as described.

Sample	Quercetin Diluted into the Ethanolic Solution of Lipoid S45 [μg/mL]
QT-LL 0	0
QT-LL 25	25
QT-LL 50	50
QT-LL 75	75
QT-LL 100	100
QT-LL 125	125
QT-LL 150	150

2.3. Quercetin Encapsulation Efficiency and Loading

The quercetin encapsulation efficiency [EE (%)] and loading [LC (%)] were conducted according to the modified procedures previously described by Souza et al. [22]. A sample of QT-LL was measured with a microplate reader (GENios, TECAN, Männedorf, Switzerland) at 373 nm, in order to obtain the total content of quercetin. The same volume of the QT-LL sample was ultracentrifuged at 180,000× *g* over 30 min and 4 °C (Micro-Ultracentrifuge Sorvall MTX 150, Thermo Fisher Scientific, Langenselbold, Germany), and the supernatant was accurately taken and measured with a microplate reader under the same conditions to obtain the amount of free quercetin. Then, the content of quercetin loaded into each QT-LL was calculated by subtracting the total content of quercetin and the amount of free quercetin. The encapsulation efficiency of quercetin [Equation (1)] and quercetin-loading (Equation (2)) were calculated by the following equations:

$$EE\ (\%) = [(W_{total} - W_{free})/W_{total}] \times 100, \quad (1)$$

$$LC\ (\%) = [(W_{total} - W_{free})/W_{vehicle}] \times 100, \quad (2)$$

where W_{free} is the amount of quercetin loaded into each QT-LL; W_{total} is the total content of quercetin into the QT-LL; and $W_{vehicle}$ is the liposome vehicle weight.

2.4. *Physicochemical Characterization of the Liposomes*

The hydrodynamic diameter (average size) and polydispersity index (PDI) of the liposomes were determined by dynamic light scattering (Zetasizer Nano ZS, Malvern Instruments, Malvern, UK). The samples were analyzed in a folded capillary cell. Detection of the scattered light was carried out at 173° (NIBS = non-invasive backscatter detection) and temperature of 25 °C. The zeta potential was measured by laser Doppler micro-electrophoresis using the same Zetasizer ZS Nano. The samples were filtered with 0.45 μm syringe filter (Minisart® High Flow, Sartorius, Gloucestershire, UK) to eliminate eventual particles above 450 nm and compared to the samples without filtering. At least five replicates were performed for each sample and the results were given as mean ± standard deviation of the obtained values.

2.5. *Antioxidant Activity Assays*

To performe the ferric reducing antioxidant power (FRAP), the method based on Benzie and Strain [23] was carried out with some modifications. Briefly, 10 μL of free quercetin and QT-LLs were mixed with 290 μL of FRAP reagent, which was freshly prepared by mixing 2,4,6-tris (1-pyridyl)-5-triazine 10 mM in 40 mM HCl with 20 mM $FeCl_3$ solution and 0.3 M acetate buffer (pH 3.6) in the proportion of 1:1:10 (*v*/*v*/*v*). The mixture containing each sample and the FRAP reagent was then incubated at 37 °C for 15 min and the absorbance determined at 593 nm (Microplate reader, Bio-Rad Laboratories, Hercules, CA, USA). The calibration curve was prepared with ascorbic acid (AA) and expressed as equivalent of AA/μL of each sample. All determinations were performed in triplicate and expressed as mean ± standard deviation.

The antioxidant capacity of QT-LLs was also determined using the ABTS (2,2′- azino-bis(3-ethylbenzothiazoline-6-sulphonic acid) method according to Torres et al. [24]. Free quercetin and QT-LLs were added to the ABTS•+ solution and the decrease in absorbance was measured at 734 nm after 15 min in the dark. The calibration curve was prepared with Trolox and the ability to eliminate ABTS radicals (%) was calculated (Equation (3)) according to the formula:

$$\text{ABTS radical elimination capacity}\ (\%) = [(A_B - A_S)/A_B] \times 100 \quad (3)$$

where A_B is the blank absorbance of Trolox and A_S is the absorbance of each sample. All determinations were performed in triplicate and the results expressed as mean ± standard deviation.

2.6. *Film Characterization*

2.6.1. Film Preparation

The galactomannan contained in *C. grandis* seeds was obtained according to Albuquerque et al. [2], while the film forming solutions (FFSs) were prepared in accordance with the methodology described by Albuquerque et al. [3]. The FFS was prepared with 0.8% (*w*/*v*) of galactomannan and 0.2% (*v*/*v*) of glycerol under magnetic stirring (500 rpm) for 12 h, at room temperature. QT-LL 75 was chosen to be immobilized at different concentrations [0–0.5% (*v*/*v*)] in the FFSs (please see Section 3.1, Liposomes' physicochemical characterization, for further explanation). The immobilization was made under magnetic stirring (500 rpm) for 2 h, at room temperature, and the obtained samples were named according to the respective amount (%, v/v) of QT-LL 75 added: A (0%), B (0.1%), C (0.2%), D (0.3%), E (0.4%), and F (0.5%). The samples were cast (15 mL) onto 90 mm diameter polystyrene Petri dishes and then dried in an oven at 33 °C for 24 h. Finally, the films were maintained at 20 °C and 54% relative humidity (RH) until further characterization.

2.6.2. Film Thickness

Film thickness was measured with a digital micrometer (No. 293–561, Mitutoyo, Nagoia, Japan). Five different randomly chosen points were performed on each film in order to obtain the values to be used in the calculations of water vapor permeability (WVP).

2.6.3. Scanning Electron Microscopy

The films' Scanning Electron Microscopy (SEM) surface scans were performed on a scanning electron microscope (Quanta FEG 650, FEI Company, ThermoFisher Scientific, Hillsboro, OR, USA) with an accelerating voltage of 5 kV under vacuum conditions. Before analysis, all samples were mounted on aluminum stubs using carbon adhesive tape and sputter-coated with gold and then shredded.

2.6.4. Fourier-Transform Infrared (FTIR) Spectroscopy

The films (A–F) were characterized by FTIR on a Bruker FT-IR VERTEX 80/80 v (Boston, MA, USA) in Attenuated Total Reflectance mode (ATR) with a platinum crystal accessory between 400 and 4000 cm^{-1}, a resolution of 4 cm^{-1} and 16 scans. Data analysis was performed with GraphPad Prism 5.00.288 (GraphPad Software, Inc., San Diego, CA, USA).

2.6.5. X-ray Patterns

An X-ray diffraction system (Malvern Panalytical Ltd., Malvern, UK) was used to evaluate the crystallographic structure. PANanalytical X'Pert HighScore Plus was the software used to gather data and analyze peak diffractions. Background noise was also measured. The powder sample was added to the glass slide through an adhesive glue and put on the sample holder for detection. The XRD diffractograms were acquired at room temperature; angular scans from 5° to 50° (2θ) were performed with a Cu source X-ray tube (° = 1.54056 Å) at 45 kV and 40 mÅ. For each sample, one measurement was performed. To determine the films' crystallinity, the total diffracted area and the area under the crystallinity peaks were evaluated by integration after correcting the data for absorption. The ratio of the crystalline area to the total area was taken as the relative crystallinity.

2.6.6. Color and Opacity

Color and opacity were determined using a digital colorimeter (Konica Minolta, model Chroma Meter CR-400, Osaka, Japan) calibrated at illuminant C with a white standard. L* (L* = 0 (black) and L* = 100 (white)), a* (−a* = greenness and +a* = redness), and b* (−b* = blueness and +b* = yellowness) were the analyzed parameters, which are the ones recommended by the International Commission on Illumination. The film's opacity (Y) was calculated as the ratio between the opacity of each sample on the black standard (Yb) and the opacity of each sample on the white standard (Yw). Five measurements were determined at random Yb and Yw and an average of them was used for calculations. The experiment was performed in triplicate and the results were expressed as a percentage and determined through Equation (4):

$$Y\ (\%) = (Yb/Yw) \times 100, \tag{4}$$

2.6.7. Moisture Content

MC was determined gravimetrically by drying the films at 105 °C in an oven with forced air circulation for 24 h. The results were expressed as the percentage of water removed from the initial mass sample. The experiments were performed in triplicate.

2.6.8. Solubility

The measurement of solubility was determined according to Albuquerque et al. [3]. Triplicates of each film were cut with a circular mold of 2 cm diameter and dried at 105 °C in an oven for 24 h until constant weight (W1). The films were placed in cups with 50 mL of distilled water, sealed over with parafilm and homogenized at 60 rpm for 24 h. The non-soluble part of each film was dried at 105 °C for 24 h and weighted again (W2). The solubility of the film samples was expressed as a percentage and calculated as follows (Equation (5)):

$$\text{Solubility (S \%)} = [(W1 - W2)/W1] \times 100, \tag{5}$$

2.6.9. Water Vapor Permeability (WVP)

WVP was determined gravimetrically based on ASTM E96-92. The film was sealed on top of a permeation cell containing distilled water (100% RH; 2337 Pa vapor pressure at 20 °C), and placed in a desiccator at 20 °C and 0% RH (0 Pa water vapor pressure) containing silica. The cells were weighed at intervals of 2 h for 10 h. Steady-state and uniform water-pressure conditions were assured by maintaining constant air circulation outside the test cell by using a miniature fan inside the desiccator. The slope of weight loss versus time was obtained by linear regression. Three replicates were obtained for each sample.

2.6.10. Contact Angle

Initially, the degree of hydrophilicity of the liposomes was measured by the sessile drop method in a face contact angle meter (OCA 20, Dataphysics, Filderstadt, Germany). The methodology was based on Mahaling and Katti [25] with some modifications. An amount of 1 mL of the liposome suspension was dropped on a glass slide to form a particle thin film and allowed to dry at room temperature in a desiccator. An amount of 500 μL of sterile water was dropped on the slides containing dried liposomes using a syringe (Hamilton, Domat/Ems, Switzerland) with a needle of 0.75 mm of diameter. Measurements were made in less than 5 s. Thirty replicates of contact angle measurements were obtained at 24.0 (±0.2) °C and ~65% RH, then calculated using the inbuilt software of the instrument. After that, contact angles for galactomannan-based films with different concentrations [0–0.5% (v/v)] of QT-LLs were evaluated as described above, in addition to a further measurement at 30 s to evaluate the pattern of hydrophobicity of the films.

2.7. *Statistical Analyses*

Statistical analyses were performed using Analysis of Variance (ANOVA) and linear regression analysis. The Tukey test ($\alpha = 0.05$) was used to determine any significance of differences between specific means (GraphPad Prism 5.00.288, GraphPad Software, Inc., San Diego, CA, USA).

3. Results and Discussion

3.1. *Liposomes' Physicochemical Characterization*

Table 1 presents the average size, PDI, and zeta potential of the produced QT-LLs. Regarding the results of average size, it can be noted that QT-LL 0 (control) presents statistically lower values ($p < 0.05$) when compared with those containing quercetin concentrations above 50 μg/mL, while for QT-LL 25 no difference ($p > 0.05$) was observed. In addition, increasing quercetin concentrations above 100 μg/mL led to higher values of average size. The results related to PDI showed that all samples presented a low PDI (≤0.29) even considering their significant differences. All QT-LLs presented zeta potential values around −45 mV with no statistical differences; regarding these values, it is possible to say that the encapsulation of quercetin does not influence the surface of the liposomes. LLs 0, 25, 50, and 75 were homogeneous; however, QT-LLs 100, 125, and 150 showed signs of instability (precipitation): QT-LL 100 and 125 after 72 h, and QT-LL 150 after 48 h of observation (Figure 1).

Table 1. Average size, polydispersity (PDI), and zeta potential for QT-LLs with different concentrations of quercetin (μg/mL).

Sample	Average Size (nm)	PDI	Zeta Potential (mV)
QT-LL 0	94.26 ± 0.823 [c]	0.27 ± 0.004 [b,d]	−44.98 ± 0.98 [a]
QT-LL 25	93.85 ± 4.343 [c]	0.28 ± 0.012 [c,d]	−46.85 ± 10.92 [a]
QT-LL 50	100.2 ± 0.567 [d]	0.23 ± 0.009 [a]	−50.83 ± 10.85 [a]
QT-LL 75	100.5 ± 4.569 [d]	0.27 ± 0.0124 [c,d]	−48.14 ± 5.64 [a]
QT-LL 100	104.1 ± 4.214 [d]	0.28 ± 0.007 [c]	−46.76 ± 1.85 [a]
QT-LL 125	114.0 ± 1.045 [b]	0.27 ± 0.006 [d]	−47.10 ± 2.59 [a]
QT-LL 150	131.8 ± 3.469 [a]	0.26 ± 0.005 [b]	−45.30 ± 0.59 [a]

[a–d] Different superscript letters in the same column indicate a statistically significant difference ($p < 0.05$).

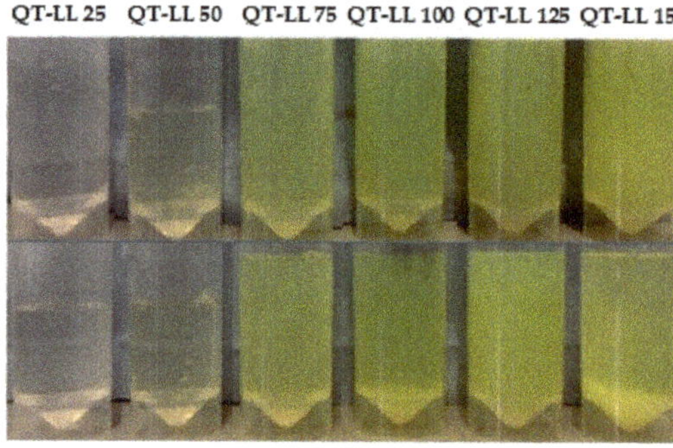

Figure 1. Images of QT-LLs with different concentrations of quercetin (μg/mL), freshly prepared and after 48 h of production.

It is known that smaller particles present a higher exposed surface area which can lead to a faster release of encapsulated drugs; however, smaller particles also have an increased risk of aggregation during storage. In view of this, it is important to produce liposomes with adequate size (<200 nm) and low values of PDI, thus reaching maximum stability by a good control of particles' size. It is also important to mention that the reproducibility of parameters such as stability and release is directly connected to a low PDI ($\leq$0.4), since a high PDI means that there is no uniformity in the size distribution of the sample [21]. In relation to the zeta potential parameter, results around ± 45 mV of surface charge can positively influence its stability through electrostatic repulsion among the liposomes. In the present study, the negative charge demonstrated by zeta potential could be associated to the phosphate groups of the mixture of phospholipids composing the lecithin (Lipoid S45).

This behavior was also reported by Gibis, Vogt, and Weiss [26] for 1% (*w*/*v*) lecithin liposomes, which had negative surface charges regardless of the exposure to different pH values; the magnitude of surface charges decreased from −21.8 at pH 3 to −57.7 at pH 11. More recently, Talón et al. [21] also reported a zeta potential value of −45.7 mV for liposomes based on lecithin and maldodextrin. This charge was inverted when chitosan was incorporated into the system, leading to a zeta potential of +61.5 mV. Our zeta potential results were not influenced by the amount of quercetin loaded in lecithin liposomes, However, a different pattern was observed for average size values; increasing the concentration of quercetin resulted in decreased stability of QT-LLs.

3.2. Antioxidant Activity Assays

Antioxidant activity is a key parameter of the antioxidant capacity of a compound; in this sense, its measure aids the understanding of the functional property of a compound, whether free-exposed or encapsulated, for example, in nanocarrier systems [24]. The antioxidant activity of QT-LLs was determined in comparison to the free quercetin samples by FRAP and ABTS methods (Table 2).

Table 2. ABTS and FRAP results for QT-LNps with different concentrations of quercetin (μg/mL) and the same concentrations for free quercetin.

Concentration of Quercetin	ABTS (%)		FRAP (AA/μL)	
	QT-LL	Free Quercetin	QT-LL	Free Quercetin
25	−4.20 ± 5.09	10.76 ± 0.33	41.61 ± 11.93	165.71 ± 12.23
50	6.53 ± 11.31	25.97 ± 3.63	100.90 ± 12.06	502.58 ± 6.82
75	28.05 ± 0.18	36.03 ± 0.18	312.59 ± 13.67	1023.47 ± 12.04
100	35.88 ± 2.75	42.34 ± 1.18	476.37 ± 24.88	2281.29 ± 15.56
125	42.53 ± 2.99	47.68 ± 0.84	731.81 ± 38.18	2579.89 ± 9.78
150	55.49 ± 3.07	52.87 ± 1.09	812.70 ± 49.80	2933.27 ± 35.19

In relation to FRAP results, free quercetin was more effective in reducing Fe^{3+} to Fe^{2+} in comparison to QT-LLs ($p < 0.05$), achieving values around four-fold higher than the reducing power of the liposomes for all of the tested concentrations. For ABTS results, the antioxidant capacities of QT-LLs 25 and 50 could be disregarded because of the low percentage when compared to free quercetin at the same concentrations. When comparing the % inhibition of quercetin at 75 μg/mL concentration between free quercetin (36.03 ± 0.18%) and QT-LL 75 (28.05 ± 0.18%), there was a decrease in this difference, and the antioxidant capacity of the liposome presented a value 1.2 times lower than the other. The results of antioxidant capacity for both free quercetin and QT-LLs containing quercetin above 100 μg/mL of concentration were quite similar. For example, QT-LL 150 showed an antioxidant capacity of 55.49 ± 3.07%, while free quercetin at 150 μg/mL concentration was 48.21 ± 3.68%. Regarding the differences in the antioxidant activities between free quercetin and QT-LLs, it is important to mention, again, our simple and quickly-prepared method of production, in which QT-LLs 75–150 reached values of antioxidant capacity very close to those of free quercetin. Additionally, higher antioxidant capacities for QT-LLs 75–150 could be explained by an improvement in the dissolution pattern of quercetin in the aqueous medium as the concentration of encapsulated quercetin increases [21].

Greater antioxidant activities of free molecules were also observed, for example, by Talón et al. [21] working with the encapsulation of eugenol by spray-drying in lecithin and maltodextrin matrices. Pure eugenol showed the best antioxidant activity when compared to emulsions based on lecithin as wall material, with or without oleic acid as a carrier agent, and chitosan as a coating. It is important to mention that the ethanol injection method used here is faster and simpler than other methods reported in the scientific literature: magnetic stirring (<1000 rpm) and approximately 20 min can produce liposomes with small particle sizes and good monodispersed properties. The lower antioxidant activities of QT-LLs than free quercetin could be due to the partial oxidation of the compound during this simple process.

3.3. *Quercetin Encapsulation Efficiency (EE %) and Loading Content (LC %)*

EE % and LC % of the compound depends on its solubility in the matrix [27]. Figure 2A shows that EE % of quercetin into liposomes presented better values for concentrations above 75 μg/mL, while Figure 2B demonstrates that LC% for QT-LLs from 0 to 75 μg/mL of quercetin were quite similar. QT-LLs containing quercetin above 100 μg/mL of concentration resulted in improved LC%. The encapsulation efficiency of quercetin in other nanocarrier systems reached higher values than the presented ones; for example, EE % of quercetin into lecithin/chitosan nanoparticles presented values higher than 95% for concentrations varying from 50 to 100 μg/mL [5], while EE % of quercetin in cholesterol liposomes coated with chitosan and sodium tripolyphosphate was 91 ± 1% [28]. This difference may be related to the structure of the mentioned systems. Regarding the results for lecithin/chitosan nanoparticles, the association of both molecules was strongly influenced by its hydrophobic nature; consequently, a larger amount of quercetin was inserted into the lipid nuclei of the nanoparticles and only a small amount of quercetin was lost in the aqueous phase during the preparation process. In relation to cholesterol liposomes, the free quercetin could be embedded in the three-dimensional polyelectrolyte shell of chitosan during the coating and cross-linking process.

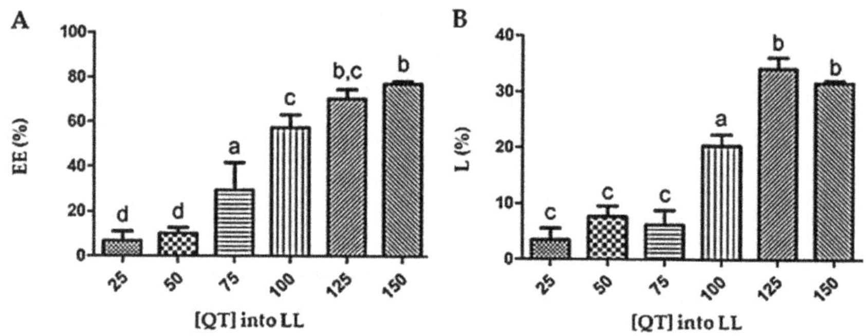

Figure 2. Influence of quercetin concentration (µg/mL) in the encapsulation efficiency (**A**) and loading (**B**) of the liposomes (LLs). All determinations were performed in triplicate, the error bars are standard deviation, and different letters represent statistical significance ($p < 0.05$).

Concerning the results of our simple and fast-prepared liposome composed only with lecithin, the system presents stability even in the absence of a coating. This is expected, since the ethanol injection method produces single bilayered liposomes with a smaller particle size and good monodispersion without sonication, extrusion or homogenization by Rotor Stator [5]. Another advantage of liposomes produced by the ethanol injection method is that they function as transfection delivery vectors. Because of this, we believe that the delivery of quercetin from liposomes can be suggested for future research, especially in relation to the nanopharmacology field.

In view of the information already demonstrated in Section 3.1 about formulations containing more than 75 µg/mL of quercetin showing signs of instability, the liposome with 75 µg/mL of quercetin (QT-LL 75) was chosen for further analyses.

3.4. *Films' Characterization*

Regarding the information already mentioned in Sections 3.1 and 3.3, all of the analyses in this very section were conducted by using QT-LL 75 immobilized at different concentrations [0–0.5% (v/v)] in galactomannan-based films.

3.4.1. Scanning Electron Microscopy (SEM)

SEM allows the visualization of the surface morphology of QT-LL 75 and the films; additionally, the distribution of the immobilized QT-LL 75 at different concentrations could be observed in the microstructure of each film matrix.

It is possible to observe in Figure 3 that QT-LL 75 presents an irregular spherical shape, while all of the film's surface is rough; in addition, the roughness is increased by increasing the concentration of immobilized QT-LL 75 from 0 to 0.5% (v/v). We hypothesize that QT-LL 75 is immobilized on the surface of the films, embedded between the chains of the galactomannan, and the higher concentrations of the immobilized liposome influenced the voluminous appearance of the films. Indeed, the presence of QT-LL 75 could be visualized even immobilized in different concentrations. Despite the absence of QT-LL 75 in the control film (sample A), a few granules are evident which could be related to contaminating residues remaining from the galactomannan extraction.

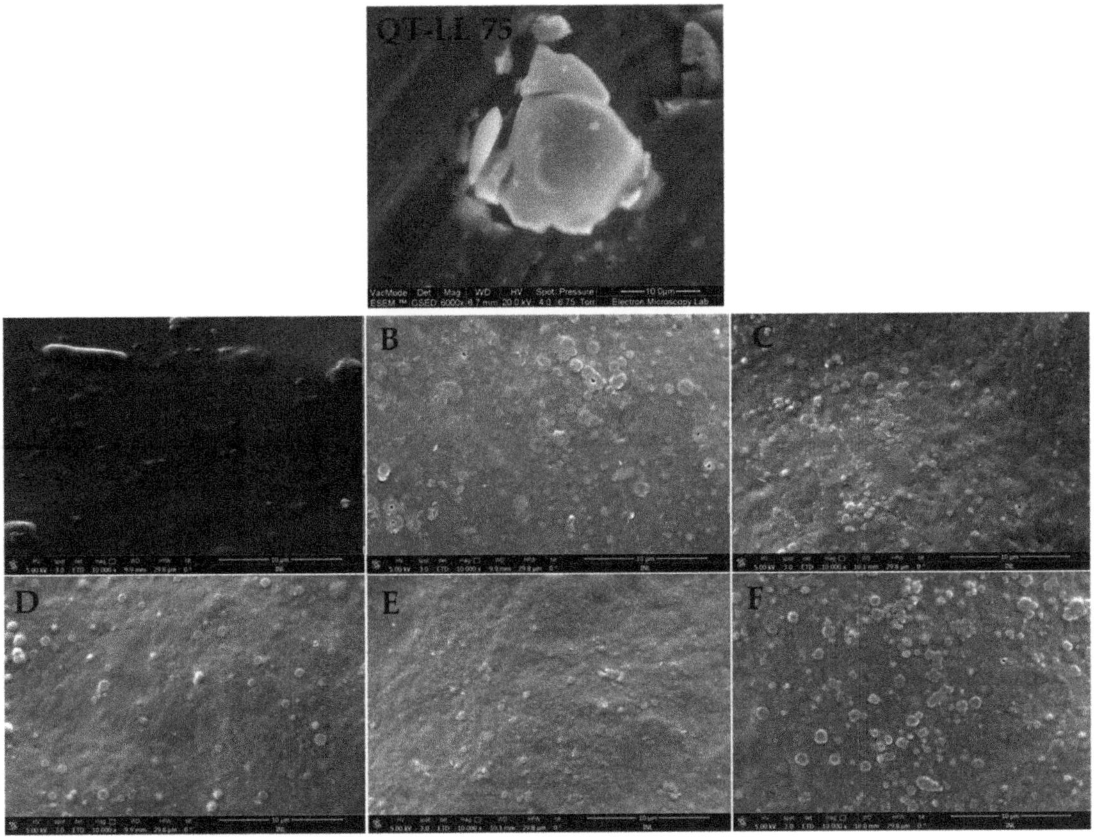

Figure 3. SEM images showing QT-LL 75 and the surface structure of galactomannan-based films containing QT-LL 75 at 0% (**A**), 0.1% (**B**), 0.2% (**C**), 0.3% (**D**), 0.4% (**E**), and 0.5% (**F**) concentration (v/v).

Regarding other research dealing with nanocarriers and polymeric films, many differences about the immobilized compounds into the microstructure of the films have been reported. For instance, working with lactoferrin, bioactive peptides, and phytosterols, Albuquerque et al. [3] reported that the immobilization of these bioactive compounds in galactomannan films derived from *C. grandis* led to films with roughness on their surface. González et al. [29] developed nanofibers obtained from soybean hulls and pods and used these low-cost agro-industrial by-products as nano-reinforcers of soy protein films; they observed that the surface for the film with the highest number of nanofibers was rougher when compared with the films without them. Bionanocomposite films of konjac glucomannan were prepared by Wu et al. [30] as immobilization matrices for chitosan/gallic acid nanoparticles. Again, the authors observed that the films' surface became less homogeneous with visible microcracks or discontinuous void when higher amounts of nanoparticles were incorporated into the matrix.

3.4.2. Fourier-Transform Infrared (FTIR) Spectroscopy

FTIR was used for the evaluation of possible chemical interactions and modifications in the structure of galactomannan-based films containing QT-LL 75 at different concentrations (Figure 4).

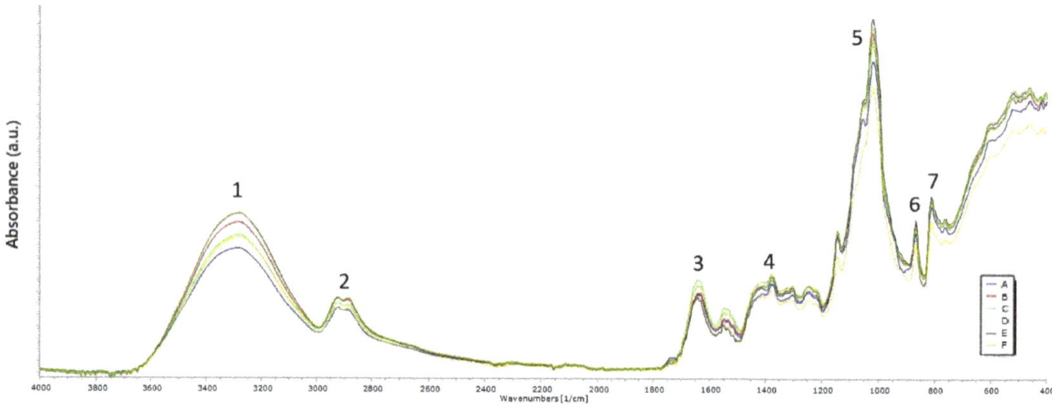

Figure 4. FTIR spectra of the galactomannan-based films containing QT-LL 75 at 0% (A), 0.1% (B), 0.2% (C), 0.3% (D), 0.4% (E), and 0.5% (F) concentration (v/v).

All spectra were quite similar to each other; indeed, the superposition of the spectra confirms that the immobilization of QT-LL 75 in different concentrations did not alter the structure of the films. The absorption peaks associated to the assigned regions are in agreement with those reported by Albuquerque et al. [2] working for the first time with the *C. grandis* galactomannan. Region 1 is related to absorption peaks between 3500–3100 cm^{-1}, which is characteristic of a carbohydrate ring and represents hydroxyl stretching vibration of polysaccharides and water involved in hydrogen bonding. Region 2 is comprised of absorptions between 3000–2800 cm^{-1} and represents the stretching vibration of the methylene group (C–H) [31]. Region 3 arose from the bending vibration of water molecules, while region 4 contains bands between 1376 and 1378 cm^{-1} related to symmetrical deformations of the CH_2 and COH groups [32]. Region 5 is associated with absorption peaks of 1021–1023 cm^{-1} that are consistent with C–H vibrations [33]. Regions 6 and 7 indicate absorption peaks at 870–874 cm^{-1} and 810–812 cm^{-1} that are associated with the presence of anomeric α and β configurations and glycosidic linkages, attributed to β-D-mannopyranose units and α-D-galactopyranose units, respectively [2,3,34,35].

The literature reports quercetin showing FTIR bands and typical molecular peaks around 1380 cm^{-1} (COH), 1610 cm^{-1} (CC), 1264 cm^{-1} (COC), 1662 cm^{-1} (CO), and 3403 cm^{-1} (OH stretch). The recorded spectra showed no characteristic peaks of quercetin, which may be indicative of quercetin encapsulation in lecithin-liposomes [22,27]. In relation to lecithin, the strong peaks reported by the literature are 1737 and 1245 cm^{-1}, which are associated to CO and PO stretching vibrations; again, they were not shown in the recorded spectra. This phenomenon revealed that quercetin might be associated with lecithin by hydrogen bonding or hydrophobic effects [36] and does not influence the structure of the films. Even considering this information, it is important to mention that the presence of QT-LL 75 slightly shifted the peak of hydrogen bonds for B–F films (region 1), implying that soft hydrogen bonding was generated among quercetin-loaded lecithin liposomes and the film matrix.

3.4.3. X-ray Patterns

In this study, all of the films (A–F) presented a broad peak around 2θ = 20°, characteristic of amorphous structures (Figure 5). The degree of crystallinity (%), or crystallinity index, was calculated by considering the total diffracted area and the area under the crystallinity peaks; after integration and correction of the data for absorption, the ratio of the crystalline area to the total area was obtained and considered as the crystallinity index (%). The obtained values for A, B, C, D, E, and F were 0.71, 2.37, 1.89, 2.33, 6.79, and 1.41%, respectively. These results are similar to those obtained by Albuquerque et al. [2] in the first study about the galactomannan obtained from the seeds of *C. grandis* collected in Pernambuco, Brazil. Other studies also showed low overall crystallinity for polysaccharide characterization by XRD; for example, water-soluble polysaccharides from potato peel [37] and chickpea flour [38].

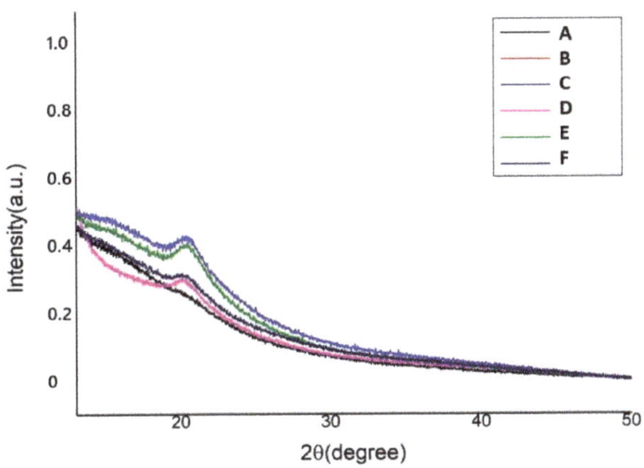

Figure 5. XRD patterns of the galactomannan-based films containing QT-LL 75 at 0% (A), 0.1% (B), 0.2% (C), 0.3% (D), 0.4% (E), and 0.5% (F) concentration (*v*/*v*).

The scientific literature reports the XRD pattern for lecithin as peak around 2θ = 20°, suggesting its amorphous characteristic [39,40]. In our case, it is not possible to claim a superimposition of the individual patterns of galactomannan and lecithin due to the low crystallinity of both of them. Concerning the crystallinity pattern expected for quercetin, one can suggest that weak interactions occur between quercetin and the film matrix. The absence of peaks corresponding to pure quercetin indicates that the flavonoid mainly exists dispersed in a non-crystalline state within the film matrix [41], which is in agreement with the results already mentioned by SEM and FTIR.

3.4.4. Color and Opacity

Color and opacity are important features for the acceptance of products by the pharmaceutical or food industry. Therefore, the measurement of the color of products potentially valuable for the industry is an important point of the physical characterization. Table 3 presents the color parameters and opacity of the films with immobilized QT-LL 75 at different concentrations. All of the films were strongly luminous, represented by L* coordinate values, with a tendency to yellowness, represented by b* coordinates; for those parameters, the films did not present statistically significant differences among them and were quite similar to the films produced with the *C. grandis* galactomannan with immobilized bioactive compounds [3]. Similar values of the L* coordinate (around 96.4) were reported for films based on *Delonix regia* galactomannan with different molecular weights and interaction with k-Carrageenan [31]; however, our films showed a greater yellowness when compared to b* values around 2.43 and 4.39, respectively, for the galactomannan films and those composed of galactomannan and k-carrageenan. The film produced with the galactomannan obtained from *Gleditsia triacanthos* was less bright (as represented by L* of 78.20 ± 0.39) and almost twice as yellow (represented by b* of 14.55 ± 0.60) as all of our films. Blended films composed of this galactomannan and soy protein led to a reduction in L* and an increase in the b* parameters, which may be associated with the natural color of the solutions prepared from the yellowish powder extract of *G. triacanthos*, and also related to Maillard reactions between the components [42].

Table 3. Color parameters L (luminosity), a* (−a* = greenness and +a* = redness), b* (−b* = blueness and +b* = yellowness), and Y (opacity) for (A) the control film and the films (B–F) with different concentrations of QT-LL 75 (values expressed as average ± standard deviation).

Film	QT-LL 75 (*v*/*v*)	L*	a*	b*	Y
A	0	95.62 ± 0.40 [a]	0.35 ± 0.06 [b,c]	7.93 ± 0.61 [a]	12.30 ± 0.32 [a]
B	0.1%	95.66 ± 0.22 [a]	0.35 ± 0.01 [b,c]	7.46 ± 0.09 [a]	12.59 ± 0.26 [a]
C	0.2%	95.77 ± 0.22 [a]	0.30 ± 0.01 [a,b,c]	7.24 ± 0.33 [a]	12.49 ± 0.24 [a]
D	0.3%	95.01 ± 1.09 [a]	0.39 ± 0.04 [b]	8.17 ± 2.08 [a]	12.11 ± 0.36 [a]
E	0.4%	96.05 ± 0.31 [a]	0.27 ± 0.03 [a,c]	7.46 ± 0.99 [a]	12.27 ± 0.29 [a]
F	0.5%	95.84 ± 0.42 [a]	0.25 ± 0.04 [a]	7.25 ± 0.36 [a]	12.62 ± 0.14 [a]

[a–c] Different superscript letters in the same column indicate a statistically significant difference ($p < 0.05$).

Regarding the a* parameter, values close to zero demonstrate that films tend to a green color besides its slight tendency to redness; also, the presence of QT-LL 75 at 0.5% (F) decreased ($p < 0.05$) the red appearance of the film when compared to the control film. These results are in agreement with those reporting the immobilization of quercetin in a chitosan-based film [43]. A tendency to greenness was also reported for the galactomannan film from *G. triacanthos* (represented by a* of 1.29 ± 0.09) [42], and Tara gum films with (a* around −2.18) or not incorporated chitosan (a* of −1.29 ± 0.09) and chitosan nanoparticles (a* around −2.23) [44].

All studied films presented low opacity values, with no significant differences between them. In comparison with Y results for films based on *D. regia* galactomannan with different molecular weights (around 9.97) [33], our films we shown to be opaquer (around 12.39) besides their evident transparency. On the other hand, the film molded with the galactomannan extracted from *G. triacanthos* [42] presented incomparable transparency with Y 18-fold lower than our films.

3.5. Film Thickness, Moisture Content (MC), Solubility (S %), Water Vapor Permeability (WVP), and Contact Angle (CA)

The moisture-binding abilities of films can affect significantly their physical and barrier properties; thus knowledge of the content and affinity of the film matrix to water is a key parameter when choosing a film for specific applications. Table 4 shows the values of thickness, moisture content (MC), solubility (S %), and water vapor permeability (WVP) of the control film and the films with QT-LL 75 immobilized in different concentrations (B–F).

The incorporation of different concentrations of QT-LL 75 in galactomannan-based films did not lead to statistical differences ($p > 0.05$) for thickness, MC, and S %; in addition, these values were similar to those reported by [3] working with similar galactomannan-based films. Other authors have reported values of thickness and discussed its importance in the characterization of galactomannan films or blends containing galactomannans. For example, films based on *D. regia* galactomannan with different molecular weight and interaction with k-carrageenan presented values of thickness between 37 and 59 μm [31], while the galactomannan extracted from *G. triacanthos* and its combination with soy protein isolates showed thickness of 76 ± 8 and 106 ± 3 μm, respectively [42]. Therefore, variations in thickness values could be observed when molecular interactions occur between the components of the blend, and also caused by acid solvents used in the preparation of the components [45].

Modifications in the MC can affect the physicochemical and thermomechanical properties of the films; the addition of glycerol is one of the main modifiers, which is explained by its hydrophilic nature and the attraction of water into the polymer matrix, thus creating more mobile regions with greater inter-chain distances in the polymer matrix [44,46]. Our results showed that the films' MC were around 22% and the addition of QT-LL 75 did not influence this parameter; however, a different behavior was observed for the same galactomannan film incorporating 0.2% of lactoferrin (15.67 ± 5.51%) [3]. According to the authors, the decreased MC values could be associated with the fact that lactoferrin presents both hydrophilic and hydrophobic parts in its conformation; the higher concentration of lactoferrin into the film may reduce the availability of hydroxyl groups to interact with water, resulting in decreased MC values. Tara gum (TG) films and nanocomposites, in this case chitosan nanoparticles incorporated in TG films, also presented lower values of MC, respectively, 14 and 13%. The authors explained that the nanocomposites presented a more compact matrix, which allowed them to occupy more free volume into the polymeric structure, thus reducing MC values [44]. Other works reported MC values similar to the ones of this work; for example, 31.01 ± 0.80% for chitosan-based films, 21–7% for chitosan films incorporated with different amounts of black soybean seed coat extract [47], and 14–7% for chitosan films neutralized with citrate and acetate buffers, added or not with glycerol [45].

Table 4. Thickness, solubility (S), moisture content (MC), and water vapour permeability (WVP) for (A) the control film and the films (B–F) with different concentrations of QT-LL 75 (values expressed as average ± standard deviation).

Film	QT-LL 75 (*v/v*)	Thickness (mm)	S (%)	MC (%)	WVP × 10^{-7} ($g{\cdot}h^{-1}{\cdot}m^{-1}{\cdot}Pa^{-1}$)
A	0	0.046 ± 0.004 ^{a}	77.0 ± 8.0 ^{a}	17.0 ± 1.0 ^{a}	7.27 ± 0.80 ^{a}
B	0.1%	0.051 ± 0.001 ^{a}	68.0 ± 10.0 ^{a}	19.0 ± 2.0 ^{a}	8.07 ± 0.63 ^{a,c}
C	0.2%	0.058 ± 0.002 ^{a}	77.0 ± 4.0 ^{a}	22.0 ± 3.0 ^{a}	10.44 ± 0.48 ^{b,c}
D	0.3%	0.056 ± 0.008 ^{a}	75.0 ± 2.0 ^{a}	27.0 ± 5.0 ^{a}	11.41 ± 1.05 ^{b}
E	0.4%	0.051 ± 0.002 ^{a}	54.0 ± 22.0 ^{a}	28.0 ± 9.0 ^{a}	9.72 ± 0.57 ^{a,b,c}
F	0.5%	0.054 ± 0.008 ^{a}	73.0 ± 13.0 ^{a}	23.0 ± 2.0 ^{a}	9.94 ± 1.46 ^{b,c}

$^{a–c}$ Different superscript letters in the same column indicate a statistically significant difference ($p < 0.05$).

The values of the films' solubility were approximately 70% and this was not influenced by the addition of QT-LL 75 in different concentrations. This great solubility behavior was also observed for films produced with *D. regia* galactomannan with different molecular weights (between 69 and 76%) [31]. In this same work, the galactomannan was blended with k-carrageenan and all of the obtained films presented 100% solubility, which was considered to be a result of the k-carrageenan sulphate functional groups which can form hydrogen bonds with water molecules helping to disrupt the film and increasing its solubility. The solubility of Tara gum films was 24% and different when compared to our results; and the addition of chitosan and chitosan nanoparticles (both at 15% w/w content) significantly decreased the films' solubility (%) to 10.37 ± 2.31 and 6.16 ± 1.17, respectively [44]. The quercetin-loaded lecithin liposomes used in this work have average sizes ranging from 93 to 131 nm, different from the chitosan nanostructures used in the above-mentioned work (180.25 nm). The average size, in addition to the strong interaction between chitosan nanoparticles and Tara gum chains, may have influenced the Tara gum film structure, justifying the differences obtained for S % when compared to our galactomannan-based films.

WVP values ranged from 7.27 ± 0.80 (A) to 11.41 ± 1.05 × 10^{-7} (D) g/(h·m·Pa). The presence of QT-LL 75 at 0.2% (C), 0.3% (D), and 0.5% (F) significantly increased ($p < 0.05$) the WVP results, leading to values up to 1.9-fold higher than the galactomannan-based film without liposome (A). On the other hand, the WVP values of B and E films were similar to the control film. According to Antoniou et al. [48], the hydrophilic pattern and the water attracted into the film contribute to create regions of higher water mobility, with greater inter-chain distances in the galactomannan film. In view of this, it is possible to consider that these WVP results are in agreement with those of MC and S, confirming the hydrophilicity of all of the films. Indeed, the differences in WVP values of films C, D, and F can be associated with the irregular surface of the film, as reported by SEM results (Section 3.4.1). Since the liposomes are immobilized on the surface of the films, the free spaces release a passage of water vapor through the chains of the galactomannan; thus, we hypothesize that the increase in WVP values occurs due to a lower compression of the film matrix, resulting in a modified diffusivity.

The water vapor pattern demonstrated by our films was also reported by other authors working with galactomannans; for instance, WVP values for films produced with the *C. grandis* galactomannan and different bioactive molecules ranged from 5.60 ± 0.39 to 22.43 ± 0.21 × 10^{-7} g/(h·m·Pa) [3], while those for films produced with *D. regia* galactomannan with different molecular weights varied between 7.55 and 12.70 × 10^{-11} g/(s.m.Pa) [31]. Easy water-vapor permeation was demonstrated for Tara gum films and associated with the high glycerol concentration used in the film-forming solution [44]; however, when chitosan and chitosan nanostructures were added to the galactomannan-based films, they decreased WVP values in a different manner than what was demonstrated for our films with incorporated quercetin-loaded lecithin liposomes. According to the authors of the above-mentioned study, the chitosan nanoparticles have a greater ability to occupy the empty spaces of the porous film matrix, thus creating a more compact structure in the films and impairing the permeation of water, which is associated with decreased WVP results. Similar decreases in WVP values were reported when chitosan film incorporated different amounts of black soybean seed coat extract [47], carvacrol extract [49], and apple polyphenol [50]; in this case, reduced WVP values could be related to the compact and dense networks formed through intermolecular interactions between the immobilized molecules and chitosan chains.

The determination of the contact angle (CA) is a simple way to evaluate the degree of hydrophobicity of a material layered on a surface or immobilized on a ready surface. Commonly, CA values lower than 90° are related to hydrophilic materials [3,51]. CA for all of the films (A–F) was measured at 5 s and additionally 30 s in order to understand the pattern of hydrophobicity of the films over time. The results demonstrate that all of the films were essentially hydrophobic with no statistical differences between them, including when analyzed over time (Table 5). The distribution of QT-LL 75 on the surface of the films was quite homogeneous and did not statistically influence CA values, as confirmed by S% and MC results (Table 4). The *C. grandis* galactomannan film reported by Albuquerque et al. [3] presented a lower CA, being considered more hydrophilic than any of the films in this study. Considering that the same source of galactomannan was used in both studies, it is possible to predict that different procedure conditions, the period of study, and environmental conditions could have influenced the final films. However, as mentioned by the authors, this difference can be useful for specific applications, since the relatively hydrophobic pattern of the film allows the graduated delivery of immobilized compounds by helping them to remain in the film until the complete solubilization of the matrix.

Other works evaluated the hydrophobic character of films produced by polymers from different sources. For example, nanofibers obtained from low-cost agro-industrial by-products (soybean hulls and pods) were used as reinforcers of soy protein films. The authors performed an extensive characterization and reported a reduction in the hydrophilicity of the reinforced films' surface, as revealed by higher CA values; they also suggested that it occurred due to the interactions between components and attributed this effect to the properties of the nanomaterials [29]. The film-forming properties of the galactomannan fraction extracted from *G. triacanthos* and its ability to improve the properties of soy protein films were also evaluated [42]. The results revealed that the galactomannan film had a more hydrophobic surface when compared to the pure soy protein film; however, a synergic effect was achieved by the combination of both components, probably derived from their interactions. In addition, the correlation of WVP with the contact angle was inferred, and they reported that the more hydrophobic surface led to a smaller WVP, as noted in our work.

Table 5. Contact angle (CA) for (A) the control film and the films (B–F) with different concentrations of QT-LL 75 (values expressed as average ± standard deviation) at 5 and 2 seconds.

Sample	CA at 5 s	CA at 30 s
A	113.5 ± 9.57 [a]	111.5 ± 2.61 [a]
B	114.2 ± 5.93 [a]	119.1 ± 3.47 [a]
C	108.3 ± 4.36 [a]	112.2 ± 3.04 [a]
D	108.2 ± 3.98 [a]	114.2 ± 6.02 [a]
E	112.8 ± 7.03 [a]	115.6 ± 6.22 [a]
F	113.8 ± 4.81 [a]	117.1 ± 3.78 [a]

[a] The same superscript letter in the same column indicates no statistical difference ($p > 0.05$).

Considering the above-mentioned results, it is possible to state that quercetin has been successfully encapsulated in lecithin liposomes by a simple technique were we used a low concentration of lecithin (Lipoid S45 at 25 mg/mL), which was injected, drop by drop, during only 2 min, in 23 mL of distilled water under magnetic stirring (600 rpm). This suspension was left in magnetic stirring for a further 8 min and then submitted to a vigorous shaking on a simple mixer for 2 min at room temperature. Other references also used ethanol injection and similar phospholipids; however, significant differences in technique could be seen:

(1) Higher concentration of phospholipids, for example, the 60 mg/mL of Lecinova® or Lipoid S75 used for the development of liposomes by Gil et al. [18];
(2) Other reagents and sophisticated equipment, as mentioned by AbouSamra, Elgohary, and Mansy [19] on the development of pirfenidone-loaded lecithin core nanocapsules, which included Span 60, Tween 80, and chloroform, in addition to a homogenizer that reached more than 20,000 rpm;
(3) Longer experimental period, for example the stirring overnight of the suspensions containing whey lecithin, maltodextrin, eugenol, oleic acid, and chitosan; these suspensions also required a high-pressure homogenizer for complete homogenization [21].

4. Conclusions

Quercetin was successfully encapsulated in lecithin liposomes using a simple and fast technique based on the ethanol injection method. From a pharmaceutical point of view, the results obtained in this study are valuable for the design and fabrication of liposomes as a delivery system for quercetin. Additionally, galactomannan-based films from *C. grandis* proved to be a promising structure for the immobilization of quercetin-loaded lecithin liposomes, thus prolonging the availability of quercetin and its biological activities. Therefore, the knowledge derived from this investigation could facilitate the encapsulation, protection, and delivery of quercetin in products suggested for the nanopharmacology field, including personalized dosage forms for skincare. However, further delivery studies, and also in vitro and in vivo evaluations, are necessary to understand transdermal delivery regarding drug permeation, drug loading, and stability of quercetin-loaded lecithin liposomes.

Author Contributions: P.B.S.d.A. was responsible for the investigation, methodology, and the original writing of the manuscript. M.P.d.S., M.A.C. and A.I.B. were responsible for the analysis and interpretation of the results, and M.A.C. supervised and contributed to the writing and the revision of the manuscript. L.P., P.J., J.A.T. and M.d.G.C.-d.-C. were responsible for the research concept, and the experimental design. J.A.T. and M.d.G.C.-d.-C. were also responsible for the funding acquisition. All authors have read and agreed to the published version of the manuscript.

Funding: This study was supported by the Coordenação de Aperfeiçoamento de Pessoal de Nível Superior (CAPES) and the Portuguese Foundation for Science and Technology (FCT) under the scope of the strategic funding of UID/BIO/04469/2013 unit and COMPETE 2020 (POCI-01-0145-FEDER-006684) and under the scope of the Project RECI/BBB-EBI/0179/2012 (FCOMP-01-0124-FEDER-027462). P.A. receipted a scholarship from CAPES. M.d.G.C.-d.-C. express her gratitude to the Conselho Nacional de Desenvolvimento Científico e Tecnológico (CNPq) for research grants and fellowship. The authors acknowledge CNPq and CAPES for financial support.

Data Availability Statement: Data supporting reported results of this work are contained within the article. The data presented in this study are available in the following articles: "Cross-linked chitosan/liposome hybrid system for the intestinal delivery of quercetin", "Immobilization of bioactive compounds in Cassia grandis galactomannan-based films: Influence on physicochemical properties", "*Delonix regia* galactomannan-based edible films: Effect of molecular weight and k-carrageenan on physicochemical properties", "Development of edible films prepared by soy protein and the galactomannan fraction extracted from *Gleditsia triacanthos* (Fabaceae) seed", and "Preparation and characterization of konjac glucomannan-based bionanocomposite film for active food packaging".

Acknowledgments: The authors acknowledge the University of Minho and University of Reading for analytical support.

Conflicts of Interest: The authors declare no conflict of interest.

References

1. Albuquerque, P.B.S.; de Oliveira, W.F.; Silva, P.M.D.S.; Correia, M.T.D.S.; Kennedy, J.F.; Coelho, L.C.B.B. Epiphanies of well-known and newly discovered macromolecular carbohydrates—A review. *Int. J. Biol. Macromol.* **2020**, *156*, 51–66. [CrossRef] [PubMed]
2. Albuquerque, P.B.; Barros, W.; Santos, G.R.; Correia, M.T.; Mourão, P.A.; Teixeira, J.A.; Carneiro-Da-Cunha, M.G. Characterization and rheological study of the galactomannan extracted from seeds of *Cassia grandis*. *Carbohydr. Polym.* **2014**, *104*, 127–134. [CrossRef] [PubMed]
3. Albuquerque, P.B.; Cerqueira, M.A.; Vicente, A.A.; Teixeira, J.A.; Carneiro-Da-Cunha, M.G. Immobilization of bioactive compounds in *Cassia grandis* galactomannan-based films: Influence on physicochemical properties. *Int. J. Biol. Macromol.* **2017**, *96*, 727–735. [CrossRef]
4. Albuquerque, P.B.; Soares, P.A.; Aragão-Neto, A.C.; Albuquerque, G.S.; Silva, L.C.; Lima-Ribeiro, M.H.; Neto, J.C.S.; Coelho, L.C.; Correia, M.T.; Teixeira, J.A.; et al. Healing activity evaluation of the galactomannan film obtained from *Cassia grandis* seeds with immobilized *Cratylia mollis* seed lectin. *Int. J. Biol. Macromol.* **2017**, *102*, 749–757. Available online: http://linkinghub.elsevier.com/retrieve/pii/S0141813017304683 (accessed on 1 February 2023). [CrossRef] [PubMed]
5. Gouda, A.; Sakr, O.S.; Nasr, M.; Sammour, O. Ethanol injection technique for liposomes formulation: An insight into development, influencing factors, challenges and applications. *J. Drug Deliv. Sci. Technol.* **2021**, *61*, 102174. [CrossRef]
6. He, Y.; Zhang, W.; Xiao, Q.; Fan, L.; Huang, D.; Chen, W.; He, W. Liposomes and liposome-like nanoparticles: From anti-fungal infection to the COVID-19 pandemic treatment. *Asian J. Pharm. Sci.* **2022**, *17*, 817–837. [CrossRef]
7. Imura, T.; Otake, K.; Hashimoto, S.; Gotoh, T.; Yuasa, M.; Yokoyama, S.; Sakai, H.; Rathman, J.F.; Abe, M. Preparation and physicochemical properties of various soybean lecithin liposomes using supercritical reverse phase evaporation method. *Colloids Surf. B Biointerfaces* **2003**, *27*, 133–140. [CrossRef]
8. Zahednezhad, F.; Zakeri-Milani, P.; Mojarrad, J.S.; Sarfraz, M.; Mahmoudian, M.; Baradaran, B.; Valizadeh, H. Acetyl carnitine modified liposomes elevate cisplatin uptake in macrophage and cancer cells. *J. Drug Deliv. Sci. Technol.* **2023**, *81*, 104198. [CrossRef]
9. Zhang, L.; Lin, Z.; Chen, Y.; Gao, D.; Wang, P.; Lin, Y.; Wang, Y.; Wang, F.; Han, Y.; Yuan, H. Co-delivery of Docetaxel and Resveratrol by liposomes synergistically boosts antitumor efficiency against prostate cancer. *Eur. J. Pharm. Sci.* **2022**, *174*, 106199. [CrossRef]
10. Yu, S.; Li, D.; Shi, A.; Long, Y.; Deng, J.; Ma, Y.; Li, X.; Wen, J.; Hu, Y.; He, X.; et al. Multidrug-loaded liposomes prevent ischemic stroke through intranasal administration. *Biomed. Pharmacother.* **2023**, *162*, 114542. [CrossRef]
11. Fereig, S.A.; El-Zaafarany, G.M.; Arafa, M.G.; Abdel-Mottaleb, M.M. Self-assembled tacrolimus-loaded lecithin-chitosan hybrid nanoparticles for in vivo management of psoriasis. *Int. J. Pharm.* **2021**, *608*, 121114. [CrossRef]
12. Hatahet, T.; Morille, M.; Hommoss, A.; Devoisselle, J.-M.; Müller, R.; Bégu, S. Liposomes, lipid nanocapsules and smartCrystals®: A comparative study for an effective quercetin delivery to the skin. *Int. J. Pharm.* **2018**, *542*, 176–185. [CrossRef] [PubMed]

13. Shah, S.; Dhawan, V.; Holm, R.; Nagarsenker, M.S.; Perrie, Y. Liposomes: Advancements and innovation in the manufacturing process. *Adv. Drug Deliv. Rev.* **2020**, *154–155*, 102–122. [CrossRef] [PubMed]
14. Kanda, H.; Katsube, T.; Wahyudiono; Goto, M. Preparation of Liposomes from Soy Lecithin Using Liquefied Dimethyl Ether. *Foods* **2021**, *10*, 1789. [CrossRef] [PubMed]
15. Płaczek, M.; Wątróbska-Świetlikowska, D.; Stefanowicz-Hajduk, J.; Drechsler, M.; Ochocka, J.R.; Sznitowska, M. Comparison of the in vitro cytotoxicity among phospholipid-based parenteral drug delivery systems: Emulsions, liposomes and aqueous lecithin dispersions (WLDs). *Eur. J. Pharm. Sci.* **2018**, *127*, 92–101. [CrossRef]
16. Savić, V.; Ilić, T.; Nikolić, I.; Marković, B.; Čalija, B.; Cekic, N.; Savic, S. Tacrolimus-loaded lecithin-based nanostructured lipid carrier and nanoemulsion with propylene glycol monocaprylate as a liquid lipid: Formulation characterization and assessment of dermal delivery compared to referent ointment. *Int. J. Pharm.* **2019**, *569*, 118624. [CrossRef]
17. Vater, C.; Hlawaty, V.; Werdenits, P.; Cichoń, M.A.; Klang, V.; Elbe-Bürger, A.; Wirth, M.; Valenta, C. Effects of lecithin-based nanoemulsions on skin: Short-time cytotoxicity MTT and BrdU studies, skin penetration of surfactants and additives and the delivery of curcumin. *Int. J. Pharm.* **2020**, *580*, 119209. [CrossRef]
18. Gil, K.A.; Jerković, I.; Marijanović, Z.; Manca, M.L.; Caddeo, C.; Tuberoso, C.I.G. Evaluation of an innovative sheep cheese with antioxidant activity enriched with different thyme essential oil lecithin liposomes. *LWT* **2021**, *154*, 112808. [CrossRef]
19. AbouSamra, M.M.; Elgohary, R.; Mansy, S.S. Innovated pirfenidone loaded lecithin nanocapsules for targeting liver fibrosis: Formulation, characterization and in vivo study. *Int. J. Pharm.* **2023**, *631*, 122539. [CrossRef]
20. Correa, V.L.R.; Martins, J.A.; de Souza, T.R.; Rincon, G.D.C.N.; Miguel, M.P.; de Menezes, L.B.; Amaral, A.C. Melatonin loaded lecithin-chitosan nanoparticles improved the wound healing in diabetic rats. *Int. J. Biol. Macromol.* **2020**, *162*, 1465–1475. [CrossRef]
21. Talón, E.; Lampi, A.-M.; Vargas, M.; Chiralt, A.; Jouppila, K.; González-Martínez, C. Encapsulation of eugenol by spray-drying using whey protein isolate or lecithin: Release kinetics, antioxidant and antimicrobial properties. *Food Chem.* **2019**, *295*, 588–598. [CrossRef] [PubMed]
22. Souza, M.P.; Vaz, A.F.M.; Correia, M.T.S.; Cerqueira, M.A.; Vicente, A.A.; Carneiro-da-Cunha, M.G. Quercetin-Loaded Lecithin/Chitosan Nanoparticles for Functional Food Applications. *Food Bioprocess Technol.* **2014**, *7*, 1149–1159. [CrossRef]
23. Benzie, I.F.F.; Strain, J.J. The ferric reducing ability of plasma (FRAP) as a measure of "antioxidant power": The FRAP assay. *Anal. Biochem.* **1996**, *239*, 70–76. Available online: http://linkinghub.elsevier.com/retrieve/pii/S0003269796902924 (accessed on 1 February 2023). [CrossRef] [PubMed]
24. Torres, E.; Marín, V.; Aburto, J.; Beltran, H.I.; Shirai, K.; Villanueva, S.; Sandoval, G. Enzymatic modification of chitosan with quercetin and its application as antioxidant edible films. *Appl. Biochem. Microbiol.* **2012**, *48*, 151–158. Available online: http://link.springer.com/10.1134/S0003683812020123 (accessed on 1 February 2023). [CrossRef]
25. Mahaling, B.; Katti, D.S. Understanding the influence of surface properties of nanoparticles and penetration enhancers for improving bioavailability in eye tissues in vivo. *Int. J. Pharm.* **2016**, *501*, 1–9. [CrossRef]
26. Gibis, M.; Vogt, E.; Weiss, J. Encapsulation of polyphenolic grape seed extract in polymer-coated liposomes. *Food Funct.* **2012**, *3*, 246–254. [CrossRef]
27. Kumari, A.; Yadav, S.K.; Pakade, Y.B.; Singh, B.; Yadav, S.C. Development of biodegradable nanoparticles for delivery of quercetin. *Colloids Surf. B Biointerfaces* **2010**, *80*, 184–192. [CrossRef]
28. Caddeo, C.; Díez-Sales, O.; Pons, R.; Carbone, C.; Ennas, G.; Puglisi, G.; Fadda, A.M.; Manconi, M. Cross-linked chitosan/liposome hybrid system for the intestinal delivery of quercetin. *J. Colloid Interface Sci.* **2016**, *461*, 69–78. [CrossRef]
29. González, A.; Gastelú, G.; Barrera, G.N.; Ribotta, P.D.; Igarzabal, C.I. Preparation and characterization of soy protein films reinforced with cellulose nanofibers obtained from soybean by-products. *Food Hydrocoll.* **2018**, *89*, 758–764. [CrossRef]
30. Wu, C.; Li, Y.; Du, Y.; Wang, L.; Tong, C.; Hu, Y.; Pang, J.; Yan, Z. Preparation and characterization of konjac glucomannan-based bionanocomposite film for active food packaging. *Food Hydrocoll.* **2018**, *89*, 682–690. [CrossRef]
31. Rodriguez-Canto, W.; Cerqueira, M.A.; Chel-Guerrero, L.; Pastrana, L.M.; Aguilar-Vega, M. Delonix regia galactomannan-based edible films: Effect of molecular weight and k-carrageenan on physicochemical properties. *Food Hydrocoll.* **2020**, *103*, 105632. [CrossRef]
32. De Sousa, F.D.; Vasconselos, P.D.; da Silva, A.F.B.; Mota, E.F.; Tomé, A.D.R.; Mendes, F.R.D.S.; Gomes, A.M.M.; Abraham, D.J.; Shiwen, X.; Owen, J.S.; et al. Hydrogel and membrane scaffold formulations of Frutalin (breadfruit lectin) within a polysaccharide galactomannan matrix have potential for wound healing. *Int. J. Biol. Macromol.* **2019**, *121*, 429–442. [CrossRef] [PubMed]
33. Rossi, B.; Campia, P.; Merlini, L.; Brasca, M.; Pastori, N.; Farris, S.; Melone, L.; Punta, C.; Galante, Y.M. An aerogel obtained from chemo-enzymatically oxidized fenugreek galactomannans as a versatile delivery system. *Carbohydr. Polym.* **2016**, *144*, 353–361. Available online: http://linkinghub.elsevier.com/retrieve/pii/S0144861716300479 (accessed on 1 February 2023). [CrossRef] [PubMed]
34. Chouaibi, M.; Rezig, L.; Lakoud, A.; Boussaid, A.; Hassouna, M.; Ferrari, G.; Hamdi, S. Exploring potential new galactomannan source of Retama reatam seeds for food, cosmetic and pharmaceuticals: Characterization and physical, emulsifying and antidiabetic properties. *Int. J. Biol. Macromol.* **2019**, *124*, 1167–1176. [CrossRef]
35. Mudgil, D.; Barak, S.; Khatkar, B. X-ray diffraction, IR spectroscopy and thermal characterization of partially hydrolyzed guar gum. *Int. J. Biol. Macromol.* **2012**, *50*, 1035–1039. [CrossRef]

36. Dai, L.; Sun, C.; Li, R.; Mao, L.; Liu, F.; Gao, Y. Structural characterization, formation mechanism and stability of curcumin in zein-lecithin composite nanoparticles fabricated by antisolvent co-precipitation. *Food Chem.* **2017**, *237*, 1163–1171. [CrossRef]
37. Ben Jeddou, K.; Chaari, F.; Maktouf, S.; Nouri-Ellouz, O.; Helbert, C.B.; Ghorbel, R.E. Structural, functional, and antioxidant properties of water-soluble polysaccharides from potatoes peels. *Food Chem.* **2016**, *205*, 97–105. Available online: http://linkinghub.elsevier.com/retrieve/pii/S0308814616302783 (accessed on 1 February 2023). [CrossRef]
38. Ghribi, A.M.; Sila, A.; Gafsi, I.M.; Blecker, C.; Danthine, S.; Attia, H.; Bougatef, A.; Besbes, S. Structural, functional, and ACE inhibitory properties of water-soluble polysaccharides from chickpea flours. *Int. J. Biol. Macromol.* **2015**, *75*, 276–282. [CrossRef]
39. Wang, X.; Luo, Z.; Xiao, Z. Preparation, characterization, and thermal stability of β-cyclodextrin/soybean lecithin inclusion complex. *Carbohydr. Polym.* **2014**, *101*, 1027–1032. [CrossRef]
40. Xie, H.; Liu, C.; Gao, J.; Shi, J.; Ni, F.; Luo, X.; He, Y.; Ren, G.; Luo, Z. Fabrication of Zein-Lecithin-EGCG complex nanoparticles: Characterization, controlled release in simulated gastrointestinal digestion. *Food Chem.* **2021**, *365*, 130542. [CrossRef]
41. Pool, H.; Quintanar, D.; Figueroa, J.D.D.; Mano, C.M.; Bechara, J.E.H.; Godínez, L.A.; Mendoza, S. Antioxidant Effects of Quercetin and Catechin Encapsulated into PLGA Nanoparticles. *J. Nanomater.* **2012**, *2012*, 145380. [CrossRef]
42. González, A.; Barrera, G.N.; Galimberti, P.I.; Ribotta, P.D.; Igarzabal, C.I.A. Development of edible films prepared by soy protein and the galactomannan fraction extracted from Gleditsia triacanthos (Fabaceae) seed. *Food Hydrocoll.* **2019**, *97*, 105227. [CrossRef]
43. Souza, M.P.; Vaz, A.F.M.; Silva, H.D.; Cerqueira, M.A.; Vicente, A.A.; Carneiro-Da-Cunha, M.G. Development and Characterization of an Active Chitosan-Based Film Containing Quercetin. *Food Bioprocess Technol.* **2015**, *8*, 2183–2191. [CrossRef]
44. Antoniou, J.; Liu, F.; Majeed, H.; Zhong, F. Characterization of tara gum edible films incorporated with bulk chitosan and chitosan nanoparticles: A comparative study. *Food Hydrocoll.* **2015**, *44*, 309–319. Available online: http://www.sciencedirect.com/science/article/pii/S0268005X14003294 (accessed on 1 February 2023). [CrossRef]
45. Libio, I.C.; Demori, R.; Ferrão, M.F.; Lionzo, M.I.; da Silveira, N.P. Films based on neutralized chitosan citrate as innovative composition for cosmetic application. *Mater. Sci. Eng. C* **2016**, *67*, 115–124. Available online: http://linkinghub.elsevier.com/retrieve/pii/S0928493116304337 (accessed on 1 February 2023). [CrossRef] [PubMed]
46. Cerqueira, M.A.; Souza, B.W.S.; Teixeira, J.A.; Vicente, A.A. Effect of glycerol and corn oil on physicochemical properties of polysaccharide films—A comparative study. *Food Hydrocoll.* **2012**, *27*, 175–184. Available online: http://linkinghub.elsevier.com/retrieve/pii/S0268005X11002086 (accessed on 1 February 2023). [CrossRef]
47. Wang, X.; Yong, H.; Gao, L.; Li, L.; Jin, M.; Liu, J. Preparation and characterization of antioxidant and pH-sensitive films based on chitosan and black soybean seed coat extract. *Food Hydrocoll.* **2018**, *89*, 56–66. [CrossRef]
48. Antoniou, J.; Liu, F.; Majeed, H.; Qazi, H.J.; Zhong, F. Physicochemical and thermomechanical characterization of tara gum edible films: Effect of polyols as plasticizers. *Carbohydr. Polym.* **2014**, *111*, 359–365. [CrossRef]
49. Rubilar, J.F.; Cruz, R.M.S.; Silva, H.D.; Vicente, A.A.; Khmelinskii, I.; Vieira, M.C. Physico-mechanical properties of chitosan films with carvacrol and grape seed extract. *J. Food Eng.* **2013**, *115*, 466–474. [CrossRef]
50. Sun, L.; Sun, J.; Chen, L.; Niu, P.; Yang, X.; Guo, Y. Preparation and characterization of chitosan film incorporated with thinned young apple polyphenols as an active packaging material. *Carbohydr. Polym.* **2017**, *163*, 81–91. [CrossRef]
51. Ma, Q.; Hu, D.; Wang, L. Preparation and physical properties of tara gum film reinforced with cellulose nanocrystals. *Int. J. Biol. Macromol.* **2016**, *86*, 606–612. Available online: http://linkinghub.elsevier.com/retrieve/pii/S0141813016301052 (accessed on 1 February 2023). [CrossRef] [PubMed]

polymers

MDPI

Article

Electromagnetic Shielding Enhancement of Butyl Rubber/Single-Walled Carbon Nanotube Composites via Water-Induced Modification

Xin Guo and Guangye Liu *

Engineering Research Center of High-Performance Polymer and Molding Technology, Ministry of Education, Qingdao University of Science and Technology, Qingdao 266042, China; 0018020008@mails.qust.edu.cn
* Correspondence: qdlgy2008@sina.com

Abstract: Electromagnetic properties of polymer composites strongly depend on the loading amount and the completeness of the filler's dispersive structure. Improving the compatibility of single-walled carbon nanotubes (SWCNTs) with isobutylene butyl rubber (IIR) is a good solution to mitigate aggregation. The change in configuration of poly-oxyethylene octyl phenol ether (OP-10) was induced using water as the exposed hydrophilic groups linking with water molecules. The SWCNT and IIR/SWCNT composites were then prepared via wetly-melt mixing at a relatively high temperature to remove water, and they were then mixed with other agents after vacuum drying and cured. The SWCNTs were dispersed uniformly to form a good network for a lower percolation threshold of the wave-absorbing property to 2 phr from 8 phr. With 8 phr SWCNTs, the tensile strength of the material improved significantly from 7.1 MPa to 15.1 MPa, and the total electromagnetic shielding effectiveness of the material was enhanced to 23.8 dB, a 3-fold increase compared to the melt-mixed material. It was demonstrated that water-induced modification achieved good dispersion of SWCNTs for electromagnetic shielding enhancement while maintaining a wide damping temperature range from −55 °C to 40 °C with a damping factor over 0.2.

Keywords: electromagnetic shielding; water-induced modification; multi-absorbing

1. Introduction

With the wide use of electrical devices, electromagnetic pollution becomes more general and even universal in our lives and workplaces, which could be harmful to the health of humans and cause interference among equipment. Traditional electromagnetic shielding materials filled with metal are too heavy to wear and not flexible enough to satisfy requirements. Rubber is a good option as a flexible and soft electromagnetic shielding material; however, rubber becomes less flexible, and damping is reduced with too much metal filler. Therefore, an excellent damping polymer and a light-weight filler are desired. With a density of 1.3 g/cm^3, single-walled carbon nanotubes (SWCNTs) are multi-functional fillers for rubber composites for use in dielectric materials [1], electrically conductive materials [2], and electromagnetic interference (EMI) shielding materials [3–10], thus improving the mechanical properties, the thermal conductivity, and the electrical conductivity of materials due to their high moduli and electrical and thermal conductivities [11–13]. Isobutylene butyl rubber (IIR) is widely used in rubber products as it exhibits high damping performance and low gas permeability since it contains a great amount of methyl groups on both sides of the main chain [14]. However, the performance of isobutylene isoprene rubber/single-walled carbon nanotubes (IIR/SWCNTs) is not enhanced as designed because the dispersion of SWCNTs with large aspect ratios is not uniform but results in a great deal of physical entanglement and aggregation in the rubber matrix. Several processes and methods have been proposed and studied to solve the entanglement and aggregation problem, such as melt mixing [15], solution mixing [16,17] and latex mixing [18,19].

Carbon nanotubes (CNTs) are dispersed in the rubber matrix through strong shear forces in the melt-mixing process, but carbon nanotubes with a diameter in the range of tens of nanometers are difficult to disperse uniformly through shearing in incompatible rubber matrices [15]. In the solution-mixing process, the rubber and carbon nanotubes are mixed together in a solvent under specified conditions, and the mixed solution is casted to films on a smooth substrate. However, the agglomeration of carbon nanotubes is a serious problem in the rubber matrix during the solvent evaporating period, which affects the performance of the composite [16,17]. In the latex-mixing process, deionized water is used to disperse carbon nanotubes in advance and cast them on glass plates or other smooth surfaces, after mixing them with latex, before they are finally placed in an oven to remove water completely [20]. This process leaves a large number of carbon nanotubes dispersed in the matrix in the form of aggregates due to the polarity difference between carbon nanotubes and rubber.

Melt mixing is the preferred process for rubber due to its shear-sensitive characteristics, and a general open-mill mixer and an internal mixer can usually be used for production on compatibility between rubber and CNTs. The pre-dispersion process and the modification of CNTs have attracted the interest of researchers in order to improve the uniformity of mixing and dispersion, which can solve the aggregation problem of carbon nanotubes to a certain extent. Physical and chemical modification and combined methods are generally applied. High-power mechanical grinding and ball milling of CNTs form lattice defects or lattice distortions to produce highly active free radicals on the surface of carbon nanotubes, thereby improving the reaction between carbon nanotubes and the rubber matrix. However, it is not easy to control during grinding, resulting in shortening the length of carbon nanotubes and compromising the original properties [21]. The chemical introduction of functional groups, such as carboxyl, hydroxyl, and amino groups with activity, can significantly change the chemical reactivity of the surface of carbon nanotubes with concentrated or diluted acid, neither of which are environmentally friendly. However, compatibility is not improved without sacrificing the original advantages of CNTs [22].

IIR/SWCNT composites do not achieve good EMI performance with fewer SWCNTs through melt mixing. Aggregates of single-walled carbon nanotubes are formed in the matrix in the mixing process because of the large aspect ratio of SWCNTs, leading to higher loading and lower performance.

In attempts to solve the aggregation problem in our research, a poly-oxyethylene octyl phenol ether aqueous solution improved dispersion and the tensile strength, and water could be removed completely. This research result encouraged us to apply this process to improve electromagnetic shielding effectiveness for IIR/SWCNT composites.

Compared to the solution and emulsion methods mentioned above, pure water and a small amount of surfactant agent were used without pollution, and the process of preparation can be conducted in a simple and green way that is easy to master. All the equipment used in the process is universal in manufacturing rubber products, resulting in the potential to scale production.

The uniform dispersion of SWCNTs improves their mechanical properties due to good compatibility with rubber and also improves their thermal conductivity, dielectric properties, and electromagnetic interference shielding effectiveness; further, the percolation threshold of electromagnetic properties is relatively low. In order to fix the entanglement problem by improving compatibility with the IIR matrix, a water-induced modification method is provided in this work to enhance the EMI shielding effectiveness of highly damping IIR/SWCNT nanocomposites at a relatively low percolation threshold concentration of wave properties.

2. Materials and Methods

2.1. Materials

Isobutylene isoprene rubber (with the product name of IIR 268) was obtained from Exxonmobil Corporation, US. Zinc oxide (ZnO, purity of 99.8%), stearic acid (purity of

97%), sulfur (purity of 99.5%), benzothiazole disulfide (accelerator DM, purity of 99%), and tetramethylthiuram disulfide (accelerator TMTD, purity of 99%) were purchased from the Weifang Qinglian Chemical agents Co., Ltd. (Weifang, China) and were of industrial grade. SWCNTs (CNT-100, 1~2 nm in diameter, 30 μm in length, purity of 90%) were provided by Beijing Deke Daojin Co., Ltd. (Beijing, China). Poly-oxyethylene octyl phenol ether (OP-10, hydrophile-lipophile balance value of 14.5) was used as an emulsifying agent from Guangdong Zhongbang Fine Chemical Co., Ltd. (Guangzhou, China). All the materials were applied in this study without further purification. Pure water was made in the lab to induce the configuration change in OP-10 in order to modify the single-walled carbon nanotubes in a good dispersion state first in OP-10 solution and then in IIR matrix.

2.2. Preparation of Samples in Traditional Melt-Mixing Process

In this phase of the experiment, 0 phr (per hundred rubber in weight), 2 phr, 4 phr, 6 phr, or 8 phr single-walled carbon nanotubes were mixed in an open two-roll mixing mill of laboratory size (Φ160 mm × 320 mm, Shanghai Shuangyi Rubber & Plastic Co., Ltd., Shanghai, China) at room temperature with 100 phr IIR, 5 phr ZnO, 3 phr stearic acid, 0.5 phr sulphur, 1 phr accelerator DM, and 0.5 phr accelerator TMTD, respectively, and the prepared samples were vulcanized at 170 °C and marked as D-IIR/0SWCNTs (D is the dry state in melt mixing), D-IIR/2SWCNTs, D-IIR/4SWCNTs, D-IIR/6SWCNTs, and D-IIR/8SWCNTs.

2.3. Preparation Samples in Wetly-Melt Mixing via Water-Induced Modification of SWCNTs

The surfactant OP-10 aqueous solution was prepared at the concentration of 2 g/L. Following this, 0 phr, 1 phr, 2 phr, 4 phr, 6 phr, and 8 phr single-walled carbon nanotubes were dispersed in 25 g OP-10 solution, respectively, to prepare the wetly-melt mixed samples marked as IIR/0SWCNTs, IIR/1SWCNTs, IIR/2SWCNTs, IIR/4SWCNTs, IIR/6SWCNTs, and IIR/8SWCNTs. These mixtures were vigorously ultrasonicated in a KH100 ultrasonic bath for 30 min at an effective power of 800 W. In the mixing process, IIR was mixed in an open two-roll mixing mill of laboratory size (Φ160 mm × 320 mm, Shanghai Shuangyi Rubber & Plastic Co., Ltd., Shanghai, China) when the mills were heated to 100–110 °C. Following this, the SWCNTs-OP-10 suspension was added to the IIR on the mills after 15 to 20 min to obtain an IIR–SWCNTs blend. After vacuum drying at 50 °C under the vacuum pressure of −0.096 MPa, the IIR and SWCNTs blend was mixed with other chemical agents. The chemical agents included 5 phr ZnO, 3 phr stearic acid, 0.5 phr sulphur, 1 phr accelerator DM and 0.5 phr accelerator TMTD. And a vulcanization process was followed to prepare the final samples at 170 °C.

2.4. Test Instruments and Methods

The dispersive network micro-morphology of SWCNTs in IIR matrix was visualized with the transmission electron microscope (TEM) JEM2100 (Electronics Co., Ltd., Kyoto, Japan) after an ultrathin sample slice was obtained using a slicer Ultramicrotome EM UC7 (Leica, Wetzlar, Germany). Vulcanization parameters of the mixed compounds were tested in RPA2000 (Alpha Technology Inc., Bellingham, WA, USA) at 170 °C under a determined pressure, and the vulcanization time was 2 min longer than t_{90} for the vulcanization of 2 mm sheets under the pressure of 10 MPa. Mechanical properties were performed in a uniaxial tensile machine (AI-7000s, GOTECH testing machine corporation, Taiwan, China) at the speed of 500 mm/min. The electric conductivity of composites was measured with a broadband dielectric resistance spectrometer NOVOCONTROL (Beijing Huidexin Technology Co., Ltd., Beijing, China) after being sputtered with gold layers to the flat surfaces on both sides of the sample with the diameter of 25 mm with a small ion-sputtering apparatus SBC-12 (Beijing Zhongke Technology Instrument Co., Ltd., Beijing, China). Magnetic, dielectric, and electromagnetic shielding effectiveness were measured with the vector network scanner R&S ZNB20 with waveguide clips (SCHWARZ, Baden-Württemberg, Germany) from 8000~12,400 MHz via a cuboid specimen of 22.86 mm in length, 10.86

mm in width, and an actual thickness measured with a vernier caliper. The real part ε' and imaginary part ε'' of the dielectric constant and the real part μ' and the imaginary part μ'' of the permeability of composites were measured using a vector network scanner, as mentioned above. As the characterization parameter of microwave absorbing performance, the reflection loss was calculated using the Matrix Lab software (Matlab R2022b, The MathWorks, Inc., Alpharetta, GA, USA) corresponding to simulation of matching thickness and frequency of incident wave according to the real part ε' and imaginary part ε'' of the dielectric constant and the real part μ' and the imaginary part μ'' of the magnetic permeability. Dynamic mechanical analysis (DMA) was conducted at the frequency of 30 Hz from −80 °C to 40 °C on DMA/SDTA86 (Mettler-Toledo, Zurich, Switzerland).

3. Results and Discussion

3.1. Dispersion Structure Characterization

The dispersion structure of SWCNTs in the IIR/SWCNT composites with eight SWCNTs via melt mixing and wetly-melt mixing are shown in Figure 1 by visualizing their ultrathin slices in transmission mode. Compared to the partially aggregated state of single-walled carbon nanotubes in D-IIR/8SWCNTs composite produced via melt mixing shown in Figure 1a, single-walled carbon nanotubes are dispersed relatively uniformly and close to the mono-dispersion in the TEM photograph in Figure 1b showing IIR/8SWCNTs treated with wetly-melt mixing. With better dispersion of the single-walled carbon nanotubes in the isobutylene-isoprene rubber matrix, the dispersion network is constructed with fewer defects and aggregations to improve electrical conductivity and mechanical properties for high performance.

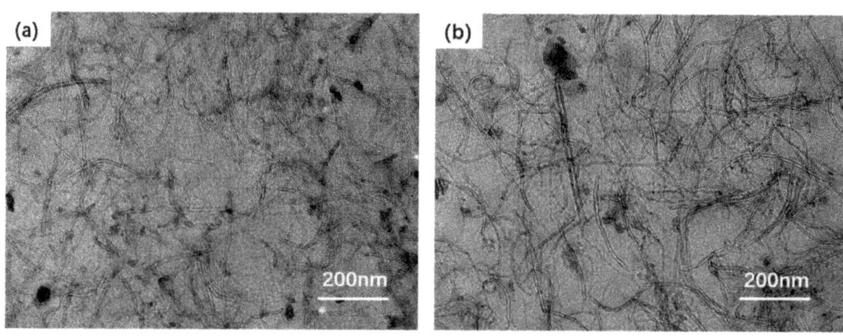

Figure 1. Transmission electron microscope photographs of isobutylene isoprene rubber composites with 8 phr SWCNTs (scale bar is 200 nm). (**a**) Melt-mixed composite. (**b**) Wetly-melt-mixed composite.

As shown in Figure 2, the OP-10 chain contains a long nonpolar group useful for its strong hydrophobic ability. In addition, its hydrophilic and lipophilic balances are low, and the hydrophilic group is hidden in the lipophilic group without water. However, the molecular spatial configuration of OP-10 is induced to change under the action of water, forming a tortuous structure when water is mixed with the OP-10 emulsifier. The hydrophilic groups of OP-10 therefore become exposed to water molecules, while the hydrophobic groups are wrapped with the hydrophilic groups. Through the connection of a hydrogen bond with the ether group of OP-10 and water molecules, SWCNTs are very well dispersed in the OP-10 emulsifier aqueous solution.

During the wetly-melt mixing process, the SWCMTs/OP-10 aqueous solution is added to the rubber matrix directly to avoid dry aggregation. Nonpolar groups of OP-10 are then released from being wrapped with the hydrophilic groups. The good compatibility of the nonpolar groups of OP-10 with IIR rubber greatly improves SWCNTs' dispersion in the rubber matrix.

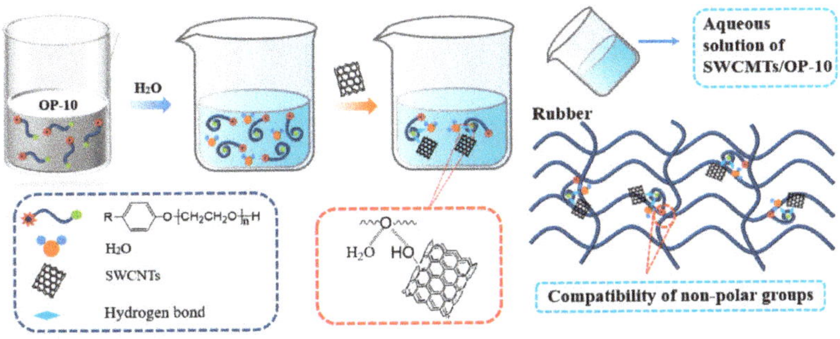

Figure 2. Schematic diagram of dispersion mechanism of single-walled carbon nanotubes in isobutylene isoprene rubber via water-induced modification of SWCNTs.

3.2. Vulcanization Properties

As shown in Figure 3, both ML and MH of isobutylene isoprene rubber composites produced by melt mixing and wetly-melt mixing improved when the amount of SWCNTs increased in the composites, as shown in Figure 3a. There are various polar groups on the surface of SWCNTs as physical crosslinking points to cause greater entanglement between isobutene–isoprene rubber molecular chains, which contributes to the total crosslinking density for higher MH and ML. Single-walled carbon nanotubes are provided more opportunities in a mono-dispersed state to contact with the rubber chains to form more physical crosslinking points, resulting in a much higher MH for IIR/8SWCNTs composites prepared via wetly-melt mixing. Both t_{10} and t_{90} of IIR/SWCNTs show a downward trend with the increasing loading phr of SWCNTs, as shown in Figure 3b. The first reason for this is that the thermal conductivity of single-walled carbon nanotubes at room temperature reaches up to 3500 W/m·K, which is far more than the thermal conductivity of 0.09 W/m·K of butyl rubber at room temperature. The more single-walled carbon nanotubes fill in the compound, the less time it needs for torque increment and vulcanization. The more uniform the dispersion of single-walled carbon nanotubes, the more perfect the heat transfer network is. Additionally, the alkalinity on the surface of SWCNTs accelerates the vulcanization process to perform shorter t_{10} and t_{90}.

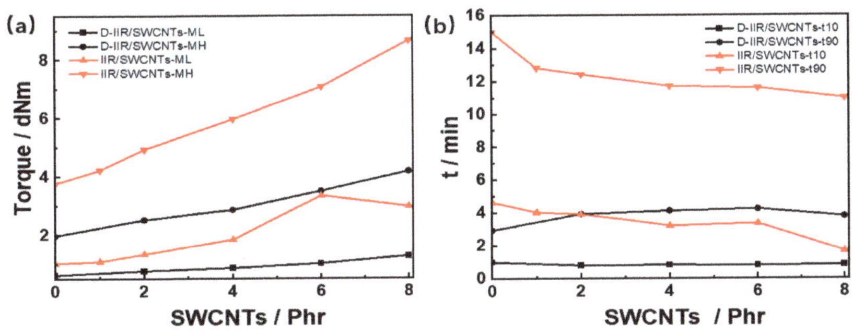

Figure 3. Vulcanization parameters of IIR/SWCNT composites via melt mixing and wetly-melt mixing. (**a**) MH and ML of composites corresponding to the amount of SWCNTs. (**b**) t_{10} and t_{90} of composites corresponding to numbers of SWCNTs. Note: ML—Lowest torque of curing curve; MH—maximum torque of curing curve; MH-ML—the difference between maximum torque and lowest torque, t_{10}—the time corresponding to a 10% increase in torque, t_{90}—the time corresponding to a 90% increase in torque.

3.3. Mechanical Properties

Single-walled carbon nanotubes can serve as an excellent reinforcing filler, especially in the uniform dispersive state, because of the high Young's modulus of 1 TPa [12]. As shown in Figure 4a, the tensile strength of IIR/SWCNT composites positively corresponds to the number of $SWCNT_S$. The composite with 8 phr SWCNTs reaches up to 15.09 MPa via wetly-melt mixing, which is 2.1 times that of IIR/8SWCNTs (7.07 MPa) via melt mixing. The elongations at the breaking points of composites prepared via wetly-melt mixing show excellent strain tolerances of over 650%, as shown in Figure 4b. The elongation at the breaking point of the IIR composite without SWCNTs reaches 760%, featuring high elasticity of isobutylene-isoprene rubber. The introduction of a small number of SWCNTs, i.e., less than 2 phr, is not enough to form a network contributing to higher elongation; yet, it causes a discontinuous network and defects in the matrix, resulting in lower elongation. With the addition of more SWCNTs from 2 phr to 6 phr, the elongations of composites increase, which can be attributed to the SWCNT-reinforced network via alkaline promotion of SWCNTs, slidable, entangled, and moveable molecular chains. When the filling of SWCNTs is further increased from 6 to 8 phr, the cross-linking density of the IIR/SWCNT composite increases further, whereas the mobility of molecular chain decreases because of more physically tangled points of SWCNTs. As a result, the elongation is shorter with more than 6 phr SWCNTs in the composite. For the composite materials prepared via melt mixing, the aggregation particles inside the isobutylene isoprene rubber matrix function as a stress concentration point for early failure, and contribute to better elongation, and even more severely when the number of SWCNTs increases.

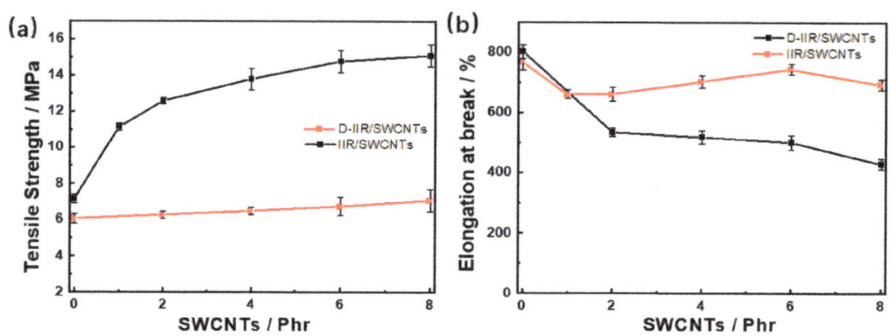

Figure 4. Mechanical properties of IIR/SWCNT composites via melt mixing and wetly-melt mixing. (**a**) Tensile strength corresponding to the loading phr of SWCNTs. (**b**) Elongation at breaking point corresponding to the loading phr of SWCNTs.

3.4. Dielectric Properties

As shown in Figure 5a,b, the dielectric properties of isobutylene isoprene rubber composites correspond to the loading phr of single-walled carbon nanotubes from 0 to 8 phr during the frequency range from 10^0 Hz to 10^7 Hz. The dielectric constant (ε'-real part) of melt-mixed composites and wetly-melt-mixed composites improve with the number of SWCNTs in the materials, owing to the conductive one-dimensional structure and the insulating butyl rubber, which together form a number of micro-capacitors improving the ability of the composites to store charge.

As the SWCNT content increases, the one-dimensional carbon nanotubes bond to each other in the IIR matrix, gradually building a conductive network. When the SWCNT content increases to a certain critical value, the conductive network can be formed, and the composite material realizes the transformation from an insulator to a semi-conductor, and the conductivity and dielectric constant rise significantly to show "percolation", as mentioned in our previous work [23]. A high dielectric constant of composite is harmful to impedance matching when used in electromagnetic shielding. It causes many electromagnetic waves to be reflected rather than being absorbed. Usually, it is a semi-conductor material, which affects the conductivity of SWCNTs, so as to reduce the dielectric constant with fewer SWCNTs and improve the effect of impedance matching. The dielectric loss factor of melt-mixed composites and wetly-melt-mixed composites are shown in Figure 5c,d. The dielectric loss of wetly-melt-mixed material is also relatively high at 1.43, while it is 0.1 for melt-mixed material when both are filled with 8 phr SWCNTs. An interconnecting network forms, as shown in TEM images, and a dielectric constant of the IIR /SWCNT composite shows a low percolation threshold value due to the high aspect ratio of SWCNTs and better dispersion. The main reason for this is that good modification dispersion induced by water in OP-10 aqueous solution provides more interface between the butyl rubber molecular chains and SWCNTs for more interfacial polarization with charge accumulation at the interfaces. It is also demonstrated that the influence of loading concentration on the dielectric property lessened when the loading phr was greater than 4.

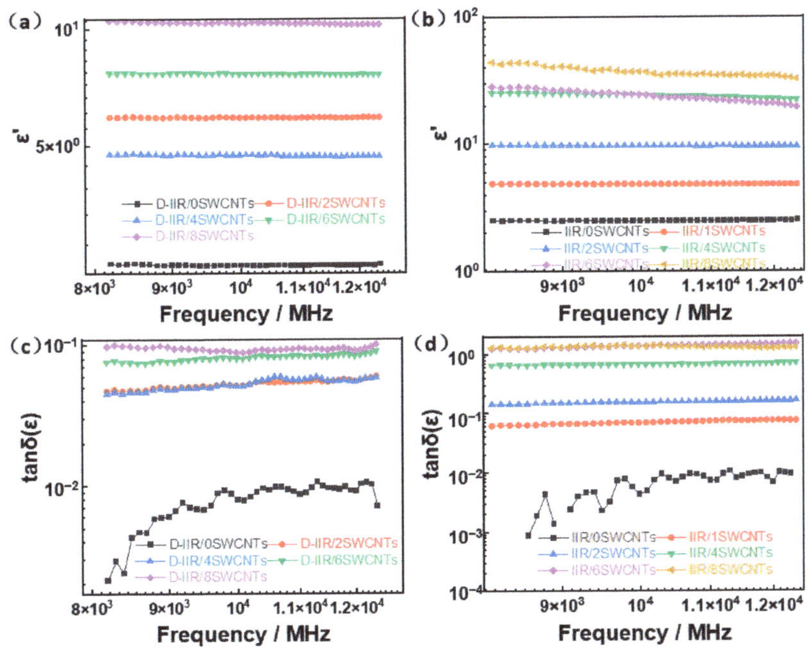

Figure 5. Dielectric properties of IIR/SWCNT composites. (**a**) Dielectric constant (ε′-real part) of melt-mixed composites. (**b**) Dielectric constant (ε′-real part) of wetly-melt-mixed composites. (**c**) Dielectric loss factor of melt-mixed composites. (**d**) Dielectric loss factor of wetly-melt-mixed composites.

3.5. Electric Conductivity

The electric conductivity of composites is enhanced with the increasing number of SWCNTs loading, as shown in Figure 6. Via a mixed process, the electric conductivity of the IIR/SWCNT composites achieves a threshold transition at 6 phr of SWCNTs, as shown in Figure 6a, while it shows a lower threshold at 2 phr in Figure 6b when a water-induced modification and wetly-melt-mixed process are applied. At the same number of 8 phr SWCNTs, the wetly-melt-mixed sample features the electric conductivity of 10^{-3} S·cm^{-1} due to more paths for electro flow in a continuous network, which is much higher than the melt-mixed one at 10^{-6} S·cm^{-1}.

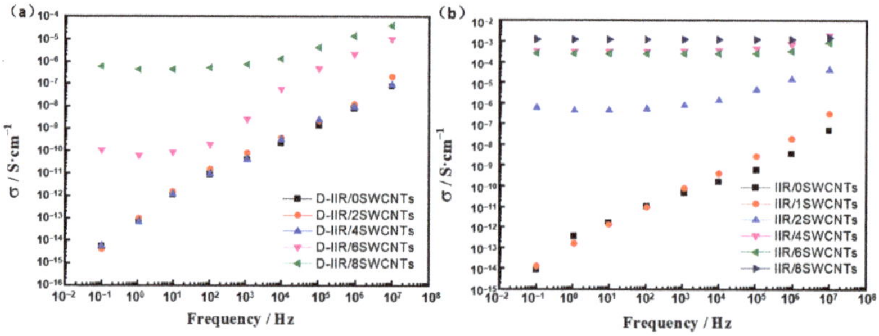

Figure 6. Electric conductivity of IIR/SWCNT composites. (**a**) Melt-mixed composites. (**b**) Wetly-melt-mixed composites.

3.6. EMI Shielding Effectiveness

The electromagnetic interference shielding effectiveness is enhanced with the increasing number of the SWCNTs loading, as shown in Figure 5. The EMI shielding efficiency of the materials increases to 6.5 dB gradually with the number of SWCNTs via the melt-mixing process for poor dispersion, as shown in Figure 7a. However, the wetly-melt-mixed materials present a sharp increase when the loading phr is added from 0 to 2 phr before maintaining a certain level, as shown in Figure 7b. As shown in Figure 7c,d, the absorption effectiveness of melt-mixed composites and wetly-melt-mixed composites improves to a different level such that the maximum value of the latter is 17 dB and ten times that of the former. Meanwhile, the total effectiveness of wetly-melt-mixed composites is enhanced to 23.8 dB at the loading phr of 8 SWCNTs, which is almost three times that of the melt-mixed one, shown in Figure 7e,f.

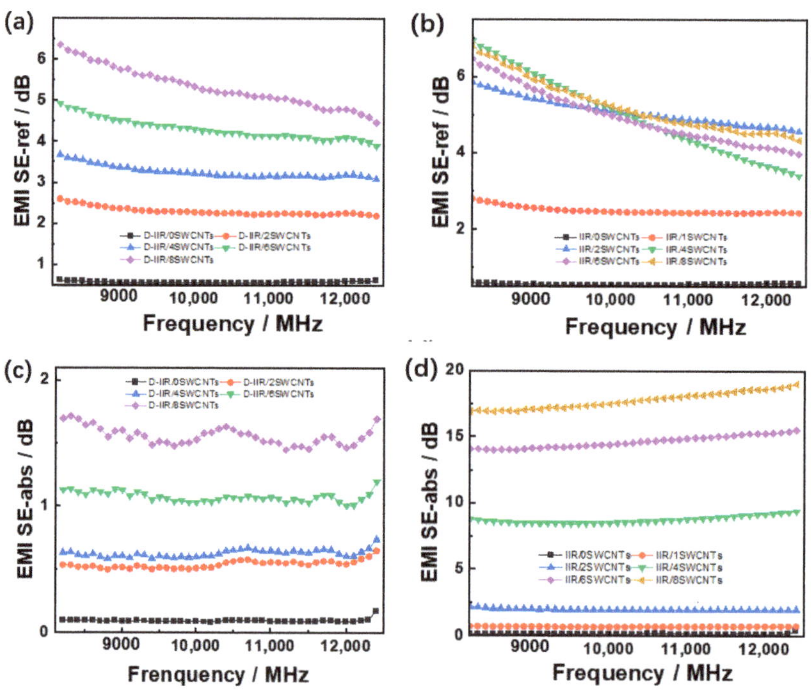

Figure 7. *Cont.*

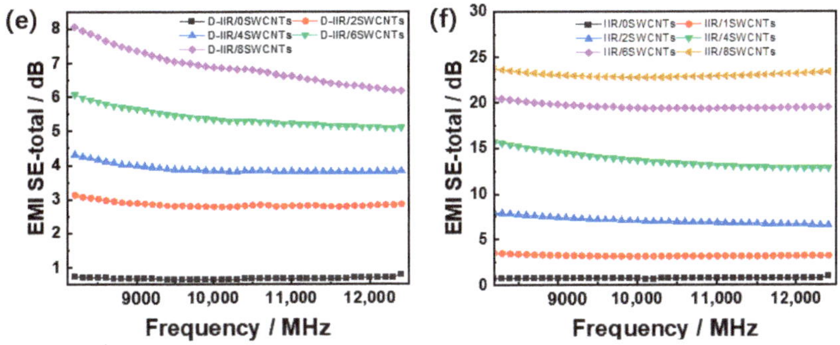

Figure 7. Electromagnetic interference shielding effectiveness of IIR/SWCNT composites. (**a**) Reflection effectiveness of melt-mixed composites. (**b**) Reflection effectiveness of wetly-melt-mixed composites. (**c**) Absorption effectiveness of melt-mixed composites. (**d**) Absorption effectiveness of wetly-melt-mixed composites. (**e**) Total effectiveness of melt-mixed composites. (**f**) Total effectiveness of wetly-melt-mixed composites.

As these composites are designed for multi-absorbing applications, the comparable examples are less effective. Nina [5] studied the EMI performance of butyl rubber-single walled carbon nanotube composites prepared via a solution-mixing process, and the shielding effectiveness was about 9~11.5 dB for the composite with 8 phr of SWCNT loading. Via water-induced modification and wetly-melt mixing, the EMI is 23.8 dB for composites prepared in our work, twice that in the mentioned research, indicating the effectiveness of the preparation method.

From Figure 8a,b, the shielding effectiveness at the frequency of 8200 MHz of composites is related to the loading phr increase of SWCNTs via both melt-mixing and wetly-melt-mixing processes. Obviously, the wetly-melt-mixing process benefits the material in forming a good dispersive state to enhance the EMI shielding total effectiveness, and the dominant contribution factor of EMI performance of the materials is the absorption effectiveness.

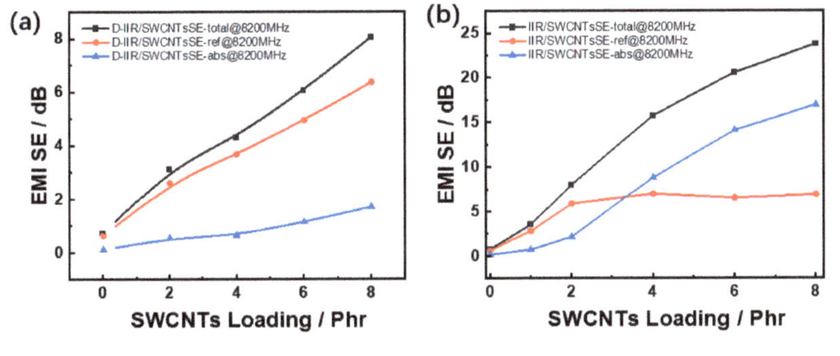

Figure 8. Electromagnetic interference shielding effectiveness of IIR/SWCNT composites with the loading phr of single-walled carbon nanotubes at the frequency of 8200 MHz. (**a**) Melt-mixed composites. (**b**) Wetly-melt-mixed composites.

3.7. Electromagnetic-Wave-Absorbing Properties

The three-dimensional image of reflection loss corresponding to thickness of melt-mixed composites and incident EM wave frequency presents an optimum value of −6.3 dB at 8 phr SWCNTs, as shown in Figure 9a. In Figure 9b, the optimum reflection loss of the wetly-melt-mixed composite is −8.7 dB with 2 phr SWCNTs. From Figure 9c, the best matching thickness of melt-mixed composite is 2.2 mm. For IIR/2SWCNTs, the matching thickness is 2.5 mm (see Figure 9d).

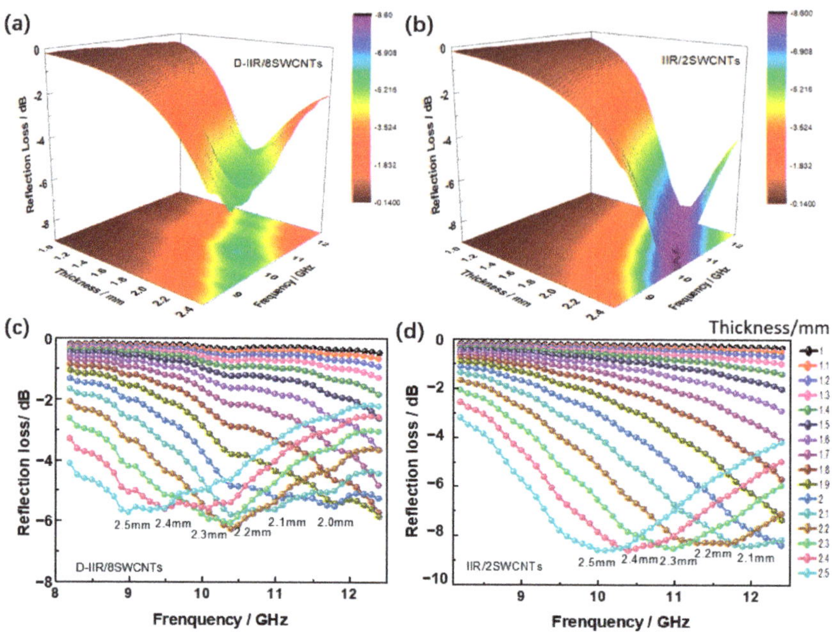

Figure 9. Electromagnetic-wave-absorbing properties of isobutylene isoprene rubber/SWCNT composites. (**a**) Three-dimensional image of reflection loss corresponding to thickness of the melt-mixed composite with 8 phr SWCNTs and incident wave frequency. (**b**) Three-dimensional image of reflection loss corresponding to thickness of the wetly-melt-mixed composite with 2 phr SWCNTs and incident wave frequency. (**c**) Reflection loss corresponding to thickness of the melt-mixed composite with 8 phr SWCNTs. (**d**) Reflection loss corresponding to thickness of the wetly-melt-mixed composite with 2 phr SWCNTs.

The water-induced modification of SWCNTs makes the wetly-melt-mixed composite construct a perfect net structure for the maximum wave absorbing performance with a lower dosage of SWCNTs.

3.8. Damping Properties

Butyl rubber composite material is an excellent option for damping applications. DMA tests of the composites were conducted at the frequency of 30 Hz in order to determine whether their damping properties were maintained and ascertain the effect of water-induced modification of SWCNTs on the damping properties as well as the effect of loading phr of SWCNTs (shown in Figure 10). As shown in Figure 10a, the main peaks appear at −18 °C and −40 °C in the curve of the damping factor. The peak temperature stays the same, and the value of the damping factor lowers as more single-walled carbon nanotubes are used.

The temperature ranges with damping factors over 0.2 of the wetly-melt-mixed composites are not as wide as that of the wetly-melt-mixed composites due to the uniform dispersion forming a continuous network of SWCNTs and rubber, as shown in Figure 10b. In shear mode, the response speed of the entanglement network of single-walled carbon nanotubes to strain is higher than that of rubber molecular segments, and the hysteresis loss of composites decreases.

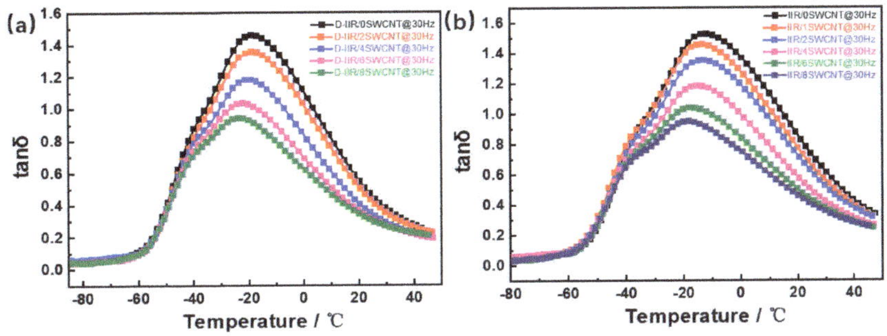

Figure 10. Effect of the amount of SWCNTs on the damping properties of IIR/SWCNT composites at the frequency of 30 Hz. (**a**) Melt-mixed composites. (**b**) Wetly-melt-mixed composites.

4. Conclusions

The IIR/SWCNT composites produced via melt-mixing could not achieve good EMI performance with fewer CNTs. The tortuous configuration of poly-oxyethylene octyl phenol ether was induced with water with the aid of ultrasonication in OP-10 solution. Under the water-induced modification of single-walled carbon nanotubes, OP-10 could be wrapped with hydrophobic non-polar groups inside with hydrophilic groups outside to link water and SWCNTs with hydrogen bonds and further improve compatibility in the wetly-melt-mixing process. Electromagnetic interference shielding effectiveness of butyl rubber/single-walled carbon nanotube composites presents a huge enhancement due to good dispersion compatibility improvement with the aid of water-induced modification of SWCNTs. The SWCNTs are dispersed uniformly to form a complete network with a lower percolation threshold of wave-absorbing properties to 2 phr from 8 phr. With 8 phr SWCNTs, the tensile strength of the material was improved significantly from 7.1 MPa to 15.1 Mpa, and total electromagnetic interference shielding effectiveness of material was enhanced to 23.8 dB, three times that of the melt-mixed one. This demonstrates that the water-induced modification achieved good dispersion of SWCNTs for EMI shielding enhancement while keeping a wide damping temperature range from −55 °C to 40 °C with the damping factor over 0.2. IIR/SWCNT composite materials obtained in this work show great potential in the field of high damping EMI material application or multi-absorbing materials including electromagnetic wave absorbing and shocking absorbing service conditions in order to mitigate the vibration and EMI for sophisticated devices. The water-induced modification of CNTs could be applied to prepare other nonpolar rubber composites for applications in the field of electrostatic resistance, EMI, EM wave absorption, and flexible sensors.

Author Contributions: Conceptualization, X.G.; methodology, X.G.; validation, X.G.; investigation, X.G.; resources, X.G.; writing—original draft preparation, X.G.; writing—review and editing, G.L.; supervision, G.L.; project administration, G.L.; funding acquisition, G.L. All authors have read and agreed to the published version of the manuscript.

Funding: This research was funded by the National Natural Science Foundation of China (NFSC no. 51972185).

Institutional Review Board Statement: Not applicable.

Data Availability Statement: Not applicable.

Conflicts of Interest: The authors declare no conflict of interest.

References

1. Zoughi, R.; Arias-Monje, P.J.; Gallion, J.; Sarkar, S.; Wang, P.-H.; Gulgunje, P.; Verghese, N.; Kumar, S. Microwave dielectric properties and Targeted heating of polypropylene nano-composites containing carbon nanotubes and carbon black. *Polymer* **2019**, *179*, 121658. [CrossRef]
2. Wang, L.; Wang, W.; Fu, Y.; Wang, J.; Lvov, Y.; Liu, J.; Lu, Y.; Zhang, L. Enhanced electrical and mechanical properties of rubber/graphene film through layer-by-layer electrostatic assembly. *Compos. Part B Eng.* **2016**, *90*, 457–464. [CrossRef]
3. Wu, H.-Y.; Zhang, Y.-P.; Jia, L.-C.; Yan, D.-X.; Gao, J.-F.; Li, Z.-M. Injection Molded Segregated Carbon Nanotube/Polypropylene Composite for Efficient Electromagnetic Interference Shielding. *Ind. Eng. Chem. Res.* **2018**, *57*, 12378–12385. [CrossRef]
4. Jia, L.-C.; Yan, D.-X.; Jiang, X.; Pang, H.; Gao, J.-F.; Ren, P.-G.; Li, Z.-M. Synergistic Effect of Graphite and Carbon Nanotubes on Improved Electromagnetic Interference Shielding Performance in Segregated Composites. *Ind. Eng. Chem. Res.* **2018**, *57*, 11929–11938. [CrossRef]
5. Joseph, N.; Janardhanan, C.; Sebastian, M.T. Electromagnetic interference shielding properties of butyl rubber-single walled carbon nanotube composites. *Compos. Sci. Technol.* **2014**, *101*, 139–144. [CrossRef]
6. Jia, L.-C.; Yan, D.-X.; Yang, Y.; Zhou, D.; Cui, C.-H.; Bianco, E.; Lou, J.; Vajtai, R.; Li, B.; Ajayan, P.M.; et al. High Strain Tolerant EMI Shielding Using Carbon Nanotube Network Stabilized Rubber Composite. *Adv. Mater. Technol.* **2017**, *2*, 1700078. [CrossRef]
7. Pawar, S.P.; Rzeczkowski, P.; Potschke, P.; Krause, B.; Bose, S. Does the Processing Method Resulting in Different States of an Interconnected Network of Multiwalled Carbon Nanotubes in Polymeric Blend Nanocomposites Affect EMI Shielding Properties? *ACS Omega* **2018**, *3*, 5771–5782. [CrossRef]
8. Nallabothula, H.; Bhattacharjee, Y.; Samantara, L.; Bose, S. Processing-Mediated Different States of Dispersion of Multiwalled Carbon Nanotubes in PDMS Nanocomposites Influence EMI Shielding Performance. *ACS Omega* **2019**, *4*, 1781–1790. [CrossRef]
9. Xu, Y.; Li, Y.; Hua, W.; Zhang, A.; Bao, J. Light-Weight Silver Plating Foam and Carbon Nanotube Hybridized Epoxy Composite Foams with Exceptional Conductivity and Electromagnetic Shielding Property. *ACS Appl. Mater. Interfaces* **2016**, *8*, 24131–24142. [CrossRef]
10. Sambyal, P.; Iqbal, A.; Hong, J.; Kim, H.; Kim, M.K.; Hong, S.M.; Han, M.; Gogotsi, Y.; Koo, C.M. Ultralight and Mechanically Robust Ti3C2Tx Hybrid Aerogel Reinforced by Carbon Nanotubes for Electromagnetic Interference Shielding. *ACS Appl. Mater. Interfaces* **2019**, *11*, 38046–38054. [CrossRef]
11. Sahoo, N.G.; Rana, S.; Cho, J.W.; Li, L.; Chan, S.H. Polymer nanocomposites based on functionalized carbon nanotubes. *Prog. Polym. Sci.* **2010**, *35*, 837–867. [CrossRef]
12. Treacy, M.M.J.; Ebbesen, T.W.; Gibson, J.M. Exceptionally high Young's modulus observed for individual carbon nanotubes. *Nature* **1996**, *381*, 678–680. [CrossRef]
13. Huang, H.; Liu, C.H.; Wu, Y.; Fan, S.S. Aligned Carbon Nanotube Composite Films for Thermal Management. *Adv. Mater.* **2010**, *17*, 1652–1656. [CrossRef]
14. Abraham, J.; Kailas, L.; Kalarikkal, N.; George, S.C.; Thomas, S. Developing highly conducting and mechanically durable styrene butadiene rubber composites with tailored microstructural properties by a green approach using ionic liquid modified MWCNTs. *RSC Adv.* **2016**, *6*, 32493–32504. [CrossRef]
15. Minot, E.D.; Yaish, Y.; Sazonova, V.; Park, J.Y.; Brink, M.; Mceuen, P.L. Tuning carbon nanotube bandgaps with strain. *Phys. Rev. Lett.* **2003**, *90*, 156401. [CrossRef] [PubMed]
16. Safadi, B.; Andrews, R.; Grulke, E.A. Multiwalled carbon nanotube polymer composites: Synthesis and characterization of thin films. *J. Appl. Polym. Sci.* **2010**, *84*, 2660–2669. [CrossRef]
17. Du, F.; Fischer, J.E.; Winey, K.I. Coagulation method for preparing single-walled carbon nanotube/poly(methyl methacrylate) composites and their modulus, electrical conductivity, and thermal stability. *J. Polym. Sci. Part B Polym. Phys.* **2003**, *41*, 3333–3338. [CrossRef]
18. Pötschke, P.; Fornes, T.D.; Paul, D.R. Rheological behavior of multiwalled carbon nanotube/polycarbonate composites. *Polymer* **2002**, *43*, 3247–3255. [CrossRef]
19. Das, A.; Stöckelhuber, K.W.; Jurk, R.; Saphiannikova, M.; Fritzsche, J.; Lorenz, H.; Klüppel, M.; Heinrich, G. Modified and unmodified multiwalled carbon nanotubes in high performance solution-styrene–butadiene and butadiene rubber blends. *Polymer* **2008**, *49*, 5276–5283. [CrossRef]
20. Nakaramontri, Y.; Nakason, C.; Kummerloewe, C.; Vennemann, N. Enhancement of electric conductivity and filler dispersion of carbon nanotube filled natural rubber composites by Latex mixing and in situ silanization. *Rubber Chem. Technol.* **2016**, *89*, 272–291. [CrossRef]
21. Bi, S.; Xiao, B.L.; Ji, Z.H.; Liu, B.S.; Liu, Z.Y.; Ma, Z.Y. Dispersion and Damage of Carbon Nanotubes in Carbon Nanotube/7055Al Composites During High-Energy Ball Milling Process. *Acta Metall. Sin. (Engl. Lett.)* **2021**, *34*, 9. [CrossRef]

22. Hirata, T.; Amiya, S.; Akiya, M.; Takei, O.; Hatakeyama, R. Chemical modification and its evaluation of CNT-based bio-nanosensor by plasma activation technique. *Am. Phys. Soc.* **2007**, *78*, 63–72. [CrossRef]
23. Guo, X.; Kang, L.; Sun, L.; Liu, L.; Liu, G. A simple and green approach to the preparation of super tough IIR/SWCNTs nanocomposites with tunable and strain responsive electrical conductivity. *J. Polym. Eng.* **2021**, *41*, 827–834. [CrossRef]

nanomaterials

MDPI

Article

Effect of Graphite Nanoplatelet Size and Dispersion on the Thermal and Mechanical Properties of Epoxy-Based Nanocomposites

Elsye Agustina, Jeung Choon Goak *, Suntae Lee, Yongse Kim, Sung Chul Hong, Yongho Seo and Naesung Lee *

Hybrid Materials Center (HMC), Department of Nanotechnology and Advanced Materials Engineering, Sejong University, 209 Neungdong-ro, Gwangjin-gu, Seoul 05006, Republic of Korea
* Correspondence: jcgoak@sejong.ac.kr (J.C.G.); nslee@sejong.ac.kr (N.L.)

Abstract: This study investigated the effect of graphite nanoplatelet (GNP) size and dispersion on the thermal conductivities and tensile strengths of epoxy-based composites. GNPs of four different platelet sizes, ranging from 1.6 to 3 μm, were derived by mechanically exfoliating and breaking expanded graphite (EG) particles using high-energy bead milling and sonication. The GNPs were used as fillers at loadings of 0–10 wt%. As the GNP size and loading amount increased, the thermal conductivities of the GNP/epoxy composites increased, but their tensile strengths decreased. However, interestingly, the tensile strength reached a maximum value at the low GNP content of 0.3% and thereafter decreased, irrespective of the GNP size. Our observations of the morphologies and dispersions of the GNPs in the composites indicated that the thermal conductivity was more likely related to the size and loading number of fillers, whereas the tensile strength was more influenced by the dispersion of fillers in the matrix.

Keywords: nanocomposite; graphite nanoplatelets; thermal conductivity; tensile strength; size-dependence; dispersion; ultrasonication; attrition mill

1. Introduction

Thermal management in advanced electronic devices has become increasingly important to extend their lifespan and performance reliability. For example, light-emitting diodes (LEDs) undergo a considerable reduction in the lifespan and quantum efficiency with increasing temperature [1], which makes heat dissipation via thermal interface materials (TIMs) and heat sinks particularly important [2,3].

Materials most frequently used for the heat sinks hitherto are metals such as aluminum or copper, which are heavy and expensive [4,5]. Therefore, there have been attempts to replace metal heat sinks with polymer composites containing thermally conductive fillers. The polymer material is attractive as the matrix of a composite due to its low weight, low cost, low thermal expansion coefficient, high chemical resistance, and easy processability [4,6]. For heat sink application, however, a polymer suffers from an extremely low thermal conductivity of 0.1–0.4 W/m·K [7]. Fillers such as alumina and silica have often been loaded to improve the thermal conductivity of polymers because they are inexpensive and lightweight. Unfortunately, because alumina and silica have low thermal conductivities of 28 and 32 W/m·K, respectively, high loading is required to dissipate the large amount of heat generated in high-power electronic devices, which in turn degrades the mechanical properties of the composite [8].

The current best filler candidates are nanocarbon materials, such as graphene and carbon nanotubes (CNTs), which have been receiving particular attention due to their following extremely high thermal conductivities: 2000–5000 W/m·K [9–11] along the in-plane direction for graphene and 2000–6000 W/m·K along the axial direction for CNTs [7]. Furthermore, they are mechanically strong, chemically inert, and light in weight. In terms of contact areas between fillers through which heat transfer occurs, however, two-dimensional (2D) graphene is a better thermally conductive filler than one-dimensional (1D) CNTs.

Single-layer graphene is difficult to prepare; it tends to agglomerate due to strong van der Waals forces via π–π interactions, as well as roll up during processing due to its thin shape. Thus, graphite nanoplatelets (GNPs) composed of tens of graphitic layers are most frequently used as fillers for practical applications. GNP-reinforced polymer nanocomposites have shown great potential as TIMs for heat dissipation [12]. Many studies have investigated the effect of adding GNPs to epoxy matrices on the thermal conductivity of resulting nanocomposites. Yu et al. demonstrated that an epoxy nanocomposite with 25 vol% exfoliated GNPs had a thermal conductivity of 6.44 W/m·K [13]. Chandrasekaran et al. studied the influence of GNP addition to an epoxy matrix using the three-roll milling technique and observed an increase in thermal conductivity of 14% at a filler loading of 2 wt% [14]. Similarly, the thermal conductivity of GNP/epoxy nanocomposites was found to increase by 627% at a filler loading of 8 wt% compared to neat epoxy [15]. The enhancement in thermal conductivity is attributed to the thermally conductive pathways formed by the 2D structure of GNPs in the polymer matrix. As highlighted in the above studies, the thermal conductivity of GNP/epoxy nanocomposites depends on various factors such as filler size, loading level, dispersion, alignment, stiffness, and thermal contact resistance between the polymer matrix and nanofillers. While much research has focused on obtaining the high thermal conductivity of GNP/epoxy nanocomposites, the mechanical properties of these materials at high GNP loadings, which directly affect their thermal management applications, have been not much concern.

TIMs are widely used in electronic devices, but they can experience mechanical failure during thermal cycling due to a mismatch between the coefficient of thermal expansion (CTE) of the device and TIM [16]. This CTE mismatch induces strain, which generates shear and stress at the interface between them. As a result, repetitive thermal cycling can gradually cause the TIM to be damaged, reducing its thermal conductivity. Thus, the reliability of mechanical properties for TIM is important to preserve its thermal conductivity. The tensile strength of GNP/epoxy composites depends on several factors, such as the interfacial adhesion, strength of the epoxy matrix, and shape and dispersion of the fillers [17]. The dimension and dispersion of the GNP in the epoxy matrix are important for improving tensile strength, and they also play a crucial role in enhancing thermal conductivity. This is because good interfacial interaction between the fillers and the epoxy matrix allows for efficient heat transfer, which in turn can help prevent thermal cycling-induced damage to the TIM. By reinforcing the epoxy matrix with well-dispersed GNPs, both the tensile strength and thermal conductivity of the composite can be improved, reducing the risk of mechanical failure during thermal cycling.

Controlling the GNP platelet size is one of the key factors in determining the thermal conductivity of the composites in which the GNPs are incorporated. As the GNPs become larger in size, the number of contacts between them decreases, resulting in higher thermal conductivity but poorer dispersion quality; thus, the mechanical properties of the composite tend to deteriorate as the GNP size increases. As such, the GNP size should be optimized to balance the trade-off between thermal conductivity and mechanical strength. This study investigated the effect of GNP size and dispersion on the thermal conductivities and tensile properties of epoxy composites. GNPs of different sizes were derived by mechanically exfoliating EG using high-energy bead milling and sonication.

2. Materials and Methods

2.1. Materials

The raw EG used herein was purchased from Hana Chemtech (Korea). Epikote 862 (Bisphenol F; Momentive) was used as the epoxy resin, HN2200 (3 or 4-methyl-1, 2, 3, 6-tetrahydrophthalic anhydride) (Hitachi Chemicals) as the curing agent, and N,N-dimethylbezylamine ($\geq$99%, Sigma–Aldrich) as the accelerating agent. Isopropanol (IPA) (ca. 99.5% purity, Samchun Pure Chemical) and acetone (ca. 99.5% purity, Samchun Pure

Chemical) were used as solvents for the mechanical treatment and EG/epoxy composite fabrication, respectively.

2.2. Mechanical Treatment

EG powders were attrition-milled in IPA to reduce their particle sizes using 1-mm-diameter zirconia beads at 2000 rpm for 90 min. Thereafter, the milled EG (denoted as "M-EG") was obtained by filtration and drying in an oven.

2.3. Preparation of the EG/Epoxy Composites

EG or M-EG powders were added to 80 mL of acetone, where the epoxy resin was dissolved and then exfoliated using a horn sonicator (ULH-700S Sonosmasher; Ulsso Hitech) at 497 W for 2 h to provide dispersions of S2- and MS2-GNPs, respectively. Here, "S" and "M" denote sonication and milling, respectively, and the number 2 indicates the sonication time in hours. An additional 2-h sonication was performed to produce the S4-GNP or MS4-GNP epoxy mixtures. The amount of EG or M-EG was fixed at 0.25 g in each sonication batch, while the amounts of epoxy resin, curing, and accelerating agents were adjusted to obtain the pre-designed filler contents in the GNP/epoxy composites. Sonication was carried out in a 100-mL bottle held in an ice bath while the sonication tip was immersed to the same depth as the solution surface. After sonication, the acetone was evaporated under vacuum while the mixture was kept at 60 °C. Subsequently, the curing and accelerating agents were added, and the mixture was stirred for 30 min at room temperature. The ratio between the epoxy resin, curing agent, and accelerating agent was 100:85.5:0.05. The prepared mixture was kept in a vacuum oven for 1 h at 80 °C for complete degassing and acetone evaporation. The GNP/epoxy mixture was transferred to a stainless-steel mould and hot-pressed at a pressure of 10 MPa during curing for 1 h at 80 °C and then for 2 h at 120 °C. Mechanical pressing was needed to prepare the GNP/epoxy composites with filler contents higher than 5 wt% because they had extremely high viscosities.

2.4. Characterisation

The morphologies and sizes of the EG particles and GNPs were observed using field-emission scanning electron microscopy (FE-SEM; model S-4700; Hitachi). SEM samples were prepared by filtering the diluted solutions of EG powders and GNPs through Pt-coated polycarbonate membranes so that the particles or platelets were individually scattered for clear observation. Atomic force microscopy (AFM; model XE-100; Park Systems) in the contact mode was used to examine the thickness of the GNPs deposited on a silicon wafer. An AFM cantilever tip having a radius less than 50 nm was operated at a force constant of ca. 7.4 N/m and at the resonance frequency of ca. 160 kHz. The AFM samples were prepared by filtering dispersions of the GNPs in IPA through polycarbonate membranes, followed by transferring the GNPs onto silicon wafers by pressing the membranes with a roller. The particle size analysis was performed on acetone dispersions of the EG and GNP powders using laser granulometry (Mastersizer 3000; Malvern Instruments). Their specific surface areas were measured by the Brunauer–Emmett–Teller method (BET; BELSORP-mini II; Microtrac) in nitrogen gas at 77 K. Crystallinity and structural defects of the EG and GNPs were characterized using Raman spectroscopy with a laser wavelength of 633 nm (Renishaw System 3000; Renishaw). The oxidation temperatures of both GNPs and composite samples were measured by thermogravimetric analysis (TGA; model TGA-50H; Shimadzu). The dispersion states of the fillers in the epoxy matrix were examined using light-transmission optical microscopy (model BX41M; Olympus). The through-plane thermal diffusivities (α, mm^2/s) of circular-shaped pellets of the GNP/epoxy composites (thickness: 1 mm; diameter: 12.7 mm) were measured by a laser flash system (Flashline 3050 System; TA Instruments) at room temperature with a laser power of 500 W. The thermal conductivity (k, W/m·K) was calculated using the following equation, $k = \alpha \times \rho \times C_P$, where the apparent densities (ρ, kg/m^3) of the samples were measured by dividing the mass over the volume, and the specific heat capacity (C_p, J/kg K) was measured by DSC

(Q200, TA instruments). Tensile tests were conducted using a tensile machine (model AGS-X; Shimadzu) at a speed of 1 mm/min. Tensile testing specimens having a 1-mm-thick dog-bone shape were prepared according to the standard ASTM-D1708.

3. Results and Discussion

EG is usually prepared by intercalating natural graphite flakes with strong acids or oxidizing agents, followed by high-temperature shock treatment [18]. EG can be easily exfoliated to produce GNPs due to the weak van der Waals interactions between the widened graphitic layers [19]. In this study, GNPs were produced by mechanical milling and sonication of EG. During the mechanical milling, EG particles experience an impact force via bead collision and a shear force via bead sliding [20]; the former force may cause fragmentation of EG particles, and the latter force may induce exfoliation along their graphitic layers, thus effectively reducing the planar size as well as the thickness of the EG particles [21]. During sonication, EG particles are exfoliated to GNPs by turbulent flow and micro-jets from acoustic cavitation (i.e., the formation, growth, and implosive collapse of bubbles) [22].

The size distribution of EG particles subjected to exfoliation processes was analyzed using SEM images and laser granulometry. Figure 1 shows SEM images of the raw EG, M-EG, and four GNPs (S2-, S4-, MS2-, and MS4-GNPs). Individual particles or platelets were observed by diluting and filtering the dispersed solutions of EG powders and GNPs through Pt-coated polycarbonate membranes. Platelet sizes of GNPs were measured for at least 150 platelets using SEM. On the other hand, laser granulometry was also used to measure particle sizes of the raw EG, M-EG, and four GNPs (see Figure S1). For the S2- and S4-GNPs given in Figure 1b,c, the average platelet sizes drastically decreased to ca. 3.0 and 2.0 μm, respectively. For the MS2- and MS4-GNPs, prepared by milling for 1.5 h and sonicating for 2 and 4 h, the average platelet size decreased to ca. 2.0 and 1.6 μm, respectively (Figure 1e,f). Referring to the platelet sizes given in Table S1 (Supplementary Information), the MS2-GNPs (2.0 ± 1.2 μm) showed smaller standard deviations than the S4-GNPs (2.0 ± 1.6 μm), although they had the same average size. This indicated that attrition milling was effective in preparing GNPs with a narrow size distribution. The specific surface areas of GNPs, measured using the BET method, are summarised in Table S1 (Supplementary Information). The specific surface areas of the GNPs expectedly increased with decreasing platelet size.

In laser granulometry, to observe particle size distribution, the D50 value represents the mass median diameter of all the particles. The raw EGs had a D50 of ca. 143.0 μm and a full-width at half-maximum (FWHM) of ca. 387.8 μm (Supplementary Information, Figure S1). Łoś et al. reported that sonication alone of EGs for a short period of time was inadequate to achieve a homogenous size distribution of GNPs because the fragmentation speed slowed down with irradiation time [23]. Our result also showed that the S2-GNPs, prepared by sonication for 2 h, had an inhomogeneous size distribution, as evidenced by the broad curve having a hump on the left side in the particle size distribution (Figure S1). For the S2-GNPs, the D50 was shifted to ca. 31.8 μm, and the FWHM was narrower at 136.5 μm. Increasing the sonication time to 4 h (S4-GNPs) moved the D50 value to a still smaller size of ca. 11.4 μm and resulted in a narrower FWHM of ca. 23.9 μm by downsizing the large particles corresponding to the right-most peak of the S2-GNP distribution (Figure S1).

Similarly, attrition milling solely was not effective in reducing graphite flake size [24]. Figure S2 gives the average particle size of the EG, measured using laser granulometry, as a function of attrition-milling time. The particle size drastically decreased from ca. 143 μm for the raw EG (0 h) to the saturated value of ca. 43 μm after 1.5 h of milling time; this was thus chosen as the optimum milling time. For the M-EG milled for 1.5 h, the D50 was ca. 43.0 μm with a narrow FWHM of ca. 55.3 μm (Figure S1). Interestingly, the M-EG had a much narrower FWHM than the S2-GNPs.

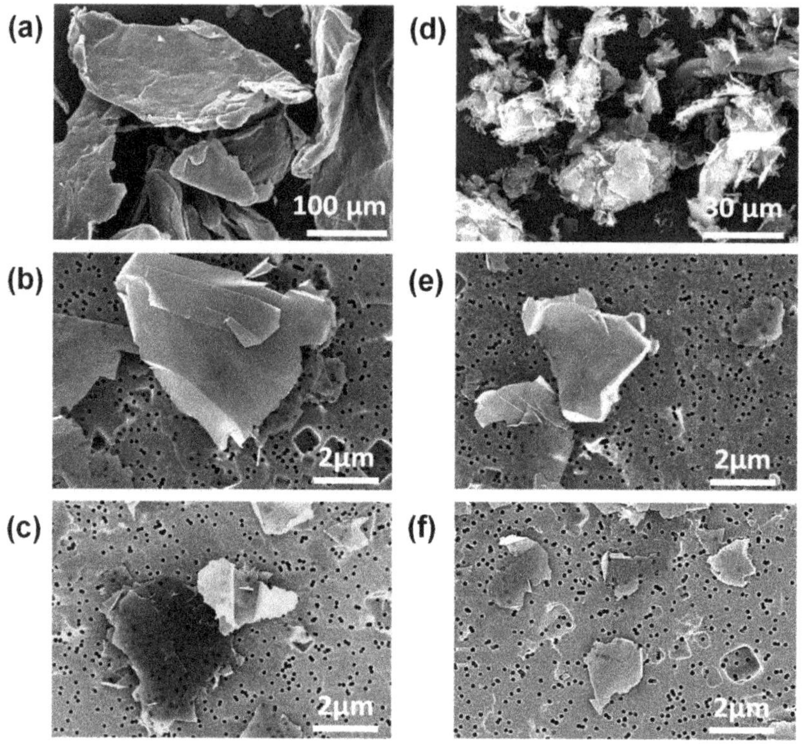

Figure 1. Scanning electron microscopy (SEM) images of (**a**) raw EG, (**b**) S2-GNPs, (**c**) S4-GNPs, (**d**) M-EG, (**e**) MS2-GNPs and (**f**) MS4-GNPs, where "S" and "M" denote sonication and milling, respectively, and the number indicates the sonication time in hours. EG: expanded graphite; GNP: graphite nanoplatelet.

This study combined high-energy attrition milling with sonication to produce GNPs with a small size and narrow distribution. Sonicating M-EG for 2 h, forming MS2-GNPs, shifted the D50 to a smaller size of ca. 20.3 μm and decreased the FWHM to ca. 51.8 μm. MS4-GNPs, produced by sonicating M-EG for 4 h, showed a further D50 shift to a smaller size of ca. 9.6 μm with a narrower FWHM of ca. 24.1 μm. Thus, a combination of milling and sonication led to more symmetric peaks in the particle size distribution, which indicated a small, homogeneous, and narrow size distribution of the GNPs.

The platelet thickness of each GNP was measured using AFM. Averaging the line profiles in Figure 2, the S2- and MS2-GNPs had platelet thicknesses of ca. 140 and 98 nm, respectively; the S4- and MS4-GNPs had smaller platelet thicknesses of ca. 35 and 17 nm, respectively. The data of Table S1 show the average thicknesses of S2-, S4-, MS2-, and MS4-GNPs from the measurements of at least 10 platelets for each sample. Attrition milling prior to sonication was effective not only in reducing platelet size but also in exfoliating layers of GNPs. Smaller graphite flakes can be exfoliated with greater ease than larger ones due to reduced van der Waals forces between their graphitic layers [21]. The number of graphene layers of a platelet was calculated from its thickness measured by AFM, using the approximation reported by Gupta et al. [25] that the distance between adjacent graphene layers is 0.35 nm. Thus, using the average thicknesses of ca. 33.4 and 21.3 nm given in Table S1 (Supplementary Information), the S4- and MS4-GNPs were composed of ca. 95 and 65 graphene layers, respectively. To minimize the surface energy, thin GNPs readily roll up into nanoscrolls after exfoliation, which can drastically decrease their aspect ratios and reduce the matrix–filler, and filler–filler interactions in the composite, hindering heat transfer between them [26,27]. According to Viculis et al. [28], GNPs thicker than ca. 20 nm would be favorable to maintaining the 2D platelet-like shape without rolling up.

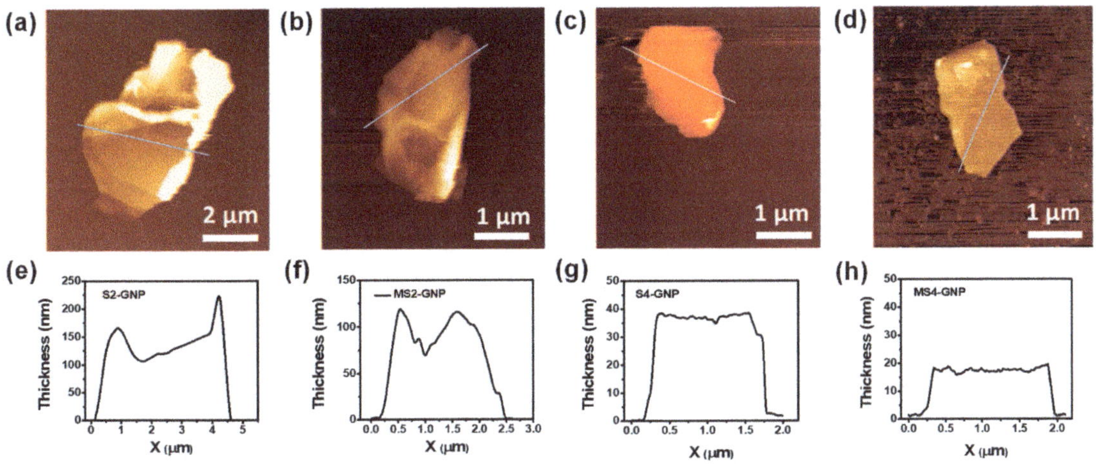

Figure 2. Atomic force microscopy (AFM) images and line profiles along the white lines of (**a**,**e**) S2-, (**b**,**f**) MS2-, (**c**,**g**) S4- and (**d**,**h**) MS4-GNPs, which had average thicknesses of 140, 98, 35, and 17 nm, respectively.

The raw EG, M-EG, and four GNPs were characterized using Raman spectroscopy (Figure 3a). Their crystalline qualities were assessed by their G-to-D peak ratios, I_G/I_D, where the D-band peak at ca. 1350 cm^{-1} is caused by structural defects and grain boundaries, and the G-band peak at ca. 1580 cm^{-1} corresponds to the planar sp^2 configuration of carbon atoms in the graphitic structure [29,30]. A decrease in the I_G/I_D value then indicates an increased defect and grain boundary density in the graphitic structure. The I_G/I_D ratio is also related to the particle size of GNPs, as Tuinstra et al. [31] previously reported; in that work, more intense D peaks were attributed to smaller flake sizes. The raw EG and M-EG had I_G/I_D ratios of 8.0 and 7.0, respectively. For the GNPs, the I_G/I_D ratios decreased in the order of S2- > MS2- > S4- > MS4-GNPs, which is in agreement with the order of their particle sizes, as shown in Table S1. The thermal stabilities of the EG and GNPs were measured using TGA in flowing air, and their derivative thermogravimetric (DTG) curves are given in Figure 3b. The DTG peak gives the rate of mass loss (dm/dT) at a maximal temperature, which can be used to clearly distinguish the difference in the thermal behavior of materials according to their composition. The oxidation temperature, T_{ox}, is defined as the peak temperature of the DTG curve and represents the thermal stability of a carbon material [32]. The T_{ox} value was 743.9 °C for the raw EG and decreased with attrition milling and longer sonication time. For the GNPs, the T_{ox} values decreased in the same order as their particle sizes and decreased I_G/I_D ratios, as presented in Table S1. The T_{ox} is a measure of the resistance to oxidation and is related to the defects and platelet sizes of the GNPs [5].

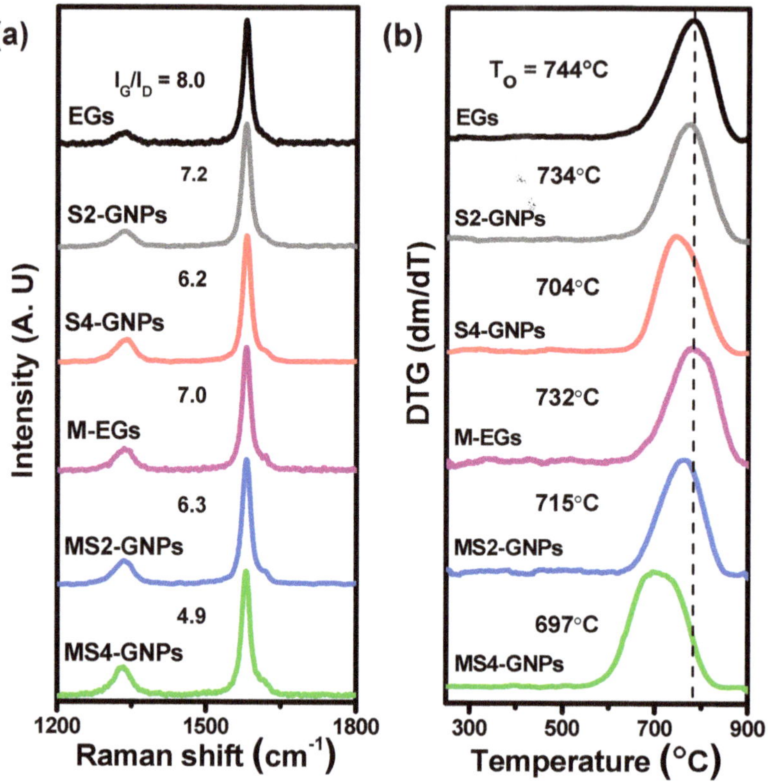

Figure 3. (**a**) Raman spectra and (**b**) derivative thermogravimetric (DTG) curves of EGs and GNPs. The G−to−D peak intensity ratios (I_G/I_D) of the Raman spectra and the oxidation temperatures (T_{ox}) from the peaks of DTG curves are shown.

The dispersion states of the GNPs in the epoxy matrix were observed using transmission optical microscopy. We exploited the light transparency of the epoxy resin, which enables clear visualization of GNP agglomeration, dispersion, and network state at the macroscopic level without interference from the epoxy matrix, as shown in Figure 4 [33]. GNP/epoxy composite films, as thin as 1.4 μm, were prepared by pressing and curing them between two glass plates. Small amounts of GNPs, 0.3 and 1 wt%, were loaded into the epoxy so that the GNPs did not overlap in the thickness direction of the films. The SEM images of Figure 1 show the individual GNPs that were used to measure their average platelet sizes. The platelets appear agglomerated in the optical microscopy images observed at a low magnification (Figure 4). Large aggregates and small particles coexisted in the S2-GNP/epoxy composite having the 0.3 wt% loading of the GNPs (Figure 4a). As the sonication time was extended to 4 h, the large aggregates were broken down but remained with reduced sizes in the S4-GNP sample (Figure 4b). Sonication alone, even for a long time of 4 h, was insufficient to completely break down the large aggregates. The MS2-GNPs, which had been attrition-milled for 1.5 h and then sonicated for 2 h, showed a decrease in the aggregate size but an increase in their number (Figure 4c) compared with the S2- and S4-GNPs. After 4 h of sonication, the MS4-GNPs exhibited a homogeneous dispersion of small GNPs of uniform size (Figure 4d). Increasing the loading amount of the GNPs to 1 wt% (Figure 4e–h) resulted in larger agglomerates of particles, and their populations also increased for each of the four GNPs.

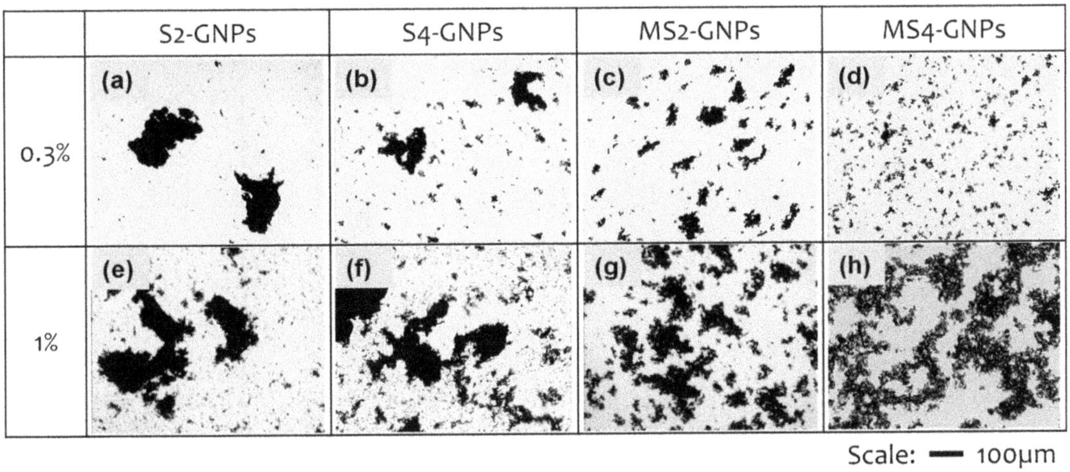

Figure 4. Transmission optical microscopy images of (**a**,**e**) S2-, (**b**,**f**) S4-, (**c**,**g**) MS2- and (**d**,**h**) MS4-GNPs in epoxy composites. The (**a**–**d**) and (**e**–**h**) contained 0.3 and 1 wt% of GNP fillers in the composites, respectively. GNP/epoxy composite films as thin as 1.4 μm were prepared by pressing and curing them between two glass plates to observe the dispersion states of the GNPs using optical microscopy. Scale bars correspond to 100 μm.

Thermal conductivities were measured for the GNP/epoxy composites loaded with 0–10 wt% of the four different GNPs (Figure 5a). In each case, the thermal conductivity linearly increased with a loading amount. At 10 wt% loading, S2-GNPs provided the highest thermal conductivity of 1.74 W/m·K, which was ca. 8 times higher than that of neat epoxy (0.22 W/m·K). At GNP concentrations greater than ca. 1 wt%, the thermal conductivities decreased in the order of S2- > S4- > MS2- > MS4-GNPs. In the high-loading region over 1 wt%, the thermal conductivities of the composites increased with increasing platelet size. The thermal conductivities of the composites having the same loading amounts of GNPs are affected by many factors of GNPs, including their platelet sizes, thicknesses, dispersion states, crystallinities, and contact resistances [34,35]. The physical connectivity between the GNPs within the epoxy matrix, which is determined by their loading amounts and morphological properties such as size, thickness, and dispersion, would make the greatest contribution to the thermal conductivities of the composites. The connectivity of the GNPs increases with the loading amount. Once they are connected within the epoxy in the high-loading region, the thermal conductivities of the composites increase as the number of physical contacts between the GNPs decreases, i.e., as the GNPs increase in platelet size. It is thus noted that the thermal conductivities increased with the platelet size of GNPs in the high-loading region (Figure 5a). This trend, however, was not followed in the low-loading region below 1 wt%. The thermal conductivities were higher at 0.3 wt% in the order of MS4- > MS2- > S4- > S2-GNPs, which was the reverse of their order in the high-loading region (Figure 5a, inset). The thermal conductivities increased with decreasing platelet size of GNPs in the low-loading region. Notably, the 0.3-wt% loading of GNPs increased the thermal conductivity of the composites relative to the neat epoxy for the S4-, MS2- and MS4-GNPs while was not of help in the case of the S2-GNPs (Figure 5a, inset). The order of the thermal conductivities of the composites was exactly reversed for the low- and high-loading regions. At the low-loading level of 0.3 wt%, GNPs were not physically connected with each other over a long distance (Figure 4a–d). It may appear that the smaller GNPs increased their connectivity over short distances to establish more heat-conducting channels.

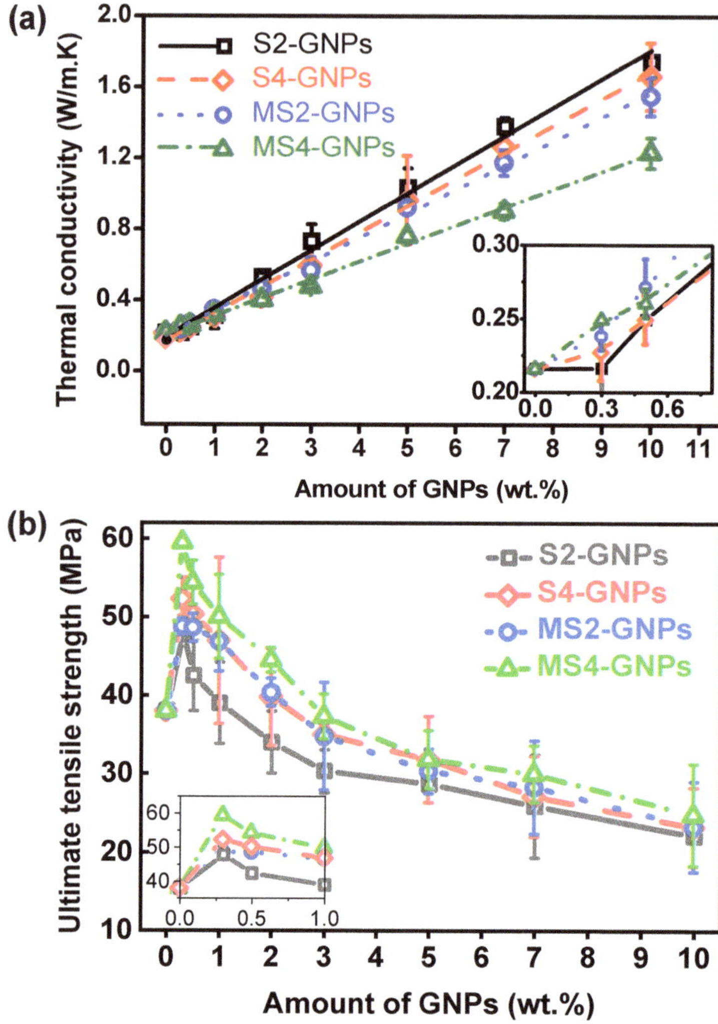

Figure 5. (**a**) Thermal conductivities and (**b**) ultimate tensile strengths of composites with varying amounts of S2-, S4-, MS2-, and MS4-GNP fillers. The inset in (**a**,**b**) shows the magnified graphs at low GNP concentrations.

Hypothetical explanation of the thermal conduction mechanism of the GNP/epoxy composites having the GNPs at a low loading of ca. 0.3 wt% is suggested in Figure 6, based on the following observation of Figure 4: the GNPs are agglomerated locally but are not connected over a long distance. The epoxy matrix behaves as the bottleneck slowing down cumulative phonon transport in the composites due to its nature of poor thermal conduction, while the GNP aggregates spread over the matrix accelerate the phonon transport in the composites. Red lines in Figure 6 represent thermal pathways in the composites, where the dotted lines illustrate slow heat conduction through the epoxy matrix, and the solid lines display fast heat conduction through the aggregated GNP fillers. As their sizes decrease from S2-GNPs to S4-, MS2- and MS4-GNPs, the distance between the GNP aggregates decreases, enhancing their connectivity over a short distance and their resultant thermal conductivities. With higher loading of GNPs, however, the GNP aggregates are in physical contact with each other, and then the other factors, such as the number of contacts between GNPs, will have greater influences on the phonon transport through the composites. Under such a circumstance, the GNPs with a larger size would have an advantage over the smaller-sized GNPs in terms of thermal conductivity.

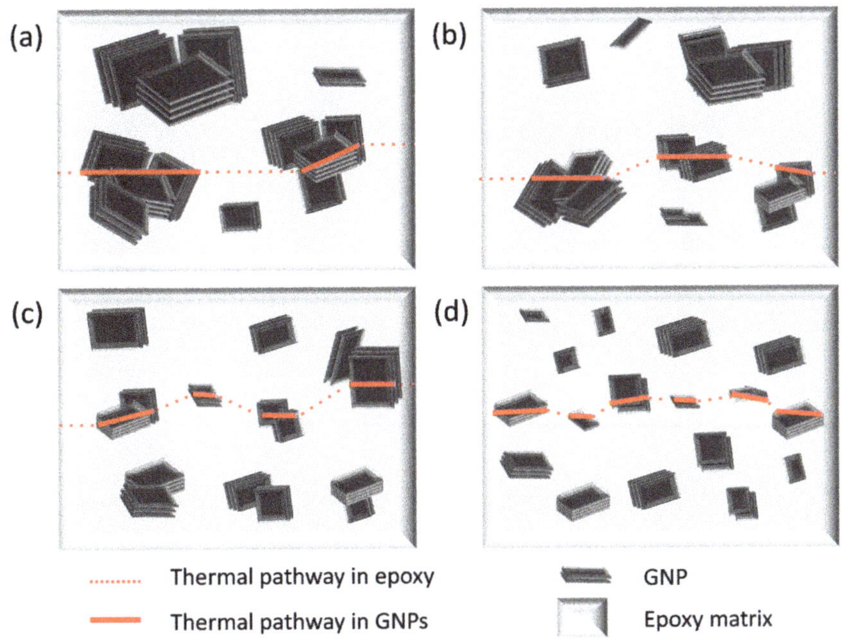

Figure 6. Pathways of thermal conduction for the composites loaded with a low concentration (ca. 0.3 wt.%) of (**a**) S2-, (**b**) S4-, (**c**) MS2-, and (**d**) MS4-GNPs.

Thermal stability of the GNP/epoxy composites having 10 wt% GNPs was examined using TGA and DTG curves, as given in Figure S4 and Figure 7, respectively. The degradation temperature of epoxy material, Td, represents the maximum weight loss rate in the DTG curve [36]. The neat epoxy showed a T_d of ca. 398 °C. The composites filled with S2-, S4-, and MS2-GNPs had T_d of ca. 399, 400, and 399 °C, respectively. Loading S2-, S4-, and MS2-GNPs hardly enhanced the thermal stability of the composites. Significant improvement, however, was achieved by loading MS4-GNPs, whose composite showed T_d of ca. 410 °C, which was 12 °C higher than that of the neat epoxy. The higher thermal stability of the composite filled with MS4-GNPs is related to more interfacial interaction between fillers and matrix due to their larger surface area [37].

Ultimate tensile strengths were measured for the GNP/epoxy composites with the four different GNPs loaded between 0 and 10 wt% (Figure 5b). In each case, the tensile strength reached a maximum of 0.3 wt% and then decreased continuously as the loading amount increased. The highest tensile strength was ca. 60 MPa for MS4-GNPs at 0.3 wt% loadings, which was improved by ca. 56% over that of neat epoxy. While other studies [34,38,39] reported that the incorporation of GNPs into polymers had a negative effect on strengthening due to their aggregation or poor dispersion, our study showed a considerable reinforcing effect of GNPs on mechanical strength when they were loaded below ca. 1 wt%. Their reinforcing effect varied according to the type and amount of GNPs. However, loading excessive amounts of GNPs into the epoxy decreased the strengths of the composites to values even below that of neat epoxy. For example, the composites containing more than 2 wt% of S2-GNPs, or more than 3 wt% of other GNPs, had lower strengths than the neat epoxy. At such high concentrations, it is likely that the GNPs were agglomerated in the composites and were easily disintegrated during the tensile test. Agglomeration of GNPs may cause voids and cracks when the polymer fails to penetrate the aggregates [40]. Such aggregated GNPs would act as structural defects rather than reinforcing fillers [41]. Fillers generally strengthen composites because they retard crack propagation through the matrix by energy absorption, crack pinning, or crack deflection [38,42]. The fillers should be fully disintegrated, evenly distributed in the matrix, and surrounded tightly by epoxy so that they can act as effective crack retardants. Figure 5a,b shows that the thermal conductivity increased while the mechanical strength decreased with increasing GNP loading. With increased loading, GNPs became more connected, which increased the number of thermal conduction pathways; however, these contact points also acted as mechanically weak points. Thus, the loading amount should be optimized to balance the trade-off between thermal conductivity and mechanical strength.

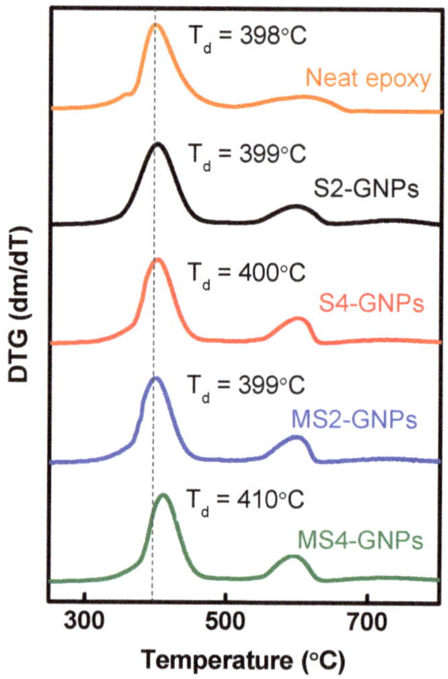

Figure 7. DTG curves of neat epoxy and GNP/epoxy composites loaded with 10 wt.% of S2-, S4, MS2-, and MS4-GNPs.

Over the range of concentrations between 0 and 10 wt%, the strength of the composite followed the order of MS4- > S4- ≈ MS2- > S2-GNPs, i.e., increasing with decreasing GNP size. The strength is mainly determined by the interfacial areas between the filler particles and the matrix where stress transfer occurs [6]. Mechanical milling and extended sonication break and exfoliate GNPs, eventually producing GNPs of smaller sizes and thicknesses, and larger surface areas, as given in Table S1. The MS4-GNPs, which were the smallest in size and thickness and the largest in surface area, consistently showed higher tensile strengths than the other GNPs. Our results agreed with those of Khan et al., who varied the GNP size by centrifugation and reported a greater reinforcement of smaller graphene flakes in the polymer matrix [43]. Additionally, thick GNPs may adversely affect the strength of composites due to the easy shearing and gliding of graphitic layers along the (001) basal plane [44].

There are additional reasons why small-sized GNPs are favored as fillers in composites. First, small GNPs may reduce the surface asperities that prevent tight surface contacts between a heat sink and a heat generator. In particular, thermal interface materials prefer small GNPs to produce unruffled composite surfaces [45]. Second, small GNPs may decrease the viscosities of composites and mitigate issues related to their anisotropic properties, thus facilitating manufacturing.

4. Conclusions

Composites were fabricated by incorporating GNPs of different sizes into an epoxy matrix. Combined attrition milling and sonication of raw EG with a large size of ca. 280 μm produced GNPs having small platelet sizes between 1.6 and 3.0 μm, i.e., S2-, S4, MS2- and MS4-GNPs. The I_G/I_D crystallinities and TGA oxidation temperatures of the GNPs decreased with attrition milling and longer sonication time, most likely due to the generation of more defects as the platelet size became smaller. The thermal conductivities of the composites linearly increased as the loading amount of the GNPs increased from 0 to 10 wt%, while the tensile strength reached a maximum at 0.3 wt% and then decreased continuously with increasing GNP amount. As the size of the GNPs decreased, the thermal conductivity decreased at high loadings over 1 wt%, but this trend was reversed at the low loading of 0.3 wt%. This behavior was attributed to changes in the GNP-to-GNP connections with a loading amount. The tensile strength increased as the size of the GNPs decreased. Thus, a higher loading amount and larger GNP size increased the thermal conductivity but decreased tensile strength. Thermal conductivity was affected significantly by the connectivity and contact number of GNPs, whereas the tensile strength was strongly related to the dispersion and filler–matrix interfacial area. The loading amount and size of the GNPs should be optimized for specific applications such as TIM and heat sink, where the trade-off relationship between the thermal conductivity and mechanical strength of the composites is important.

Supplementary Materials: The following supporting information can be downloaded at: https://www.mdpi.com/article/10.3390/nano13081328/s1, Figure S1: Particle size analyses of raw EG, S2-, and S4-GNPs, M-EG, and MS2- and MS4-GNPs; Figure S2: Average particle sizes of M-EG attrition-milled in IPA; Figure S3: Thermogravimetric analysis of raw EG, S2- and S4-GNPs, M-EG, and MS2- and MS4-GNPs; Figure S4: Thermogravimetric analysis of neat epoxy and GNP/epoxy nanocomposites; Table S1: Platelet sizes and thicknesses, surface areas, G-to-D peak intensity ratios (IG/ID), and oxidation temperatures (Tox) of EG and GNPs.

Author Contributions: Conceptualization, E.A., J.C.G. and N.L.; data curation, E.A., S.L. and Y.K.; formal analysis, E.A., J.C.G., S.C.H. and Y.S.; funding acquisition, N.L.; investigation, E.A., S.L. and Y.K.; methodology, E.A., J.C.G., S.C.H., Y.S. and N.L.; project administration, J.C.G. and N.L.; resources, E.A., J.C.G. and N.L.; supervision, N.L.; validation, E.A., J.C.G. and N.L.; visualization, E.A.; writing—original draft, E.A.; writing—review and editing, N.L. All authors have read and agreed to the published version of the manuscript.

Funding: This research was supported by Basic Science Research Program through the National Research Foundation of Korea (NRF) funded by the Ministry of Education (NRF-2020R1A6A1A03043435) and the Ministry of Science and ICT (NRF-2021R1A2C1008798).

Data Availability Statement: The data presented in this study are available on request from the corresponding author.

Conflicts of Interest: The authors declare no conflict of interest.

References

1. Christensen, A.; Graham, S. Thermal effects in packaging high power light emitting diode arrays. *Appl. Therm. Eng.* **2009**, *29*, 364–371. [CrossRef]
2. Zhou, S.; Xu, J.; Yang, Q.-H.; Chiang, S.; Li, B.; Du, H.; Xu, C.; Kang, F. Experiments and modeling of thermal conductivity of flake graphite/polymer composites affected by adding carbon-based nano-fillers. *Carbon* **2013**, *57*, 452–459. [CrossRef]
3. Xing, W.; Xu, Y.; Song, C.; Deng, T. Recent Advances in Thermal Interface Materials for Thermal Management of High-Power Electronics. *Nanomaterials* **2022**, *12*, 3365. [CrossRef] [PubMed]
4. Chung, D.D.L. Materials for thermal conduction. *Appl. Therm. Eng.* **2001**, *21*, 1593–1605. [CrossRef]
5. Shtein, M.; Nadiv, R.; Buzaglo, M.; Kahil, K.; Regev, O. Thermally Conductive Graphene-Polymer Composites: Size, Percolation, and Synergy Effects. *Chem. Mater.* **2015**, *27*, 2100–2106. [CrossRef]

6. Chatterjee, S.; Nafezarefi, F.; Tai, N.H.; Schlagenhauf, L.; Nüesch, F.A.; Chu, B.T.T. Size and synergy effects of nanofiller hybrids including graphene nanoplatelets and carbon nanotubes in mechanical properties of epoxy composites. *Carbon* **2012**, *50*, 5380–5386. [CrossRef]
7. Han, Z.; Fina, A. Thermal conductivity of carbon nanotubes and their polymer nanocomposites: A review. *Prog. Polym. Sci.* **2011**, *36*, 914–944. [CrossRef]
8. Wang, J.-W.; Wang, Y.; Wang, F.; Li, S.-Q.; Xiao, J.; Shen, Q.-D. A large enhancement in dielectric properties of poly(vinylidene fluoride) based all-organic nanocomposite. *Polymer* **2009**, *50*, 679–684. [CrossRef]
9. Ghosh, S.; Bao, W.; Nika, D.L.; Subrina, S.; Pokatilov, E.P.; Lau, C.N.; Balandin, A.A. Dimensional crossover of thermal transport in few-layer graphene. *Nat. Mater.* **2010**, *9*, 555–558. [CrossRef]
10. Ghosh, S.; Calizo, I.; Teweldebrhan, D.; Pokatilov, E.P.; Nika, D.L.; Balandin, A.A.; Bao, W.; Miao, F.; Lau, C.N. Extremely high thermal conductivity of graphene: Prospects for thermal management applications in nanoelectronic circuits. *Appl. Phys. Lett.* **2008**, *92*, 151911. [CrossRef]
11. Balandin, A.A.; Ghosh, S.; Bao, W.; Calizo, I.; Teweldebrhan, D.; Miao, F.; Lau, C.N. Superior Thermal Conductivity of Single-Layer Graphene. *Nano Lett.* **2008**, *8*, 902–907. [CrossRef] [PubMed]
12. Bilisik, K.; Akter, M. Graphene nanoplatelets/epoxy nanocomposites: A review on functionalization, characterization techniques, properties, and applications. *J. Reinf. Plast. Compos.* **2022**, *41*, 99–129. [CrossRef]
13. Yu, A.; Ramesh, P.; Itkis, M.E.; Bekyarova, E.; Haddon, R.C. Graphite Nanoplatelet−Epoxy Composite Thermal Interface Materials. *J. Phys. Chem. C* **2007**, *111*, 7565–7569. [CrossRef]
14. Chandrasekaran, S.; Seidel, C.; Schulte, K. Preparation and characterization of graphite nano-platelet (GNP)/epoxy nano-composite: Mechanical, electrical and thermal properties. *Eur. Polym. J.* **2013**, *49*, 3878–3888. [CrossRef]
15. Wang, Y.; Yu, J.; Dai, W.; Song, Y.; Wang, D.; Zeng, L.; Jiang, N. Enhanced thermal and electrical properties of epoxy composites reinforced with graphene nanoplatelets. *Polym. Compos.* **2015**, *36*, 556–565. [CrossRef]
16. Due, J.; Robinson, A.J. Reliability of thermal interface materials: A review. *Appl. Therm. Eng.* **2013**, *50*, 455–463. [CrossRef]
17. Zakaria, M.R.; Abdul Kudus, M.H.; Akil, H.M.; Mohd Thirmizir, M.Z. Comparative study of graphene nanoparticle and multiwall carbon nanotube filled epoxy nanocomposites based on mechanical, thermal and dielectric properties. *Compos. Part B Eng.* **2017**, *119*, 57–66. [CrossRef]
18. Li, B.; Zhong, W.-H. Review on polymer/graphite nanoplatelet nanocomposites. *J. Mater. Sci.* **2011**, *46*, 5595–5614. [CrossRef]
19. Du, W.; Jiang, X.; Zhu, L. From graphite to graphene: Direct liquid-phase exfoliation of graphite to produce single- and few-layered pristine graphene. *J. Mater. Chem. A* **2013**, *1*, 10592–10606. [CrossRef]
20. Suryanarayana, C. Mechanical alloying and milling. *Prog. Mater. Sci.* **2001**, *46*, 1–184. [CrossRef]
21. Yi, M.; Shen, Z. A review on mechanical exfoliation for the scalable production of graphene. *J. Mater. Chem. A* **2015**, *3*, 11700–11715. [CrossRef]
22. Doktycz, S.; Suslick, K. Interparticle collisions driven by ultrasound. *Science* **1990**, *247*, 1067–1069. [CrossRef] [PubMed]
23. Łoś, S.; Duclaux, L.; Alvarez, L.; Hawełek, Ł.; Duber, S.; Kempiński, W. Cleavage and size reduction of graphite crystal using ultrasound radiation. *Carbon* **2013**, *55*, 53–61. [CrossRef]
24. Kim, B.G.; Park, C.L.; Jeon, H.S.; Jang, H.D. Fabrication of Platelet-Like Natural Crystalline Graphite with Nano-Scale Thickness via Layer Distance Expansion. *Mater. Trans. JIM* **2009**, *50*, 2323–2328. [CrossRef]
25. Gupta, A.; Chen, G.; Joshi, P.; Tadigadapa, S.; Eklund. Raman Scattering from High-Frequency Phonons in Supported n-Graphene Layer Films. *Nano Lett.* **2006**, *6*, 2667–2673. [CrossRef] [PubMed]
26. Shioyama, H. Cleavage of graphite to graphene. *J. Mater. Sci. Lett.* **2001**, *20*, 499–500. [CrossRef]
27. Meyer, J.C.; Geim, A.K.; Katsnelson, M.I.; Novoselov, K.S.; Booth, T.J.; Roth, S. The structure of suspended graphene sheets. *Nature* **2007**, *446*, 60–63. [CrossRef]
28. Viculis, L.M.; Mack, J.J.; Kaner, R.B. A Chemical Route to Carbon Nanoscrolls. *Science* **2003**, *299*, 1361. [CrossRef]
29. Wang, G.; Yang, J.; Park, J.; Gou, X.; Wang, B.; Liu, H.; Yao, J. Facile Synthesis and Characterization of Graphene Nanosheets. *J. Phys. Chem. C* **2008**, *112*, 8192–8195. [CrossRef]
30. Prolongo, S.G.; Jiménez-Suárez, A.; Moriche, R.; Ureña, A. Graphene nanoplatelets thickness and lateral size influence on the morphology and behavior of epoxy composites. *Eur. Polym. J.* **2014**, *53*, 292–301. [CrossRef]
31. Tuinstra, F.; Koenig, J.L. Raman Spectrum of Graphite. *J. Chem. Phys.* **1970**, *53*, 1126–1130. [CrossRef]
32. Mansfield, E.; Kar, A.; Hooker, S.A. Applications of TGA in quality control of SWCNTs. *Anal. Bioanal. Chem.* **2010**, *396*, 1071–1077. [CrossRef] [PubMed]
33. Kilic, U.; Sherif, M.M.; Ozbulut, O.E. Tensile properties of graphene nanoplatelets/epoxy composites fabricated by various dispersion techniques. *Polym. Test.* **2019**, *76*, 181–191. [CrossRef]
34. Ghaleb, Z.A.; Mariatti, M.; Ariff, Z.M. Properties of graphene nanopowder and multi-walled carbon nanotube-filled epoxy thin-film nanocomposites for electronic applications: The effect of sonication time and filler loading. *Compos. Part A Appl. Sci. Manuf.* **2014**, *58*, 77–83. [CrossRef]
35. Shtein, M.; Pri-Bar, I.; Varenik, M.; Regev, O. Characterization of Graphene-Nanoplatelets Structure via Thermogravimetry. *Anal. Chem.* **2015**, *87*, 4076–4080. [CrossRef]
36. Wu, C.S.; Liu, Y.L.; Chiu, Y.C.; Chiu, Y.S. Thermal stability of epoxy resins containing flame retardant components: An evaluation with thermogravimetric analysis. *Polym. Degrad. Stabil.* **2002**, *78*, 41–48. [CrossRef]

37. Chen, G.-X.; Kim, H.-S.; Park, B.H.; Yoon, J.-S. Multi-walled carbon nanotubes reinforced nylon 6 composites. *Polymer* **2006**, *47*, 4760–4767. [CrossRef]
38. Zaman, I.; Phan, T.T.; Kuan, H.-C.; Meng, Q.; Bao La, L.T.; Luong, L.; Youssf, O.; Ma, J. Epoxy/graphene platelets nanocomposites with two levels of interface strength. *Polymer* **2011**, *52*, 1603–1611. [CrossRef]
39. Ma, J.; Meng, Q.; Zaman, I.; Zhu, S.; Michelmore, A.; Kawashima, N.; Wang, C.H.; Kuan, H.-C. Development of polymer composites using modified, high-structural integrity graphene platelets. *Compos. Sci. Technol.* **2014**, *91*, 82–90. [CrossRef]
40. Yang, S.-Y.; Lin, W.-N.; Huang, Y.-L.; Tien, H.-W.; Wang, J.-Y.; Ma, C.-C.M.; Li, S.-M.; Wang, Y.-S. Synergetic effects of graphene platelets and carbon nanotubes on the mechanical and thermal properties of epoxy composites. *Carbon* **2011**, *49*, 793–803. [CrossRef]
41. Chandrasekaran, S.; Sato, N.; Tölle, F.; Mülhaupt, R.; Fiedler, B.; Schulte, K. Fracture toughness and failure mechanism of graphene based epoxy composites. *Compos. Sci. Technol.* **2014**, *97*, 90–99. [CrossRef]
42. Rafiee, M.A.; Rafiee, J.; Wang, Z.; Song, H.; Yu, Z.-Z.; Koratkar, N. Enhanced Mechanical Properties of Nanocomposites at Low Graphene Content. *ACS Nano* **2009**, *3*, 3884–3890. [CrossRef] [PubMed]
43. Khan, U.; May, P.; O'Neill, A.; Coleman, J.N. Development of stiff, strong, yet tough composites by the addition of solvent exfoliated graphene to polyurethane. *Carbon* **2010**, *48*, 4035–4041. [CrossRef]
44. Debelak, B.; Lafdi, K. Use of exfoliated graphite filler to enhance polymer physical properties. *Carbon* **2007**, *45*, 1727–1734. [CrossRef]
45. Sun, X.; Ramesh, P.; Itkis, M.E.; Bekyarova, E.; Haddon, R.C. Dependence of the thermal conductivity of two-dimensional graphite nanoplatelet-based composites on the nanoparticle size distribution. *J. Phys. Condens. Matter* **2010**, *22*, 334216. [CrossRef] [PubMed]

MDPI AG
Grosspeteranlage 5
4052 Basel
Switzerland
Tel.: +41 61 683 77 34

MDPI Books Editorial Office
E-mail: books@mdpi.com
www.mdpi.com/books

www.ingramcontent.com/pod-product-compliance
Lightning Source LLC
Chambersburg PA
CBHW061757210926
51850CB00006B/578
9783725870646